Improving integrated pest management in horticulture

It is widely recognised that agriculture is a significant contributor to global warming and climate change. Agriculture needs to reduce its environmental impact and adapt to current climate change whilst still feeding a growing population, i.e. become more 'climate-smart'. Burleigh Dodds Science Publishing is playing its part in achieving this by bringing together key research on making the production of the world's most important crops and livestock products more sustainable.

Based on extensive research, our publications specifically target the challenge of climate-smart agriculture. In this way we are using 'smart publishing' to help achieve climate-smart agriculture.

Burleigh Dodds Science Publishing is an independent and innovative publisher delivering high quality customer-focused agricultural science content in both print and online formats for the academic and research communities. Our aim is to build a foundation of knowledge on which researchers can build to meet the challenge of climate-smart agriculture.

For more information about Burleigh Dodds Science Publishing simply call us on +44 (0) 1223 839365, email info@bdspublishing.com or alternatively please visit our website at www.bdspublishing.com.

Related titles:

Critical issues in plant health: 50 years of research in African agriculture
Print (ISBN 978-1-78676-232-0) Online (ISBN 978-1-78676-234-4; 978-1-78676-235-1)

Pesticides and agriculture
Profit, politics and policy
Print (ISBN 978-1-78676-276-4); Online (ISBN 978-1-78676-279-5; 978-1-78676-278-8)

Integrated management of diseases and insect pests of tree fruit
Print (ISBN 978-1-78676-256-6) Online (ISBN 978-1-78676-258-0; 978-1-78676-259-7)

Integrated management of insect pests: Current and future developments
Print (ISBN 978-1-78676-260-3) Online (ISBN 978-1-78676-262-7; 978-1-78676-263-4)

Chapters are available individually from our online bookshop:
https://shop.bdspublishing.com

climate•SMART•publishing
IN AGRICULTURAL SCIENCE

BURLEIGH DODDS SERIES IN AGRICULTURAL SCIENCE
NUMBER 110

Improving integrated pest management in horticulture

Edited by Professor Rosemary Collier, Warwick University, UK

burleigh dodds
SCIENCE PUBLISHING

Published by Burleigh Dodds Science Publishing Limited
82 High Street, Sawston, Cambridge CB22 3HJ, UK
www.bdspublishing.com

Burleigh Dodds Science Publishing, 1518 Walnut Street, Suite 900, Philadelphia, PA 19102-3406, USA

First published 2021 by Burleigh Dodds Science Publishing Limited

Library of Congress Control Number: 2021940789

British Library Cataloguing in Publication Data
A catalogue record for this book is available from the British Library

ISBN 978-1-78676-753-0 (Print)
ISBN 978-1-78676-756-1 (PDF)
ISBN 978-1-78676-755-4 (ePub)
ISSN 2059-6936 (print)
ISSN 2059-6944 (online)

DOI: 10.19103/AS.2021.0095

Typeset by Deanta Global Publishing Services, Dublin, Ireland

Contents

Series list

Title	Series number
Achieving sustainable cultivation of maize - Vol 1 From improved varieties to local applications Edited by: Dr Dave Watson, CGIAR Maize Research Program Manager, CIMMYT, Mexico	001
Achieving sustainable cultivation of maize - Vol 2 Cultivation techniques, pest and disease control Edited by: Dr Dave Watson, CGIAR Maize Research Program Manager, CIMMYT, Mexico	002
Achieving sustainable cultivation of rice - Vol 1 Breeding for higher yield and quality Edited by: Prof. Takuji Sasaki, Tokyo University of Agriculture, Japan	003
Achieving sustainable cultivation of rice - Vol 2 Cultivation, pest and disease management Edited by: Prof. Takuji Sasaki, Tokyo University of Agriculture, Japan	004
Achieving sustainable cultivation of wheat - Vol 1 Breeding, quality traits, pests and diseases Edited by: Prof. Peter Langridge, The University of Adelaide, Australia	005
Achieving sustainable cultivation of wheat - Vol 2 Cultivation techniques Edited by: Prof. Peter Langridge, The University of Adelaide, Australia	006
Achieving sustainable cultivation of tomatoes Edited by: Dr Autar Mattoo, USDA-ARS, USA and Prof. Avtar Handa, Purdue University, USA	007
Achieving sustainable production of milk - Vol 1 Milk composition, genetics and breeding Edited by: Dr Nico van Belzen, International Dairy Federation (IDF), Belgium	008
Achieving sustainable production of milk - Vol 2 Safety, quality and sustainability Edited by: Dr Nico van Belzen, International Dairy Federation (IDF), Belgium	009
Achieving sustainable production of milk - Vol 3 Dairy herd management and welfare Edited by: Prof. John Webster, University of Bristol, UK	010
Ensuring safety and quality in the production of beef - Vol 1 Safety Edited by: Prof. Gary Acuff, Texas A&M University, USA and Prof. James Dickson, Iowa State University, USA	011
Ensuring safety and quality in the production of beef - Vol 2 Quality Edited by: Prof. Michael Dikeman, Kansas State University, USA	012
Achieving sustainable production of poultry meat - Vol 1 Safety, quality and sustainability Edited by: Prof. Steven C. Ricke, University of Arkansas, USA	013
Achieving sustainable production of poultry meat - Vol 2 Breeding and nutrition Edited by: Prof. Todd Applegate, University of Georgia, USA	014

Introduction

Pests and diseases remain a significant threat to crop yields and quality worldwide. With concerns about the environmental impact of synthetic pesticides, there remains a need to develop more environmentally friendly biological methods of control that can be combined synergistically within integrated pest management strategies.

This volume provides a comprehensive review of the recent developments in integrated pest management for horticultural crops and is broken down into four parts. Chapters in Part 1 focus on using biological agents in IPM, drawing particular attention to the use of biopesticides/bioprotectants for both plant disease and insect control. Chapters also focus on using biostimulants as an IPM tool as well as improving application systems for bioprotectants in IPM programmes. Chapters in Part 2 discuss using decision support systems in IPM, reviewing advances in pest insect monitoring and forecasting, proximal sensors and in particular, using decision support systems. Part 3 examines improving IPM techniques and implementation, the chapters address the use of agronomic practices and conservation biological control in IPM programmes. Chapters also review the economics of IPM and approaches to encouraging the take up of IPM in horticultural programmes. Part 4 provides three case studies on the practical application of IPM in greenhouses/protected cultivation, tomato and outdoor *Brassica* production and other vegetables.

Part 1 Using biological agents in integrated pest management

Part 1 begins with a chapter that reviews the advances in biopesticides for insect control in horticulture. Chapter 1 starts by giving a brief introduction to biopesticides then goes on to discuss their use in horticulture practices. The chapter also examines the key challenges in the successful use of biopesticides in IPM programmes, focusing specifically on agent selection, bioassays, ecological suitability, production, formulation and application. It also examines the other control methods that can be used for targeting pests and methods of testing their compatibility with biopesticides. The chapter then goes on to analyse the integration of biopesticide use into already established IPM strategies. There is a case study on developing *Brevibacillus laterosporus* for diamondback moth control in New Zealand, before the chapter concludes with an analysis of new approaches that are being considered, as well as potential future research trends in the area.

The next chapter focuses on advances in bioprotectants for plant disease control in horticulture. Microbial bioprotectants and natural substances have been increasingly available and used for disease management in horticultural crops. Chapter 2 examines the status of registered products and their current insufficiency to fulfil farmers' needs for disease control. It highlights possible advances to meet those needs, which include extending the spectrum of activity of existing products, finding new active substances, and facilitating registration of bioprotectants. The chapter also addresses issues on enhancing the field efficacy of bioprotectants on horticultural crops, through the expansion of knowledge on their modes of action, the identification of key factors of efficacy, the improvement of delivery and the complementation of bioprotectant use with other solutions from the IPM toolbox. Finally, the chapter provides data on the adoption of bioprotectants by horticulturists in France, reviews possible obstacles for wider adoption and provides perspectives on ways to help farmers cope with the technical complexity of deployment.

Expanding on topics previously discussed in Chapter 2, Chapter 3 reviews the advances in the application of biostimulants as part of integrated pest management approaches in horticulture. The chapter begins by discussing IPM in general and how biostimulants are being integrated into IPM strategies. It then goes on to discuss how various biostimulants can be used to enhance crop growth, crop health and yields. The chapter also highlights how biostimulants can be used to enhance soil health as well as inducing natural resistance in plants to various stresses. Sections on using biostimulants to suppress both plant diseases and arthropod pests, as well as improving a plant's tolerance to abiotic stresses are also provided. The chapter then goes on to review multitrophic interactions and the factors that can affect the effectiveness of biostimulants as well as the various strategies that can be put in place for using biostimulants. The chapter concludes by highlighting the challenges of and future needs for integrating biostimulants into crop production and IPM practices.

The final chapter of Part 1 examines approaches to improving application systems for bioprotectants in IPM programmes in horticulture. Chapter 4 begins by reviewing application technologies appropriate for integrated pest management, then goes on to examine the importance of achieving the correct dose of active substance during application. The chapter also highlights the importance of finding the 'right place' to deliver the substance as well as establishing the 'right time', together with the benefits of ensuring the right form of substance is applied. Sections on adjuvants and co-formulants and the use of efficacy trials are also provided. The chapter concludes by emphasising the importance of ensuring good application practice and regular calibration in application systems.

Part 2 Using decision support systems in integrated pest management

Part 2 opens with a chapter on advances in pest insect and disease monitoring and forecasting in horticulture. Chapter 5 begins by examining the key issues and challenges associated with pest monitoring and forecasting. It then goes on to address how these issues and challenges can be addressed, highlighting methods such as automated identification of insects from traps and plants and improving detection by being more selective with the plants that are used. A case study on white fly sampling, monitoring and forecasting is also provided. The chapter then concludes with an analysis of how developing new ways to detect pests and diseases is becoming more important and highlights the potential areas of research that can be developed in the future.

Chapter 6 discusses advances in proximal sensors to detect crop health status in horticultural crops. The chapter begins by reviewing the use of optoelectronic devices for detecting disease in vegetable plants, focusing specifically on hyperspectral, multispectral and RGB sensors. The chapter also examines the use of thermographic and chlorophyll fluorescence probes. A section on the use of sensors for the detection of micro-environmental parameters related to disease outbreaks is provided. The chapter includes a case study which analyses the use of digital and mechatronic applications in the baby leaf vegetable quality chain, before concluding with an overview of the importance of developing proximal sensors and integrating their use into IPM programmes.

The next chapter of Part 2 focuses on advances in decision support systems (DSS) for IPM in horticultural crops. Chapter 7 first discusses the availability of integrated pest management decision support systems in horticulture, drawing specific attention to tactical systems and strategic and operational DSS. The chapter then goes on to highlight what to look for in decision support systems for pest management whilst also providing case studies on the use of specific systems. It discusses how online platforms can be used to help standardise outputs to improve user experiences. The chapter concludes by highlighting why the use of decision support systems is important for boosting horticultural crop production.

Part 3 Improving integrated pest management techniques and implementation

The first chapter of Part 3 examines the use of agronomic practices in integrated pest management programmes in horticulture. Chapter 8 first reviews the IPM practices related to the crop and its annual management, such as tree architecture, plant nutrition and fertilisation, water supply, soil and groundcover management and harvest, damaged fruit and plant residue management. The chapter then goes on to address the IPM practices related to the enhancement of plant diversity in the cropping system at a field scale, focusing specifically on intercropping with companion plants and diversification with cash crops. A section on designing cropping systems that can combine agronomic IPM practices at field, farm and agri-food scales is also provided, giving two examples of experiments that have previously been conducted.

Chapter 9 discusses advancing conservation biological control (CBC) as a component of integrated pest management of horticultural crops. The chapter first discusses the effect of scale on the efficacy of CBC, focusing on spatial and temporal considerations. The chapter then examines the effectiveness of CBC management in multiple dimensions. There is a section on the benefits of CBC beyond integrated pest management, before the presentation of two case studies on the use of CBC in commercial practice. The chapter concludes by emphasising the importance of adopting CBC in the near future as its uptake is currently quite low.

Assessing the economics of integrated pest management for horticultural crops is the subject of Chapter 10. The chapter begins by introducing the concepts of economic thinking, then goes on to discuss the challenges associated with different stakeholder goals. Case studies are included to illustrate key principles of economic thinking. The chapter emphasises the importance of considering agricultural system design and utilising novel control tactics. The chapter concludes by highlighting how understanding IPM within an economic framework can help both growers and society to create sustainable value over time.

The final chapter of Part 3 focuses on encouraging integrated pest management uptake in horticultural crop production. Chapter 11 first reviews extension theories and methodologies and identifies useful resources to aid in the diffusion of IPM information to increase the adoption of IPM practices in horticultural crops. The chapter discusses Rogers (1963) 'Diffusion of Innovation Theory' and research related to the diffusion of IPM around the globe. It also examines how the use of logic models can improve the planning and efficiency of extension programs. Finally, the chapter discusses E-extension and virtual communication and outline possible barriers to the adoption of IPM innovations and how these barriers can be overcome.

Part 4 Case studies

The final part of the book opens with a chapter that focuses on the practical application of integrated pest management in greenhouses and protected cultivation. Chapter 12 provides several case studies based on the production of ornamental plants in greenhouse horticulture, focusing specifically on how good communication and education of growers is important for facilitating transition to IPM and how increased hygiene measures reduce disease pressure and the need for additional disease control. The case studies cover how the success of predatory mites increased confidence in further use of IPM programmes, especially when adapted to specific greenhouse circumstances, and how the management of broad mite, a difficult and elusive pest, benefitted from biological control. The two final case studies focus on how adding food to the crop helped establish a standing army of beneficials and how mealy bug control still poses a major challenge, but that biocontrol shows promise. The chapter concludes by emphasising how increased use of dynamic spectral lighting, increased automation and multilayer growing will pose future challenges that should be addressed in ongoing research.

Moving on from Chapter 12, Chapter 13 focuses on the practical applications of integrated pest management in horticultural cultivation: using case studies about the production of protected tomato and outdoor *Brassica* crops. The chapter first discusses integrated pest management in protected tomato production, focusing specifically on IPM strategies such as establishing resistant varieties, as well as physical and cultural control. The chapter also covers chemical and biological control strategies in protected tomato crops, before going on to discuss IPM strategies to control particular pest insects. It then moves on to review IPM in outdoor *Brassica* production, focusing on methods to control various pest insects. The chapter concludes by providing an overview of potential future research trends research for IPM in horticulture.

The final chapter of the book examines the practical application of integrated pest management to control cabbage root fly in the economics of integrated pest management for horticultral crops *Brassica* vegetables. Chapter 14 first focuses on development of a decision-support system to predict pest outbreaks, specifically focusing on the monitoring of cabbage root fly. The chapter goes on to review the use of row covers in crop protection, specifically highlighting the importance of the choice of material, mesh size of insect nets and timing of covering with these row covers. It then goes on to discuss the use of entomopathogenic nematodes as biocontrol agents, before moving on to review the key issues in chemical control. The chapter also provides three case studies on *Delia radicum*, focusing specifically on the use of row covers, entomopathogenic nematodes and chemical control.

Part 1

Using biological agents in integrated pest management

Chapter 1

Advances in biopesticides for insect control in horticulture

Travis R. Glare, Bio-Protection Research Centre, Lincoln University, New Zealand; and Aimee C. McKinnon, La Trobe University, Australia

1 Introduction

Horticulture, from the production through to the sale of fruit, vegetables and flowers, is an industry that is crucial for food production and human wellbeing. The science of horticulture is critical to the prosperity and health of people and the environment. With an increasing population, it is estimated that food production will need to increase by 70% by the middle of the century to meet demand (Davies, 2015). Opportunities exist to develop new high-value and high-nutrition horticultural produce on the decreasing area of land available for food production. However, horticulture is a more intensive farming approach than that used for broad-acre crops such as wheat and corn, which means that poor sustainability practices will have more severe consequences for future generations, although the area impacted is less than for some other crops. This includes approaches toward invertebrate pest control, as insects and other invertebrates cause significant damage and production losses across the horticulture sector around the world.

The application of synthetic pesticides has been the main approach to controlling pest insects in horticulture. However, applications of certain

http://dx.doi.org/10.19103/AS.2021.0095.01

pesticides on fruit crops, such as stone fruits, have pre-harvest intervals ranging between 7 and 28 days that preclude the use of treatments prior to fruit ripening, to minimise residues, during a period when the risk of pest outbreak is high (Hossain et al., 2006; Boston et al., 2020). Frequent use of broad-spectrum insecticides can also promote secondary pest outbreaks of insect pests such as mites, which are otherwise regulated by natural invertebrate predators (Epstein et al., 2000). Furthermore, the pest insect species of some horticultural crops occupy cryptic ecological niches. For example, *Carpophilus truncatus* beetles, pests in Australian almond orchards, live inside mummified nuts and in new nuts, so that insecticide sprays are often unable to reach them and are therefore ineffective (Boston et al., 2020). In the USA, pecan weevils are major pests that cause significant crop damage; the larvae and pupae occur inside the pecan nuts and in soil, so that control of the weevils using contact insecticides is restricted only to the adults (Mulder et al., 2012).

The development of insecticide resistance from the continual use of pesticides is also a key challenge. Development of new synthetic pesticides is now costly and difficult, partly due to the strict environmental and safety regulations involved when converting an active compound into a product (Qadri et al., 2020). For instance, a recent report estimated that in the USA, the development of a novel synthetic pesticide would cost in excess of $300 million and can take approximately 12 years (McDougall, 2016).

The withdrawal of pesticides, combined with health and environmental concerns and the rise in the resistance to current pesticides, is driving the need for new methods of control, including the growing use of biopesticides.

2 What are biopesticides?

Biopesticides have been loosely defined as biologically based pesticides (Glare, 2015). This can be quite a broad definition, including the incorporation of live organisms and their derivatives (e.g. metabolites), through to the use of semiochemicals in formulations. The use of plant extracts, for example, can fit into a broad definition of biopesticides, but authors often use the term to refer to a subgroup of biopesticides such as microbial-based biopesticides.

Kienwick (2007) limited the definition of biopesticides to only those based on microbes, while Copping and Menn (2000) included semiochemicals. The USA EPA defined biopesticides as 'certain types of pesticides derived from such natural materials as animals, plants, bacteria, and certain minerals' (http://www.epa.gov/pesticides/biopesticides). However, biopesticides are divided into three groups - microbial-based, plant-incorporated protectants (e.g. Bt corn) and biochemical - which encompass formulations of naturally occurring substances including plant extracts and pheromones.

Several novel approaches to using microbes for the control of pest insects have expanded the basic definition of biopesticides. Endophytes, microbes which can live inside plants without causing disease, can often confer resistance to pests and pathogens (McKinnon et al., 2017) and can be classed as a new biopesticide (Glare et al., 2012). Other novel biopesticide approaches include the use of RNAi, which is the use of the antiviral immune system of insects for post-transcriptional gene silencing (Vogel et al., 2019).

It is more common for researchers to define the term 'biopesticide' based on their own area of expertise and focus. In this chapter, we will focus on microbial-based biopesticides for use against pest insects. Paradoxically, 'microbial' in the area of entomopathogens generally includes nematode entomopathogens, which are not microorganisms *per se*, although, as the most commonly studied entomopathogenic nematodes that use symbiotic bacteria as an agent for lethality, this is understandable.

The term 'biopesticide' has also been questioned from a viewpoint of public understanding. The term 'pesticide' has become associated with chemical insecticides and the connotation of broad-spectrum killing. As such, it can be viewed negatively, even for biopesticides. The term 'biorationals' has been used (e.g. Rosell et al., 2008) to encompass biopesticides and other biologically based approaches.

The history of biopesticides is as long as the history of agriculture. The use of plant extracts is probably the earliest example of biopesticides, dating from about some 4000 years ago, with regard to extracts of the neem tree (El-Wakeil, 2013). The use of microbes to control insects is relatively recent; the first demonstration of a disease caused by a microbial infection in insects was for silkworms infected by *Beauveria bassiana*, published in 1835 by Agostino Bassi. It was not until the nineteenth century, after the discovery that microbes cause disease, that the idea of using microbes to control insects was proposed by Pasteur, with the first practical attempts in Russia in the 1890s (Zimmermann et al., 1995). The product Sporéine, based on the bacterium *Bacillus thuringiensis* (Bt), is thought to be the first microbial-based insecticidal biopesticide that was commercialised.

Since that time, biopesticide use has increased incrementally, especially for those based on Bt. However, in the last decade, with the reduction in available synthetic pesticides and more concern around sustainable plant production, the use of biopesticides and the availability of products have increased dramatically. There are several areas where biopesticide development has advanced significantly, which will be reviewed below.

3 Biopesticide use in horticulture

The global market for biopesticides has an estimated value of $3 billion USD and is expected to increase by 15% in the next few years, outpacing market growth of novel synthetic pesticides by 10-fold (Damalas and Koutroubas, 2018). Although biopesticide sales have been increasing year on year, they still make up only a small percentage of the insecticide market worldwide. When looking at the current, most exploited, microbes used as biopesticides in horticulture, there is an indication of where some of the key challenges lie.

Microbial-based biopesticides for insect control have been based mainly on bacterial, nematode, viral and fungal entomopathogens. Although there have been products based on protozoa and microsporidia, the lack of rapid-kill and *in vitro* production methods has greatly limited their development. For viruses and nematodes, *in vitro* production limitations have restricted biopesticide development to only a few examples, such as nucleopolyhedroviruses (Nguyen et al., 2016) and the nematodes in *Heterorhabditis* and *Steinernema* which are associated with the entomopathogenic bacteria in the genera *Xenorhabdus* and *Photorhabdus* (Shapiro-Ilan et al., 2012).

Fungi in the genera *Beauveria* and *Metarhizium* are the most exploited entomopathogenic fungi that are currently formulated into biopesticides (Quesada-Moraga et al., 2020). Strains of these genera are the most common to infect insects naturally, with host ranges in the hundreds of species across a range of insect orders. These species are largely asexual, with just conidia produced in bulk on cadavers. They can also be grown easily on simple media, including grains (Grzywacz et al., 2014), which makes mass production possible. In general, strains of *Metarhizium* and *Beauveria* are safe for non-target organisms, such as predators and parasites, when applied in the field. This does not mean these other beneficial agents may not be infected in the laboratory in some cases, but these other organisms are far more resistant to infection, meaning they have a little impact in the field. The compatibility of these beneficial fungi with other environmentally sensitive control approaches has been exploited in integrated pest management (IPM) programmes (Glare et al., 2020b).

Insect-killing bacteria are known from diverse taxonomic groups, although only a few groups have been exploited as commercially available biopesticides (Jurat-Fuentes and Jackson, 2012), and even fewer used in horticulture. Spore-forming bacteria in the genera *Bacillus, Paenibacillus, Lysinibacillus* and *Brevibacillus* are the most commonly used for insect-control products, the majority of which target caterpillars or Diptera. Among the Proteobacteria, species of *Serratia, Pseudomonas, Chromobacterium* and *Burkholderia* have been used. Bt strains and subspecies have been the mainstay of biopesticides used in horticulture for over 70 years, with the first commercialised Bt sold in 1938, and still accounting for around 60–70% of all microbial-based biopesticide sales (Olson, 2015). Some bacteria act through infection, but in many cases, including Bt, most target mortality is due to toxins. The use of toxins, as opposed to microbes with an infectious mode of action, results in very different commercial products. For example, in the case of Bt, in which there is no requirement for live cells in the final product. This greatly simplifies formulation and storage.

Entomopathogenic nematodes can be found in several major groups, but the most exploited for biopesticides are in the Heterorhabditidae and Steinernematidae families. These nematodes have symbiotic bacteria involved in the infection cycle. The two groups are not particularly closely related but do share common characteristics, since they are obligate parasites, horizontally transmitted, have an infective juvenile stage and are always associated with bacteria that are involved in infection (Koppenhőfer et al., 2020).

This group of nematodes has been the subject of commercial development, with 13 species now developed. Commercialisation has been aided by the often broad host range of the main nematode families and the development of *in vitro* mass-production methods (Shapiro-Ilan et al., 2012). Although the range of hosts can be broad in the laboratory, with over 200 species for *S. carpocapsae* across 10 insect orders (Koppenhőfer et al., 2020), for many species, the host range has not been investigated in the field. Infective juveniles are able to disperse actively and passively, which can greatly aid effectiveness.

Entomopathogenic viruses are known from 15 families, but most viruses used in biopesticides have been from the family Baculoviridae, containing the nucleopolyhedroviruses (NPV) and granuloviruses (GV), which target Lepidoptera (Srinivasan et al., 2019). These viruses have several attractive characteristics, principally that the infective virons are encased in a proteinaceous coat, ensuring more robust particles in the environment and suitability for formulation into biopesticides. The known viruses in these groups are host-specific, sometimes only known to infect a single lepidopteran species, meaning they are safe for non-target organisms. However, growing these viruses outside living cells is not currently possible, which greatly increases the cost of production and limits their use as biopesticides in many areas.

Some successful biopesticides have been developed based on baculoviruses. In Africa, commercial formulations of the *Maruca vitrata* Mavi NPV have been used in Benin, Niger and Nigeria (Srinivasan et al., 2019). NPV formulations have also been used against *Spodoptera*, *Plutella xylostella, Mamestra brassicae* and *Helicoverpa armigera,* including in citrus orchards in South Africa (Moore and Kirkman, 2010). In the Canary Islands, a *Chrysodeixis chalcites* NPV is under development as a commercial biopesticide for the pest of banana crops, *Chrysodeixis chalcites* (Bernal et al., 2018).

4 Key challenges in the successful use of biopesticides in integrated pest management programmes

There is a common list of challenges that have been reported to limit the uptake of microbial-based biopesticides. Some are based on the limitations of using live organisms, such as lack of highly virulent agents, lack of environmental persistence, difficulties in economic mass production and application, short shelf life of formulations, lack of compatibility with other control options and limitations of the host range (either too broad or too narrow); but others are to do with regulation and registration of non-chemical pesticides in many regions (Table 1). One area that has been vexing for many proponents has been the reluctance of growers to use biopesticides. Also, this area has received less attention than the development of biopesticides, partly as it requires different skills from the biologists and technologists working with product development. These are the key challenges to overcome, and the specific issues for any biopesticide will be a mixture of these topics.

Furthermore, most bacteria-, virus-, nematode- and protozoa-based agents need to be ingested for action, as opposed to fungal entomopathogens and some nematode-transmitted bacteria. The need for ingestion of the toxin or cells places restrictions on the method of application and on formulation requirements. Agents used in biopesticides need to be easy to culture and have a shelf life suitable for the target pest.

Table 1 Common barriers to uptake of biopesticides

Barrier	Suggested solution
Lack of efficacy	Improve strain selection or use multiple agents in IPM
Short shelf life of biopesticides	Formulation improvements
Required to be easy to culture	In most cases, this is necessary for mass production to be economically viable
A common tendency to view biopesticides as chemical pesticide replacements - 'silver bullet' beliefs	More education for users with respect to benefits and limitations
Cryptic or soil-dwelling pests difficult to treat	Improvements in application methods informed by knowledge of pest ecology
Lack of environmental persistence of BCA	Formulation improvements
Lack of compatibility with other control options	Strain selection of BCA
Limitations of host range (either too broad or too narrow)	Improve and extend selection of agents
Difficulties in economic mass production	Selection of agents, or genetic modifications and/or research into production approaches
Regulation and registration barriers	Region-specific and can be addressed at that level
Slow to kill, so pest continues to damage plant	Strain selection, adjuvant/formulation to enhance mortality rate or new approaches such as mixed species of BCA or transgenic approaches
Reluctance of growers to use biopesticides	Field demonstration of efficacy

4.1 Agent selection

Selection of prospective entomopathogenic agents has traditionally relied on target invertebrate mortality rates observed under laboratory conditions (Waage, 1997). Frequently, efficacy is determined using a fundamentally reductionist approach, where both the agent and dose rate are tested in otherwise sterile and optimal conditions that favour the agent. This emphasis on mortality-by-dose arises from the type of biological control typically aimed for when using microbial-based biopesticides; that is, 'inundation' biological control (Eilenberg et al., 2001). Inundation biological control was defined by Eilenberg et al. (2001) as *'The use of living organisms to control pests when control is achieved exclusively by the released organisms themselves.'* In contrast with augmentation (or inoculation) biological control, which relies on growing and sustaining a viable population of the agent after release, the agent applied by inundation is not required to reproduce in order to achieve pest control (Eilenberg et al., 2001). The appealing aspect of this strategy is that the efficacy of the biological control agent (BCA) or its products is not dependent on the density of the target host population, as is the case with, for example, parasitoid wasps or other arthropod predators adopted as BCAs. There is, however, a common tendency to view BCAs in biopesticides as chemical pesticide replacements, offering an alternative but still providing a 'silver bullet' approach to the pest problem. If anything, this common perception has limited the development and use of otherwise effective agents in horticultural practices, because it is rare to achieve consistently high mortality rates in the field comparable to chemical pesticides (Chandler et al., 2011).

To improve field efficacy and use, novel biopesticides should be intentionally developed to contribute to an IPM 'toolkit,' where effective pest control relies on a combination of complementary or synergistic methods (Glare et al., 2020b). Part of the issue with the silver bullet mindset is that the full benefits and effectiveness of BCAs have not been incorporated in efficacy evaluations with focus on high virulence alone. For example, research indicates the need to consider ecological suitability, particularly with respect to biotic and abiotic interactions, during the initial process of agent selection (Hoy, 1992; Whipps, 1997; Deacon and Berry, 1993; Eilenberg et al., 2001; Meyling and Eilenberg, 2007; Bruck, 2010; Jaronski, 2010; Scheepmaker and Butt, 2010; Campos-Herrera et al., 2019; Martínez-Barrera et al., 2019; Glare et al., 2020b). Future biopesticides are therefore likely to be those developed for, and used in, a combination of biological control-type systems and/or will possess the capacity to confer multiple beneficial functions, such as biological control activity and plant growth promotion (Kowalska et al., 2020).

4.2 Bioassay

Typically, to assess the mortality rate of target pests by candidate biopesticide strains, laboratory bioassays are conducted. Bioassays are experiments that involve challenging the target host with varying dose concentrations of the candidate inoculum to determine the median lethal dose and mortality time after inoculation (Inglis et al., 2012; Thomas and Jenkins, 1997). Infective propagules of candidate entomopathogens are usually prepared in liquid suspensions and the concentrations calculated using histological techniques. To ensure consistent results, testing of potential candidates requires a method to deliver reproducible doses of the test substances (Spence et al., 2020). How the bioassay experiment is formally designed is also of critical importance, as the design can influence the consistency of results and therefore the success of the candidate(s) (Inglis et al., 2012). Bioassays require stringent experimental design with true replication because variables such as host species and age, host condition, size of bioassay container, incubation conditions and pathogen species/strain cultivation conditions can all affect results. Specific discussions of bioassays with entomopathogenic fungi can be found in Butt and Goettel (2000) and O'Callaghan et al. (2012) for experiments involving various entomopathogenic bacteria.

Even when robust laboratory bioassays are conducted, the results are often not reproducible when candidate agents are applied under field conditions to control the target pest (Roberts and St. Leger, 2004). This is because field conditions introduce more variables that can directly affect the host-pathogen interaction, the viability of the inoculum or the persistence of the pathogenic agent in the target environment. However, conventional dose-mortality bioassays may be useful in evaluating the effect of many different variables as treatments (Keyser, 2010). Using bioassays involving entomopathogenic fungi, studies have tested the effects of temperature on virulence (Fargues et al., 1997; Thomas and Jenkins, 1997; Brooks et al., 2004), the effect of repeated sub-culturing on virulence, and, if there is an effect, whether it can be improved by passing the pathogen through an insect host (Daoust and Roberts, 1982), the effect of insect thermal regulation on virulence (Blanford and Thomas, 2001; Ouedraogo et al., 2004) and the effects of propagule formulation (Prior et al., 1995; Thomas et al., 1995; Caudwell and Gatehouse, 1996).

4.3 Ecological suitability

While bioassays remain important to determine virulent entomopathogenic species or strains, simulating field conditions to further test virulence and ecological suitability in laboratory settings is challenging. Nevertheless, expanding screening criteria for entomopathogens to test certain factors that are likely to be encountered under field conditions is crucial to increase the success of prospective biopesticides in horticultural systems (Onsongo et al., 2019; Blond et al., 2018). At a minimum, this may include heat tolerance, UV tolerance and vegetative growth-rate tests, which may be useful for shortlisting species or isolates to identify those more likely to be suitable in field conditions (Keyser, 2010). Primarily, it is important to understand the key factors that may affect the persistence of the agent once applied in the target environment. Where candidate isolates are required for application to soil, additional bioassays can be conducted using (non-sterile) soil taken from the target environment to study interactions that may occur with soil microbiota, soil structure or chemistry (Cory and Ericsson, 2010; Jackson et al., 2010; Jaronski, 2010). To date, there are few studies that have directly assessed the effect of the soil microbial community on inoculum or *vice versa* (Mayerhofer et al., 2017; Zitlalpopoca-Hernandez et al., 2017; McKinnon et al., 2018), yet the importance of plant-associated microbial communities for modulating plant health, including with respect to inducing insect-herbivore resistance, is currently highly topical (Dastogeer et al., 2020).

Conversely, plant species and plant rhizosphere systems can also directly influence the prevalence and persistence of entomopathogenic bacteria, fungi and nematodes. For example, plants have been shown to 'recruit' root-associated microbes following attack by insects or pathogens through the modulation of specific hormone-signalling pathways (Rudrappa et al., 2008; Lakshmanan et al., 2012; Lee et al., 2012). Rasmann et al. (2005) reported that an insect-induced plant root signal, (E)-β-caryophyllene, was emitted by maize roots in response to herbivory by beetle larvae *Diabrotica virgifera virgifera*. The (E)-β-caryophyllene was shown to strongly attract an entomopathogenic nematode. Some entomopathogenic fungi, particularly *Metarhizium anisopliae*, have been reported to be common in agricultural (tilled) soils compared to natural habitats, even though fungal diversity and abundance are normally higher in undisturbed soils (Bidochka et al., 2001; Meyling and Eilenberg, 2007; Inglis et al., 2008; Meyling et al., 2009).

Additionally, under field conditions, the presence of non-target insect hosts may affect the persistence of the biopesticide agent (Blond et al., 2018), as it can alter the concentration of the inoculum applied or its distribution in the environment. The presence of plants and plant microbiomes can also impact fungal persistence.

Aside from natural biotic and abiotic factors, biopesticides intended for use in horticulture need to be reasonably compatible with other components of the IPM toolkit that are used. For example, fungus-based biopesticides may be susceptible to certain fungicides used to control plant fungal pathogens, which can be problematic when these are used in the same IPM strategy (Chandler et al., 2011). Furthermore, IPM programmes that include combinations of natural enemies to achieve pest control, such as different entomopathogens with invertebrate predators, must also consider and test for possible interactions between these biocontrol agents. Apart from testing prospective entomopathogens for pathogenicity to the invertebrate predators or parasitoids intended for use within the same IPM programme (Ekesi et al., 2007), certain combinations of natural enemies can result in 'non-additive effects,' where competition for the host as a resource reduces the overall efficacy of the pest management strategy – compared with when a single biocontrol agent is used (Ferguson and Stiling, 1996).

4.4 Production, formulation and application

Another essential aspect of agent selection for biopesticide development is the ease of production. Research has frequently involved and emphasised the optimisation of the mass production of inoculum, as well as agent stability and ease of delivery for application (Charnley and Collins, 2007). Accordingly, production systems have been devised to maximise inoculum yield at lower cost for use in inundative biocontrol applications (Wraight et al., 2001; Cliquet and Jackson, 2005; Jackson et al., 2010; Jaronski, 2010; Jaronski and Jackson, 2012). The underlying assumption has been that effective insect control may be achieved if copious amounts of inoculum can be feasibly produced to be applied at sufficiently high rates (Brownbridge et al., 2008; Jaronski, 2010). The variables encountered in the environment of application that may impact biocontrol activity have not been prioritised during product development, and thus in the formulation strategies (Jackson et al., 2010). However, recent research indicates that there is scope to modify selection and production to improve product quality (Lacey et al., 2015).

Nucelopolyhedroviruses and other agents that are obligate pathogens lead to particular problems around producing enough inoculum for use in biopesticides which are both economical and able to survive formulation and application. Large-scale production has proved elusive, with many efforts to grow pathogens *in vitro*, generally in insect cell lines. NPVs have been commercially produced based on infecting insects (e.g. Gychek, TM-Biocontrol (Bell et al., 1981; Reardon and Podgwaite, 1994; USDA, 2020)) or collecting infected cadavers from the field (Moscardi, 2007), but for many promising viral control agents, production is a limiting step (Bernal et al., 2018; Srinivasan et al., 2019). For the MaviMNPV for the control of *M. vitrata* in parts of Africa, community-based production is under development, where the mass production of the host insect and infection is completed at the village level (Srinivasan et al., 2019).

The current delivery methods are generally based on spray applications, similar to those used for synthetic pesticides or surface-applied granules for the control of some soil-dwelling stages. These methods have been reviewed in detail (e.g. Bateman, 2003; Chapple et al., 2000). The main issue of any application process is getting sufficient viable inoculum to the target insect or site, and ensuring maximum persistence after application. In an IPM situation, it may include avoidance of application directly on other natural enemies or at certain times.

Attract-and-kill approaches have been developed for some insects, where a combination of semiochemical attraction to pathogen-laced traps is used (e.g. Yousef et al., 2018). The use of pathogen-impregnated traps, such as with entomopathogenic fungi, increases the subsequent chance of horizontal transfer to conspecifics and also has the advantages of using less of the pathogen than inundative approaches, with less residue in the crop (Quesada-Moraga et al., 2020).

The discovery and potential exploitation of any additional ecological roles of fungal and bacterial entomopathogens, as plant endophytes or rhizo- and phyllophane colonisers, is also under study as a mode of application (see Section 6.1).

4.5 Commercialisation: integration and safety

The most challenging aspects, that are to be considered early on, are the other control methods that are used for the target pest and whether the biopesticide is compatible (e.g. other biocontrol agents such as parasitoid wasps, lure traps, SIT, nematodes and fungi, compatibility with IPM-friendly pesticides). Is the biopesticide easy to use?

The compatibility of entomopathogens with other, especially biological agents, has been well documented (e.g. Glare et al., 2020b). Although all combinations need to be tested, generally parasitoids and predators are unaffected by field rates of biopesticides. For example, a recent study by Garriga et al. (2019) found that four predators and four parasitoids commonly used in horticulture were unaffected by *Heterorhabditis bacteriophora* or *Steinernema feltiae* treatments; and *Steinernema carpocapsae*, despite some indication of negative effects in the laboratory, did not impact the natural enemies in the field.

In most countries, biopesticides need to be registered under some form of pesticide regulations. For many years, biopesticides suffered under regulatory regimes which were established for chemical-based pesticides, which meant that some of the requirements for registration were inappropriate for biological-based products. This is slowly improving, with countries such as the USA and UK adopting specific registration requirements for biopesticides. However, a precautionary approach is often taken, which means unfamiliar products such as novel microbial-based biopesticides can face regulatory challenges.

Regulation can be tailored to biologicals, providing rigorous safe assessments, while not resulting in undue transactional costs (Grant and Gwynn, 2020). Regulators often have high field-efficacy targets of over 80% for pesticides, which, as discussed by Grant and Gwynn (2020), may not be appropriate for biopesticides. However, they can have lower efficacy, as long as the effect is to reduce a population below the economic damage threshold.

5 Integrated pest management

One of the trends in biopesticide use in horticulture is an increased focus on the integration of biopesticides into wider pest management strategies. The concept and application of IPM is not new, and there are good examples of commercially applied programmes that incorporate biopesticides. However, the use of multiple compatible pest control approaches into a single programme is increasing. Several recent reviews cover the potential and implementation of IPM with the inclusion of biopesticides (e.g. Glare et al., 2020b; Kiss et al., 2020). The compatibility of microbial pathogens with other organisms such as those providing ecosystem services (Srinivasan et al., 2019), including predators and parasites, has made IPM an attractive approach. There are various approaches that can be taken, from the rotational use of biopesticides with other softer control options, or of mixing agents together. As discussed by Kiss et al. (2020), implementation of IPM is easier in controlled situations such as glasshouses, than in open fields. IPM approaches can take advantage of the synergism that can occur between multiple agents. For example, neem has been shown to synergistically enhance NPV infections of *Spodoptera* and *Helicoverpa armigera* (Nathan and Kalaivani, 2006; Senthilkumar et al., 2008).

There are many examples of IPM using biopesticides in combinations. For example, Aristizabal et al. (2017) used the microbial biopesticides based on *Beauveria bassiana* GHA (BotaniGard) and *Metarhizium brunneum* F52 (Met-52 EC) in combinations with a horticultural oil (SuffOil-X) and azadirachtin (Molt-X) for the control of chilli thrips (*Scirtothrips dorsalis*) in Florida. In rotation, the agents gave as good a level of control as each alone, but tank-mixing the agents provided no additional benefit. Kiss et al. (2020) use the example of Ecophyto, a series of programmes by the French government to reduce the use of pesticides and plant protection products. One aim of the programmes is to shift from synthetic pesticides to biocontrol products. They quote a 33% reduction in the use of pesticides in horticultural production over 1900 test farms over 8 years.

Effective management of pest tephritid fruit fly species in various horticultural systems requires implementation of multiple techniques in an IPM program to reduce pest populations, minimise yield losses and enable growers to satisfy export market regulations (Aluja, 1996; Vargas et al., 2015;

Lezama-Gutierrez et al., 2000; Quesada-Moraga et al., 2006; Cossentine et al., 2010; Ekesi et al., 2011; Toledo et al., 2017; Yousef et al., 2017). Adult females cause damage by puncturing or 'stinging' ripening fruits, to lay eggs under the fruit skin. The eggs then hatch into larvae that feed on the flesh and these infested fruits subsequently rot (Aluja, 1996). Common techniques to achieve suppression of adult fruit flies and reduce produce losses include using food baits or attractants laced with insecticides. Food baits usually provide a protein source suitable for both female and male fruit flies, whereas pheromone attractants typically target male adult flies for a male annihilation technique (MAT) (Yousef et al., 2018). The MAT can also be used in combination with the sterile insect technique (SIT), where sterile male fruit flies are released to mate with females in order to limit damage from egg laying (Enkerlin et al., 2015). A biopesticide product, GF-120 (Dow AgroSciences, Indianapolis, IN), is an effective protein bait used for fruit flies which contains an insecticidal toxin, spinosad, that is derived from metabolites of the bacterium *Saccharopolyspora spinosa* (Sparks et al., 1995). In Hawaii, a GF-120 spinosad bait spray has been used effectively in combination with MAT and SIT, and with biological control (Mau et al., 2007). The efficacy of an application of GF-120 spinosad bait combined with an application of the entomopathogenic fungus *Metarhizium anisopliae* to soil was demonstrated by Ekesi et al. (2011) for the suppression of the tephritid fruit fly species, *Bactrocera dorsalis,* in Kenya. Prior studies also showed a reduction in adult emergence of different species of fruit flies in both laboratory and semi-field cage experiments after exposure to soil inoculated with *M. anisopliae* (Ekesi et al., 2007; Ouna, 2010).

Beauveria bassiana and *M. anisopliae* have also demonstrated pathogenicity against the Mediterranean fruit fly *Ceratitis capitata* adults, larvae, or pupae (Castillo et al. 1996; Quesada-Moraga et al., 2006; Soliman et al., 2020) and there have been studies conducted on the control of *Anastrepha fraterculus*, *A. ludens*, *B. oleae* and *B. tryoni* with different entomopathogenic fungi (Carswell et al., 1998; Lezama-Gutierrez et al., 2000;; Konstantopoulou and Mazomenos, 2005). The use of entomopathogenic nematodes against different tephritid species' larval stages has also been investigated (Langford et al., 2014; Soliman et al., 2014). Indeed, the application of these fungi or nematodes as biocontrol agents to soil can provide an additional control measure in IPM, because most control measures target adult flies and not larval stages.

5.1 *Case study: Developing* **Brevibacillus laterosporus** *for diamondback moth control in New Zealand*

Drawing on the authors' own experience, the development of the spore-forming bacterium, *Brevibacillus laterosporus* for diamondback moth (DBM) larval control highlights several of the factors that need to be considered in biopesticide development. The bacterium is a well-known entomopathogen, although historic publications are limited (Ruiu, 2013). Among species in the genus, *B. laterosporus* is the only truly entomopathogenic species. Although strains can contain parasporal crystals and in some cases produce Cry toxins (Marche et al., 2018; Glare et al., 2020a), strains without parasporal crystals can also kill insects (Ruiu et al., 2007).

In New Zealand, screening was undertaken using a range of bacterial isolates recovered from *Brassica* plants, which included bioassays for activity against DBM larvae. The screening uncovered several *B. laterosporus* isolates that were highly toxic to larvae (van Zijll de Jong et al., 2016). Genome sequencing was conducted and several Cry-encoding toxin genes were found to be present (Glare et al., 2020a). However, these were later demonstrated not to be the main toxins for DBM (Ormskirk, 2017). As the species is also used as a probiotic for human consumption, safety was expected, and limited host range studies showed no activity against bees or other beneficial insects. *B. laterosporus* had also been previously isolated from the digestive tract of healthy foraging adult honey bees as a non-pathogenic endosymbiont in Europe (Gilliam and Valentine, 1976). Given the toxicity to a major insect pest such as DBM, and a mode of action that was not based on Cry toxin, the biological control activity differed sufficiently from that of Bt as a biological product. Additionally, the ease of mass production of the isolate enabled the initiation of a project to develop a novel biopesticide based on this New Zealand *B. laterosporus* strain. The New Zealand strains were determined to be an identifiable genotype compared to other *B. laterosporus* (Glare et al., 2020a). Much of the research for the development of the biopesticide is unpublished, but some can be found in the PhD thesis of Marsha Ormskirk (2017).

Following the standard requirements of a biopesticide, we investigated efficacy in bioassays, host range, production, formulation and field efficacy. Given the origin of the strains was the *Brassica* vegetable seed, we also investigated the endophytic potential of *B. laterosporus* in brassica. The bacteria were able to colonise brassica plants at a low level, and the presence of the bacteria endophytically had a small but significant effect on pupation of DBM (Ormskirk et al., 2019). The project team were investigating a range of potential BCAs for DBM control, so research was included that looked at the effect of combined agents in the laboratory and in the field. Combined applications with fungal and another bacterial entomopathogen demonstrated occasional synergistic effects in the laboratory. However, in the field, the effect was not significantly different from the activity observed of the single agents alone (Narciso et al., 2019).

The use of the novel New Zealand strains was patented, which gave a commercial partner confidence in taking the development to the pre-commercial stage. Field trials were conducted in New Zealand (Ormskirk et al., in preparation) and overseas, which indicated efficacy in a variety of situations. Currently, the company Ecolibrium Biologicals is registering a product in New Zealand, called Lateral, which is based on the bacteria. The development to the point of a product registration covers the classic steps in biopesticide development – efficacy against a target, safety, economic mass production, field efficacy and registration.

This case study demonstrates the general process and work involved to ensure the success of biopesticide product. However, candidate entomopathogenic agents, even when shown to be highly virulent to a target pest, do not often result in the development of a biopesticide product. Table 1 outlines some of the typical barriers that hinder the success of candidate biocontrol agents.

6 New approaches

6.1 The microbiome

Additional properties of some entomopathogenic fungi and their ability to colonise plants and contribute to resistance to plant diseases are leading to new considerations of the use of these agents. Many studies have shown how entomopathogenic fungi, in particular, and some bacteria can colonise plants, leading to a new consideration of control opportunities. This can even occur during inundative treatments, as spray application of some fungi for insect control led to transient endophytic colonisation that provided additional mortality to that caused by the direct spray (Garrido-Jurado et al., 2017; Resquin-Romero et al., 2016). Vega (2018) reviewed 85 papers covering 12 entomopathogenic fungal species in 6 genera, colonising 38 plant species in 19 families as endophytes. He concluded that colonisation was often localised and ephemeral, and a lack of understanding of the endophytism and plant responses, as well as the effect on insects, is holding back exploitation of this ecological niche.

The microbiomes of soils, plants and insects have all become topics of a lot of research focus, as techniques to directly evaluate components have become refined and cheaper. New approaches to biocontrol using microbes include consideration of the microbiome, whether to modify the insect microbiome for pest control or on/in plants (Lefort et al., 2017). For insects which depend on components of their microflora for deriving essential nutrients, knocking out specific microbes might act as a biocontrol approach. However, this requires the identification of the key components of the microbiome and then an understanding of how to specifically target certain microbes.

As more information accumulates, it is clear that interactions between microbes in the community are crucial to successful biocontrol, whether in terms of interactions with an applied microbe, or regarding how the whole community might be manipulated for improved pest control outcomes. Microbes are known to provide nutrients to insects. Examples are *Buchnera aphidicola* in aphids and *Baumannia cicadellinicola* and *Sulcia muelleri* in the glassy-winged sharpshooter, *Homalodisca vitripennis* (reviewed in Gurung et al., 2019). Similar fungi can be involved in nutrient acquisition and host defence. One demonstrated benefit to plants of the *Metarhizium* infection of soil-dwelling insects was the transfer of nitrogen (Behie et al., 2012) and carbon (Behie et al., 2017) between the plant and the insect.

Viruses can sometimes manipulate host behaviour, in addition to being pathogenic. However, for many viruses, their interaction with insects is simply unknown. Similarly, protozoa and archaea are largely unexplored in this context (Gurung et al., 2019).

A review of patents in the area of beneficial endophytes (Ortega et al., 2020) listed 16 patents that mention 'pests,' including control of nematodes, although a few additional patents claim multiple activities of endophytes. The species included in the patents include well-published species with inverticidal properties (e.g. *Beauveria bassiana, Neotyphodium* sp., *Acremonium* spp., *Clonostachys rosea* and *Purpureocillium* sp.), but also groups better known as plant pathogens or saprophytes (*Aspergillus niger, Epichloë, Guignardia mangiferae, Muscodor* spp., *Sarocladium brachiariae, Cladosporium* sp., *Sphaerulina* sp., *Epicoccum* sp., *Stagonospora* sp., *Alternaria* sp., *Bipolaris* sp., *Stemphylium* sp., *Preussia* sp., *Curvularia* sp., *Penicillium* sp., *Phomopsis* sp., *Gibellulopsis* sp., *Fusarium* sp., *Chaetomium* sp. and *Cryptococcus* sp. (Ortega et al., 2020).

6.2 Genetic modification of microbes

Genetic modification of microbes has advanced greatly over the past decades and is seeing a step-change in possibilities with the use of CRISPR-Cas technologies. Modifying entomopathogens to overcome the current limitations to their use has been proposed and investigated for many years, although few genetically modified organisms have ever reached the field release and commercialisation stages. This is partly due to the current regulations in many countries.

Several studies have reported the expression of spider and scorpion toxin genes in *Metarhizium* and *Beauveria* that increase the virulence, even to the level of a single spore causing host death. Bilgo et al. (2017) reported that a range of toxins under a haemolymph-specific promoter, expressed in *Metarhizium pingshaense,* increased the virulence and speed-of-kill of mosquitoes, and a combination of two toxins, one from the North African desert scorpion *Androctonus australis (*AaIT) and another from the Australian Blue Mountains funnel-web spider *Hadronyche versuta* (hexatoxin-Hv1a) acted synergistically. This combination was used in *M. pingshaense* in successful field trials against mosquitoes in Burkina Faso (Lovett et al., 2019). *Beauveria bassiana* was also modified to express scorpion toxin AaIT, for use against the adult mosquito *Aedes albopictus*. The lethal concentration required was 313-fold lower on day 4 and killed faster compared to the wild type (Deng et al., 2017). Similarly, *B. bassiana* expressing *AaIT* and *Pri*, a gene from *Metarhizium*, was more virulent to the Masson's pine caterpillar *Dendrolimus punctatus* and the wax moth *Galleria mellonella*, and killed 36–40% faster (Lu et al., 2008). Expressing a trypsin-modulating oostatic factor in *B. bassiana* increased virulence and reduced time-to-kill against *Anopheles gambiae* mosquitoes (Kamareddine et al., 2013).

Metarhizium anisopliae has been modified to overexpress a toxic protease, Pr1, which was coupled with a haemolymph-induced promoter. This overexpression led to enhanced virulence in the haemolymph of *Manduca sexta*, exhibited a 25% reduction in time-of-death, and reduced food consumption by 40%, compared to infections by the wild-type fungus (St. Leger et al., 1996). In *Metarhizium robertsii*, overexpression of the esterase gene *mest1* increased virulence against caterpillars, while expression of the same gene in *Metarhizium acridum*, where is it otherwise absent in the wild type, allowed infection of caterpillar hosts similar to *M. robertsii* (Wang et al., 2011b).

Increased persistence has also been the target of modification efforts. With *M. anisopliae*, melanin biosynthetic genes from *Alternaria alternata* were expressed, showing improvements in antistress capabilities and virulence against *Plutella xylostella* (Tseng et al., 2011).

Combining entomopathogenic components has also shown promise. Deng et al. (2019) reported that the expression of Bt Cyt2Ba toxin in *B. bassiana* showed increased virulence against *Aedes* mosquitoes. The lethal time-to-kill was reduced by up to 47%, although in other studies, expression of Bt toxins in NPVs did not improve the toxicity (Ribeiro and Crook, 1993; Martens et al., 1995).

Similarly, new approaches include combining entomopathogenicity with plant disease antagonists. Sorokan et al. (2020) reported that an endophytic strain of *Bacillus subtilis* that can inhibit mycelial growth of the potato pathogen *Phytophthora infestans* can also be modified to express the Cry1Ia insecticidal endotoxin from Bt. When tested, the modified strain showed insecticidal activity against the Colorado potato beetle *Leptinotarsa decemlineata*

The potential of modification has been demonstrated in areas such as increased virulence, reduced time-to-kill, altered host range, and in the case of some NPVs, reduced plant damage before insect death. Around 20 baculoviruses have been genetically modified to enhance performance (Scheepmaker et al., 2016). One area of focus in the 1990s was the viral *egt* gene, which blocks insect moulting to prolong caterpillar feeding. This is a method for the virus to enhance the reproduction of viral particles in a fatter caterpillar but also results in more damage to the crop by infected caterpillars. Deletion in the *Autographa californica* nucleopolyhedrovirus (AcMNPV) of *egt* resulted in ~405× less feeding damage before insect death (Tomalski and Miller, 1992). This virus was also modified to express the insect toxin gene (Androctonus australis hector insect toxin [AaHIT]) from the scorpion *Androclonus australis,* which resulted in a 25% reduction in LT_{50} against *Trichoplusia ni*, compared to the wild-type virus, but with unaltered pathogenicity and host range. This combination was used in field trials in the 1990s, resulting in reduced crop damage and faster mortality of the caterpillars (Cory et al., 1994). Expression of an algal virus pyrimidine dimer-specific glycosylase cv-PDG in AcMNPV made the virus less susceptible to UV (3-fold decrease) and conferred a 16× increase in virulence (Petrika et al., 2003).

Toxins from straw itch mites (Tomalski and Miller, 1992), scorpions (Maeda et al., 1991; Elazar et al., 2001), sea anemone, and spiders have also been expressed in a range of NPVs, generally showing reductions in the time-to-kill by transgenic viruses compared to the wild type (reviewed in Kroemer et al., 2015). However not all gene deletions seem to affect the activity of NPVs (Bianchi et al., 2000; Kroemer et al., 2015).

Other modifications have included the deletion of a polyhedral envelope gene which resulted in a 6× increase in infectivity against *Trichoplusia* ni by AcMNPV (Ignoffo et al., 1995). The discovery of RNA interference as a gene-specific regulation mechanism has led to its use as a direct pest control approach (discussed below), but it has also been used to improve baculovirus efficacy (Kroemer et al., 2015).

The genes encoding Bt Cry toxins were identified many years ago, and some of these genes have been transgenically expressed in the most important crop plants in the world, such as maize, cotton and soybean. Bt-plants are the most common transgenic plants grown in the world. Similarly, expression of Cry toxin genes in other bacteria has been common, such as the expression of Bt cry1Ac in *Pseudomonas fluorescens,* with the recombinants highly toxic to *Helicoverpa armigera* (Ding et al., 2000). Recombinant *B. thuringiensis israelensis* with toxins from *B. sphaericus* were up to 5× more potent against *Culex* mosquitoes than wild-type strains (Park et al., 2003). The expression of genes encoding insect toxins in other beneficial bacteria has been explored. For example, a dipteran-active toxin gene from *B. thuringiensis israelensis* was expressed in the nitrogen-fixing bacterium *Bradyrhizobium* in nodules of pigeonpea, which provided protection against larvae of *Rivellia angulate* (Diptera) (Nambiar et al., 1990). Other potentially beneficial metabolites and enzymes can be expressed in insecticidal species for pest control enhancement. A Bt-expressing *Serratia marcescens* chitinase gene can help weaken the integrity of the insect gut epithelium, giving better access for bacterial toxin genes to the insect, for stronger biocontrol activity (Okay et al., 2008).

The potential of genetic modification for the improvement of the active agents in biopesticides has been shown for decades. However, few commercially available products have been brought to market. Two Bt-based biopesticides were commercialised; CrymaxTM, based on a Bt strain with an additional Cry1C protein expressed, was commercialised by Certis; and LepinoxTM, another Bt which was modified to produce Cry 1Ac/1F proteins (Scheepmaker et al., 2016). The lack of GMO-based products is likely due to a combination of lack of public acceptance, restrictive regulations and insufficient benefits demonstrated to go through the regulatory process. There is also concern about non-target impacts (Azizoglu et al., 2020). With ever-increasing knowledge and development of novel techniques, and pressure for new control approaches, it is likely that GMO- based biopesticides will be sold in some countries in the near future.

6.3 The use of RNAi for insect control

A growing area of development for new biopesticide approaches is the use of RNAi for insect control. While not strictly a microbial-based approach, the approach can be used in genetically modified crops (expressing dsRNA) or applied as a formulation through a spray application, plant injection or irrigation (Yan et al., 2020). RNAi can also be transgenically expressed by microbes for enhanced effects. Expressing dsRNA in bacteria is considered the most efficacious production method (Joga et al., 2016). Sprayable RNAi has shown potential in pest management and has been trialled in horticulture. Topically applied RNAi can cause insect death, especially in insects such as aphids and mosquitoes, and has been demonstrated in greenhouse control of *L. decemlineata* (San Miguel et al., 2016). However, this approach has not been effective against all insects (Yan et al., 2020). In horticulture, application through irrigation has been used, such as for citrus tree protection (Hunter et al., 2012).

Delivery remains an issue, as formulation improvements are still needed to protect the RNAi from degradation. Nanoparticles have been investigated for this purpose, which can also promote translocation of the dsRNA across the peritrophic membrane or insect cuticle (Yan et al., 2020). There are still issues with RNAi as a pest management tool, including high production cost, sublethal effects against non-target species, resistance development and efficiency (Yan et al., 2020).

6.4 Production and formulation advances

Production is a major step in the development of biopesticides, when an organism or its bioactive metabolites are used inundatively. Scale-up is expensive and so not all potential biopesticides are mass-produced, often because the market size is too small to justify the expense of production set-up. Growing massive amounts of cells can be challenging, and for some entomopathogens, *in vitro* growth is not possible currently (e.g. viruses, some nematodes and protozoa). Solid- and liquid-state fermentation are the basic methods to produce microbes for biopesticides, and in general, production has been incrementally improving. However, most of the technology has not changed much for decades.

For entomopathogenic fungi, local production has been a successful approach, especially where labour costs are low. Using production on solid substrates such as grains, fungi including *Beauveria* and *Metarhizium* can be produced cheaply in the locality of need. This approach has been widely used in the sugarcane-growing regions of Central and South America (Lacey et al., 2015), and there are other examples. However, in general, current mass solid fermentation methods for conidia production are not considered cost effective (Singhania et al., 2016).

For bacteria, liquid fermentation has been the most common approach, in the standard stirred batch fermentation. This has been particularly cost effective for the spore-forming bacilli, including Bt. For non-spore-forming bacteria, liquid fermentation has also been used, but post-production formulation and storage are more problematic without a spore stage (Johnson et al., 2001).

There are new production methods under investigation, but there must be a significant cost-saving incentive for companies to change processes, as capital equipment costs can be high. For example, approaches such as liquid interface production of fungi (e.g. Lopes et al., 2018) and production of microbes for biopesticides on waste materials have been investigated, as a method to reduce costs (e.g. Ballardo et al., 2016).

One area showing promise for the production of entomopathogenic fungi is the production of microsclerotia in liquid culture. Many fungi produce sclerotia - pigmented hyphal aggregates - which can survive environmental stresses better than most spores. It has been demonstrated that a number of entomopathogenic fungi can produce what are termed 'microsclerotia,' small sclerotia that can produce conidia after storage. The production of these microsclerotia was first demonstrated with *Metarhizium* spp. (Jackson and Jaronski, 2009; Song et al., 2014) and suggested as a useful new approach to producing propagules for biocontrol. Subsequent studies have also shown that *Beauveria* spp. can produce microsclerotia in submerged culture (Wang et al., 2011a; Villamizar et al., 2018). Huarte-Bonnet et al. (2019) have recently demonstrated that aggregation is dependent on FE2+ presence. Microsclerotia can be dried for formulation and storage, and then produce hyphae and sporulate. This makes microsclerotia a promising approach to liquid culture production of fungal biopesticides, which would be a cheaper approach than the use of solid substrates for large-scale commercial production (Jackson and Jaronski, 2009; Song et al., 2014; Song, 2018).

7 Conclusion and future trends

Biopesticides are moving from niche products to more widespread use in large horticultural industries. Integrated pest management has been implemented in horticultural systems for many years, and in some cases, entomopathogens are highly effective tools in the arsenal of pest management systems. Integrated pest management systems are going to become much more common as synthetic pesticide use declines, and entomopathogens are well placed to be major components of new systems.

The increase in the use of IPM and of biopesticides is also due to improved technologies. The advent of new monitoring technologies, particularly automated systems, is benefiting the implementation of IPM, allowing more accurate prediction of pest peaks. In this review, we have outlined both a range of traditional approaches to IPM, such as combinations with other natural enemies and attractants, and examples of new technologies that have potential. Figure 1 provides a concept map that considers the various ideas for research and improvement of microbial-based biopesticides, which includes how candidate agents may fit within IPM programs.

In addition to the use of microbiome approaches, genetic engineering and RNAi, other new technologies are emerging. One method that has been used for a number of years is 'lure-and-kill' (El-Sayed et al., 2009), in which pheromone insect traps are laced with a microbial pathogen to kill

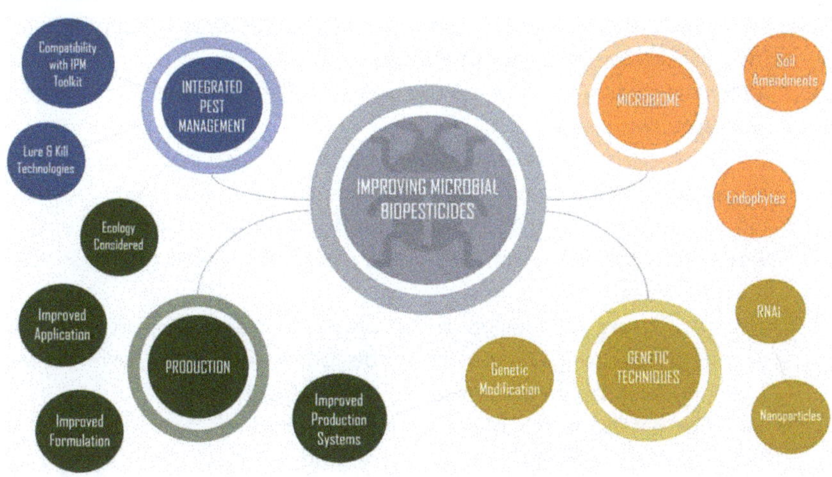

Figure 1 Concept map to consider how to improve biopesticide agent outcomes for insect pest management.

adults. A new approach by the American biotech company Pheronym is to use a novel nematode pheromone extract that increases the effectiveness of entomopathogenic nematodes, by increasing dispersal of the nematodes to hunt for insect targets. In a patent, the inventors claim that the use of the pheromone also expands the effective temperature range for the nematode pathogens. These Ascaroside pheromones stimulate dispersal of nematodes such as *Steinernema* spp., which both increase the dispersal of nematodes and reduce the quiescent period, which was reducing *S. carpocapsae* effectiveness at lower temperatures (Kaplan et al., 2020a). The team has also demonstrated the effect of the pheromone in microgravity (Kaplan et al., 2020b). This is an example of how a new understanding of the fundamentals of ecology and behaviour can be manipulated to improve biopesticide use in IPM.

Biopesticide use in horticulture will continue to increase, especially as the multinational pesticide companies move into these product opportunities and consumer demand drives the need for more environmentally friendly and safe pest control methods.

8 Where to look for further information

Further information on biopesticides can be found in many reviews and books devoted to the subject:

1. The Society for Invertebrate Pathology (http://www.sipweb.org/) is devoted to the study of entomopathogens.
2. The International Biocontrol manufacturers and the Research Institute of Organic Agriculture jointly run an Annual Biocontrol Industry meeting (https://www.abim.ch/about-abim.html).
3. International Organisation for Biological and Integrated Control: https://www.iobc-wprs.org/.
4. OECD biological pesticides website: http://www.oecd.org/chemicalsafety/pesticides-biocides/biological-pesticides.htm.
5. Information on regulations can be found on registration entities' websites, particularly the EPA USA (https://www.epa.gov/pesticides/biopesticides).
6. The FAO works to develop and promote IPM and support regulation of biocontrol products. Details can be found here: http://www.fao.org/agriculture/crops/thematic-sitemap/theme/pests/en/.
7. The EU database for plant protection products and guidance for biopesticides can be found here: http://ec.europa.eu/food/plant/pesticides/eu-pesticides-database/public/?event=homepage&language=EN.

8.1 Books

- Hall, F. R. and Menn, J. J. (Eds). (1999). *Biopesticides: Use and Delivery*. Springer.
- Birch, N. and Glare, T. R. (Eds). (2020). *Biopesticides for Sustainable Agriculture*. Burleigh Dodds Science Publishing Ltd., UK. 366 pp.

9 References

Aluja, M. (1996). Future trends in fruit fly management. In: McPheron, B. A. and Steck, G. J. (Eds), *Fruit Fly Pests: A World Assessment of Their Biology and Management*, St Lucie Press, Delray Beach, pp. 309–320.

Aristizabal, L. F., Chen, Y., Cherry, R. H., Cave, R. D. and Arthurs, S. P. (2017). Efficacy of biorational insecticides against chilli thrips, *Scirtothrips dorsalis* (Thysanoptera: Thripidae), infesting roses under nursery conditions. *J. Appl. Entomol.* 141(4), 274–284.

Azizoglu, U., Salehi Jouzani, G., Yilmaza, N., Baz, E. and Ozkok, D. (2020). Genetically modified entomopathogenic bacteria, recent developments, benefits and impacts: A review. *Sci. Total Environ.* 734, 139–169.

Ballardo, C., Abraham, J., Barrena, R., Artola, A., Gea, T. and Sánchez, A. (2016). Valorization of soy waste through SSF for the production of compost enriched with *Bacillus thuringiensis* with biopesticide properties. *J. Environ. Manage.* 169, 126–131.

Bateman, R. P. (2003). Rational pesticide use: Spatially and temporally targeted application of specific products. In: Wilson, M. (Ed.), *Optimising Pesticide Use*, John Wiley & Sons Ltd, Chichester, pp. 129–157.

Behie, S. W., Moreira, C. C., Sementchoukova, I., Barelli, L., Zelisko, P. M. and Bidochka, M. J. (2017). Carbon translocation from a plant to an insect-pathogenic endophytic fungus. *Nat. Commun.* 8, 14245.

Behie, S. W., Zelisko, P. M. and Bidochka, M. J. (2012). Endophytic insect-parasitic fungi translocate nitrogen directly from insects to plants. *Science.* 336(6088), 1576–1577.

Bell, R. A., Owens, C. D., Shapiro, M. and Tardif, J. R. (1981). Development of mass-rearing technology. In: Doane, C. C. and McManus, M. L. (Eds), *The Gypsy Moth: Research Towards Integrated Pest Management*. *For. Serv. Techn. Bull.* 1584, 599–632.

Bernal, A., Simón, O., Williams, T., Muñoz, D. and Caballero, P. (2018). Remarkably efficient production of a highly insecticidal *Chrysodeixis chalcites* nucleopolyhedrovirus (ChchNPV) isolate in its homologous host. *Pest Manage. Sci.* 74(7), 1586–1592.

Bianchi, F. J., Snoeijing, I., van der Werf, W., Mans, R. M., Smits, P. H. and Vlak, J. M. (2000). Biological activity of SeMNPV, AcMNPV, and three AcMNPV deletion mutants against *Spodoptera exigua* larvae (Lepidoptera: Noctuidae). *J. Invertebr. Pathol.* 75(1), 28–35.

Bidochka, M. J., Kamp, A. M., Lavender, T. M., Dekoning, J. and De Croos, J. N. A. (2001). Habitat association in two genetic groups of the insect-pathogenic fungus *Metarhizium anisopliae*: Uncovering cryptic species? *Appl. Environ. Microbiol.* 67(3), 1335–1342.

Bilgo, E., Lovett, B., Fang, W., Bende, N., King, G. F., Diabate, A. and St. Leger, R. J. (2017). Improved efficacy of an arthropod toxin expressing fungus against insecticide-resistant malaria-vector mosquitoes. *Sci. Rep.* 7(1), 3433.

Blanford, S. and Thomas, M. B. (2001). Adult survival, maturation, and reproduction of the desert locust *Schistocerca gregaria* infected with the fungus *Metarhizium anisopliae* var *acridum*. *J. Invertebr. Pathol.* 78(1), 1-8.

Blond, C. L. H., Ridgway, H. J., Brownbridge, M., Chapman, R. B., Condron, L. M., Saville, D. J. and Glare, T. R. (2018). Non-host larvae negatively impact persistence of the entomopathogen *Beauveria bassiana* in soil. *J. Invertebr. Pathol.* 156, 19-28.

Boston, W., Leemon, D. and Cunningham, J. P. (2020). Virulence screen of *Beauveria bassiana* isolates for Australian *Carpophilus* (Coleoptera: Nitidulidae) beetle biocontrol. *Agronomy.* 10(8), 1207.

Brooks, A. J., Muro, M. A., Burree, E., Moore, D., Taylor, M. A. and Wall, R. (2004). Growth and pathogenicity of isolates of the fungus *Metarhizium anisopliae* against the parasitic mite, *Psoroptes ovis*: Effects of temperature and formulation. *Pest Manage. Sci.* 60(10), 1043-1049.

Brownbridge, M., Townsend, R. J., O'Callaghan, M., Bell, N. L., Mander, C. and Jackson, T. A. (2008). Developing opportunities for entomopathogenic microbes and nematodes in crop protection. In: Butcher, M. R., Walker, J. T. S. and Zydenbos, S. M. (Eds), *Future Challenges in Crop Protection: Repositioning New Zealand's Primary Industries for the Future*, New Zealand Plant Protection Society Inc., Auckland, New Zealand, pp. 129-142.

Bruck, D. J. (2010). Fungal entomopathogens in the rhizosphere. *BioControl.* 55(1), 103-112.

Butt, T. M. and Goettel, M. S. (2000). Bioassays of entomogenous fungi. In: Navon, A. and Ascher, K. R. S. (Eds), *Bioassays of Entomopathogenic Microbes and Nematodes*, CAB International, Wallingford, pp. 141-195.

Campos-Herrera, R., Stuart, R. J., Pathak, E., El-Borai, F. E. and Duncan, L. W. (2019). Temporal patterns of entomopathogenic nematodes in Florida citrus orchards: Evidence of natural regulation by microorganisms and nematode competitors. *Soil Biol. Biochem.* 128, 193-204.

Carswell, I., Spooner-Hart, R. and Milner, R. J. (1998). Laboratory susceptibility of *Musca domestica* L. (Diptera: Muscidae) *and Bactrocera tryoni* (Frogatt) (Diptera: Tephritidae) to an isolate of *Metarhizium anisopliae* (Metsch.) Sorokin. *Aust. J. Entomol.* 37, 281-284. doi: 10.1111/j.1440-6055.1998.tb01584.x.

Castillo, M. A., Moya, P., Hernándeza, E. and Primo-Yúferab, E. (1996). Susceptibility of *Ceratitis capitata* Wiedemann (Diptera: Tephritidae) to entomopathogenic fungi and their extracts. *Biol. Control.* 19, 274-282.

Caudwell, R. W. and Gatehouse, A. G. (1996). Formulation of grasshopper and locust entomopathogens in baits using starch extrusion technology. *Crop Prot.* 15(1), 33-37.

Chandler, D., Bailey, A. S., Tatchell, G. M., Davidson, G., Greaves, J. and Grant, W. P. (2011). The development, regulation and use of biopesticides for integrated pest management. *Phil. Trans. Roy. Soc. Lond. B Biol. Sci.* 366(1573), 1987-1998.

Chapple, A. C., Downer, R. A. and Bateman, R. P. (2000). Theory and practice of microbial insecticide application. In: Lacey, L. and Kaya, H. (Eds), *Field Manual of Techniques in Invertebrate Pathology*, Kluwer Publishers, NLE, Dordrecht, pp. 5-37.

Charnley, A. K. and Collins, S. A. (2007). Entomopathogenic fungi and their role in pest control. In: Kubicek, C. P. and Druzhinina, I. S. (Eds), *Environmental and Microbial Relationships: The Mycota IV* (2nd edn.), Springer-Verlag, Berlin, pp. 159-187.

Cliquet, S. and Jackson, M. A. (2005). Impact of carbon and nitrogen nutrition on the quality, yield and composition of blastospores of the bioinsecticidal fungus *Paecilomyces fumosoroseus*. *J. Ind. Microbiol. Biotechnol.* 32(5), 204-210.

Copping, L. G. and Menn, J. J. (2000). Biopesticides: A review of their action, applications and efficacy. *Pest Manage. Sci.* 56(8), 651–676.

Cory, J. S. and Ericsson, J. D. (2010). Fungal entomopathogens in a tritrophic context. *BioControl.* 55(1), 75–88.

Cory, J. S., Hirst, M. L., Williams, T., Hails, R. S., Goulson, D., Green, B. M., Carty, T. M., Possee, R. D., Cayley, P. J. and Bishop, D. H. L. (1994). Field trial of a genetically improved baculovirus insecticide. *Nature.* 370(6485), 138–140.

Cossentine, J., Thistlewood, H., Goettel, M. and Jaronski, S. (2010). Susceptibility of preimaginal western cherry fruit fly, *Rhagoletis indifferens* (Diptera: Tephritidae) to *Beauveria bassiana* (Balsamo) Vuillemin Clavicipitaceae (Hypocreales). *J. Invertebr. Pathol.* 104(2), 105–109.

Damalas, C. A. and Koutroubas, S. D. (2018). Current status and recent developments in biopesticide use. *Agriculture.* 8(1), 13.

Daoust, R. A. and Roberts, D. W. (1982). Virulence of natural and insect-passaged strains of *Metarhizium-anisopliae* to mosquito larvae. *J. Invertebr. Pathol.* 40(1), 107–117.

Dastogeer, K. M. G., Tumpa, F. H., Sultana, A., Akter, M. A. and Chakraborty, A. (2020). Plant microbiome: An account of the factors that shape community composition and diversity. *Curr. Plant Biol.* 23, 100161.

Davies, F. T. (2015). Opportunities for horticulture to feed the world©. *Acta Hortic.* 1085, 455–458.

Deacon, J. W. and Berry, L. A. (1993). Biocontrol of soil-borne plant-pathogens: Concepts and their application. *Pestic. Sci.* 37(4), 417–426.

Deng, S. Q., Cai, Q. D., Deng, M. Z., Huang, Q. and Peng, H. J. (2017). Scorpion neurotoxin AaIT-expressing *Beauveria bassiana* enhances the virulence against *Aedes albopictus* mosquitoes. *AMB Exp.* 7(1), 121.

Deng, S. Q., Zou, W. H., Li, D. L., Chen, J. T., Huang, Q., Zhou, L. J., Tian, X. X., Chen, Y. J. and Peng, H. J. (2019). Expression of *Bacillus thuringiensis* toxin Cyt2Ba in the entomopathogenic fungus *Beauveria bassiana* increases its virulence towards *Aedes* mosquitoes. *PLoS Negl. Trop. Dis.* 13(7), e0007590.

Ding, Z., Zhang, J., Song, F., Huang, D. and Li, J. (2000). [Expression and synergism of two Cry insecticidal protein genes in *Pseudomonas fluorescens*]. *Wei Sheng Wu Xue Bao* 40(6), 573–578. [Chinese].

Eilenberg, J., Hajek, A. and Lomer, C. (2001). Suggestions for unifying the terminology in biological control. *BioControl.* 46(4), 387–400.

Ekesi, S., Dimbi, S. and Maniania, N. K. (2007). The role of entomopathogenic fungi in the integrated management of fruit flies (Diptera: Tephritidae) with emphasis on species occurring in Africa. In: Ekesi, S. and Maniania, N. K. (Eds), *Use of Entomopathogenic Fungi in Biological Pest Management*, Research Signpost, Trivandrum, India, pp. 239–374.

Ekesi, S., Maniania, N. K. and Mohamed, S. A. (2011). Efficacy of soil application of *Metarhizium anisopliae* and the use of GF-120 Spinosad bait spray for suppression of *Bactrocera invadens* (Diptera: Tephritidae) in mango orchards. *Biocon. Sci. Technol.* 21, 299–316.

Elazar, M., Levi, R.. and Zlotkin, E. (2001). Targeting of an expressed neurotoxin by its recombinant baculovirus. *J. Exp. Biol.* 204(15), 2637–2645.

El-Sayed, A. M., Suckling, D. M., Byers, J. A., Jang, E. B. and Wearing, C. H. (2009). Potential of "lure and kill" in long-term pest management and eradication of invasive species. *J. Econ. Entomol.* 102(3), 815–835.

El-Wakeil, N. E. (2013). Retracted article: Botanical pesticides and their mode of action. *Gesunde Pflanzen.* 65(4), 125–149.

Enkerlin, W., Gutiérrez-Ruelas, J. M., Villaseñor Cortes, A. V., Roldan, E. C., Midgarden, D., Lira, E., López, J. L. Z., Hendrichs, J., Liedo, P. and Arriaga, F. J. T. (2015). Area freedom in Mexico from Mediterranean fruit fly (Diptera: Tephritidae): A review of over 30 years of a successful containment program using an integrated area-wide SIT approach. *Fla Entomol.* 98(2), 665–681.

Epstein, D. L., Zack, R. S., Brunner, J. F., Gut, L. and Brown, J. J. (2000). Effects of broad-spectrum insecticides on epigeal arthropod biodiversity in Pacific Northwest apple orchards. *Environ. Entomol.* 29(2), 340–348.

Fargues, J., Ouedraogo, A., Goettel, M. S. and Lomer, C. J. (1997). Effects of temperature, humidity and inoculation method on susceptibility of *Schistocerca gregaria* to *Metarhizium flavoviride*. *Sci. Technol.* 7(3), 345–356.

Ferguson, K. I. and Stiling, P. (1996). Non-additive effects of multiple natural enemies on aphid populations. *Oecologia.* 108(2), 375–379.

Garrido-Jurado, I., Resquın-Romero, G., Amarilla, S. P., Rıos-Moreno, A., Carrasco, L. and Quesada-Moraga, E. (2017). Transient endophytic colonization of melon plants by entomopathogenic fungi after foliar application for the control of *Bemisia tabaci* Gennadius (Hemiptera: Aleyrodidae). *J. Pest Sci.* 90(1), 319–330.

Garriga, A., Morton, A., García-López, D. and García-del-Pino, F. (2019). Compatibility of entomopathogenic nematodes with natural enemies for horticultural pest control. *Biol. Control.* 138, 104050.

Gilliam, M. and Valentine, D. K. (1976). Bacteria isolated from the intestinal contents of foraging worker honey bees, *Apis mellifera*: The Genus *Bacillus*. *J. Invertebr. Pathol.* 28(2), 275–276.

Glare, T. R. (2015). Types of biopesticides. In: Nollet, L. and Rathore, H. S. (Eds), *Handbook of Biopesticides*, CRC Press, Boca Raton, pp. 7–24.

Glare, T. R., Caradus, J., Gelernter, W. D., Jackson, T., Keyhani, N., Köhl, J., Marrone, P., Morin, L. and Stewart, A. (2012). Have biopesticides come of age? *Trends Biotechnol.* 30(5), 250–258.

Glare, T. R., Durrant, A., Berry, C., Palma, L., Ormskirk, M. M. and Cox, M. P. (2020a). Phylogenetic determinants of toxin gene distribution in genomes of *Brevibacillus laterosporus*. *Genomics.* 112(1), 1042–1053.

Glare, T. R., McKinnon, A. C., Gwynn, R. L. and Brownbridge, M. (2020b). Advances in microbial control in integrated pest management: Entomopathogenic fungi. In: Kogan, M. and Higley, L. (Eds), *Integrated Management of Pest Insects: Current and Future Developments*, Burleigh Dodds Science Publishing, Cambridge, pp. 549–592.

Grant, W. and Gwynn, R. (2020). Improving regulatory approval process for biopesticides and other new biological technologies in agriculture. In: Birch, N. and Glare, T. R. (Eds), *Biopesticides for Sustainable Agriculture*, Burleigh Dodds Science Publishing, Cambridge, pp. 45–62.

Grzywacz, D., Moore, D. and Rabindra, R. J. (2014). Mass production of entomopathogens in less industrialized countries. In: Morales-Ramos, J. A., Guadalupe Rojas, M. and Shapiro-Ilan, D. I. (Eds), *Mass Production of Beneficial Organisms: Invertebrates and Entomopathogens*, Academic Press/Elsevier Inc., London, pp. 519–561.

Gurung, K., Wertheim, B. and Falcao Salles, J. (2019). The microbiome of pest insects: It is not just bacteria. *Entomol. Exp. Appl.* 167(3), 156–170.

Hossain, M. S., Williams, D. G., Mansfield, C., Bartelt, R. J., Callinan, L. and Il'ichev, A. L. (2006). An attract-and-kill system to control *Carpophilus* spp. in Australian stone fruit orchards. *Entomol. Exp. Appl.* 118(1), 11-19.

Hoy, M. A. (1992). Biological control of arthropods: Genetic engineering and environmental risks. *Biol. Control.* 2(2), 166–170.

Huarte-Bonnet, C., Paixão, F. R. S., Mascarin, G. M., Santana, M., Fernandes, ÉK. K. and Pedrini, N. (2019). The entomopathogenic fungus *Beauveria bassiana* produces microsclerotia-like pellets mediated by oxidative stress and peroxisome biogenesis. *Environ. Microbiol. Rep.* 11(4), 518–524.

Hunter, W. B., Glick, E., Paldi, N. and Bextine, B. R. (2012). Advances in RNA interference: dsRNA treatment in trees and grapevines for insect pest population suppression. Southwestern Entomol. 37, 85–87.

Ignoffo, C. M., Garcia, C., Zuidema, D. and Vlak, J. M. (1995). Relative in vivo activity and simulated sunlight-UV stability of inclusion bodies of a wild-type and an engineered polyhedral envelope-negative isolate of the nucleopolyhedrosis virus of *Autographa californica*. *J. Invertebr. Pathol.* 66(2), 212–213.

Inglis, G. D., Duke, G. M., Goettel, M. S. and Kabaluk, J. T. (2008). Genetic diversity of *Metarhizium anisopliae* var. *anisopliae* in southwestern British Columbia. *J. Invertebr. Pathol.* 98(1), 101–113.

Inglis, G. D., Enkerli, J. and Goettel, M. S. (2012). Laboratory techniques used for entomopathogenic fungi: Hypocreales. In: Lacey, L. A. (Ed.), *Manual of Techniques in Invertebrate Pathology* (2nd edn.), Academic Press, New York, pp. 189–253.

Jackson, M. A., Dunlap, C. A. and Jaronski, S. T. (2010). Ecological considerations in producing and formulating fungal entomopathogens for use in insect biocontrol. *BioControl.* 55(1), 129–145.

Jackson, M. A. and Jaronski, S. T. (2009). Production of microsclerotia of the fungal entomopathogen *Metarhizium anisopliae* and their potential for use as a biocontrol agent for soil-inhabiting insects. *Mycol. Res.* 113(8), 842–850.

Jaronski, S. T. (2010). Ecological factors in the inundative use of fungal entomopathogens. *BioControl.* 55(1), 159–185.

Jaronski, S. T. and Jackson, M. A. (2012). Mass production of entomopathogenic Hypocreales. In: Lacey, L. A. (Ed.), *Manual of Techniques in Invertebrate Pathology*, Academic Press, San Diego, pp. 257–286.

Joga, M. R., Zotti, M. J., Smagghe, G. and Christiaens, O. (2016). RNAi efficiency, systemic properties, and novel delivery methods for pest insect control: What we know so far. Front. Physiol. 7. doi: 10.3389/fphys.2016.00553.

Johnson, V. W., Pearson, J. F. and Jackson, T. A. (2001). Formulation of *Serratia entomophila* for biological control of grass grub. *N. Z. Plant Prot.* 54, 125–127.

Jurat-Fuentes, J. L. and Jackson, T. A. (2012). Bacterial entomopathogens. In: Vega, F. and Kaya, H. (Eds), *Insect Pathology* (2nd edn.), Elsevier, San Diego, CA, pp. 265–349.

Kamareddine, L., Fan, Y., Osta, M. A. and Keyhani, N. O. (2013). Expression of trypsin-modulating oostatic factor (TMOF) in an entomopathogenic fungus increases its virulence towards *Anopheles gambiae* and reduces fecundity in the target mosquito. *Parasit. Vectors.* 6, 22.

Kaplan, F., Perret-Gentil, A., Giurintano, J., Stevens, G., Erdogan, H., Schiller, K. C., Mirti, A., Sampson, E., Torres, C., Sun, J., Lewis, E. E. and Shapiro-Ilan, D. (2020a). Conspecific and heterospecific pheromones stimulate dispersal of entomopathogenic nematodes during quiescence. *Sci. Rep.* 10(1), 5738.

Kaplan, F., Shapiro-Ilan, D. and Schiller, K. C. (2020b). Dynamics of entomopathogenic nematode foraging and infectivity in microgravity. *NPJ Microgravity.* 6, 20.

Keyser, C. A. (2010). *Development of a Laboratory Based System for Selecting Insect Pathogenic Fungi with Greatest Potential for Success in the Field*, Vol. 604. All Graduate Theses and Dissertation. Utah State University, Utah.

Kienwick, S. (2007). Practicalities of developing and registering microbial biological control agents. *CAB Rev.* 2(13), 1–11.

Kiss, J., Istvan, S. and Delos, M. (2020). Implementing biopesticides as part of an integrated pest management (IPM) programme. In: Birch, N. and Glare, T. R. (Eds), *Biopesticides for Sustainable Agriculture*, Burleigh Dodds Science Publishing, Cambridge, pp. 21–44.

Konstantopoulou, M. and Mazomenos, B. E. (2005). Evaluation of Beauveria bassiana and B. brongniartii strains and four wild-type fungal species against adults of Bactrocera oleae and Ceratitis capitate. BioControl 50(2), 293–305. doi: 10.1007/s10526-004-0458-4.

Koppenhőfer, A. M., Shapiro-Illan, I. and Hiltpold, I. (2020). Advances in the use of entomopathogenic nematodes (EPNs) in suppressing crop pest insects. In: Birch, N. and Glare, T. R. (Eds), *Biopesticides for Sustainable Agriculture*, Burleigh Dodds Science Publishing, Cambridge, pp. 195–232.

Kowalska, J., Tyburski, J., Matysiak, K., Tylkowski, B. and Malusá, E. (2020). Field exploitation of multiple functions of beneficial microorganisms for plant nutrition and protection: Real possibility or just a hope? *Front. Microbiol.* 11, 1904–1904.

Kroemer, J. A., Bonning, B. C. and Harrison, R. L. (2015). Expression, delivery and function of insecticidal proteins expressed by recombinant baculoviruses. *Viruses.* 7(1), 422–455.

Lacey, L. A., Grzywacz, D., Shapiro-Ilan, D. I., Frutos, R., Brownbridge, M. and Goettel, M. S. (2015). Insect pathogens as biological control agents: Back to the future. J. Invertebr. Pathol. 132, 1–41. doi: 10.1016/j.jip.2015.07.009.

Lakshmanan, V., Kitto, S. L., Caplan, J. L., Hsueh, Y. H., Kearns, D. B., Wu, Y. S. and Bais, H. P. (2012). Microbe-associated molecular patterns-triggered root responses mediate beneficial rhizobacterial recruitment in *Arabidopsis*. *Plant Physiol.* 160(3), 1642–1661.

Langford, E. A., Nielsen, U. N., Johnson, S. N. and Riegler, M. (2014). Susceptibility of Queensland fruit fly, *Bactrocera tryoni* (Froggatt) (Diptera: Tephritidae), to entomopathogenic nematodes. *Biol. Control.* 69, 34–39.

Lee, B., Lee, S. and Ryu, C. M. (2012). Foliar aphid feeding recruits rhizosphere bacteria and primes plant immunity against pathogenic and non-pathogenic bacteria in pepper. *Ann. Bot.* 110(2), 281–290.

Lefort, M.-C., Boyer, S. and Glare, T. R. (2017). A response to Pennisi: "How do gut microbiomes help herbivores," a hint into next-generation biocontrol solutions. *Rethink. Ecol.* 1, 9–13.

Lezama-Gutierrez, R., Trujillo-de la Luz, A., Molina-Ochoa, J., Rebolledo-Dominguez, O., Pescador, A. R., López-Edwards, M. and Aluja, M. (2000). Virulence of *Metarhizium anisopliae* (Deuteromycotina: Hyphomycetes) on *Anastrepha ludens* (Diptera: Tephritidae): Laboratory and field trials. *J. Econ. Entomol.* 93(4), 1080–1084.

Lopes, R. B., Faria, M. and Glare, T. R. (2018). A novel biphasic fermentation system for the production of aerial conidia of entomopathogenic fungi utilizing surface tension. *Appl. Environ. Microbiol.* 126, 155–164.

Lovett, B., Bilgo, E., Millogo, S. A., Ouattarra, A. K., Sare, I., Gnambani, E. J., Dabire, R. K., Diabate, A. and St. Leger, R. J. (2019). Transgenic *Metarhizium* rapidly kills mosquitoes in a malaria-endemic region of Burkina Faso. *Science*. 364(6443), 894–897.

Lu, D., Pava-Ripoll, M., Li, Z. and Wang, C. (2008). Insecticidal evaluation of *Beauveria bassiana* engineered to express a scorpion neurotoxin and a cuticle-degrading protease. *Appl. Microbiol. Biotechnol.* 81(3), 515–522.

Maeda, S., Volrath, S. L., Hanzlik, T. N., Harper, S. A., Majima, K., Maddox, D. W., Hammock, B. D. and Fowler, E. (1991). Insecticidal effects of an insect-specific neurotoxin expressed by a recombinant baculovirus. *Virology*. 184(2), 777–780.

Marche, M. G., Camiolo, S., Porceddu, A. and Ruiu, L. (2018). Survey of Brevibacillus laterosporus insecticidal protein genes and virulence factors. *J. Invertebr. Pathol.* 155, 38–43.

Martens, J. W., Knoester, M., Weijts, F., Groffen, S. J., Hu, Z., Bosch, D. and Vlak, J. M. (1995). Characterization of baculovirus insecticides expressing tailored *Bacillus thuringiensis* CryIA(B) crystal proteins. *J. Invertebr. Pathol.* 66(3), 249–257.

Martínez-Barrera, O. Y., Toledo, J., Liedo, P., Gómez, J., Valle-Mora, J., Cancino, J. and Montoya, P. (2019). Does *Beauveria bassiana* (Hypocreales: Cordycipitaceae) affect the survival and fecundity of the parasitoid *Coptera haywardi* (Hymenoptera: Diapriidae)? *Environ. Entomol.* 48(1), 156–162.

Mau, R., Jang, E. B. and Vargas, R. (2007). The Hawaii area-wide fruit fly pest management programme: Influence of partnerships and a good education programme. In: Vreysen, M. J. B., Robinson, A. S. and Hendrich, J. (Eds), *Area-Wide Control of Pest Insects: From Research to Field Implementation*, Springer, The Netherlands, pp. 671–683.

Mayerhofer, J., Eckard, S., Hartmann, M., Grabenweger, G., Widmer, F., Leuchtmann, A. and Enkerli, J. (2017). Assessing effects of the entomopathogenic fungus *Metarhizium brunneum* on soil microbial communities in *Agriotes* spp. biological pest control. *FEMS Microbiol. Ecol.* 93(10), 117.

McDougall, P. (2016). The cost of new agrochemical product discovery, development and registration consultancy study. Available at: https://croplife.org/wp-content/uploads/2016/04/Cost-of-CP-report-FINAL.pdf (accessed August 2020).

McKinnon, A. C., Glare, T. R., Ridgway, H. J., Mendoza Mendoza, A., Holyoake, A., Godsoe, W. K. and Bufford, J. L. (2018). Detection of the entomopathogenic fungus *Beauveria bassiana* in the rhizosphere of wound-stressed *Zea mays* plants. Front. Microbiol. 9, 1161. doi: 10.3389/fmicb.2018.01161.

McKinnon, A. C., Saari, S., Moran-Diez, M. E., Meyling, N. V., Raad, M. and Glare, T. R. (2017). *Beauveria bassiana* as an endophyte: A critical review on associated methodology and biocontrol potential. *BioControl*. 62(1), 1–17.

Meyling, N. V. and Eilenberg, J. (2007). Ecology of the entomopathogenic fungi *Beauveria bassiana* and *Metarhizium anisopliae* in temperate agroecosystems: Potential for conservation biological control. *Biol. Control*. 43(2), 145–155.

Meyling, N. V., Lübeck, M., Buckley, E. P., Eilenberg, J. and Rehner, S. A. (2009). Community composition, host range and genetic structure of the fungal entomopathogen Beauveria in adjoining agricultural and seminatural habitats. *Mol. Ecol.* 18(6), 1282–1293.

Moore, S. and Kirkman, W. (2010). Helicovir™: A virus for the biological control of bollworm. South Afr. Fruit J. 9, 63–67.

Moscardi, F. (2007). A nucleopolyhedrovirus for control of the velvetbean caterpillar in Brazilian soybeans. In: Goettel, M. S. and Lazarovits, G. (Eds), *Biological Control: A Global Perspective*, CABI Publishing, Wallingford, UK, pp. 344–361.

Mulder, P. G., Harris, M. K. and Grantham, R. A. (2012). Biology and management of the Pecan Weevil (Coleoptera: Curculionidae). *J. Integr. Pest Manage.* 3(1), A1–A9.

Nambiar, P. T. C., Ma, S. W. and Iyer, V. N. (1990). Limiting an insect infestation of nitrogen-fixing root nodules of the pigeon pea (*Cajanus cajan*) by engineering the expression of an entomocidal gene in its root nodules. *Appl. Environ. Microbiol.* 56(9), 2866–2869.

Narciso, J. O., Ormskirk, M. M., Jones, S. A., Rolston, P., Moran-Diez, M. E., Hurst, M. R. H., Brookes, J. J. and Glare, T. R. (2019). Using multiple insecticidal microbial agents against diamondback moth larvae: Does it increase toxicity? *N. Z. J. Agric. Res.* 64, 178–193.

Nathan, S. S. and Kalaivani, K. (2006). Combined effects of azadirachtin and nucleopolyhedrovirus (SpltNPV) on *Spodoptera litura* Fabricius (Lepidoptera: Noctuidae) larvae. *Biol. Control.* 39(1), 96–104.

Nguyen, Q., Tran, T., Chan, L. C., Nielsen, L. K. and Reid, S. (2016). In vitro production of baculoviruses: Identifying host and virus genes associated with high productivity. *Appl. Microbiol. Biotechnol.* 100(21), 9239–9253.

O'Callaghan, M., Glare, T. R. and Lacey, L. A. (2012). Bioassay of bacterial entomopathogens against insect larvae. In: Lacey, L. A. (Ed.), *Manual of Techniques in Invertebrate Pathology* (2nd edn.), Academic Press, New York, pp. 101–127.

Okay, S., Tefon, B. E., Okan, M. and Ocengiz, G. (2008). Expression of chitinase A (chiA) gene from a local isolate of *Serratia marcescens* in Coleoptera-specific *Bacillus thuringiensis*. *J. Appl. Microbiol.* 104(1), 161–170.

Olson, S. (2015). An analysis of the biopesticide market now and where it is going. *Outlooks Pest Manage.* 26(5), 203–206.

Onsongo, S. K., Gichimu, B. M., Akutse, K. S., Dubois, T. and Mohamed, S. A. (2019). Performance of three isolates of *Metarhizium anisopliae* and their virulence against *Zeugodacus cucurbitae* under different temperature regimes, with global extrapolation of their efficiency. *Insects.* 10(9), 270.

Ormskirk, M. M. (2017). *Brevibacillus laterosporus as a Potential Bio-Control Agent of the Diamondback Moth and Other Insects*. PhD Thesis. Lincoln University, New Zealand.

Ormskirk, M. M., Narciso, J., Hampton, J. G. and Glare, T. R. (2019). Endophytic ability of the insecticidal bacterium *Brevibacillus laterosporus* in Brassica. *PLoS ONE* 14(5), e0216341.

Ortega, H. E., Torres-Mendoza, D. and Cubilla-Rios, L. (2020). Patents on endophytic fungi for agriculture and bio- and phytoremediation applications. *Microorganisms.* 8(8), 1237.

Ouedraogo, R. M., Goettel, M. S. and Brodeur, J. (2004). Behavioral thermoregulation in the migratory locust: A therapy to overcome fungal infection. *Oecologia.* 138(2), 312–319.

Ouna, E. A. (2010). *Entomopathogenicity of Hyphomycte Fungi to Fruit Fly Bactrocera invadens (Diptera: Tephiritidae) and Their Potential for Biological Control on Mango*. MSc Thesis. Kenyatta University, Nairobi, Kenya, p. 134.

Park, H. W., Bideshi, D. K. and Federici, B. A. (2003). Recombinant strain of *Bacillus thuringiensis* producing Cyt1A, Cry11B, and the *Bacillus sphaericus* binary toxin. *Appl. Environ. Microbiol.* 69(2), 1331–1334.

Petrika, D. T., Iseli, A.,Montelone, B. A., Van Etten, J. L. and Clem, R. J. (2003). Improving baculovirus resistance to UV inactivation: Increased virulence resulting from expression of a DNA repair enzyme. *J. Invertebr. Pathol.* 82(1), 50–56.

Prior, C., Carey, M., Abraham, Y. J., Moore, D. and Bateman, R. P. (1995). Development of a bioassay method for the selection of entomopathogenic fungi virulent to the desert locust, *Schistocerca gregaria* (Forskal). *J. Appl. Entomol.* 119(1–5), 567–573.

Qadri, M., Short, S., Gast, K., Hernandez, J. and Wong, A. C. (2020). Microbiome innovation in agriculture: Development of microbial based tools for insect pest management. *Front. Sustain. Food Syst.* 4, 547751.

Quesada-Moraga, E., Ruiz-Garcia, A. and Santiago-Alvarez, C. (2006). Laboratory evaluation of entomopathogenic fungi *Beauveria bassiana* and *Metarhizium anisopliae* against puparia and adults of *Ceratitis capitata* (Diptera: Tephritidae). *J. Econ. Entomol.* 99(6), 1955–1966.

Quesada-Moraga, E., Yousef-Naef, M. and Garrido-Jurado, I. (2020). Advances in the use of entomopathogenic fungi as biopesticides in suppressing crop pest insects. In: Birch, N. and Glare, T. (Eds), *Biopesticides for Sustainable Agriculture*, Burleigh Dodds Science Publishing, Cambridge, UK, pp. 63–98.

Rasmann, S., Köllner, T. G., Degenhardt, J., Hiltpold, I., Toepfer, S., Kuhlmann, U., Gershenzon, J. and Turlings, T. C. (2005). Recruitment of entomopathogenic nematodes by insect-damaged maize roots. *Nature.* 434(7034), 732–737.

Reardon, R. C. and Podgwaite, J. D. (1994). Summary of efficacy evaluations using aerially applied Gypchek against gypsy moth in the USA. *J. Environ. Sci. Health B.* 29(4), 739–756.

Resquín-Romero, G., Garrido-Jurado, I., Delso, C., Ríos-Moreno, A. and Quesada-Moraga, E. (2016). Transient endophytic colonizations of plants improve the outcome of foliar applications of mycoinsecticides against chewing insects. *J. Invertebr. Pathol.* 136, 23–31.

Ribeiro, B. M. and Crook, N. E. (1993). Expression of full-length and truncated forms of crystal protein genes from *Bacillus thuringiensis* subsp. *kurstaki* in a baculovirus and pathogenicity of the recombinant viruses. *J. Invertebr. Pathol.* 62(2), 121–130.

Roberts, D. W. and St. Leger, R. J. (2004). *Metarhizium* spp., cosmopolitan insect-pathogenic fungi: Mycological aspects. *Adv. Appl. Microbiol.* 54, 1–70.

Rosell, G., Quero, C., Coll, J. and Guerrero, A. (2008). Biorational Insecticides in pest management. *J. Pestic. Sci.* 33(2), 103–121.

Rudrappa, T., Czymmek, K. J., Pare, P. W. and Bais, H. P. (2008). Root-secreted malic acid recruits beneficial soil bacteria. *Plant Physiol.* 148(3), 1547–1556.

Ruiu, L. (2013). *Brevibacillus laterosporus*, a pathogen of invertebrates and a broad-spectrum antimicrobial species. *Insects.* 4(3), 476–492.

Ruiu, L., Floris, I., Satta, A. and Ellar, D. J. (2007). Toxicity of a *Brevibacillus laterosporus* strain lacking parasporal crystals against *Musca domestica* and *Aedes aegypti*. *Biol. Control.* 43(1), 136–143.

San Miguel, K. and Scott, J. G. (2016). The next generation of insecticides: dsRNA is stable as a foliar-applied insecticide. Pest Manag. Sci. 72, 801–809. doi: 10.1002/ps.4056.

Scheepmaker, J. W. A. and Butt, T. M. (2010). Natural and released inoculum levels of entomopathogenic fungal biocontrol agents in soil in relation to risk assessment and in accordance with EU regulations. *Biocont. Sci. Technol.* 20(5), 503–552.

Scheepmaker, J. W. A., Hogervorst, P. A. M. and Glandor, D. C. M. (2016). *Future Introductions of Genetically Modified Microbial Biocontrol Agents in the EU Are*

Current EU Legislation and Risk Assessment Fit for Purpose? RIVM Letter Report, Bilthoven, The Netherlands, pp. 2016–0057.
Senthilkumar, N., Murugan, K. and Zhang, W. (2008). Additive interaction of *Helicoverpa armigera* nucleopolyhedrovirus and azadirachtin. *BioControl.* 53(6), 869–880.
Shapiro-Ilan, D. I., Han, R. and Dolinksi, C. (2012). Entomopathogenic nematode production and application technology. *J. Nematol.* 44(2), 206–217.
Singhania, R. R., Patel, A. K., Thomas, L. and Pandey, A. (2016). Solid-state fermentation. In: Wittman, C. and Liao, J. C. (Eds), *Solid-State Fermentation. Industrial Biotechnology*, Wiley-VCH Verlag GmbH and Co. KGaA, Weinheim, pp. 187–204.
Soliman, N. A., Al-amin, S. M., Mesbah, A. E., Ibrahim, A. M. A. and Mahmoud, A. M. A. (2020). Pathogenicity of three entomopathogenic fungi against the Mediterranean fruit fly, *Ceratitis capitata* (Wiedemann) (Diptera: Tephritidae). *Egypt. J. Biol. Pest. Control.* 30(1), 49.
Soliman, N. A., Ibrahim, A. A., Shams El-Deen, M. M., Shams Ramadan, N. F. and Farag, S. R. (2014). Entomopathogenic nematodes and fungi as biocontrol agents for the peach fruit fly, *Bactrocera zonata* (Saunders) and the Mediterranean fruit fly, *Ceratitis capitata* (Wiedemann) soil-borne stages. *Egypt. J. Biol. Pest.* Contol. 24, 497–502.
Song, Z. (2018). Fungal microsclerotia development: Essential prerequisites, influencing factors, and molecular mechanism. *Appl. Microbiol. Biotechnol.* 102(23), 9873–9880.
Song, Z. Y., Yin, Y. P., Jiang, S. S., Liu, J. J. and Wang, Z. K. (2014). Optimization of culture medium for microsclerotia production by *Nomuraea rileyi* and analysis of their viability for use as a mycoinsecticide. *BioControl.* 59(5), 597–605.
Sorokan, A., Benkovskaya, G., Burkhanova, G., Blagova, D. and Maksimov, I. (2020). Endophytic strain *Bacillus subtilis* 26DCryChS producing Cry1Ia toxin from *Bacillus thuringiensis* promotes multifaceted potato defense against *Phytophthora infestans* (Mont.) de Bary and pest *Leptinotarsa decemlineata* Say. *Plants.* 9(9), 1115.
Sparks, T. C., Thompson, G. D., Larson, L. L., Kirst, H. A., Jantz, O. K., Worden, T. V., Hertlein, M. B. and Busacca, J. D. (1995). Biological characteristics of the spinosyns: A new naturally derived insect control agent. *Proc. Beltwide Cotton Conf.* 2, 903–907.
Spence, E. L., Chandler, D., Edgington, S., Berry, S. D., Martin, G., O'Sullivan, C., Svendsen, C. and Hesketh, H. (2020). A standardised bioassay method using a bench-top spray tower to evaluate entomopathogenic fungi for control of the greenhouse whitefly, *Trialeurodes vaporariorum. Pest. Manage. Sci.* 76(7), 2513–2524.
Srinivasan, R., Sevgan, S., Ekesi, S. and Tamòc, M. (2019). Biopesticide-based sustainable pest management for safer production of vegetable legumes and brassicas in Asia and Africa. *Pest. Manage. Sci.* 75(9), 2446–2454.
St. Leger, R., Joshi, L., Bidochka, M. J. and Roberts, D. W. (1996). Construction of an improved mycoinsecticide overexpressing a toxic protease. *Proc. Natl. Acad. Sci. USA.* 93(13), 6349–6354.
Thomas, M. B.and Jenkins, N. E. (1997). Effects of temperature on growth of *Metarhizium flavoviride* and virulence to the variegated grasshopper, *Zonocerus variegatus. Mycol. Res.* 101(12), 1469–1474.
Thomas, M. B., Wood, S. N. and Lomer, C. J. (1995). Biological control of locusts and grasshoppers using a fungal pathogen - the importance of secondary cycling. *Proc. R. Soc. Lond. B.* 259(1356), 265–270.
Toledo, J., Flores, S., Campos, S., Villaseñor, A., Enkerlin, W., Liedo, P., Valle, Á and Montoya, P. (2017). Pathogenicity of three formulations of *Beauveria bassiana* and efficacy of

autoinoculation devices and sterile fruit fly males for dissemination of conidia for the control of *Ceratitis capitata*. *Entomol. Exp. Appl.* 164(3), 340–349.

Tomalski, M. D. and Miller, L. K. (1992). Expression of a paralytic neurotoxin gene to improve insect baculoviruses as biopesticides'. *Nat. Biotechnol.* 10(5), 545–549.

Tseng, M. N., Chung, P. C. and Tzean, S. S. (2011). Enhancing the stress tolerance and virulence of an entomopathogen by metabolic engineering of dihydroxynaphthalene melanin biosynthesis genes. *Appl. Environ. Microbiol.* 77(13), 4508–4519.

USDA Forest Service (2020). Pesticide management & coordination. Available at: https://www.fs.fed.us/foresthealth/protecting-forest/integrated-pest-management/pesticide-management/pesticide-virus-products.shtml (accessed 1 February 2021).

van Zijll de Jong, E., Roush, T. L., Glare, T. R. and Hampton, J. G. (2016). Discovery of two *Brevibacillus laterosporus* isolates from brassica with insecticidal properties against diamondback moth. *Biocont. Sci. Technol.* 26, 426–431.

Vargas, R. I., Piñero, J. C. and Leblanc, L. (2015). An overview of pest species of *Bactrocera* fruit flies (Diptera: Tephritidae) and the integration of biopesticides with other biological approaches for their management with a focus on the Pacific Region. *Insects.* 6(2), 297–318.

Vega, F. E. (2018). The use of fungal entomopathogens as endophytes in biological control: A review. *Mycologia.* 110(1), 4–30.

Villamizar, L. F., Nelson, T. L., Jones, S. A., Jackson, T. A., Hurst, M. R. H. and Marshall, S. D. G. (2018). Formation of microsclerotia in three species of *Beauveria* and storage stability of a prototype granular formulation. *Biocont. Sci. Technol.* 28(12), 1097–1113.

Vogel, E., Santos, D., Mingels, L., Verdonckt, T. W.. and Broeck, J. V. (2019). RNA interference in insects: Protecting beneficials and controlling pests. *Front. Physiol.* 9, 1912.

Waage, J. (1997). Global developments in biological control and the implications for Europe. *EPPO Bull.* 27(1), 5–13.

Wang, H. H., Wang, J.-L., Li, Y.-P., Liu, X., Wen, J.-Z. and Lei, Z.-R. (2011a). Liquid culturing of microsclerotia of *Beauveria bassiana*, an entomopathogenic fungus to control western flower thrip, *Frankliniella occidentalis*. *J. Appl. Ent.* 48, 588–595.

Wang, S., Fang, W., Wang, C. and St. Leger, R. J. (2011b). Insertion of an esterase gene into a specific locust pathogen (*Metarhizium acridum*) enables it to infect caterpillars. *PLoS Pathog.* 7(6), e1002097.

Whipps, J. M. (1997). Developments in the biological control of soil-borne plant pathogens. *Adv. Bot. Res.* 26, 1–134.

Wraight, S. P., Jackson, M. A. and de Lock, S. L. (2001). Production, stabilization and formulation of fungal biocontrol agents. In: Butt, T. M., Jackson, C. W. and Magan, N. (Eds) Fungi as Biocontrol Agents, CABI Publishing, pp. 253–287.

Yan, S., Ren, B., Zeng, B. and Shen, J. (2020). Improving RNAi efficiency for pest control in crop species. *BioTechniques.* 68(5), 283–290.

Yousef, M., Aranda-Valera, E. and Quesada-Moraga, E. (2018). Lure-and-infect and lure-and-kill devices based on *Metarhizium brunneum* for spotted wing *Drosophila* control. *J. Pest. Sci.* 91(1), 227–235.

Yousef, M., Garrido-Jurado, I., Ruíz-Torres, M. and Quesada-Moraga, E. (2017). Reduction of adult olive fruit fly populations by targeting preimaginals in the soil with the entomopathogenic fungus *Metarhizium brunneum*. *J. Pest. Sci.* 90(1), 345–354.

Zimmermann, G., Papierok, B. and Glare, T. (1995). Elias Metschnikoff, Elie Metchnikoff or Ilya Ilich Mechnikov (1845–1916): A pioneer in insect pathology, the first describer of the entomopathogenic fungus *Metarhizium anisopliae* and how to translate a Russian name. *Biocon. Sci. Technol.* 5(4), 527–530.

Zitlalpopoca-Hernandez, G., Najera-Rincon, M. B., del-Val, E., Alarcon, A., Jackson, T. and Larsen, J. (2017). Multitrophic interactions between maize mycorrhizas, the root-feeding insect *Phyllophaga vetula* and the entomopathogenic fungus *Beauveria bassiana*. *Appl. Soil Ecol.* 115, 38–43.

Chapter 2

Advances in bioprotectants for plant disease control in horticulture

Philippe C. Nicot, Thomas Pressecq and Marc Bardin, INRAE, Pathologie Végétale, France

1 Introduction

Horticultural crops (as defined by the International Society for Horticultural Science; https://www.ishs.org/defining-horticulture; accessed in January 2021) encompass a particularly high number of plant species which are each challenged by a large gamut of plant pathogens. Disease control is necessary to ensure adequate production and product quality. For many of the crops (ornamentals, but also fruits and vegetables), the consumer expects a high level of visual quality (possibly 'zero-defect'), particularly when the produce does not undergo substantial processing before its commercialization (Porat et al., 2018; Leus, 2018). Farmers often rely on chemical control for disease management in these crops, but bioprotectants are increasingly available and may contribute to more sustainable pest management strategies.

Among the four types of commercial bioprotectants described by the International Biocontrol Manufacturers Association (IBMA, 2019), two may be available to farmers for use against diseases of horticultural crops: microbial bioprotectants and natural substances. Microbial bioprotectants are preparations based on microorganism active substances. According to a definition jointly proposed by the Food and Agricultural Organization of the United Nations and the World Health Organization, a microorganism active

http://dx.doi.org/10.19103/AS.2021.0095.02

substance may contain viable and/or non-viable microorganisms. It can contain relevant metabolites/toxins produced during cell proliferation (growth), and material from the growth medium, provided none of these components have been intentionally altered (FAO and WHO, 2017). *Natural substances*, as defined by IBMA (2019), 'consist of one or more components that originate from nature, including but not limited to: plants, algae/micro algae, animals, minerals, bacteria, fungi, protozoans, viruses, viroids and mycoplasmas' and 'they can either be sourced from nature or are nature-identical if synthetized'. These two types of bioprotectants are sometimes referred to as 'biopesticides' (Marrone, 2019). In most countries, their commercialization is regulated by a formal registration process (FAO and WHO, 2017), in the same manner as synthetic chemical pesticides (see for example the 'List of databases on registered plant protection products in the EPPO region' provided by the European and Mediterranean Plant Protection Organization; https://www.eppo.int/ accessed in January 2021).

Considering the increasing concerns about negative side effects of chemical pesticides in many countries, bioprotectants present a favorable profile in terms of public health and environmental issues (van Lenteren et al., 2018). Their use may thus present a number of advantages for the horticulturist. In fact, it is estimated that 80% of the use of biopesticides at a global scale (both against pests and diseases) is on fruits and vegetables (Marrone, 2019). Increasingly, the commercialization of horticultural crops benefits from quality labels or eco-labels that allow the consumer to identify those produced without any, or with limited, synthetic chemical pesticides (Chen et al., 2018; Nguyen and Le, 2020; Thøgersen et al., 2019). In the production phase on the farm, regulations in numerous countries allow crop treatments with bioprotectants at periods of the growing season when protection is needed against diseases, but chemical control is restricted to limit the presence of toxic residues on the plants. This could be the case, for instance, close to harvest or postharvest and close to the selling time and consumption of the produce.

However, the protection of horticultural crops against diseases with the help of natural substances or microbial bioprotectants also presents a number of challenges, which will be addressed in the present chapter. One key challenge is a large gap between the needs of horticulturists and the limited number of bioprotectants available to them, even for major diseases of economically important crops. The issue is thus how to increase the number of registered products. Another challenge, for products that are available to farmers, is the level and consistency of the protection that can be achieved. Bioprotectants are often considered to be less efficacious and less reliable than synthetic chemical pesticides. The issues are thus to enhance the field efficacy of bioprotectants and to help farmers optimize their deployment to improve reliability.

2 Availability of bioprotectants against diseases of horticultural crops

One feature of horticultural crops is the very large number of plant species they include, most of which are considered as 'minor' crops on the basis of their cultivation area, production and consumption (SANCO, 2017; OECD, 2009). Many different plant pathogens may cause diseases in those crops, and most of these pathogens have narrow host ranges. This results in a vast range of possible host–pathogen combinations, each corresponding to a potential need for a plant protection solution.

The number of host–pathogen combinations (or 'uses') for which at least one registered bioprotectant is available to horticulturists may vary greatly from one country to another, depending on national regulations. However, it is likely to represent a small fraction of all possible needs by farmers. For example, among bioprotectants registered in France, microbial biocontrol agents covered 163 uses in 2020 (Table 1), with 36 commercial products based on 26 microbial preparations (16 with fungal strains, 8 with bacterial strains and 2 with viruses). Most of these products are registered for application on above-ground plant parts in the field or after harvest (Table 1A), but several may be used for treatment of seeds and transplants (Table 1B) and a few are registered for the treatment of soil or growth substrates (Table 1C).

Without considering sulfur- and copper-based products, bioprotectants based on natural substances cover 61 horticultural uses in France (Table 2), 34 of which are in common with microbial biocontrol agents. These natural substances are mostly plant or microbial extracts, but the majority of commercial products are based on phosphonates (20 products) and potassium carbonate (3 products). In France, several natural substances (and one microbial bioprotectant) benefited from a general registration for use as 'stimulation of plant defenses', without the mention of any specific disease or plant pathogen. Sulfur-based active substances (55 commercial products) cover 41 uses, mostly against powdery mildew of different crops. Copper-based active substances (10 products) cover 54 uses, mostly against downy mildews or late blight and bacterial diseases (Table 2). Overall, bioprotectants registered in France in 2020 (if one includes sulfur- and copper-based products) cover 244 distinct uses on horticultural crops, of which 31 are specific to copper-based products and 13 to sulfur-based products.

Despite these three-digit figures on uses covered by bioprotectants, which tend to increase year after year, a large part of the potential needs of horticulturists is not well covered at a global scale. In fact, many of the needs are not covered by any plant protection product at all, complicating the management of diseases by farmers. The situation has been made more severe by the evolution of the safety standards set by regulatory authorities in the last

Table 1 Examples of microbial strains used as active substances in bioprotectants registered in France for disease control on horticultural crops (from the E-phy database https://ephy.anses.fr/ consulted in October 2020)

Microbial strains	Registered use: target disease/pathogen and crops
A: Products which can be applied as foliar sprays or post-harvest treatments	
Ampelomyces quisqualis (strain AQ10)	**Powdery mildew** (on cucumber, melon, pepper, strawberry, tomato, rose)
Aureobasidium pullulans (strains DSM 14940 & DSM 14941)	**Gray mold** (grapes, strawberry, tomato); **fireblight** (apple); **post-harvest** (apple, citrus)
Bacillus amyloliquefaciens (strain FZB24)	***Botrytis* and *Sclerotinia*** (lettuce); **brown spot** (potato, tomato); **downy mildew or late blight** (cucumber, grapes, lettuce, potato, tomato); **gray mold** (grapes); **powdery mildew** (cucumber, grapes, tomato); **unspecified bacteria** (tomato)
Bacillus amyloliquefaciens (strain MBI600)	***Botrytis* and *Sclerotinia*** (lettuce, pepper, strawberry, tobacco, tomato); **gray mold** (grapes, raspberry); **unspecified fungi** (chicory, mushroom)
Bacillus amyloliquefaciens ssp. plantarum (strain D747)	***Botrytis* and *Sclerotinia*** (cucumber, lettuce, melon); **fireblight** (apple); **gray mold** (blackcurrant, grapes, pepper, raspberry, strawberry, tomato); ***Monilinia* brown rot** (cherry, peach, plum); **post-harvest rot** (apple); **powdery mildew** (blackcurrant, pepper, raspberry, strawberry, tomato); **Stemphylium** (apple); **unspecified bacteria** (kiwi)
Bacillus pumilus (strain QST 2808)	**Powdery mildew** (aromatic and medicinal plants, black currant, chicory, cucumber, flowers and potted plants, grapes, herbs, lettuce, melon, pepper, raspberry, strawberry, tobacco, tomato, vegetable crops)
Bacillus subtilis (strain IAB/BS03)	**Powdery mildew** (cucumber, lettuce); **scab** (apple)
Bacillus subtilis (strain QST 713)	Bacterial diseases: **fire blight** (apple, pear) and **various bacteria** (almond, aromatic and medicinal plants, artichoke, cardoon, cherry, eggplant, jujube, kiwi, nectarine, nuts, olive, ornamental plants, peach, pepper, plum, tomato) Fungal diseases: **anthracnose** (avocado); ***Botrytis* and *Sclerotinia*** (beans, chicory, cowpea, courgette, cucumber, eggplant, fava beans, lamb's lettuce, lentils, lettuce, melon, peas, pepper, rocket, strawberry, tobacco, tomato, watermelon); **gray mold** (aromatic and medicinal plants, blackberry, blackcurrant, grapes, lingonberry, ornamental plants, raspberry, tobacco, tomato); ***Monilinia* brown rot** (almond, cherry, jujube, nectarine, peach, plum; stone fruits); **Mycosphaerella leaf streak** (banana); **peacock eye** and other olive fruit diseases (olive); **powdery mildew** (aromatic and medicinal plants, blackberry, blackcurrant, cucumber, flowers and potted plants, lingonberry, mango, melon, ornamental plants, raspberry, rose, small fruits, trees and shrubs); **scab** (apple); ***Sclerotinia* collar rot** (lettuce); **Stemphylium** (pear); **unspecified fungi** (aromatic and medicinal plants) Post-harvest rot (kiwi)

(*Continued*)

Table 1 (*Continued*)

Microbial strains	Registered use: target disease/pathogen and crops
Candida oleophila (strain O)	**Post-harvest rot** (apple, banana, citrus)
Clonostachys rosea (*Gliocladium catenulatum*) (strain J1446)	***Botrytis* and *Sclerotinia*** (cucumber, flowers and green plants, pepper, raspberry, strawberry, tomato) and **other fungal diseases on leaves** (cucumber, flowers and potted plants, pepper, tomato)
Metschnikowia fructicola (strain NRRL Y-27328)	***Botrytis* and *Sclerotinia*** (blackcurrant, grapes, raspberry, strawberry); ***Monilinia* brown rot** (cherry, peach, plum)
Phlebiopsis gigantea	**Armilaria** on tree stumps
Saccharomyces cerevisiae (strain LAS02)	***Botrytis* and *Sclerotinia*** (blackcurrant, grapes, pepper, raspberry, strawberry, tomato); ***Monilinia* brown rot** (peach, plum)
Trichoderma atroviride (strain SC1)	***Botrytis* and *Sclerotinia*** (tomato, grapes); **wood-decay fungi** (grapes)
Trichoderma atroviride (strain I-1237)	**Oomycetes** (carrot); **unspecified fungi** (lettuce); **wood-decay fungi** (grapes)
Trichoderma gamsii (strain ICC080) + *Trichoderma asperellum* (strains ICC012 T25 & TV1)	**Wood-decay fungi** (grapes)
B: Products which can be applied to seeds and nursery transplants	
Bacillus amyloliquefaciens (strain MBI600)	**Unspecified fungi** (witloof chicory)
Pseudomonas sp. (strain DSMZ 13134)	**Unspecified fungi** (potato)
Trichoderma asperellum (strain T34)	**Wilt diseases** (ornamental plants)
Trichoderma atroviride (strain I-1237)	**Unspecified fungi** (potato)
Trichoderma atroviride (strain SC1)	**Wood-decay fungi** (grapes)
Trichoderma harzianum Rifai (strains T-22 & ITEM-908)	**Unspecified fungi** (aromatic and medicinal plants)
Mild pepino mosaic virus (isolates VC1 & VX1)	**Pepino mosaic virus** (tomato)
Pepino Mosaic Virus (strain CH2, isolate 1906)	**Pepino mosaic virus** (tomato)
C: Products which can be applied as soil or substrate treatments	
Bacillus amyloliquefaciens (strain MBI600)	**Unspecified fungi** (mushroom)
Bacillus amyloliquefaciens ssp. plantarum (strain D747)	**Substrate disinfestation** (mushroom)

(*Continued*)

Table 1 *(Continued)*

Microbial strains	Registered use: target disease/pathogen and crops
Bacillus subtilis (strain QST 713)	**Soil oomycetes** (carrot); **various soil fungi** (cucumber, lettuce, melon, pepper, potato, tomato); **competing fungi and substrate disinfestation** (mushroom)
Clonostachys rosea (*Gliocladium catenulatum*) (strain J1446)	**General soil treatment** against soil fungi and oomycetes
Paraphaeosphaeria minitans (*Coniothyrium minitans*) (strain CON/M/91-08)	**General soil treatment** against *Sclerotinia*
Pseudomonas sp. (strain DSMZ 13134)	**Various soil fungi** (aromatic and medicinal plants, beet, cabbage, chicory, cucumber, lettuce, melon, pepper, poppy, potato, spinach, strawberry, tomato, turnip, watercress)
Streptomyces griseoviridis (strain K61)	**General soil treatment** against soil fungi and oomycetes
Trichoderma asperellum (strain T34)	**Wilt-causing fungi** (ornamentals); **soil oomycetes** (pepper, tomato)
Trichoderma asperellum(strains ICC012 T25 & TV1)	**Soil oomycetes** (vegetable crops); **general soil treatment** against soil fungi and oomycetes
Trichoderma asperellum (strains ICC012 T25 & TV1) + *Trichoderma atroviride* (formerly *T. harzianum*) (strainT11)	**Soil fungi and oomycetes** (trees and shrubs, cucumber, flowers and potted plants, melon, ornamental crops, pepper, vegetable crops, tomato); **soil oomycetes** (strawberry); **soil fungi** (lettuce)
Trichoderma atroviride (strain I-1237)	**Soil oomycetes** (carrot); **soil fungi** (ornamental crops, lettuce, potato)
Trichoderma gamsii (strain ICC080) + *Trichoderma asperellum* (strains ICC012 T25 & TV1)	**Armilaria** (ornamental crops); **soil oomycetes** (ornamental and vegetable crops); **soil fungi** (strawberry); **soil disinfestation** (fruit trees)
Trichoderma harzianum Rifai (strains T-22 & ITEM-908)	**General soil treatment** against soil fungi and oomycetes; **soil fungi and oomycetes** (carrot, cabbage, cauliflower, broccoli, cucumber, ornamental crops, herbs, strawberry, grasses, beans, lettuce, turnip, onion, melon, leek, pepper, tomato); **soil oomycetes** (celery, fruit and vegetable crops); **soil fungi** (aromatic and medicinal plants)

decades. The reassessment of toxicological issues with previously registered plant protection products has resulted in a drastic reduction in the number of synthetic chemical pesticides that are authorized for commercialization in the European Union and many other countries (Donley, 2019). The problems posed by requirements for which no plant protection product is available have long been a concern for farmers, including horticulturists, and are dealt with by expert groups at national and international levels. One example is the European Union Minor Uses Co-ordination Facility, which maintains a database compiling all unfulfilled needs reported by EU member states. Horticultural crops represent a high proportion of recorded needs, amounting to 2862 of the 3177 minor uses compiled in this database (http://www.eumuda.eu/index; accessed in October 2020). The great majority of these needs are for vegetable crops (covering over 100 plant species) and for fruits and nuts (over 50 species) (Fig. 1). For all horticultural crops, plant diseases account for 34% of the reported needs, with animal pests and weeds each representing 52% and 14%, respectively. The range of implicated pathogens of horticultural crops is wide, encompassing over 300 species of fungi or oomycetes, 30 species or pathovars of bacteria and 24 viruses.

Thus, there is a globally strong need for more active substances to be registered as bioprotectants for a wide range of uses.

3 Advances in increasing the number of bioprotectants for disease control

Several approaches may contribute to increasing the number of active substances registered as bioprotectants for a wider range of horticultural uses.

3.1 Extending the spectrum of activity of existing products

Among bioprotectants currently available for horticulturists, most cover a limited number of uses, as illustrated for France in Tables 1 and 2. This situation may be due to the inability of the registered microbial strains or natural substances to provide plant protection beyond a narrow range of host-pathogen combinations. This is the case, for example, for the mycoparasite *Ampelomyces quisqualis*, whose biological activity is limited to the fungi responsible for powdery mildew diseases (Pintye et al., 2015).

However, it may also be due to a lack of information on the spectrum of activity of the bioprotectants, as the acquisition of such knowledge requires time-consuming and costly experimental research and field trials. Over the years, the range of registered uses for certain bioprotectants has markedly increased, possibly in relation with the growth of the original biocontrol companies that developed them or with the increasing involvement of large international plant protection businesses with strong investment capacities in research and development (Marrone, 2019; Pelaez and Mizukawa, 2017). For example, the strain QST713 of *Bacillus subtilis*, originally registered by the company Agraquest in the USA, was commercialized in France with a single use in 2006 (Table 3). In 2020, its registration in France covered over 60 uses on horticultural crops and several on arable crops.

Table 2 Examples of natural substances used in bioprotectants registered in France for disease control on horticultural crops (from the E-phy database https://ephy.anses.fr/ consulted in October 2020)

Natural substance[a]	Registered use (target disease/pathogen and crops)[b]
Cerevisiane (2 products)	**Stimulation of plant defenses** (cucumber, strawberry, lettuce, melon, tomato, grapes)
Clove oil + plant oils	**Post-harvest rot** of apple
Copper (54 uses ; 10 products)	**Anthracnose** (hazelnut); **Bacteria** (apple, artichoke, bean, cabbage, celery, chicory, grapes, kiwi, leek, lettuce, melon, nut trees, olive, onion, peach, pea, plum, strawberry, tomato); **brown spot** (strawberry); **Coryneum blight and Polystigma red leaf spot** (cherry, peach, plum); **downy mildew and late blight** (artichoke, cabbage, cucumber, grapes, leek, lettuce, melon, onion, potato, potted plants, rose, tomato, trees & shrubs); **fungal canker** (almond); **fungal dieback** (hazelnut); **leaf curl** (cherry, peach, plum); **leaf spots** (potted plants, trees & shrubs); **Mycosphaerella leafspot** (chestnut); **Nectria canker** (apple); **oomycetes** (carrot); **peacock eye** (olive); **scab** (apple); **various unspecified diseases** (citrus)
COS-OGA (2 products)	**Stimulation of plant defenses** (grapes, vegetable crops)
Essential oil from orange (2 products)	**Black spot** (rose); **brown spot** (tomato); **downy mildew or late blight** (cucumber, flowers & potted plants, grapes, lettuce, melon, rose, tobacco, tomato, turnip); **powdery mildew** (apple, blackcurrant, carrot, chicory, cucumber, flowers & potted plants, grapes, herbs, hops, lettuce, mango, melon, papaya, peach, rose, raspberry, salsify, strawberry, tobacco, tomato, trees and shrubs); **rusts** (cauliflower, flowers & potted plants, rose, trees and shrubs); **unspecified fungi** (herbal tea plants)
Fenugreek extract	**Stimulation of plant defenses** (grapes, melon)
Laminarin	**Fire blight** (apple); **powdery mildew** (strawberry); **scab** (apple); **stimulation of plant defenses** (apple, grapes, lettuce, strawberry)
Potassium carbonate (3 products)	**Gray mold** (grapes); ***Monilinia* brown rot** (peach); **powdery mildew** (apple, artichoke, blackcurrant, cucumber, flowers & potted plants, grapes, hops, melon, pepper, raspberry, rose, strawberry, tomato, trees and shrubs); scab (apple); **unspecified fungi** (aromatic & medicinal plants)
Potassium or sodium phosphonates (20 products)	**Collar rot and oomycetes** (citrus, fruit trees); **downy mildew** (grapes); **scab** (apple); **stimulation of plant defenses** (chicory, cucumber, lettuce, melon, pepper, tomato) + **post-harvest** (citrus) + **soil treatment against collar rot** of fruit trees
Sulfur (42 uses ; 55 products)	**Black spot** (rose); **Cercospora leaf spot** (beet); **downy mildew** (grapes); **leaf spots** (potted plants); **Phomopsis cane and leaf spot** (grapes); **powdery mildew** (apple, artichoke, beans, beet, blackcurrant, carrot, celery, chicory, cucumber, flowers & potted plants, garden legumes, grapes, hazelnut, hops, lettuce, mango, melon, papaya, peach, peas, pepper, rose, raspberry, salsify, spinach, strawberry, tomato, trees & shrubs); **scab** (apple); **unspecified fungi** (aromatic & medicinal plants)
Thymol	**Gray mold** (grapes)

[a] Commercialized as one single product unless otherwise specified.
[b] Products are applicable as foliar sprays unless otherwise stated.

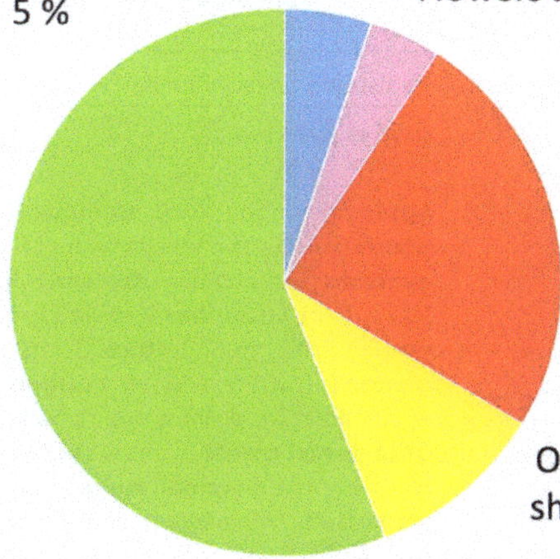

Figure 1 Distribution, according to main types of crops, of horticultural uses compiled in the 'Minor Uses Table of Needs' of the European Union Minor Uses Co-ordination Facility (accessed in October 2020).

For many existing bioprotectants, further progress may be expected on the identification of new protective effects against diseases of horticultural crops. This trend could be stimulated by information already available for the active substances of existing products at a global scale, including those with current uses registered on non-horticultural crops or in different regions of the world with various climates and agricultural practices. A recent review of microbial biological control agents registered in Australia, Brazil, Canada, the European Union, Japan, New Zealand and the USA (van Lenteren et al., 2018) shows a potential for such data mining. It presents a number of microbial strains (26 bacteria, 31 fungi and 2 viruses) and 28 species that are not listed in Table 1 of the present chapter (Fig. 2). Published scientific literature is generally available for registered strains, or for strains of similar species, and may provide useful guidance for the expansion of registered uses. It could provide evidence or assist in devising hypotheses on the protective potential of registered strains on new plant-pathogen combinations. Hypotheses could, for instance, result from careful extrapolations of known protective activity of a strain on horticultural crops to that of closely related strains of the same species. Advanced knowledge on the modes of action of a bioprotectant may also help to identify additional target pathogens or crop-pathogen combinations that could be responsive to the same protective effects.

One promising development of bioprotectants currently registered against invertebrate pests, especially when their mode of action relies on the stimulation of plant defenses, could also result from the evaluation of possible protective

Table 3 Recent evolution of uses for bioprotectants. Example of *Bacillus subtilis* QST 713 in France (sources: Index Phytosanitaire ACTA. ACTA Editions, France & the E-phy database https://ephy.anses.fr/ consulted in October 2020)

Commercialization[a]	Year	Registered uses in France[b]
Nufarm SAS (Serenade Biofungicide)	2006	**grapes** (gray mold)
BASF Agro SAS (Serenade Max)	2010	**apple** (fire blight, scab); **aromatic & medicinal plants** (various bacteria, gray mold, powdery mildew); **avocado** (anthracnose); **banana** (Mycosphaerella leaf streak); **cucumber & melon** (powdery mildew); **grapes** (gray mold); **lettuce** (*Sclerotinia* collar rot); **mango** (powdery mildew); **mushrooms** (treatment of substrate for disinfestation); **nuts** (bacteria); **ornamental plants** (various bacteria, gray mold, powdery mildew) **small fruits** (powdery mildew); **stone fruits** (*Monilinia* brown rot); **tomato** (bacteria, gray mold)
Bayer Crop Science (Serenade Max)	2018	2010 uses + **almond** (bacteria, *Monilinia* brown rot); **apple & pear** (fire blight); **aromatic and medicinal plants** (various bacterial and fungal diseases); **artichoke & cardoon** (bacteria); **blackberry, blackcurrant, lingonberry & raspberry** (gray mold; powdery mildew); **carrot & parsley** (soil treatment against oomycetes); **cherry** (bacteria, *Monilinia* brown rot); **chicory, lamb's lettuce, lettuce & rocket** (*Botrytis* and *Sclerotinia*; soil treatments against soil-borne fungi); **cucumber** (*Botrytis* and *Sclerotinia*); **eggplant & tomato** (bacteria, *Botrytis* and *Sclerotinia*, soil treatments against soil-borne fungi); **flowers and potted plants** (powdery mildew); **mushrooms** (treatment of substrate against competing fungi); **peach & nectarine** (bacteria, *Monilinia* brown rot); **pear** (Stemphylium); **pepper** (bacteria, *Botrytis* and *Sclerotinia*, soil treatment against soil-borne fungi); **plum & jujube** (bacteria, *Monilinia* brown rot); **rose** (powdery mildew); **strawberry** (*Botrytis* and *Sclerotinia*); trees & shrubs (powdery mildew); **table grapes** (gray mold); **tobacco** (*Botrytis* and *Sclerotinia*); **walnut** (bacteria)
Bayer Crop Science (Rhapsody, Serenade ASO, Serenade Max)	2020	2018 uses + **beans, cowpeas, fava beans, lentils & peas** (*Botrytis* and *Sclerotinia*), **courgette** (*Botrytis* and *Sclerotinia*); **courgette & cucumber** (soil treatment against soil-borne fungi); **kiwi** (bacteria, post-harvest rot); **melon & watermelon** (*Botrytis* and *Sclerotinia*; soil treatment against soil-borne fungi); **olive** (bacteria, peacock eye and various fruit diseases)

[a] Company (+ name of commercial product).
[b] Registered uses are for foliar application unless otherwise stated. All uses registered in 2010 were for stimulation of plant defenses, with the exception of substrate treatment for mushroom production.

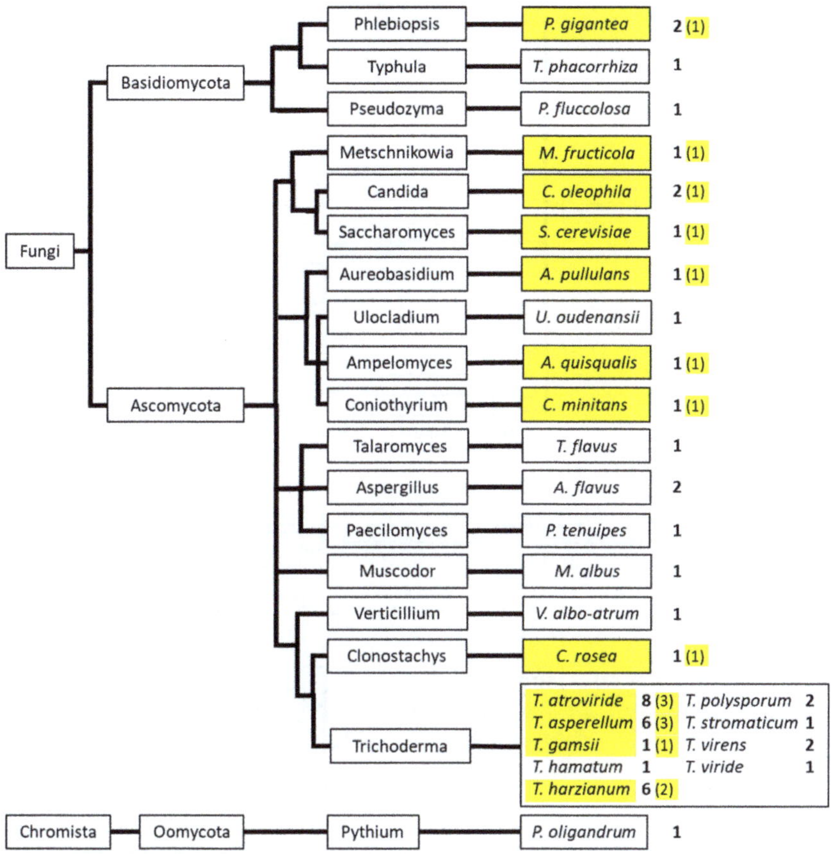

Figure 2 Diversity of genera and species of microorganisms registered as bioprotectants in Australia, Brazil, Canada, the European Union, Japan, New Zealand and the USA, based on the review of (van Lenteren et al., 2018). The dendrograms show key taxonomic links between genera. Yellow highlights indicate species registered in France on horticultural crops (listed in Table 1). For each species, bold-faced numbers show the total counts of registered strains (the subset of those listed in Table 1 is indicated between brackets).

effects against diseases. One plant extract listed in Table 2 (essential oil from orange) is also registered in various countries for the protection of numerous horticultural crops against pests such as leafhoppers, psyllids, scale insects, thrips, white flies, and mites. Several species of *Trichoderma*, a genus mostly known for its bioactivity against plant diseases, have been reported to harbor strains that can protect plants against aphids, thrips and caterpillars (Coppola et al., 2019). Vice versa, microbial species known for their entomoparasitic effects were shown to have the ability to protect horticultural crops against plant pathogens. This is the case, for example, of *Metarhizium anisopliae* against gray mold of tomato (Sarven et al., 2020) and *Verticillium lecanii* against

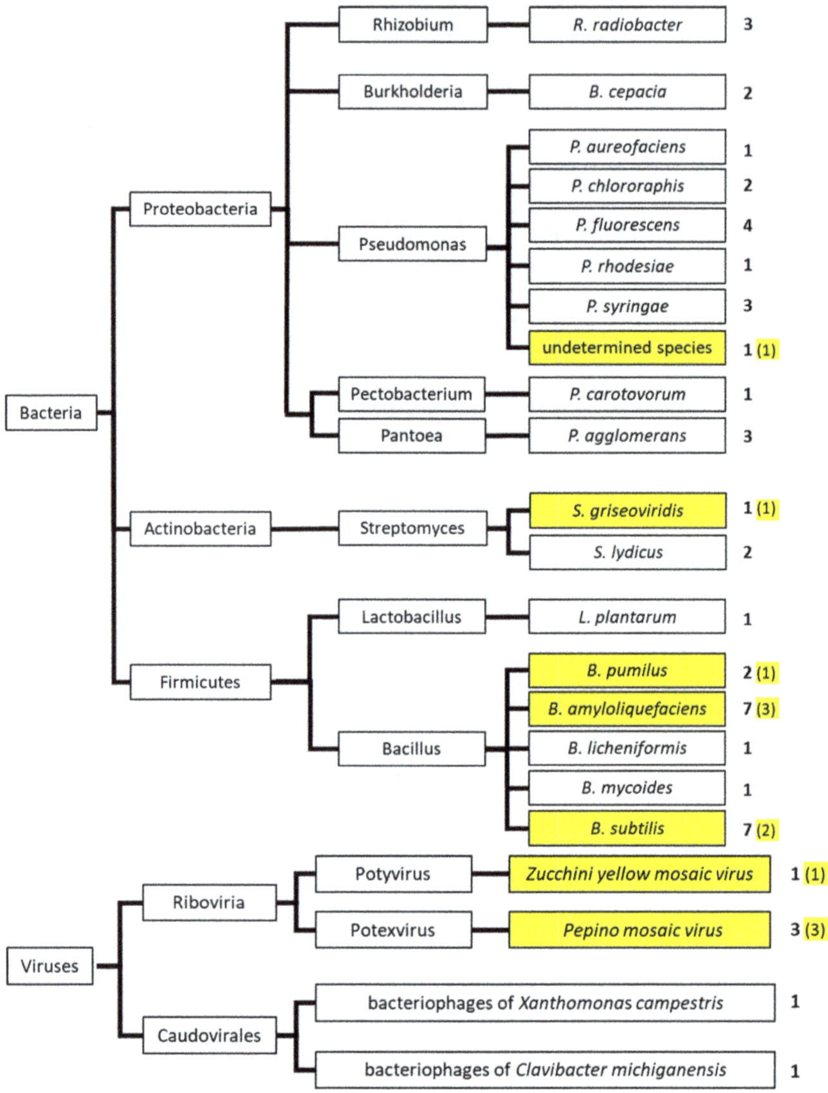

Figure 2 (*Continued*).

several powdery mildew fungi, rust fungi and *Penicillium digitatum*, the agent of post-harvest green mold on fruits (Benhamou, 2004).

3.2 Finding new active substances: where and how?

There is a high potential for escalating the discovery of new active substances with protective effects against diseases of horticultural crops. Two approaches may stimulate advances in this domain: (i) expanding the sourcing of new candidate microbial strains or natural substances and (ii) rationalizing and upscaling screening methods for the selection of candidates.

The microorganisms currently registered as bioprotectants are limited to a small number of species and genera, suggesting that the range of sourcing for new strains could be expanded. For example, among the 47 fungal strains and 43 bacterial strains accounted for in Fig. 2, 28 belong to the genus *Trichoderma* (nearly 60% of fungal strains), 18 to *Bacillus* and 12 to *Pseudomonas* (nearly 42% and 28% of bacterial strains, respectively). In contrast, estimates of global microbial biodiversity on earth are usually in the range of millions of species (Vitorino and Bessa, 2018). A wider exploration of this vast pool of potential bioprotectants has been stimulated by the recent surge of studies on microbiomes that were fostered by advances in multiple 'omics' tools and bioinformatics (Massart et al., 2015). In the agricultural context, a special focus is placed on soil- or plant-associated microbiota (Berg et al., 2017; Ab Rahman et al., 2018; Schlatter et al., 2017; Massart et al., 2015; Compant et al., 2019). Concretely, the large amounts of generated data may present challenges, and new methods need to be devised for the interpretation of microbiome networks with the objective of identifying candidate microbes that affect plant health (Poudel et al., 2016).

Another issue with the sourcing of new bioprotectants is the choice of ecological niches where microbiomes are most likely to be interesting. Formerly, recommendations to seek potential bioprotectant strains were to focus on agricultural environments, in areas where the disease caused by a given pathogen did not occur or had declined despite the presence of a susceptible host, rather than on areas where the disease occurred (Ellis, 2017). This applies well, for example, to the study of microorganisms in suppressive soils (Schlatter et al., 2017). These early recommendations have been questioned and a wider range of possible niches is increasingly being considered, including endophytic microbiota and microbiota on diseased plants, on survival structures of the plant pathogens and in river or marine environments (Ellis, 2017; Mülner et al., 2019; Sahu et al., 2020; Hu et al., 2019; Kong, 2018; Marimuthu et al., 2020; De Silva et al., 2019). Marine samples, consisting of fish, shellfish and shrimp, as well as intertidal sand and mud collected along the seashore, proved to be a valuable source of actinomycetes for the control of blue mold of citrus (Hu et al., 2019). Exploring wider ranges of climatic conditions may also be valuable. Psychrotrophic yeasts were successfully selected from water and soil samples collected in Antarctica, for use as biocontrol agents for the post-harvest diseases of apples stored at low temperature (Arrarte et al., 2017).

Much effort is also dedicated to the exploration of new sources of natural substances, for example as extracts from a variety of terrestrial or sea plants (Raveau et al., 2020; Aguilar-Veloz et al., 2020) or from a wide variety of microorganisms. Marine-derived bacterial lipopeptides were shown to inhibit gray mold caused by *Botrytis cinerea* on potted cucumber plants and on Chinese rose (*Rosa chinensis*) in field experiments (Gu et al., 2017). Natural products from mangrove rhizospheric microorganisms were recently reported to provide control of anthracnose, caused by *Colletotrichum fructicola* on pear fruits (Hao et al., 2019). The microbiome studies mentioned may also lead to the identification of natural substances of microbial origin (Hao et al., 2019).

Improved screening strategies and advanced phenotyping tools are also likely to accelerate the selection and industrial development of new bioprotectants. In many cases, screening of candidate microorganisms or natural substances was initially focused on criteria that pertain to their protective efficacy against one or several plant pathogens. For microorganisms, the bioassays in early studies generally relied on *in vitro* confrontation, on nutrient agar media, of microbial candidate strains with a plant pathogen to assess their biocontrol potential. This method presents many logistical advantages. Its implementation in the laboratory is simple, fast, cheap and compatible with high throughput screening. However, it focuses the selection criteria on antagonism by the candidate strains against the pathogen and the production of antimicrobial compounds, while disregarding possible protective effects that could result from interactions with the plant to be protected or with its microbiome (Berg et al., 2017). Additionally, there has been mounting evidence for a possible lack of correlation between the *in vitro* performance of a strain and its subsequent protection efficacy on the host plants, leading to the conception of more sophisticated screening procedures that implicate *in vivo* tests on plants or detached plant organs (Bouaoud et al., 2018; Raymaekers et al., 2020; Köhl et al., 2020; Besset-Manzoni et al., 2019; Köhl et al., 2019b). Increased throughput for *in vivo* tests can be expected with the involvement of automated phenotyping platforms, possibly assisted by artificial intelligence (Ozyilmaz, 2020). The evaluation of complex microbial communities may also benefit from high throughput systems such as the kChip screening platform (Kehe et al., 2019).

Aside from criteria concerning the protective efficacy of candidate microorganisms or natural substances, many other qualities need to be evaluated in a screening process leading to the industrial and commercial development of a bioprotectant. They may include a variety of biological and ecological traits such as UV and drought resistance, cost-effective mass production and compatibility with other bioprotectants or chemical pesticides. These additional selection criteria may address requirements related to industrial production and to the registration of a future product. Integrative screening strategies have been proposed to prioritize this whole broad range of criteria in relation to their cost and logistical weight (Köhl et al., 2011; Marrone, 2019; Köhl et al., 2019b).

3.3 Facilitating registration of bioprotectants

In most countries, the commercialization of a bioprotectant necessitates its registration with regulatory authorities. For this, the company needs to provide a dossier with extensive information covering the chemical, physical and biological properties of the product or active substance, as well as toxicological and environmental issues (Gotor-Vila et al., 2017; Gu et al., 2017; Marrone, 2019; Korsten, 2004). In many countries, an additional dossier must be submitted to document the field efficacy of the product (Frederiks and Wesseler, 2019). Overall, the procedure may be complex, it requires costly investments for the companies, and it generally involves a lengthy evaluation period by the regulatory authorities (Marrone, 2019; Frederiks and Wesseler, 2019). This has often been viewed as an impediment to the development of bioprotectants, particularly by small companies or when the product targets a small market (Zaki et al., 2020; Robin and Marchand, 2019; Korsten, 2004; Bailey et al., 2010; Glare et al., 2012; Chandler et al., 2011; Abbey et al., 2019; Aguilar-Veloz et al., 2020). Improvement of this situation can be expected when regulatory authorities provide incentives, such as preferential reduced administrative fees or a reduced delay for dossier evaluation, that may be granted for the registration of bioprotectants by small companies (Marrone, 2019). However, an evolution of the data requirements for the registration of bioprotectants remains to be carefully considered in many countries, as they were initially devised for the registration of chemical pesticides and may not be fully adapted for biological compounds (Köhl et al., 2019a,b; Koch et al., 2018; Chandler et al., 2011). Developing fully adapted toxicological criteria for the registration of bioprotectants to be used for post-harvest protection of horticultural crops may be particularly challenging (Aguilar-Veloz et al., 2020).

4 Enhancing the field efficacy of bioprotectants on horticultural crops

The field performance of bioprotectants is usually compared to, and expected by the farmers, to match that of chemical pesticides (Marian and Shimizu, 2019). In past decades, however, commercialized bioprotectants have shown success, but also partial success, failures and lack of economic viability on the farm, which have affected the perception of end-users about their protective efficacy (Glare et al., 2012). This assessment fits well with reviews of the scientific literature, which have shown a gap between an abundance of reports on high protective efficacy observed in the laboratory in controlled conditions, and a smaller number of reports on effective protection in the field (Nicot et al., 2012; Ojiambo and Scherm, 2006; Marian and Shimizu, 2019). Several advances, which contribute to improving this situation, will be discussed below.

4.1 Expanding knowledge on their modes of action

Knowledge on the modes of action of a bioprotectant is essential, because they affect its protective efficacy, its efficiency of use, consistency of response, and host target and non-target susceptibility (Hubbard et al., 2014; Köhl et al., 2019b; Aguilar-Veloz et al., 2020). In terms of industrial production of the bioprotectants, they may also influence the choice of manufacturing method and the final cost of production (Aguilar-Veloz et al., 2020;, Hubbard et al., 2014). Information on the modes of action may also be required in registration dossiers of bioprotectants (Köhl et al., 2019b).

The modes of action of bioprotectants can involve a direct impact on the pathogen, through antibiosis (by means of non-volatile or volatile antimicrobial compounds), competition or hyperparasitism (Fig. 3). Competition and hyperparasitism (the direct competitive interaction between two organisms in which one organism [the bioprotectant] is gaining nutrients from the other [the plant pathogen]), are specific to bioprotectants based on living microorganisms. The action of bioprotectants can also result from indirect effects through the stimulation of plant defenses or through impacts on the microbiota of the plant or the soil (Guimarães et al., 2020; Gao et al., 2017; Elsayed et al., 2020; Antoniou et al., 2017). There are excellent reviews describing all these modes of action (Köhl et al., 2019b; Singh et al., 2020; De Silva et al., 2019; Aguilar-Veloz et al., 2020; Ab Rahman et al., 2018; Rao et al., 2019) and thus they will not be presented in detail here.

One important property of bioprotectants, particularly those based on living microorganisms, is that their protective effect is rarely limited to a single mode of action. This is exemplified by species of *Trichoderma*, which have long been studied as biocontrol agents on horticultural crops (Verma et al., 2007). All of the modes of action included in Fig. 3 have been reported for these species, and a number of strains are known for their multiple modes of actions (Zaki et al., 2020; Sood et al., 2020; Verma et al., 2007). Their stimulatory effects on plant defenses implicate complex mechanisms. These can result in a localized or a systemic plant response, and depending on strains and host plants, they may mediate directly-induced defenses or defense priming (Zaki et al., 2020; Conrath et al., 2015; Mauch-Mani et al., 2017; Lorito et al., 2010).

Natural substances are often used as antimicrobial compounds, possibly also through a volatile effect (Aguilar-Veloz et al., 2020). Many of them are known for stimulating plant defenses, a property which may be officially recognized in the registered usage, as for many of the substances listed in Table 2. A combination of both properties is not rare, and the protective effect of natural substances may also result from additional modes of action. One good example is chitin and chitin-based polymers of D-glucosamine from crab shells, such as chitosan, which have long been known for their ability to elicit plant defenses against plant pathogens, including on horticultural crops (Malerba and Cerana, 2019).

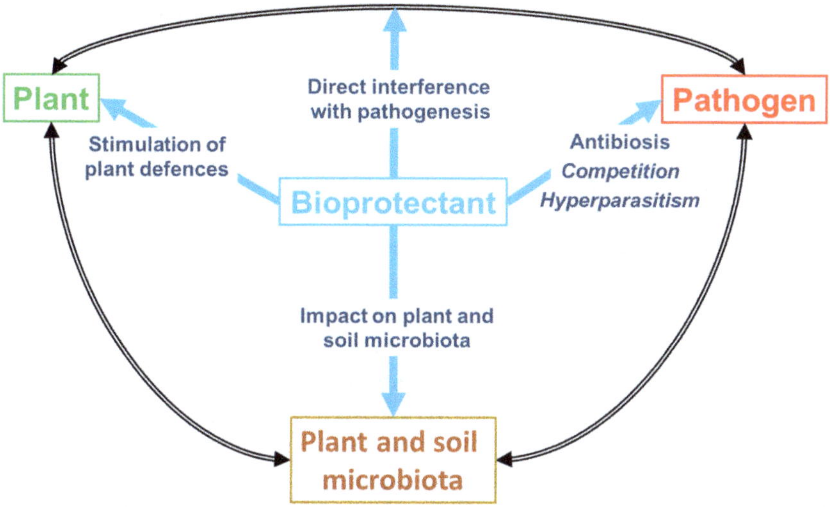

Figure 3 Possible modes of action of bioprotectants used against plant pathogens, either with direct effects, or through an impact on the plant or on its microbiota and that of the soil. Italics indicate those modes of action that are specific to living microorganisms.

In addition, chitosan may provide plant protection through two other possible effects: (i) antimicrobial activity and (ii) film formation on the treated surface, leading to the constitution of a physical barrier against pathogen penetration (Romanazzi et al., 2019). In Europe, it was the first basic substance approved by the European Union for plant protection (Reg. EU 2014/563), for both organic agriculture and integrated pest management (IPM).

In the last decade, spectacular progress has been made in the elucidation of complex mechanisms implicated in the modes of actions of an increasing number of bioprotectants (Köhl et al., 2019b; Sharma et al., 2017). However, one remaining question for those bioprotectants with the capacity to mobilize multiple modes of action, is whether they are all systematically implicated in the protective effect on the plants. One of the rare reports on this issue proposes that defence stimulation represents 30–40% of the activity of chitosan, antibiosis 35–45%, and film-forming activity 20–30%, in terms of its effectiveness in the control of postharvest decay of fresh fruit (Romanazzi et al., 2019). One key challenge for improving field efficacy of bioprotectants will be to identify factors that determine which modes of action are mobilized, singly or in combination and under which conditions.

4.2 Identifying key factors of efficacy

The term 'factor of efficacy' refers to factors that may arise before, during or after the application of a bioprotectant in the field, which could have a positive or a negative impact on the extent of protection it provides to the crop. The factors of efficacy can affect differently the roles of a bioprotectant in reducing the incidence of the targeted disease, avoiding damage to the plant and securing the yield of the crop. Environmental conditions encountered in the field, farming practices, ecology of the targeted plant pathogen, epidemiology of the disease it causes, and biological or biochemical properties and quality of the bioprotectant are among the factors that may impact the efficacy of protection (Aguilar-Veloz et al., 2020; Bardin and Pugliese, 2020). Examples of the main factors of efficacy and their interactions are presented in Fig. 4.

There is abundant scientific literature describing new microbial strains and new compounds with bioprotectant potential against plant diseases, or evaluating their protective efficacy in controlled conditions. However, detailed information on factors governing the efficacy of bioprotectants in the field is still scarce.

Microclimatic conditions (temperature and relative humidity) in the canopy, pH of the targeted area, and quality and intensity of light (ranges of wavelength, photoperiod) have been known for years for their possible impacts on the efficacy of chemical pesticides (Cress, 2006). For bioprotectants, the same main environmental factors can be considered (Abbey et al., 2019). Temperature is probably the most important factor, and one of the most studied. It directly puts to the test the environmental resilience of microbial bioprotectants and can strongly influence their efficacy by affecting their growth rate (Wei et al., 2017), their survival (Landa et al., 2004) and their mechanism of action (Humair et al.,

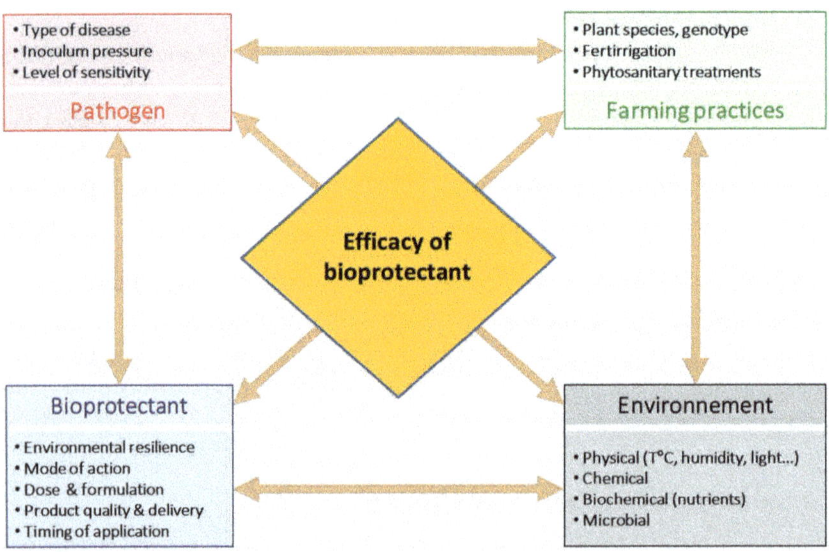

Figure 4 Examples of factors of efficacy of bioprotectants and their possible interactions.

2009). For instance, the production of antibiotic compounds and hydrogen cyanide by the root-colonizing bacterial strain *Pseudomonas fluorescens* CHA0 was significantly lower at 35°C than at 30°C (Humair et al., 2009). Temperature also interacts with the effect of other factors such as water activity or relative air humidity, making it important to determine ranges that are optimal, permissive or deleterious for protective efficacy. For example, it was shown for six commercialized bioprotectants that the effect against *B. cinerea* was reduced by 60% if relative humidity decreased below 90%. The optimal temperatures were between 22°C and 25°C, with a strong loss of efficacy below 20°C (Fedele et al., 2020). Environmental conditions are also known to play a major role in the efficacy of natural substances (Aguilar-Veloz et al., 2020). They strongly affect the development of the pathogen (Roger et al., 1999; Elad and Pertot, 2014) and its virulence (Hardwick, 2002), which can ultimately impact the efficacy of a bioprotectant.

The timing of application of the bioprotectants may be critical for successful protection, and can be guided by knowledge of the properties of the bioprotectant, but also of the biology of the pathogen and the epidemiology of the disease. For example, the widely used mycoparasite *Paraphaeosphaeria minitans* (formerly *Coniothyrium minitans*) was found to have a greater effect against sclerotia of *Sclerotinia sclerotiorum* in soil samples collected from field plots treated immediately after a disease outbreak, than from plots treated at a later stage (Gerlagh et al., 2003). A possible explanation for this outcome was that mycoparasitism may be easier on freshly produced sclerotia than on more mature ones. Optimized field application schedules for microbial bioprotectants such as *Bacillus cereus* (SC-1 and P-1) and *Bacillus subtilis* (W-67) against infections by ascospores of *S. sclerotiorum* were focused on the flowering stage of the crop, although they targeted stem rot (Kamal et al., 2015). Appropriate timing for the application of natural substances is also essential (Abbey et al., 2019). For all bioprotectants, their optimization requires knowledge about the persistence of the product on its target, and needs to take into account the microclimatic conditions encountered in the field.

Various factors linked to cultural practices have been studied. For instance, the effect of bioprotectants may be variety-dependent, as was observed for the protection of tomato against *B. cinerea* with resistance-inducing strains of *Trichoderma* (Tucci et al., 2011) or against the seed pathogen *Pythium torulosum* with *Bacillus cereus* UW85 (Smith et al., 1999). Tomato genotypes also had distinct but overlapping root bacterial microbiomes and responded differently to specific bacterial endophytes (French et al., 2020). Plant fertilization may affect the efficacy of bioprotectants. Nitrogen nutrition of the plant was shown to affect the protective effects against *B. cinerea* of two fungal bioprotectants applied to tomato, and of two resistance-inducing products (one based on strain QST713 of *B. subtilis* and one on chitosan) applied to strawberry (Nicot et al., 2013; Abro et al., 2014).

Even though the complexity of factors affecting the efficacy of bioprotectants is often suggested in the scientific literature, most studies only focus on one or two factors at a time or hardly consider what is actually happening in the field. For instance, the effect of temperature, or that of relative humidity, is generally investigated in controlled conditions as static factors (e.g., a constant temperature throughout the experiment) and not as fluctuating factors (e.g., frequent and potentially significant modification during 24 h). This limitation is shared by a large proportion of the publications on bioprotectants, causing possible discrepancies between promising results obtained under controlled conditions in the laboratory, real efficacy observed in the field and feedback from farmers (Ojiambo and Scherm, 2006). Identifying the key factors of efficacy and their effects, alone or in interaction, is an essential step in ensuring the integration of bioprotectants into complex farming systems and thus improving their reliability.

4.3 Improving delivery

Delivery is a critical phase for the protective effect of a bioprotectant, and it can be a major obstacle to its commercial success (Glare et al., 2012). Advances on several issues will be considered here, as delivery is not limited to the context of field application of a bioprotectant by the farmer as presented in section 4.2, but also entails industrial steps such as mass production, formulation, packaging and organization of the supply chain of the product before it reaches the farmers. Throughout this series of steps, one key issue is the maintenance of product stability and preservation of optimal properties of the bioprotectant, to ensure that it will have the capacity to deliver expected bioactive functions when used by the farmer. Other issues are to facilitate effective application and to enhance post-application persistence of the active substance in its target niche. For foliar applications, it was estimated that achieving persistence for more than 3 weeks would constitute a difficult but key goal (Glare et al., 2012). Specific technological details on production and formulation are usually kept as trade secrets by companies, but increasing numbers of publications provide a glimpse into some of the major issues and advances.

At the production stage, optimized technology is needed to ensure that sufficient amounts of bioactive substances are delivered in a cost-effective manner. For microbial bioprotectants, the type of fermentation, its conditions and the composition of the medium are critical. They may also determine the types of propagules that are produced, which in turn influence the stability and the effectiveness of the final product (Zaki et al., 2020). Production conditions may also be manipulated to increase the environmental resilience of the microbial protectant after its application to the crop (Cabrefiga et al., 2011; Daranas et al., 2018). For natural substances such as essential oils and other plant extracts, a high level of standardization is also needed in commercial production. Key factors that may cause variability in the quality of the product were reviewed by Aguilar-Veloz et al. (2020). They include factors pertaining to the plant material, such as the variety and cultivation conditions influenced by climatic factors (photoperiod, shade, temperature, etc.); the plant age (maturity); and agronomic practices (soils, time of planting and time of harvest, application of fertilizers, irrigation, etc.). The type of plant part used in the biological evaluation may also be influential. Other key factors pertain to the experimental procedures used for extraction and conservation and to the methods of chemical characterization (Aguilar-Veloz et al., 2020).

Much progress has been made in optimizing liquid as well as dry formulations, to foster the shelf life and the environmental resilience of microbial bioprotectants (Carbó et al., 2018; Zaki et al., 2020; De La Cruz Quiroz et al., 2019). Inclusion of specific ingredients may play a key role (Marian and Shimizu, 2019). For example, the survival and efficacy of liquid formulations of a strain of *Candida sake* was improved by adding a commercial fatty acid-based product called Fungicover (Carbó et al., 2018). The identification of useful additives may benefit from nutrient profiling of the microbial strains (Gramisci et al., 2018). Further improvements, both for liquid and for dry formulations, should be facilitated by the development of rapid screening tools (combining temperature, water activity and pH conditions) to optimize the formulation of microorganisms for environmental resilience and to predict their ecological niches (Carbó et al., 2018).

Coating agents are increasingly included in the formulations of microbial bioprotectants and natural substances to be used on post-harvest fruits (Marín et al., 2017; Barragán-Menéndez et al., 2020; Aguilar-Veloz et al., 2020). They enhance the protective effect of the active substance, while also providing physiological benefits for the preservation of the produce. For example, the incorporation of the yeast *Meyerozyma caribbica* in a sodium alginate matrix greatly increased its protective effect against anthracnose caused by *Colletotrichum gloeosporioides* on avocado fruits, reaching 100% protection at 6°C, and it reduced weight loss due to dehydration (Iñiguez-Moreno et al., 2020). Lactic acid bacteria such as *Lactobacillus plantarum*, incorporated into various edible coatings, were shown to provide control of decay fungi on grape berries and to maintain post-harvest quality (Marín et al., 2019). Besides microbial bioprotectants, coatings with a range of molecules and natural products incorporated into chitosan-based polymers were shown to provide protection of blueberry, jujube and strawberry fruits against post-harvest diseases (Malerba and Cerana, 2019).

Other reported additives for the formulation of microbial bioprotectants include dextrin, milk, molasses, oils, pectin, starch and sucrose (Zaki et al.,

2020; De La Cruz Quiroz et al., 2019). Some of these compounds contribute to the microencapsulation of the bioprotectants with biopolymers, a technology for which tremendous advances have been made in recent years. Microencapsulation consists of creating a physical barrier that provides protection of the active ingredient against external factors such as mechanical stress, ultraviolet radiation, oxidation and high temperatures. As a result, the microbial bioprotectants are able to survive for a longer period and their metabolic activity is maintained (Fraceto et al., 2018). Advances in microencapsulation have also been significant for the formulation of natural substances such as plant extracts, as reviewed by Aguilar-Veloz et al. (2020). Improved formulations have contributed to better control of post-harvest rot of horticultural produce, as well as reducing the mycotoxin contents. Enhanced control of post-harvest rot of sweet potato, caused by fungal pathogens such as *Rhizopus stolonifer* and *Ceratocystis fimbriata*, was recently achieved using the lipopeptide iturin A with a new microencapsulation technology based on chitosan (Yuan et al., 2020). After 50 days of storage, the incidence of rot was 8% in tubers treated with the micro-encapsulated iturin, compared to 14% with non-encapsulated iturin, and nearly 56% for the untreated control. It also significantly decreased the tuber content of ipomeamarone, a toxic furanoterpenoid which accumulates in response to microbial attack in rotting sweet potatoes (Wamalwa et al., 2015). The content of this compound was 2 mg/100g in tubers treated with micro-encapsulated iturin, compared to 19 mg/100 g for tubers treated with the non-encapsulated iturin and 135 mg/100 g for the untreated control (Yuan et al., 2020).

There is increasing interest in formulations based on nanotechnologies both for microbial bioprotectants (Fraceto et al., 2018) and for natural substances (Aguilar-Veloz et al., 2020). Recent application of chitosan-based polymers and nanoparticles complexed with a range of molecules and natural substances were shown to increase the resistance of horticultural crops such as tomato, pepper and lettuce against various soil-borne pathogens (Malerba and Cerana, 2019). However, further research will be necessary to optimize and confirm the potential of these technologies, and to support decision-makers in the development of adapted regulations and safety assessment procedures (Aguilar-Veloz et al., 2020; Fraceto et al., 2018).

Advances in packaging technology may also contribute to the stability of bioprotectants and the extension of their shelf life, by controlling gas exchange, preventing the loss or gain of moisture and avoiding contamination of the product (Gotor-Vila et al., 2019). Successful packaging optimization was recently reported for formulated preparations of the strain CPA-8 of *Bacillus amyloliquefaciens*, with significant positive effects for long-term storability at 22°C while providing suitable product delivery and field application, as well as the effective control of *Monilinia* brown rot of nectarines (Gotor-Vila et al., 2019).

4.4 Complementing bioprotectant use with other solutions from the integrated pest management toolbox

IPM approaches have long been investigated for the management of plant health in horticultural crops. As early as 1959, a 'Working Group for Integrated Plant Protection in Fruit Orchards' was established by the International Organization for Biological and Integrated Control of Noxious Animals and Plants (IOBC), followed by working groups on integrated protection for greenhouse crops in 1968 and vegetable crops in 1970 (Freier and Boller, 2009; Boller et al., 2006). While early work on IPM was mostly focused on pest animals (particularly invertebrates), the concept was gradually extended to include the control of plant diseases, and the word 'pest' is increasingly understood to include weeds and causal agents of diseases as well as pest animals (van Lenteren and Nicot, 2020). Bioprotectants are particularly well suited for use in IPM programs and their application can be combined with a range of other plant protection practices that can enhance their efficacy (van Lenteren and Nicot, 2020; Glare et al., 2012).

The bioprotection-enhancing effect of agricultural practices may result from a reduction of disease pressure on the crop, for example by avoiding conditions that foster severe forms of disease expression or epidemic development, or by decreasing plant susceptibility (through the planting of less susceptible varieties or through interventions that stimulate plant defenses). This may be achieved by combining the application of bioprotectants with abiotic or biotic stresses. The application of different abiotic and biotic stresses on the roots of tomato plants, for example, was shown to enhance chemotropism and the growth of a biocontrol strain of *T. harzianum* in the rhizosphere (Lombardi et al., 2018). Fertilization may be a particularly interesting lever. For example, appropriate levels of nitrogen fertilization have a potential to reduce the susceptibility of various horticultural crops to *B. cinerea* and *S. sclerotiorum*, but also to increase the protective effect of bioprotectants (Nicot et al., 2013; Lecompte et al., 2010; Abro et al., 2014; Abro et al., 2013; Lecompte et al., 2013).

A large body of literature describes possible bioprotection-enhancing effects of soil organic amendments, such as composts, manures and organic wastes on the efficacy of microbial bioprotectants (Marian and Shimizu, 2019). Composts amended with strains of microbial bioprotectants have shown high potential for disease control. Recent examples include the efficient reduction in the incidence and severity of Fusarium wilt of lettuce, provided by incorporating into the soil a compost-based biofertilizer amended with the strain TW2 of *Trichoderma* sp. (Gilardi et al., 2019), and the highly efficient control of *Pythium* infection in tomato by strain 4.4.1 of *Pseudomonas chlororaphis*, and various organic amendments (Postma and Nijhuis, 2019). Organic amendments may act as nutrients that are preferentially used by specific microbial bioprotectants, but there is also increasing interest in their global effects on indigenous beneficial microorganisms present in the soil, to foster soil suppressiveness against soil-borne plant pathogens (Tao et al., 2020; De Corato, 2020; Mazzola and Freilich, 2017).

Enhanced disease control may also be achieved above ground by combining natural substances with microbial bioprotectants or by combining several microbial strains with synergistic or complementary capacities. Chitosan has been combined successfully with various microorganisms for the post-harvest protection of horticultural produce (Zhou et al., 2016; Godana et al., 2020). For example, enhanced protection of grape berries against *Penicillium expansum* was achieved by the addition of 1% chitosan to preparations of biocontrol yeast *Pichia anomala* (Godana et al., 2020). The increased protection was associated with increased pullulation of yeast cells in the wounds in the presence of chitosan. The combined application of chitosan with extracts from the brown alga *Ascophyllum nodosum* provided significantly reduced development of pea powdery mildew by a synergistic activation of the plant's defenses (Patel et al., 2020). Several strains of microbial bioprotectants are already commercialized in combinations (Table 1). Combinations of microbial strains belonging to the same species or to species of the same genus are already commercialized for the protection of horticultural crops (Table 1). More may come in the future, with increasing numbers of reports of successful multispecies assemblages as reviewed by Marian and Shimizu (2019), such as mixtures of Bacillus, Pseudomonas and Rhizobium that were shown to provide yield improvement of bean together with control of Fusarium wilt (Kalantari et al., 2018). Ultimately, microbiome research may lead to the screening or engineering of more complex consortia of microbial bioprotectants (Kehe et al., 2019; Mitter et al., 2019; Woo and Pepe, 2018).

5 Fostering the adoption of biocontrol

The reticence of farmers to use bioprotectants has long been considered to be an obstacle to the wide adoption of bioprotectants for disease management, both in horticultural and other production systems (Barratt et al., 2018; Lamichhane et al., 2017; Abbey et al., 2019; Villemaine et al., 2020; Marrone, 2019). This section will present an example of current use of bioprotectants by horticulturists in France, considerations on possible obstacles for wider adoption, and outlooks for helping farmers cope with the technical complexity of deploying bioprotectants in IPM strategies.

5.1 Case study: adoption of bioprotectants by horticulturists in France

To gain a better perspective on bioprotectant use in horticultural crops, we recently carried out a survey of farmers, farm advisors and farm suppliers. A questionnaire was widely disseminated through farmers' associations, advisory services and in technical journals. Among 170 responses, 90 were received from farm advisors, 53 from horticulturists, (essentially fruit and vegetable farmers) and 27 from suppliers of bioprotectants. Among responding horticulturists, 41% (22 of the 53) declared that they used one or more microbial bioprotectants for disease management. This figure may overestimate the frequency of use of microbial bioprotectants, as the survey was clearly focused on bioprotection and non-users may have chosen not to respond. Nearly half of the responding horticulturists (26 of the 53) had some organic production, and among them 16 used both microbial bioprotectants and natural products for disease control, while 10 used only natural products. Microbial bioprotectants were used by 6 of the 27 conventional farmers. The survey generated a greater sample size for farm advisors. Among the 90 respondents, 77 declared that they recommended microbial bioprotectants for disease control. Again, this high number may be biased by the fact that the questionnaire was focused on bioprotection. Microbial bioprotectants were most often recommended for the control of *Sclerotinia* white mold, powdery mildews and *Botrytis* gray mold, with reported recommendations by 32–39% of the farm advisors (Fig. 5). The recommendation of microbial bioprotectants appeared to be somewhat limited for the control of bacterial diseases, and viral diseases were the least targeted, with recommendations of bioprotectant use by only 4 % of the farm advisors.

Twenty microbial bioprotectants commercialized in France were recommended by the farm advisors for disease management of horticultural crops (Fig. 6). They included 13 of the 16 microbial preparations based on fungal strains shown in Fig. 1: 6 Trichoderma-based products (labeled A, C, E, H, M and P in Fig. 6); 4 yeasts (labeled B, K, N and O) and 3 other fungal preparations (labeled D, Q and S). The

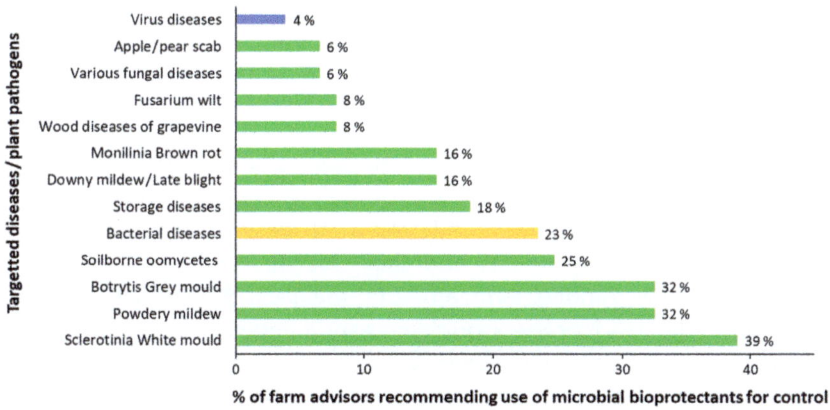

Figure 5 Key diseases and plant pathogens targeted by recommendations of microbial bioprotectant use made by 77 French farm advisors. Blue, green and yellow bars indicate viral, fungal and bacterial targets, respectively.

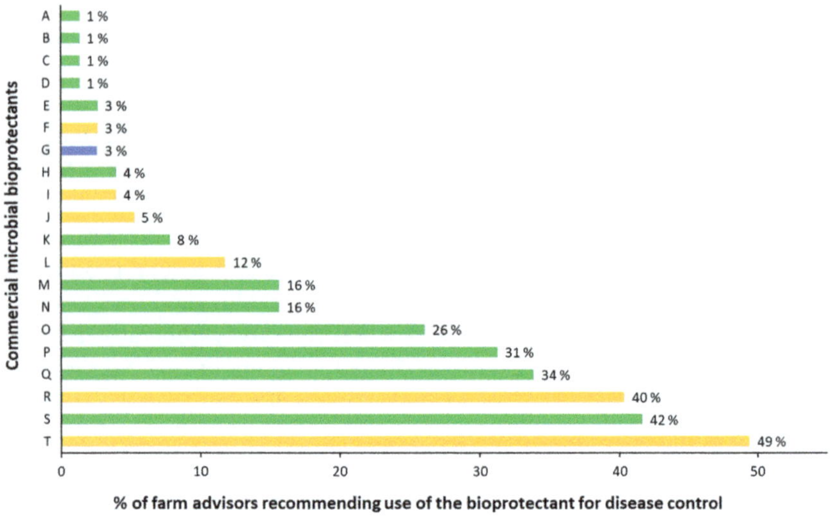

Figure 6 Frequency of recommendation of 20 commercial microbial bioprotectants (anonymously labeled from A to T) for disease control of horticultural crops by 77 French farm advisors. The colors of the bars indicate the type of microorganisms present as active substance in the bioprotectants: green = fungus; yellow = bacterium; blue = virus.

six bacterium-based bioprotectants prescribed by the farm advisors (out of eight shown in Fig. 1) included five *Bacillus*-based preparations (labeled F, J, L, R and T in Fig. 6) and one other bacterium (labeled I).

Although some products are recommended by nearly half of the farm advisors, there is clearly room for wider adoption of microbial bioprotectants, as 5 products were not prescribed at all and 11 were recommended by fewer than 10% of the farm advisors.

5.2 Possible obstacles for wider adoption

In the survey mentioned, certain respondents spontaneously offered comments to express their reticence about the use or the prescription of bioprotectants. Among farmers, the most commonly reported reason was the absence of knowledge about the products and their use. Products were sometimes unavailable locally. Among farm advisors, the lack of information on the products and the need for users' guidelines and references from experimental stations were also frequently evoked. Lack of reliable field efficacy and high costs were also given as reasons.

These survey returns corroborate, for French horticulturists, previously reported analyses of possible impediments to the adoption of bioprotectants at a more global scale (Marrone, 2009, 2019; Moser et al., 2008; Abdollahzadeh et al., 2017; Lamichhane et al., 2017). Marrone (2019) recently described several challenges to the adoption of biopesticides: (i) lack of awareness and education on how to deploy their unique modes of action in integrated programs, (ii) testing products alone versus in integrated programs and (iii) lingering perceptions of high cost and low efficacy.

Socio-economic factors may also play a role, as well as possible issues with risk aversion and acceptance of novelty, as shown from a survey-based study on the prescription of bioprotectants for wine-grapes in the Burgundy region of France (Villemaine et al., 2020) and other studies (Moser et al., 2008; Abdollahzadeh et al., 2017).

The limited number of bioprotectants currently available for disease control, together with the general situation with minor uses presented in part 2 of this chapter, may be complicating factors. In the absence of available bioprotectants for the control of a key disease, a farmer will be likely to resort to chemical control. This, in turn, may interfere with the activity of other microbial bioprotectants and reduce their efficacy. This hypothetical situation highlights the need to provide farmers with detailed information on the compatibility of bioprotectants with other plant protection products.

One key stumbling block for widening the adoption of bioprotectants by farmers may be their perceived or actually encountered lack of reliable efficacy in the field (Bardin and Pugliese, 2020; Abbey et al., 2019). Bioprotectants, and especially microbials, are generally more sensitive than synthetic chemicals to environmental factors (Bailey et al., 2010; Marrone, 2009). They are also much more complex to use, in relation to the intricate modes of actions and factors of efficacy described in this chapter. Comprehending this complexity is a major challenge for the research community, as it requires advanced multidisciplinary expertise in biology, physics and (bio)chemistry, among others. For the farmers, harnessing this complexity to make optimal decisions for the deployment of bioprotectants is, understandably, also a tremendous challenge. This is compounded by the fact that many farmers have ample experience with the application of synthetic pesticides, but no or very limited experience with the use of bioprotectants.

5.3 Coping with the technical complexity of deployment

One possible way to help farmers overcome this challenge is the development of decision support systems (DSS) that would translate scientific knowledge on the biology and modes of action of bioprotectants into simple and hierarchized decision trees. In relation to the digitalization of agriculture, DSS are becoming increasingly available to farmers, for example for the optimization of fertilization and irrigation (Zhai et al., 2020). Excellent reviews have highlighted the potential (and complexity) of DSS for IPM, based on models that forecast the risk of disease development (Damos, 2015; Rossi et al., 2019; Guimapi et al., 2020). To date however, there is a shortage of tools to support decisions on the deployment of bioprotectants for disease control. To our knowledge, only two have been reported. One is 'BOTMAN', a weather-based tool designed to assist decisions on the utilization of a Trichoderma bioprotectant, or alternatively chemical pesticides, for the control of *B. cinerea* in tomato and cucumber greenhouses (Shtienberg and Elad, 1997). More recently, a weather-based disease forecasting model, designed to help farmers schedule copper sprays against Cercospora spot of avocado, was extended to include elements about the behavior of a microbial bioprotectant (Van Eeden and Korsten, 2003).

Work has been initiated in our group to develop knowledge-driven DSS to facilitate the deployment of microbial bioprotectants against diseases of horticultural crops. The first objective is to identify and prioritize the factors governing the efficacy of microbial biocontrol agents and then to identify those that could be modulated by farmers' practices. As a second step, a generic DSS prototype will be designed to integrate key factors of efficacy and their interactions with each other. The final step will be to parameterize and evaluate a DSS for a specific crop-bioprotectant-pathogen combination.

6 Conclusion

Microbial bioprotectants and natural substances have been increasingly available and deployed for disease management, contributing to enhanced sustainability of horticultural production. Because of the large number of horticultural plant species, and of pathogens that can affect them, currently registered products cover only a fraction of farmers' needs for disease control solutions. However, biocontrol and IPM constitute a highly dynamic field of research, both in academia and in the plant protection industry, and high numbers of new bioprotectants are likely to be developed in the coming years, covering an increasing proportion of the needs. The exploration of a broader range of possible sources can be expected to provide a much wider diversity of microbial species and natural substances. Meanwhile, expanding knowledge about the properties of bioprotectants currently on the market should result in significant widening of their spectrum of action, to cover more disease targets on more horticultural plant species. Even entomopathogenic microorganisms traditionally used against pest insects may prove to be useful for disease control (Sarven et al., 2020).

7 Future trends in research

One key challenge to further increase adoption of bioprotectants by farmers is to enhance and stabilize their field efficacy. Continued improvements in their formulation can play a determinant role, for example in sheltering the bioprotectants from detrimental environmental conditions. Novel formulations may also increasingly foster their bioactivity, with increased knowledge about the modes of action of the bioprotectants. This knowledge is also likely to support the design of bioprotectants combining microbial strains or natural substances with complementary or synergistic modes of action. One promising avenue derived from microbiome research is the engineering of microbial consortia to support bioprotectants that are applied, or with their own protective effect (Eckshtain-Levi et al., 2020; Ab Rahman et al., 2018). This research is also likely to lead to innovative tools to recruit indigenous rhizosphere microbiomes to develop suppressiveness against soil-borne diseases (Schlatter et al., 2017; Mazzola and Freilich, 2017).

Improved field efficacy will also result from the inclusion of bioprotectants in IPM strategies. Bioprotectants for plant disease control have long been regarded as substitution tools, to be applied as sustainable alternatives to synthetic pesticides. This concept is conveyed by the word 'biopesticide', which is commonly used to designate certain bioprotectants. Although bioprotectants may often have the ability to provide full protection of crops on their own (van Lenteren et al., 2018), they are decreasingly considered to be stand-alone solutions, as has been expected with chemical pesticides. Rather, bioprotectants are increasingly regarded as components of a toolbox of IPM solutions, which can be combined to design strategies for plant health management (Abbey et al., 2019). This is particularly true in the context of horticultural crops, for which IPM is already a quite common practice and which benefits from a wide range of possible plant health management tools.

Finally, a key challenge for the future will be to enable farmers to cope with the complexity attached to achieving optimized integration of bioprotectants in IPM schemes. For this, it will be necessary to acquire and translate detailed and multifaceted knowledge into simple decision rules that can be implemented at field level, and to deliver them within user-friendly DSS.

8 Where to look for further information

The following articles provide a good overview of key aspects of the subject:

- Ab Rahman, S. F. S., Singh, E., Pieterse, C. M. J. and Schenk, P. M. (2018), Emerging microbial biocontrol strategies for plant pathogens. *Plant Science*, 267, 102–111.
- Bardin, M. and Pugliese, M. (2020), Biocontrol agents against diseases. In: Gullino, M., Albajes , R. and Nicot, P. (Eds) *Integrated Pest and Disease Management in Greenhouse Crops*. Springer.

- Barratt, B., Moran, V., Bigler, F. and van Lenteren, J. (2018), The status of biological control and recommendations for improving uptake for the future. *BioControl*, 63, 155–167.
- Compant, S., Samad, A., Faist, H. and Sessitsch, A. (2019), A review on the plant microbiome: Ecology, functions, and emerging trends in microbial application. *Journal of Advanced Research*, 19, 29–37.
- De Silva, N. I., Brooks, S., Lumyong, S. and Hyde, K. D. (2019), Use of endophytes as biocontrol agents. *Fungal Biology Reviews*, 33, 133–148.
- Köhl, J., Kolnaar, R. and Ravensberg, W. J. (2019), Mode of action of microbial biological control agents against plant diseases: relevance beyond efficacy. *Frontiers in Plant Science*, 10, 845. doi: 10.3389/fpls.2019.00845.
- Raveau, R., Fontaine, J. and Sahraoui, A. L. H. (2020), Essential oils as potential alternative biocontrol products against plant pathogens and weeds: A Review. *Foods*, 9, 31.
- Raymaekers, K., Ponet, L., Holtappels, D., Berckmans, B. and Cammue, B. P. A. (2020), Screening for novel biocontrol agents applicable in plant disease management – A review. *Biological Control*, 144, 104240.
- Zhai, Z., Martínez, J. F., Beltran, V. and Martínez, N. L. (2020), Decision support systems for agriculture 4.0: Survey and challenges. *Computers and Electronics in Agriculture*, 170, 105256.

Key research in this area can be found at member institutions of the International Organisation of Biological Control (https://www.iobc-global.org/) and those of scientists who participate in the activities of Working Group of its West Palaearctic Regional Section on the following topics:

- Biological and integrated control of Plant Pathogens (https://www.iobc-wprs.org/expert_groups/06_wg_plant_pathogens.html).
- Induced resistance in plants against insects and diseases (https://www.iobc-wprs.org/expert_groups/17_wg_induced_resistance.html).
- Integrated control in Protected crops, Temperate climate (https://www.iobc-wprs.org/expert_groups/08_wg_protected_crops_temperate.html) and Mediterranean climate (https://www.iobc-wprs.org/expert_groups/09_wg_protected_crops_mediterranean.html).
- Integrated control in Citrus fruit crops (https://www.iobc-wprs.org/expert_groups/16_wg_citrus_crops.html).
- Integrated protection of Field vegetable Crops (https://www.iobc-wprs.org/expert_groups/12_wg_vegetables.html).
- Integrated protection of Fruit crops (https://www.iobc-wprs.org/expert_groups/02_wg_fruit_crops.html).
- Integrated protection of Olive crops (https://www.iobc-wprs.org/expert_groups/15_wg_olive_crops.html).

9 References

Ab Rahman, S. F. S., Singh, E., Pieterse, C. M. J. and Schenk, P. M. 2018. Emerging microbial biocontrol strategies for plant pathogens. *Plant Science: An International Journal of Experimental Plant Biology* 267, 102-111.

Abbey, J. A., Percival, D., Abbey, L., Asiedu, S. K., Prithiviraj, B. and Schilder, A. 2019. Biofungicides as alternative to synthetic fungicide control of grey mould (*Botrytis cinerea*)–prospects and challenges. *Biocontrol Science and Technology* 29(3), 207–228.

Abdollahzadeh, G., Damalas, C. A. and Sharifzadeh, M. S. 2017. Understanding adoption, non-adoption, and discontinuance of biological control in rice fields of northern Iran. *Crop Protection* 93, 60–68.

Abro, M. A., Lecompte, F., Bryone, F. and Nicot, P. C. 2013. Nitrogen fertilization of the host plant influences production and pathogenicity of *Botrytis cinerea* secondary inoculum. *Phytopathology* 103(3), 261-267.

Abro, M. A., Lecompte, F., Bardin, M. and Nicot, P. C. 2014. Nitrogen fertilization impacts biocontrol of tomato gray *Agronomy for Sustainable Development* 34(3), 641-648.

Aguilar-Veloz, L. M., Calderón-Santoyo, M., Vázquez González, Y. and Ragazzo-Sánchez, J. A. 2020. Application of essential oils and polyphenols as natural antimicrobial agents in postharvest treatments: advances and challenges. *Food Science and Nutrition* 8(6), 2555–2568.

Antoniou, A., Tsolakidou, M. D., Stringlis, I. A. and Pantelides, I. S. 2017. Rhizosphere microbiome recruited from a suppressive compost improves plant fitness and increases protection against vascular wilt pathogens of tomato. *Frontiers in Plant Science* 8, 2022.

Arrarte, E., Garmendia, G., Rossini, C., Wisniewski, M. and Vero, S. 2017. Volatile organic compounds produced by Antarctic strains of *Candida sake* play a role in the control of postharvest pathogens of apples. *Biological Control* 109, 14–20.

Bailey, K. L., Boyetchko, S. M. and Längle, T. 2010. Social and economic drivers shaping the future of biological control: a Canadian perspective on the factors affecting the development and use of microbial biopesticides. *Biological Control* 52(3), 221-229.

Bardin, M. and Pugliese, M. 2020. Biocontrol agents against diseases. In: Gullino, M., Albajes , R. and Nicot, P. (Eds) *Integrated Pest and Disease Management in Greenhouse Crops*, pp. 385–407. Springer. Available at: https://www.springer.com/gp/book/9783030223038.

Barragán-Menéndez, C., Gálvez-López, D., Rosas-Quijano, R., Salvador-Figueroa, M., Ovando-Medina, I. and Vázquez-Ovando, A. 2020. Films of chitosan and aloe vera for maintaining the viability and antifungal activity of *Lactobacillus paracasei* TEP6. *Coatings* 10(3), 259. https://doi.org/10.3390/coatings10030259

Barratt, B. I. P., Moran, V. C., Bigler, F. and van Lenteren, J. C. 2018. The status of biological control and recommendations for improving uptake for the future. *BioControl* 63(1), 155–167.

Benhamou, N. 2004. Potential of the mycoparasite, *Verticillium lecanii*, to protect citrus fruit against *Penicillium digitatum*, the causal agent of green mold: a comparison with the effect of chitosan. *Phytopathology* 94(7), 693–705.

Berg, G., Koberl, M., Rybakova, D., Muller, H., Grosch, R. and Smalla, K. 2017. Plant microbial diversity is suggested as the key to future biocontrol and health trends. *FEMS Microbiology Ecology* 93(5), 9.

Besset-Manzoni, Y., Joly, P., Brutel, A., Gerin, F., Soudiere, O., Langin, T. and Prigent-Combaret, C. 2019. Does in vitro selection of biocontrol agents guarantee success in planta? A study case of wheat protection against Fusarium seedling blight by soil bacteria. *PLoS ONE* 14(12), e0225655.

Boller, E. F., van Lenteren, J. C. and Delucchi, V. (Eds) 2006. International Organization for Biological Control of Noxious Animals and Plants (IOBC): History of the first 50 years (1956–2006). Available at: http://www.iobc-wprs.org/members/shop_en.cfm?mod_Shop_detail_produkte=210.

Bouaoud, Y., Troulet, C., Foughalia, A., Berge, O., Aissat, K. and Bardin, M. 2018. A multi-criteria approach for the selection of efficient biocontrol agents against *Botrytis cinerea* on tomato in Algeria. *BioControl* 63(2), 299–311.

Cabrefiga, J., Francés, J., Montesinos, E. and Bonaterra, A. 2011. Improvement of fitness and efficacy of a fire blight biocontrol agent via nutritional enhancement combined with osmoadaptation. *Applied and Environmental Microbiology* 77(10), 3174–3181.

Carbó, A., Torres, R., Teixidó, N., Usall, J., Magan, N. and Medina, A. 2018. Predicted ecological niches and environmental resilience of different formulations of the biocontrol yeast *Candida sake* CPA-1 using the Bioscreen C. *BioControl* 63(6), 855–866.

Chandler, D., Bailey, A. S., Tatchell, G. M., Davidson, G., Greaves, J. and Grant, W. P. 2011. The development, regulation and use of biopesticides for integrated pest management. *Philosophical Transactions of the Royal Society of London. Series B, Biological Sciences* 366(1573), 1987–1998.

Chen, X., Gao, Z., Swisher, M., House, L. and Zhao, X. 2018. Eco-labeling in the fresh produce market: not all environmentally friendly labels are equally valued. *Ecological Economics* 154, 201–210.

Compant, S., Samad, A., Faist, H. and Sessitsch, A. 2019. A review on the plant microbiome: ecology, functions, and emerging trends in microbial application. *Journal of Advanced Research* 19, 29–37.

Conrath, C., Beckers, G. J., Langenbach, C. J. and Jaskiewicz, M. R. 2015. Priming for enhanced defense. *Annual Review of Phytopathology* 53, 97–119.

Coppola, M., Cascone, P., Lelio, I. D., Woo, S. L., Lorito, M., Rao, R., Pennacchio, F., Guerrieri, E. and Digilio, M. C. 2019. *Trichoderma atroviride* P1 colonization of tomato plants enhances both direct and indirect defense barriers against insects. *Frontiers in Physiology* 10, 813.

Cress, D. 2006. Factors affecting pesticide behavior. *Greenhouse Product News*, September issue, 28–33.

Damos, P. 2015. Modular structure of web-based decision support systems for integrated pest management: a review. *Agronomy for Sustainable Development* 35(4), 1347–1372.

Daranas, N., Badosa, E., Francés, J., Montesinos, E. and Bonaterra, A. 2018. Enhancing water stress tolerance improves fitness in biological control strains of *Lactobacillus plantarum* in plant environments. *PLoS ONE* 13(1), e0190931.

De Corato, U. 2020. Disease-suppressive compost enhances natural soil suppressiveness against soil-borne plant pathogens: a critical review. *Rhizosphere* 13, 100192.

De La Cruz Quiroz, R., Cruz Maldonado, J. J., Rostro Alanis, MdJ., Torres, J. A. and Parra Saldívar, R. 2019. Fungi-based biopesticides: shelf-life preservation technologies used in commercial products. *Journal of Pest Science* 92(3), 1003–1015.

De Silva, N. I., Brooks, S., Lumyong, S. and Hyde, K. D. 2019. Use of endophytes as biocontrol agents. *Fungal Biology Reviews* 33(2), 133–148.

Donley, N. 2019. The USA lags behind other agricultural nations in banning harmful pesticides. *Environmental Health: A Global Access Science Source* 18(1), 44.

Eckshtain-Levi, N., Harris, S. L., Roscios, R. Q. and Shank, E. A. 2020. Bacterial community members increase bacillus subtilis maintenance on the roots of *Arabidopsis thaliana*. *Phytobiomes Journal* 4(4), 303–313.

Elad, Y. and Pertot, I. 2014. Climate change impacts on plant pathogens and plant diseases. *Journal of Crop Improvement* 28(1), 99–139.

Ellis, J. G. 2017. Can plant microbiome studies lead to effective biocontrol of plant diseases? *Molecular Plant-Microbe Interactions: MPMI* 30(3), 190–193.

Elsayed, T. R., Jacquiod, S., Nour, E. H., Sorensen, S. J. and Smalla, K. 2020. Biocontrol of bacterial wilt disease through complex interaction between tomato plant, antagonists, the indigenous rhizosphere microbiota, and *Ralstonia solanacearum*. *Frontiers in Microbiology* 10, 2835.

FAO and WHO 2017. International Code of Conduct on Pesticide Management: Guidelines for the Registration of Microbial, Botanical and Semiochemical Pest Control Agents for Plant Protection and Public Health Uses. The Food and Agriculture Organization of the United Nations (FAO) and the World Health Organization (WHO).

Fedele, G., Brischetto, C. and Rossi, V. 2020. Biocontrol of *Botrytis cinerea* on grape berries as influenced by temperature and humidity. *Frontiers in Plant Science* 11, 1232.

Fraceto, L. F., Maruyama, C. R., Guilger, M., Mishra, S., Keswani, C., Singh, H. B. and De Lima, R. 2018. Trichoderma harzianum-based novel formulations: potential applications for management of Next-Gen agricultural challenges. *Journal of Chemical Technology and Biotechnology* 93(8), 2056–2063.

Frederiks, C. and Wesseler, J. H. H. 2019. A comparison of the EU and US regulatory frameworks for the active substance registration of microbial biological control agents. *Pest Management Science* 75(1), 87–103.

Freier, B. and Boller, E. F. 2009. Integrated pest management in Europe-history, policy, achievements and implementation. In: Peshin R. and Dhawan A. K. (Eds) *Integrated Pest Management: Dissemination and Impact*, pp. 435–454. Dordrecht: Springer. https://doi.org/10.1007/978-1-4020-8990-9_14.

French, E., Tran, T. and Iyer-Pascuzzi, A. S. 2020. Tomato genotype modulates selection and responses to root microbiota. *Phytobiomes Journal* 4(4), 314–326.

Gao, Z., Zhang, B., Liu, H., Han, J. and Zhang, Y. 2017. Identification of endophytic *Bacillus velezensis* ZSY-1 strain and antifungal activity of its volatile compounds against *Alternaria solani* and *Botrytis cinerea*. *Biological Control* 105, 27–39.

Gerlagh, M., Goossen-Van De Geijn, H. M., Hoogland, A. E. and Vereijken, P. F. G. 2003. Quantitative aspects of infection of *Sclerotinia sclerotiorum* sclerotia by *Coniothyrium minitans*–timing of application, concentration and quality of conidial suspension of the mycoparasite. *European Journal of Plant Pathology* 109(5), 489–502.

Gilardi, G., Pugliese, M., Gullino, M. L. and Garibaldi, A. 2019. Nursery treatments with resistant inducers, soil amendments and biocontrol agents for the management of the Fusarium wilt of lettuce under glasshouse and field conditions. *Journal of Phytopathology* 167(2), 98–110.

Glare, T., Caradus, J., Gelernter, W., Jackson, T., Keyhani, N., Köhl, J., Marrone, P., Morin, L. and Stewart, A. 2012. Have biopesticides come of age? *Trends in Biotechnology* 30(5), 250–258.

Godana, E. A., Yang, Q., Wang, K., Zhang, H., Zhang, X., Zhao, L., Abdelhai, M. H. and Guillaume Legrand, N. N. 2020. Bio-control activity of *Pichia anomala* supplemented with chitosan against *Penicillium expansum* in postharvest grapes and its possible inhibition mechanism. *LWT* 124, 109188.

Gotor-Vila, A., Teixido, N., Sisquella, M., Torres, R. and Usall, J. 2017. Biological characterization of the biocontrol agent bacillus amyloliquefaciens CPA-8: the effect of temperature, pH and water activity on growth, susceptibility to antibiotics and detection of enterotoxic genes. *Current Microbiology* 74(9), 1089–1099.

Gotor-Vila, A., Usall, J., Torres, R., Solsona, C. and Teixido, N. 2019. Enhanced shelf-life of the formulated biocontrol agent *Bacillus amyloliquefaciens* CPA-8 combining diverse packaging strategies and storage conditions. *International Journal of Food Microbiology* 290, 205–213.

Gramisci, B. R., Lutz, M. C., Lopes, C. A. and Sangorrín, M. P. 2018. Enhancing the efficacy of yeast biocontrol agents against postharvest pathogens through nutrient profiling and the use of other additives. *Biological Control* 121, 151–158.

Gu, K. B., Zhang, D. J., Guan, C., Xu, J. H., Li, S. L., Shen, G. M., Luo, Y. C. and Li, Y. G. 2017. Safe antifungal lipopeptides derived from *Bacillus marinus* B-9987 against grey mold caused by *Botrytis cinerea*. *Journal of Integrative Agriculture* 16(9), 1999 -2008.

Guimapi, R. A., Mohamed, S. A., Biber-Freudenberger, L., Mwangi, W., Ekesi, S., Borgemeister, C. and Tonnang, H. E. Z. 2020. Decision support system for fitting and mapping nonlinear functions with application to insect pest management in the biological control context. *Algorithms* 13, 20.

Guimarães, R. A., Pherez-Perrony, P. E., Müller, H., Berg, G., Medeiros, F. H. V. and Cernava, T. 2020. Microbiome-guided evaluation of *Bacillus subtilis* BIOUFLA2 application to reduce mycotoxins in maize kernels. *Biological Control* 150, 104370.

Hao, L., Wang, Y., Chen, X., Zheng, X., Chen, S., Li, S., Zhang, Y. and Xu, Y. 2019. Exploring the potential of natural products from mangrove rhizosphere bacteria as biopesticides against plant diseases. *Plant Disease* 103(11), 2925–2932.

Hardwick, N. V. 2002. Weather and plant diseases. *Weather* 57(5), 184–190.

Hu, X., Cheng, B., Du, D., Huang, Z., Pu, Z., Chen, G., Peng, A. and Lu, L. 2019. Isolation and identification of a marine actinomycete strain and its control efficacy against citrus green and blue moulds. *Biotechnology and Biotechnological Equipment* 33(1), 719–729.

Hubbard, M., Hynes, R. K., Erlandson, M. and Bailey, K. L. 2014. The biochemistry behind biopesticide efficacy. *Sustainable Chemical Processes* 2(1), 18.

Humair, B., González, N., Mossialos, D., Reimmann, C. and Haas, D. 2009. Temperature-responsive sensing regulates biocontrol factor expression in *Pseudomonas fluorescens* CHA0. *The ISME Journal* 3(8), 955–965.

IBMA 2019. Definition: bioprotection as the global term for all biocontrol technologies. In: IBMA (Ed.) International Biocontrol Manufacturers Association. Available at: https://ibma-global.org/wp-content/uploads/2020/12/ibmadefinitionleafletweb.pdf.

Iñiguez-Moreno, M., Ragazzo-Sánchez, J. A., Barros-Castillo, J. C., Sandoval-Contreras, T. and Calderón-Santoyo, M. 2020. Sodium alginate coatings added with *Meyerozyma caribbica*: postharvest biocontrol of *Colletotrichum gloeosporioides* in avocado (*Persea americana* Mill. cv. Hass). *Postharvest Biology and Technology* 163, 111123.

Kalantari, S., Marefat, A., Naseri, B. and Hemmati, R. 2018. Improvement of bean yield and Fusarium root rot biocontrol using mixtures of *Bacillus, Pseudomonas* and *Rhizobium*. *Tropical Plant Pathology* 43(6), 499–505.

Kamal, M. M., Lindbeck, K. D., Savocchia, S. and Ash, G. J. 2015. Biological control of Sclerotinia stem rot of canola using antagonistic bacteria. *Plant Pathology* 64(6), 1375–1384.

Kehe, J., Kulesa, A., Ortiz, A., Ackerman, C. M., Thakku, S. G., Sellers, D., Kuehn, S., Gore, J., Friedman, J. and Blainey, P. C. 2019. Massively parallel screening of synthetic microbial communities. *Proceedings of the National Academy of Sciences of the United States of America* 116(26), 12804–12809.

Koch, E., Becker, J. O., Berg, G., Hauschild, R., Jehle, J., Kohl, J. and Smalla, K. 2018. Biocontrol of plant diseases is not an unsafe technology! *Journal of Plant Diseases and Protection* 125(2), 121–125.

Köhl, J., Booij, K., Kolnaar, R. and Ravensberg, W. J. 2019a. Ecological arguments to reconsider data requirements regarding the environmental fate of microbial biocontrol agents in the registration procedure in the European Union. *BioControl* 64(5), 469–487.

Köhl, J., Kolnaar, R. and Ravensberg, W. J. 2019b. Mode of action of microbial biological control agents against plant diseases: relevance beyond efficacy. *Frontiers in Plant Science* 10, 845. doi: 10.3389/fpls.2019.00845.

Köhl, J., Medeiros, F. H. V., Lombaers-Van Der Plas, C., Groenenboom-De Haas, L. and Van Den Bosch, T. 2020. Efficacies of bacterial and fungal isolates in biocontrol of *Botrytis cinerea* and *Pseudomonas syringae* pv. *tomato* and growth promotion in tomato do not correlate. *Biological Control* 150, 104375.

Köhl, J., Postma, J., Nicot, P., Ruocco, M. and Blum, B. 2011. Stepwise screening of microorganisms for commercial use in biological control of plant-pathogenic fungi and bacteria. *Biological Control* 57(1), 1–12.

Kong, Q. 2018. Marine microorganisms as biocontrol agents against fungal phytopathogens and mycotoxins. *Biocontrol Science and Technology* 28(1), 77–93.

Korsten, L. 2004. Biological control in Africa: can it provide a sustainable solution for control of fruit diseases? *South African Journal of Botany* 70(1), 128–139.

Lamichhane, J. R., Bischoff-Schaefer, M., Bluemel, S., Dachbrodt-Saaydeh, S., Dreux, L., Jansen, J. P., Kiss, J., Köhl, J., Kudsk, P., Malausa, T., Messéan, A., Nicot, P. C., Ricci, P., Thibierge, J. and Villeneuve, F. 2017. Identifying obstacles and ranking common biological control research priorities for Europe to manage most economically important pests in arable, vegetable and perennial crops. *Pest Management Science* 73(1), 14–21.

Landa, B. B., Navas-Cortés, J. A. and Jiménez-Díaz, R. M. 2004. Influence of temperature on plant-rhizobacteria interactions related to biocontrol potential for suppression of fusarium wilt of chickpea. *Plant Pathology* 53(3), 341–352.

Lecompte, F., Abro, M. A. and Nicot, P. C. 2010. Contrasted responses of *Botrytis cinerea* isolates developing on tomato plants grown under different nitrogen nutrition regimes. *Plant Pathology* 59(5), 891–899.

Lecompte, F., Abro, M. A. and Nicot, P. C. 2013. Can plant sugars mediate the effect of nitrogen fertilization on lettuce susceptibility to two necrotrophic pathogens: *Botrytis cinerea* and *Sclerotinia sclerotiorum*? *Plant and Soil* 369(1–2), 387–401.

Leus, L. 2018. Breeding for disease resistance in ornamentals. In: Van Huylenbroeck J. (Ed.) Ornamental Crops. Handbook of Plant Breeding, Vol. 11, pp. 97-125. Cham: Springer. https://doi.org/10.1007/978-3-319-90698-0_5

Lombardi, N., Vitale, S., Turrà, D., Reverberi, M., Fanelli, C., Vinale, F., Marra, R., Ruocco, M., Pascale, A., D'errico, G., Woo, S. L. and Lorito, M. 2018. Root exudates of stressed plants stimulate and attract Trichoderma soil fungi. *Molecular Plant-Microbe Interactions: MPMI* 31(10), 982-994.

Lorito, M., Woo, S. L., Harman, G. E. and Monte, E. 2010. Translational research on Trichoderma: from'omics to the field. *Annual Review of Phytopathology* 48, 395-417.

Malerba, M. and Cerana, R. 2019. Recent applications of chitin- and chitosan-based polymers in plants. *Polymers* 11(5), 839.

Marian, M. and Shimizu, M. 2019. Improving performance of microbial biocontrol agents against plant diseases. *Journal of General Plant Pathology* 85(5), 329-336.

Marimuthu, S., Karthic, C., Mostafa, A. A., Mohammed Al-Enazi, N., Abdel-Raouf, N. and Nageh Sholkamy, E. 2020. Antifungal activity of *Streptomyces* sp. SLR03 against tea fungal plant pathogen *Pestalotiopsis theae*. *Journal of King Saud University–Science* 32(8), 3258-3264.

Marín, A., Atarés, L. and Chiralt, A. 2017. Improving function of biocontrol agents incorporated in antifungal fruit coatings: a review. *Biocontrol Science and Technology* 27(10), 1220-1241.

Marín, A., Plotto, A., Atarés, L. and Chiralt, A. 2019. Lactic acid bacteria incorporated into edible coatings to control fungal growth and maintain postharvest quality of grapes. *Hortscience* 54(2), 337-343.

Marrone, P. G. 2009. Barriers to adoption of biological control agents and biological pesticides. *CAB Reviews: Perspectives in Agriculture, Veterinary Science, Nutrition and Natural Resources* 2(51), 163-178.

Marrone, P. G. 2019. Pesticidal natural products–status and future potential. *Pest Management Science* 75(9), 2325-2340.

Massart, S., Martinez-Medina, M. and Jijakli, M. H. 2015. Biological control in the microbiome era: challenges and opportunities. *Biological Control* 89, 98-108.

Mauch-Mani, B., Baccelli, I., Luna, E. and Flors, V. 2017. Defense priming: an adaptive part of induced resistance. *Annual Review of Plant Biology* 68, 485-512.

Mazzola, M. and Freilich, S. 2017. Prospects for biological soilborne disease control: application of indigenous versus synthetic microbiomes. *Phytopathology* 107(3), 256-263.

Mitter, B., Brader, G., Pfaffenbichler, N. and Sessitsch, A. 2019. Next generation microbiome applications for crop production–limitations and the need of knowledge–based solutions. *Current Opinion in Microbiology* 49, 59-65.

Moser, R., Pertot, I., Elad, Y. and Raffaelli, R. 2008. Farmers' attitudes toward the use of biocontrol agents in IPM strawberry production in three countries. *Biological Control* 47(2), 125-132.

Mülner, P., Bergna, A., Wagner, P., Sarajlić, D., Gstöttenmayr, B., Dietel, K., Grosch, R., Cernava, T. and Berg, G. 2019. Microbiota associated with sclerotia of soilborne fungal pathogens–a novel source of biocontrol agents producing bioactive volatiles. *Phytobiomes Journal* 3(2), 125-136.

Nguyen, H. T. and Le, H. T. 2020. The effect of agricultural product eco-labelling on green purchase intention. *Management Science Letters* 10, 2813-2820.

Nicot, P., Alabouvette, C., Bardin, M., Blum, B., Kohl, J. and Ruocco, M. 2012. Review of factors influencing the success or failure of biocontrol: technical, industrial and socio-economic perspectives. *IOBC WPRS Bulletin* 78, 95–98.

Nicot, P. C., Bardin, M., Debruyne, F., Duffand, M., Lecompte, F., Neu, L. and Pascal, M. 2013. Effect of nitrogen fertilisation of strawberry plants on the efficacy of defence-stimulating biocontrol products against Botrytis cinerea. *IOBC WPRS Bulletin* 88, 39–42.

OECD 2009. OECD guidance document on defining minor uses of pesticides. ENV/JM/MONO(2009)39.

Ojiambo, P. S. and Scherm, H. 2006. Biological and application-oriented factors influencing plant disease suppression by biological control: a meta-analytical review. *Phytopathology* 96(11), 1168–1174.

Ozyilmaz, U. 2020. Evaluation of the effectiveness of antagonistic bacteria against Phytophthora blight disease in pepper with artificial intelligence. *Biological Control* 151, 104379.

Patel, J. S., Selvaraj, V., Gunupuru, L. R., Rathor, P. K. and Prithiviraj, B. 2020. Combined application of Ascophyllum nodosum extract and chitosan synergistically activates host-defense of peas against powdery mildew. *BMC Plant Biology* 20(1), 113.

Pelaez, V. and Mizukawa, G. 2017. Diversification strategies in the pesticide industry: from seeds to biopesticides. *Ciência Rural* 47(2). http://www.scielo.br/scielo.php?script=sci_arttext&pid=S0103-84782017000200152&nrm=iso.

Pintye, A., Ropars, J., Harvey, N., Shin, H. D., Leyronas, C., Nicot, P. C., Giraud, T. and Kiss, L. 2015. Host phenology and geography as drivers of differentiation in generalist fungal mycoparasites. *PLoS ONE* 10(3), e0120703.

Porat, R., Lichter, A., Terry, L. A., Harker, R. and Buzby, J. 2018. Postharvest losses of fruit and vegetables during retail and in consumers' homes: quantifications, causes, and means of prevention. *Postharvest Biology and Technology* 139, 135–149.

Postma, J. and Nijhuis, E. H. 2019. Pseudomonas chlororaphis and organic amendments controlling Pythium infection in tomato. *European Journal of Plant Pathology* 154(1), 91-107.

Poudel, R., Jumpponen, A., Schlatter, D. C., Paulitz, T. C., Gardener, B. B. M., Kinkel, L. L. and Garrett, K. A. 2016. Microbiome networks: a systems framework for identifying candidate microbial assemblages for disease management. *Phytopathology* 106(10), 1083-1096.

Rao, J., Chen, B. and McClements, D. J. 2019. Improving the efficacy of essential oils as antimicrobials in foods: mechanisms of action. *Annual Review of Food Science and Technology* 10, 365–387.

Raveau, R., Fontaine, J. and Sahraoui, A. L. H. 2020. Essential oils as potential alternative biocontrol products against plant pathogens and weeds: a review. *Foods* 9(3), 31.

Raymaekers, K., Ponet, L., Holtappels, D., Berckmans, B. and Cammue, B. P. A. 2020. Screening for novel biocontrol agents applicable in plant disease management–a review. *Biological Control* 144, 104240.

Robin, D. C. and Marchand, P. A. 2019. Evolution of the biocontrol active substances in the framework of the European pesticide Regulation (EC) No. 1107/2009. *Pest Management Science* 75(4), 950–958.

Roger, C., Tivoli, B. and Huber, L. 1999. Effects of temperature and moisture on disease and fruit body development of *Mycosphaerella pinodes* on pea (*Pisum sativum*). *Plant Pathology* 48(1), 1-9.

Romanazzi, G., Feliziani, E. and Sivakumar, D. 2019. Chitosan, a biopolymer with triple action on postharvest decay of fruit and vegetables: eliciting, antimicrobial and film-forming properties. *Frontiers in Microbiology* 9, 2745.

Rossi, V., Sperandio, G., Caffi, T., Simonetto, A. and Gilioli, G. 2019. Critical success factors for the adoption of decision tools in IPM. *Agronomy-Basel* 9, 19.

Sahu, P. K., Singh, S., Gupta, A. R., Gupta, A., Singh, U. B., Manzar, N., Bhowmik, A., Singh, H. V. and Saxena, A. K. 2020. Endophytic bacilli from medicinal-aromatic perennial Holy basil (*Ocimum tenuiflorum* L.) modulate plant growth promotion and induced systemic resistance against *Rhizoctonia solani* in rice (*Oryza sativa* L.). *Biological Control* 150, 104353.

SANCO 2017. Guidelines on comparability, extrapolation, group tolerances and data requirements for setting MRLs. *Guidance document.* European Commission, Directorate-General for Health and Food Safety.

Sarven, M. S., Hao, Q., Deng, J., Yang, F., Wang, G., Xiao, Y. and Xiao, X. 2020. Biological control of tomato gray mold caused by *Botrytis cinerea* with the entomopathogenic fungus *Metarhizium anisopliae*. *Pathogens* 9(3), 213.

Schlatter, D., Kinkel, L., Thomashow, L., Weller, D. and Paulitz, T. 2017. Disease suppressive soils: new insights from the soil microbiome. *Phytopathology* 107(11), 1284–1297.

Sharma, V., Salwan, R., Sharma, P. N. and Gulati, A. 2017. Integrated translatome and proteome: approach for accurate portraying of widespread multifunctional aspects of Trichoderma. *Frontiers in Microbiology* 8, 1602.

Shtienberg, D. and Elad, Y. 1997. Incorporation of weather forecasting in integrated, biological-chemical management of Botrytis cinerea. *Phytopathology* 87(3), 332–340.

Singh, S., Kumar, V., Dhanjal, D. S. and Singh, J. 2020. Biological control agents: diversity, ecological significances, and biotechnological applications. In: Singh, J. and Yadav, A. N. (Eds) *Natural Bioactive Products in Sustainable Agriculture*, pp. 31–44. Singapore: Springer.

Smith, K. P., Handelsman, J. and Goodman, R. M. 1999. Genetic basis in plants for interactions with disease-suppressive bacteria. *Proceedings of the National Academy of Sciences of the United States of America* 96(9), 4786–4790.

Sood, M., Kapoor, D., Kumar, V., Sheteiwy, M. S., Ramakrishnan, M., Landi, M., Araniti, F. and Sharma, A. 2020. Trichoderma: the "secrets" of a multitalented biocontrol agent. *Plants-Basel* 9(6), 25.

Tao, C., Li, R., Xiong, W., Shen, Z., Liu, S., Wang, B., Ruan, Y., Geisen, S., Shen, Q. and Kowalchuk, G. A. 2020. Bio-organic fertilizers stimulate indigenous soil Pseudomonas populations to enhance plant disease suppression. *Microbiome* 8(1), 137.

Thøgersen, J., Pedersen, S. and Aschemann-Witzel, J. 2019. The impact of organic certification and country of origin on consumer food choice in developed and emerging economies. *Food Quality and Preference* 72, 10–30.

Tucci, M., Ruocco, M., De Masi, L., De Palma, M. and Lorito, M. 2011. The beneficial effect of Trichoderma spp. on tomato is modulated by the plant genotype. *Molecular Plant Pathology* 12(4), 341-354.

Van Eeden, M. and Korsten, L. 2003. Determining appropriate parameters for an integrated disease control system for biological predictive modelling. *South African Avocado Growers' Association Yearbook* 26, 83–95.

van Lenteren, J. C., Bolckmans, K., Köhl, J., Ravensberg, W. J. and Urbaneja, A. 2018. Biological control using invertebrates and microorganisms: plenty of new opportunities. *BioControl* 63(1), 39–59.

van Lenteren, J. C. and Nicot, P. C. 2020. Integrated pest management methods and considerations concerning implementation in greenhouses. In: Gullino, M. L., Albajes, R. and Nicot, P. C. (Eds) *Integrated Pest and Disease Management in Greenhouse Crops*, pp. 1–20. Cham: Springer International Publishing. Available at: https://www.sciencedirect.com/science/article/abs/pii/S1369703X07001994.

Verma, M., Brar, S. K., Tyagi, R. D., Surampalli, R. Y. and Valero, J. R. 2007. Antagonistic fungi, Trichoderma spp.: panoply of biological control. *Biochemical Engineering Journal* 37(1), 1–20.

Villemaine, R., Compagnone, C. and Falconnet, C. 2020. The social construction of alternatives to pesticide use: A study of biocontrol in Burgundian viticulture. *Sociologia Ruralis* 61(1), 74–95, doi: 10.1111/soru.12320.

Vitorino, L. C. and Bessa, L. A. 2018. Microbial Diversity: the gap between the estimated and the known. *Diversity* 10(2), 29.

Wamalwa, L. N., Cheseto, X., Ouna, E., Kaplan, F., Maniania, N. K., Machuka, J., Torto, B. and Ghislain, M. 2015. Toxic ipomeamarone accumulation in healthy parts of sweetpotato (*Ipomoea batatas* L. Lam) storage roots upon infection by *Rhizopus stolonifer*. *Journal of Agricultural and Food Chemistry* 63(1), 335–342.

Wei, Z., Huang, J., Yang, T., Jousset, A., Xu, Y., Shen, Q. and Friman, V. P. 2017. Seasonal variation in the biocontrol efficiency of bacterial wilt is driven by temperature-mediated changes in bacterial competitive interactions. *Journal of Applied Ecology* 54(5), 1440–1448.

Woo, S. L. and Pepe, O. 2018. Microbial consortia: promising probiotics as plant biostimulants for sustainable agriculture. *Frontiers in Plant Science* 9, 1801.

Yuan, B., Jia, H., Bu, W., Yang, T., Han-Meng, L., Ju, X. Y. and Li, R. P. 2020. A new chitosan sub-micron and encapsulated Iturin A with enhanced antifungal activity against *Ceratocystis fimbriata* and *Rhizopus strolonifer*. *International Journal of Biological Macromolecules* 159, 995–1003.

Zaki, O., Weekers, F., Thonart, P., Tesch, E., Kuenemann, P. and Jacques, P. 2020. Limiting factors of mycopesticide development. *Biological Control* 144, 104220.

Zhai, Z., Martínez, J. F., Beltran, V. and Martínez, N. L. 2020. Decision support systems for agriculture 4.0: survey and challenges. *Computers and Electronics in Agriculture* 170, 105256.

Zhou, Y., Zhang, L. and Zeng, K. 2016. Efficacy of pichia membranaefaciens combined with chitosan against Colletotrichum gloeosporioides in citrus fruits and possible modes of action. *Biological Control* 96, 39–47.

Chapter 3

Advances in biostimulants as an integrated pest management tool in horticulture

Surendra K. Dara, University of California Cooperative Extension, USA

1 Introduction

Plant biostimulants are microorganisms, microbial derivatives, and materials of mineral, marine, or botanical origin that stimulate plant physiological responses including induced systemic resistance (ISR), and improve nutrient or water uptake, promote plant growth and health, or impart tolerance to biotic and abiotic stressors, without having nutritional or pesticidal properties. This definition covers both the technical and the regulatory aspects of biostimulants. However, there are some exceptions to the biostimulant products that are available in the market, due to the inherent nutritional or pesticidal properties of some of the active ingredients or their sources. For example, silicon has biostimulant properties, but when applied as calcium or potassium silicate,

http://dx.doi.org/10.19103/AS.2021.0095.03

plants receive calcium and potassium as additional nutrients along with silicon (Savvas and Ntatsi, 2015). Some protein hydrolysates have several nutrient components in addition to biostimulant activity (Baglieri et al., 2014; Visconi et al., 2015). Similarly, certain microorganisms have both biostimulatory and biopesticidal properties and depending on the strain or application procedure, they are registered or used for one of these purposes (Kim et al., 2017; Fiorentino et al., 2018; Dara, 2019a). Crop habitat is a dynamic environment with complex interactions among various biotic and abiotic elements. Sometimes, biostimulatory and biopesticidal properties overlap, interchange, or are activated or synergized by their interactions with other influencing factors. To avoid under-exploring the full potential of biostimulants, this chapter does not involve the regulatory aspects but focuses only on the biostimulatory properties of various active ingredients and how they can be used in integrated pest management (IPM) programs as a part of cultural control. The IPM section briefly describes the new IPM model (Dara, 2019b), the role of biostimulants in improving crop production and protection, and lists various categories of biostimulants. The following sections focus on the role of biostimulants in enhancing crop growth, health, and yield, improving soil health, inducing natural resistance, suppressing diseases and arthropod pests, and imparting tolerance to abiotic stresses. Various interactions of biostimulants with other biotic and abiotic factors in the crop environment, the different strategies for successfully using biostimulants in food production, and the challenges and future needs in promoting the use of biostimulants will be discussed in the concluding sections.

2 Integrated pest management

The traditional definition of IPM was limited to pest monitoring, various control options, and making treatment decisions based on economic thresholds with a focus on protecting the ecosystem (Peterson et al., 2018). In many cases, the rotational use of pesticides with different modes of action, mainly intended for managing pesticide resistance, is considered as practicing IPM. In general, a heavy emphasis on minimizing pesticide applications with a focus on environmental safety, a lack of consideration for factors that influence pest management practices at the farm level, and various other reasons have limited the widescale implementation of IPM strategies. The transformation of agriculture as a global enterprise, the technological advances in multiple disciplines of crop production, the growing world populations and their food needs, and an increasing demand for sustainably produced food warranted a practical IPM model that is economically viable, environmentally sustainable, and socially acceptable. As a result, the new IPM model has been developed to meet the needs of modern-day agriculture (Dara, 2019b). The new model provides a template for an integrated crop production system that makes the best use of traditional and modern technologies, incorporates information management and effective decision-making, and emphasizes the importance of research and outreach to maintain farm profitability and food affordability (Fig. 1). The knowledge of pests, the vulnerable stages of their life cycle, and the influence of various agronomic practices on pest populations and their damage are critical in the new IPM model as they help devise and implement effective strategies. Good agronomic practices such as using biostimulants not only improve crop health but also lessen the burden on crop protection directly or indirectly. Biostimulants help crop production in one or more of the following ways (Fig. 2):

1 Stimulate ISR and prime the plants for biotic and abiotic stressors.
2 Improve water and nutrient use efficiency and thus reduce the input costs and associated environmental risks from leaching or runoff,

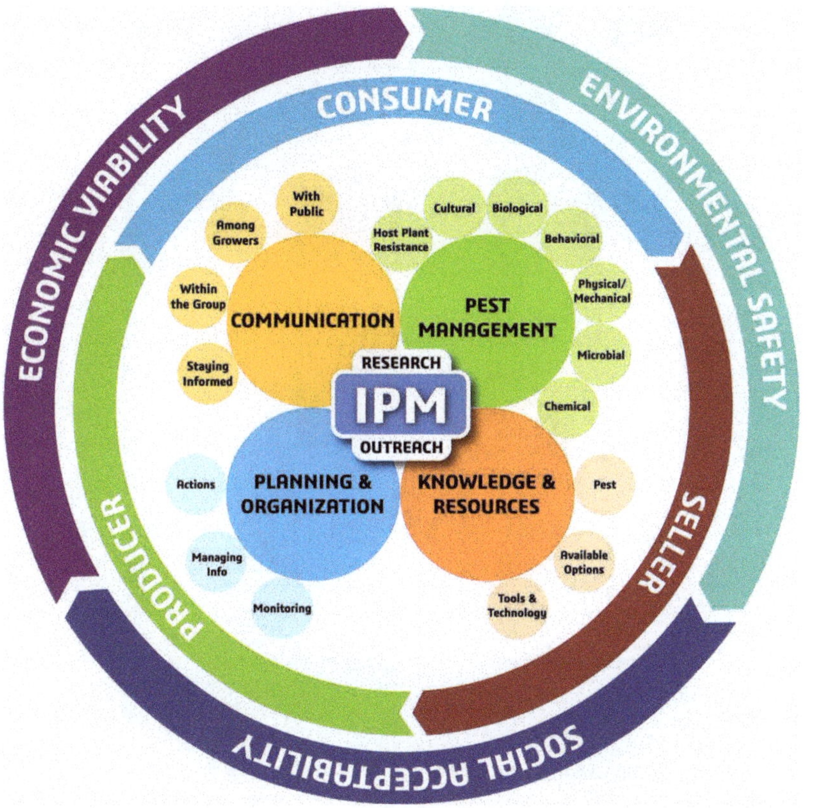

Figure 1 The new integrated pest management (IPM) model provides a template for various crop and pest situations to produce crops in an economically viable, socially acceptable, and environmentally sustainable manner (Dara, 2019b).

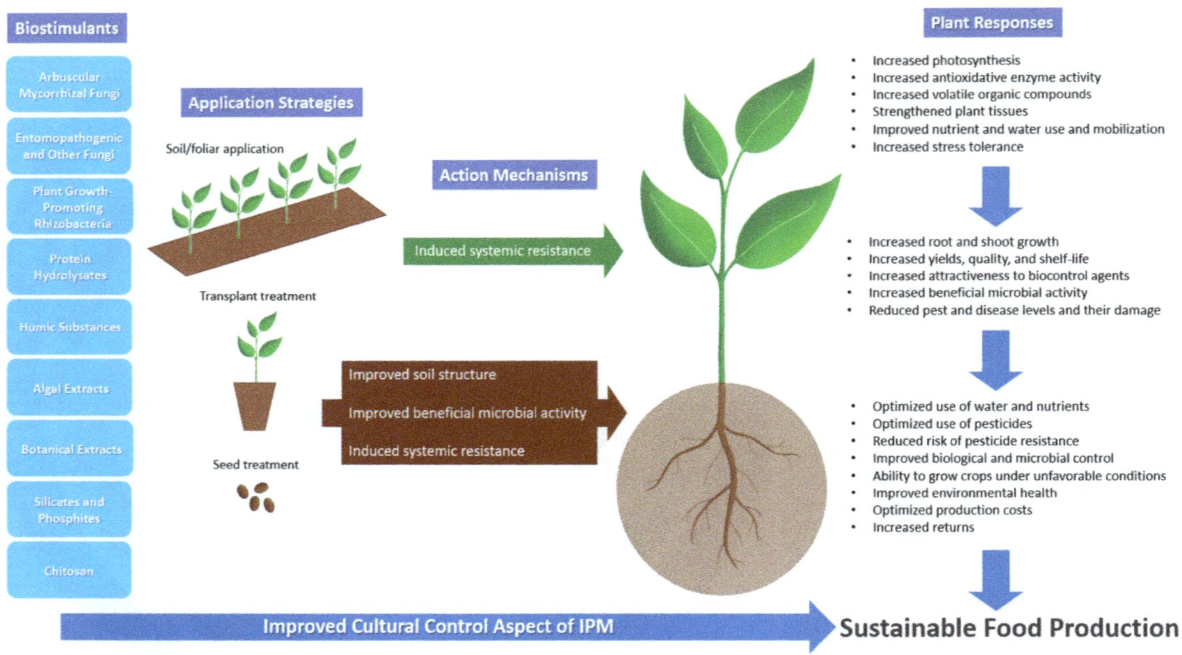

Figure 2 Biostimulant categories, application strategies, mechanisms of action, and their impact on crop production and protection.

enhance plant vigor and yields and thus compensate for losses due to pest damage.

3 Help optimize fertilizer inputs and prevent overuse of certain nutrients that increase a plant's attractiveness to pests.
4 Strengthen plant tissues and reduce pest damage or proliferation.
5 Release metabolites or enhance the production of plant defense chemicals that negatively influence pest populations or attract natural enemies.
6 Contribute to the reduction of pest management practices, pesticide resistance, and the potential negative impact on the environment.
7 Build soil structure and promote natural beneficial microbial populations that further improve crop health in the short term and improve soil health in the long term.
8 Contribute to carbon sequestration and a reduction in the use of synthetic agricultural inputs.

One key area of IPM is focusing on preventive measures and minimizing the reliance on curative measures. Biostimulants are good agronomic inputs that contribute to these preventive measures as a part of cultural control. Biostimulant use has increased in recent years as a result of increased market availability, research and outreach efforts, the need for improved crop production practices, and the interest in sustainably produced food items. Examples of some research studies with commercially available or developmental products that have demonstrated the benefits of various categories of biostimulants, including arbuscular mycorrhizal fungi (AMF), entomopathogenic and other fungi, plant growth-promoting rhizobacteria (PGPR), protein hydrolysates, humic substances, algal and botanical extracts, and minerals are discussed in the following sections.

3 Enhancement of crop growth, health and yield

Biostimulants are primarily used for improving crop growth, health, and yields through one or more mechanisms. Increased nutrient and water uptake through enhanced root growth, increased solubility or bioavailability of nutrients, higher beneficial microbial activity, and improved metabolic activity are among various mechanisms that aid plant growth and health.

3.1 Arbuscular mycorrhizal fungi (AMF)

AMF have been used in crop production for a long time for their biostimulatory properties. *Glomus aggregatum* significantly enhanced the growth of grape rootstocks with improved root morphology (Aguín et al., 2004). Inoculation of *Glomus mosseae* resulted in higher biomass and concentration of some nutrients in shoots, and higher single-fruit weight in cucumbers (Wang et al., 2008). *Glomus versiforme* inoculation improved osmotic adjustment, leaf water potential, and soluble sugar content, and increased the transpiration and photosynthetic rates among other changes, resulting in improved plant growth of tangerine (Wu and Xia, 2006). A review by Rouphael et al. (2015) discussed yield and growth improvement of several horticultural crops growing under nutrient-deficient conditions with the help of AMF.

3.2 Entomopathogenic and other fungi

Although not primarily used as biostimulants, entomopathogenic fungi such as *Beauveria bassiana*, *Cordyceps fumosorosea*, and *Metarhizium* spp. appear to have biostimulatory properties such as promoting plant growth through improved nutrient and water uptake and alleviating plant stresses. A strain of *Beauveria bassiana* is currently available in the United States as a soil amendment, for its biostimulant properties. A recent review of entomopathogenic fungi described various non-entomopathogenic roles and how they can be exploited as holistic tools (Dara, 2019a). When entomopathogenic fungi are applied for arthropod pest control, they also benefit plants through their endophytic and mycorrhiza-like roles and improve plant growth and health. Additionally, entomopathogenic fungi could also serve as the substitutes of AMF for crops that have a poor relationship with AMF.

The mycoparasitic fungal species of *Trichoderma* are commonly used as biofungicides, but some of them also have biostimulatory properties and are sold as biofertilizers. A biofertilizer enriched with *Trichoderma harzianum* significantly increased tomato yields, quality, and nutritional value and contributed to the reduction of synthetic fertilizers (Molla et al., 2012). A biofungicidal strain of *T. harzianum* increased the access of AMF to non-host brassica plants and prompted plant growth (Poveda et al., 2019). A *Trichoderma*-based biofertilizer enhanced the soil enzyme activity that improved nutrient availability and promoted the growth, quality, and yield of flowering Chinese cabbage (Ji et al., 2020). Similarly, positive results were seen in yield or other parameters from field studies conducted in California strawberries with biostimulants that have *T. harzianum* and *T. virens* or *T. harzianum* and *Bacillus amyloliquefaciens* (Dara, 2020, 2021), and a tomato study with a combination of products where one of them contained *T. harzianum* and some PGPR (Dara and Lewis, 2019).

3.3 Plant growth-promoting rhizobacteria (PGPR)

The rhizosphere environment supports an abundance of microorganisms due to the presence of organic matter, root exudates, nutrients, and moisture.

Among these microorganisms are the PGPR, which regulate plant hormones, produce antibiotics and organic compounds, fix nitrogen, solubilize phosphate, improve soil structure, increase nutrient and water uptake, and alleviate plant stresses (Glick, 2012; Backer et al., 2018). Similarly, root exudates also influence soil microbial communities (Bakker et al., 2012). Thus, both plants and soil microorganisms influence and depend on each other. Examples of PGPR genera that are commonly used in crop production include *Agrobacterium*, *Azospirillum*, *Azotobacter*, *Bacillus*, *Burkholderia*, *Delfitia*, *Paenibacillus*, *Pseudomonas*, *Rhizobium*, *Serratia*, and *Streptomyces* that enhanced plant growth and yields in multiple studies. For example, *Azospirillum brasilense* improved root and shoot growth in strawberry (Pedraza et al., 2010), *Bacillus subtilis* increased shoot development and yields in apricot (Esitken et al., 2002), *Pseudomonas fluorescens* improved nutrient uptake and promoted plant growth and yields in broccoli (Tanwar et al., 2014), a consortium of *Pseudomonas putida*, *Camanomonas testosterone*, *Citrobacter freundii*, and *Enterobacter cloacae* and a nutrient program with a product containing *Azotobacter chroococcum*, *B. subtilis*, *Bacillus megaterium*, *Bacillus mycoides*, and *T. harzianaum* improved tomato yields (Dara and Lewis, 2019).

3.4 Protein hydrolystates

Protein hydrolysates contain peptides and amino acids derived from enzymatic or chemical hydrolysis of animal- or plant-derived materials and have biostimulant effects that help plants in multiple ways (Colla et al., 2015). Significant improvement in marketable strawberry yields was seen when a hydrolysate of grocery store food waste was used as a substitute for the grower-standard nutrient program at 50% or 100% through drip application in a commercial field in California (Dara, 2017). An Italian study showed that protein hydrolysates improved the biomass of the aerial parts and fruit diameter, both under well-watered and water-stressed conditions (Meggio et al., 2020). In another Italian study, a protein hydrolysate derived from legume seeds increased antioxidant activities, total soluble solids, mineral and nutritional value, photosynthetic activity, and yields of tomato (Rouphael et al., 2017). Such a positive impact on tomato yields and quality parameters was also seen in another study from a legume-based protein hydrolysate, seaweed extract, and a tropical plant extract (Colla et al., 2017). Similarly, a soy protein hydrolysate improved strawberry fruit yields comparable to synthetic fertilizers in a Californian study (Dara, 2020).

3.5 Humic substances

Another group of biostimulants comprises humic substances, which are biologically derived organic matter in soil and water that regulate critical ecological and environmental processes and have a major impact on plant growth and metabolism (Nardi et al., 2002; Canellas et al., 2015). Reducing soil compaction, increasing nutrient use efficiency and microbial activity, and improving plant biomass are some of the benefits of humic substances with their hormone (auxin)-like activity (Nardi et al., 2002; Rose et al., 2014). In a Cuban study, humates from vermicompost decreased total carbohydrate content, increased protein content, nitrate uptake, and crop yields, and shorted the crop cycle by 21 days in lettuce (Hernandez et al., 2015). In a potted plant study in Italy, humic-like substances derived from alkaline hydrolysis of tomato plant residue increased the nitrogen assimilation in kidney bean (Baglieri et al., 2014). A meta-analysis by Rose et al. (2014) showed that exogenously applied humic substances increased the root biomass by 21% and the shoot biomass by 22%. Similarly, Seyedbagheri (2010) reported from 17 years of research studies in Idaho, USA, that a 11–20% yield increase was seen in potato from various humic acid products. An extensive review by Canellas et al. (2015) summarized the positive effects of humic substances in several horticultural crops including fruits such as apricots, grapes, pineapple, and strawberry, and vegetables such as broccoli, beans, eggplant, garlic, lettuce, onion, pepper, potato, and tomato with an increase in root dry weight by up to 124% and increase in yield by 80% in some of these crops.

3.6 Algal extracts

Extracts from several species of micro- and macroalgae have been used in crop production. Species of *Aphanizomenon*, *Chlorella*, *Dunaliella*, and *Spirulina* are among the microalgae used for their biostimulant and biofertilizer properties. Species of *Ascophyllum*, *Durvillea*, *Ecklonia*, *Saragassum* (brown algae), *Ulva* (green algae), and *Gelidium* (red algae) are some of the common macroalgae or seaweeds used for improving plant growth or relieving them from stresses. A bioassay conducted in Brazil by immersing onion seedlings in a combination of the microalga *Scendesmus subspicatus* and humic acid showed synergism and promoted plant growth and improved bulb caliber, sugar and protein content, and yields (Gemin et al., 2019). An Indian study showed faster germination and plant growth, and improved root and shoot development of tomato from seed treatment; and higher chlorophyll content, and increased root and shoot development from foliar application of a microalgal consortium (Supraja et al., 2020). In an Italian study, Barone et al. (2018) showed that extracts of the microalgae *Chlorella vulgaris* and *Scenedesmus quadricauda* upregulated several genes of sugar beets putatively involved in sulfate starvation, and triggered primary and secondary metabolic pathways and intracellular transport, resulting in improved root development and nutrient uptake. Similarly, treating tomato and pepper plants with the polysaccharide extract of *Spirulina platensis* resulted in as much as a 30% increase in plant size, 230% increase in root weight, and 100% increase in size and number of nodes, in a Moroccan study (Elarroussi et al., 2016). Microalgae also have various plant hormones such as auxins, cytokinins, and gibberellins and influence several physiological processes related to plant growth, development, maturity, yields, and abiotic stress tolerance (Ronga et al., 2019).

Several macroalgal biostimulants are available around the world, and the majority are based on *Ascophyllum nodosum* (Sharma et al., 2014). Improved plant growth, chlorophyll content, nutritional value, yield, or other benefits were seen from macroalgal biostimulants in apple (Spinelli et al., 2009), onion (Araujo et al., 2012), orange (Fornes et al., 2002), peanut (Featonby-Smith and van Staden, 1987), strawberry (Dara and Peck, 2018), and various other crops. A review by Battacharyya et al. (2015) summarized the benefits of macroalgal extracts in several fruit and vegetable crops, with increased germination, plant growth, chlorophyll content, fruit yield and quality, nutritional value, and early maturity among the several parameters measured.

3.7 Botanical extracts

Several botanical extracts are commonly used as biopesticides, and some of them have biostimulatory properties. Field studies conducted in California in strawberry and tomato with cold-pressed neem extract containing neem oil and azadirachtin, carboxylic acids from rice hulls, and extracts of black walnut and willow showed yield improvements (Dara and Peck, 2018; Dara, 2019c; Dara, 2021). Godlewska et al. (2019) evaluated raw extracts of dandelion, mugwort, and other plants applied to white head cabbage and observed an increase of shoot and root growth by 246% and 106%, respectively. Ali et al. (2019) reported that a single foliar pre-transplant application of garlic bulb extract improved plant growth and antioxidant activity of eggplant.

3.8 Silicates and phosphites

In addition to macronutrients and micronutrients that ensure optimal plant growth and health, products based on silicon and phosphorus have biostimulatory effects of improving mechanical strength, modifying leaf architecture for improved exposure to sunlight, enhancing the activity of antioxidative enzymes to reduce oxidative damage, and improving water use efficiency, mineral uptake, and photosynthesis, among others (Gómez-Merino and Trejo-Téllez, 2015; Savvas and Ntatsi, 2015). Several studies have demonstrated the benefits of using silica-based products in improving plant growth or yields. For example, the application of silica clay improved the fresh weight of onion leaves by 46% (Araujo et al., 2012). Potassium silicate

Published by Burleigh Dodds Science Publishing Limited, 2022.

application resulted in a numerical increase of tomato yields by 27% (Dara and Lewis, 2019). Substrate application of potassium silicate and foliar application of sodium silicate improved plant and flower characters of gerbera (Kamenidou et al., 2010). A combination of silicon- and seaweed-based products improved strawberry yields by 16% (Dara and Peck, 2018). Similarly, phosphite improved the yield parameters in celery, lettuce, onion, pepper, potato, strawberry, and tomato when applied as phosphorous acid (Rickard, 2000; Bertsch et al., 2009; Estrada-Ortiz et al., 2012) and emergence, early growth, and mycorrhizal colonization in potato seed tubers and root and shoot growth in strawberry when applied as potassium phosphite (Glinicki et al., 2010; Tambascio et al., 2014).

3.9 Chitosan

Chitosan is a polysaccharide derived from chitin in the exoskeleton of shellfish. It is also present in the arthropod exoskeleton and in fungi. Similar to other biostimulants, chitosan also induces plant defenses and imparts tolerance to various stresses. A review by Pichyangkura and Chadchawan (2015) discussed increased antioxidant activity in apricot, lychee, grapes, navel oranges, sweet basil, and tomato, and improved growth parameters in artichoke, chilli pepper, Indian spinach, okra, watermelon, and several ornamental crops, with chitosan application.

4 Enhancement of soil health

Soil is more than dirt and is full of life. Organic matter and microbial communities are the major components of soil, which influence its water holding capacity and nutrient content. Soils rich in organic matter and beneficial microorganisms improve the physical, chemical, and biological properties of soil, which in turn influence plant growth and health. Healthy soils are also important for nutrient cycling and water conservation. Most biostimulant materials enrich soils with organic molecules, beneficial microorganisms, or minerals. For example, humic substances, being a major component of soil organic matter, have a major impact on soil structure, microbial activity, and overall soil health. Humic substances also influence heavy metal mobilization in soil with improved leachability or plant uptake (Shahid et al., 2012; Canellas et al., 2015). Certain microorganisms such as *B. subtilis* can also be used for the bioremediation of soils (Nadhirawaty and Titah, 2019). Microalgae can remove antibiotics, pharmaceuticals, and personal care products from water and such benefits can be extended for improving soil health as well (Xiong et al., 2017; Kiki et al., 2019; Hena et al., 2021). AMF, PGPR, and other biostimulants fix, mobilize, solubilize, or increase the bioavailability of nutrients in soils and thus improve soil health (Pindi and Satyanarayana, 2012; De Pascale et al., 2017). Soil health and soilborne disease incidence are closely related, and improved soil health through organic and microbial amendments and other cultural practices is important for disease suppression (Janvier et al., 2007). Similarly, improved soil quality and health through organic amendments and other practices not only improves water and nutrient availability, but also increases the resilience of soil against adverse environmental conditions and contributes to disease suppression (Lal, 2016).

5 Inducing natural resistance

Plants respond to various beneficial and harmful stimulants of both a biotic and an abiotic nature and trigger defense mechanisms to naturally induce resistance to stresses. The mechanism where the salicylic acid pathway is activated in response to virulent and avirulent pathogens, nonpathogenic microorganisms, chemicals, and other stressors is called systemic acquired resistance (SAR) (Dara, 2021). The mechanism where jasmonic acid and ethylene pathways are activated in response to beneficial microorganisms is called ISR. Pathogenesis-related proteins produced in both mechanisms also help plants perform well under stress. While SAR is more of a curative or reactive strategy, ISR is a proactive strategy. Many biostimulants activate ISR and prime plants for potential stressors. Several studies have demonstrated induced resistance from biostimulants including algal extracts (Raghavendra et al., 2007; Sharma et al., 2014), AMF (Jung et al., 2012; Cameron et al., 2013), *Trichoderma* (Yoshioka et al., 2012), PGPR (Zehnder et al., 2001), protein hydrolysates and humic substances (Nardi et al., 2016), and chitosan (Nandeeshkumar et al., 2008). Primed plants have enhanced responses to stressors and thus mitigate their negative impact (Pastor et al., 2014). Although some biostimulatory active ingredients may also exhibit pesticidal properties, ISR is the primary mechanism in crop protection and in their role in IPM.

6 Suppression of diseases

6.1 Arbuscular mycorrhizal fungi and other microorganisms

Numerous studies have demonstrated the role of biostimulants in disease suppression through ISR and other mechanisms. Larkin (2008) reported that an AMF inoculant containing six *Glomus* spp., *Gigospora margarita*, and *Paraglomus brazianum* reduced stem canker and black scurf caused by *Rhizoctonia solani* in potato by 17–28%. Larkin (2008) also found that when aerobically composted tea was used alone and in combination with a mixture of six *Bacillus* spp., *Streptomyces griseoviridis*, *T. harzianum*, humic acids, and seaweed and yeast extracts in a crop rotation with barley/ryegrass, the incidence of stem canker and black scurf, and the common scab caused by *Streptomyces scabiei* declined by 18–33%, with a 20–23% yield increase. In another study, the *T. harzianum* strain T39 reduced symptoms of downy mildew caused by *Plasmopara viticola* by triggering resistance locally and systemically (Perazzolli et al., 2008), similar to that seen in resistant genotypes (Perazzolli et al., 2012). Similarly, a strain of *T. harzianum*, in the presence of root rot-causing *Fusarium solani*, upregulated several genes involved in plant defenses in olive (Amira et al., 2017).

6.2 Plant growth-promoting rhizobacteria

Wei et al. (2011) demonstrated that fortifying an organic fertilizer with two strains of *B. amyloliquefaciens* reduced the incidence of tomato wilt caused by *Ralstonia solanacearum* in greenhouse and field studies. Omara et al. (2017) reported that treating soybean seedlings with PGPR *Methylobacterium aminovorans* and *M. rhodinum*, *Bradyrhizobium japonicum*, *B. megaterium* var. *phosphaticum*, and the mycoparasitic fungus *Trichoderma viride* suppressed damping off caused by *R. solani*. These treatments also increased the nodule numbers; dry weight of nodules; nitrogen, phosphorus, and potassium content; and seed yield. Recep et al. (2009) observed that 17 strains of *B. amyloliquefaciens*, *Bacillus atrophaeus*, *Bacillus pumilus*, *Bacillus macerans*, *Burkholderia cepacia*, *Flavobacter balastinium*, and *Pseudomonas putida* inhibited some of the dry rot-causing *Fusarium oxysporum*, *Fusarium culmorum*, and *Fusarium sambucinum, in vitro*. However, only the OSU-7 strain of *B. cepacia* controlled the disease under storage conditions. Other studies demonstrated similar disease suppression including the control of shot-hole disease (*Coryneum blight*) in apricot with *B. subtilis* (Esitken et al., 2002), tomato yellow leaf curl virus with *Enterobacter asburiae* (Li et al., 2016), *F. oxysporum* and *Fusarium guttiforme* with chitinase from *Chromobacterium violaceum* (Sousa et al., 2019), and *Verticillium dahliae* with cyclodextrin glycosyltransferase, an antimicrobial protein, from *Bacillus cereus* YUPP-10 (Zhou et al., 2021).

6.3 Protein hydrolysates

Soybean and casein hydrolysates reduced downy mildew infection in grape by 76% and 63%, respectively (Lachhab et al., 2014). Both hydrolysates boosted the plant defenses by upregulating the genes responsible for pathogenesis-related proteins and the stilbene synthase enzyme involved in the production of resveratrol, a defensive phytoalexin. In another study, guar hydrolysate suppressed *Pdosphaera xanthii*, the causal agent of powdery mildew, in zucchini (Cappelletti et al., 2017).

6.4 Humic substances

Disease suppression is one of the benefits of humic substances in addition to their impact on improving crop health and yields. Hernandez et al. (2015) observed reduced disease incidence in lettuce when vermicompost was applied. Increased phenylalanine ammonia lyase activity, which improves plant defenses, was thought to be responsible for disease suppression. Suppression of *Pythium* damping off in cucumber, *Rhizoctonia* root rot in radish, and *Verticillum* wilt in strawberry were observed from vermicompost application (Chaoui et al., 2002). Similarly, the combination of vermicompost and wood vinegar (pyroligneous acid) reduced *V. dahliae* in cucumber (Saberi et al., 2013). Several other studies demonstrated the impact of humic substances in improving crop health through disease suppression (Olivares et al., 2017; Jindo et al., 2020)

6.5 Algal and botanical extracts

A review by Shukla et al. (2021) explained the role of seaweed extracts in priming plants and triggering defense mechanisms against plant pathogens. Seaweed polysaccharides and their derivate oligosaccharides activate the salicylic, the jasmonic acid, and the ethylene signaling pathways that increase the production of pathogenesis-related proteins and other compounds with antimicrobial activity, and offer protection from pathogens (Vera et al., 2011). For example, the seaweed polysaccharides ulvan, carrageenan, alginate, and laminarin improved total polyphenol and lignin contents, inhibited fungal growth, and reduced Verticillium wilt in olives (Salah et al., 2018). Extract of *A. nodosum* increased the activity of peroxidase, polyphenoloxidase and other enzymes, and several pathogenesis-related proteins in carrots, resulting in enhanced resistance to *Alternaria radicina* and *Botrytis cinerea* (Jayaraj et al., 2008). In a different study, laminarin suppressed scab (*Venturia inaequalis*) in apple by inducing plant resistance, but supplemental fungicidal applications were necessary to achieve the desired control (Van Hemelrijck et al., 2013). Foliar application or seed treatment with the extracts of brown algae *Cystoseira myriophylloides*, *Fucus spiralis*, and *Laminaria digitata* significantly reduced disease severity by *V. dahliae* and *Agrobacterium tumefaciens* in tomato through induced resistance (El Modafar et al., 2012).

Many botanical extracts are primarily used as biopesticides rather than biostimulants. Some products based on soybean and other vegetable oils are used as both biostimulants and biopesticides. However, in general, there are fewer examples where botanicals have been used exclusively as biostimulants and have indirectly contributed to pest or disease suppression. Willow bark extract, which is used as a plant growth regulator and biostimulant, also has fungicidal properties (Deniau et al., 2019). Treating potato with Canada milkvetch extract, *P. fluorescens* DF37, and *B. pumulis* M1 provided up to 84% of *Verticillium* control with up to 24% of yield increase (Uppal et al., 2008), but it was not clear whether the disease suppression was a result of biostimulation or the pesticidal activity of these materials.

6.6 Silicates and phosphites

Sodium silicate inhibited conidial germination and mycelial growth of *Fusarium sulphureum* in controlled dry rot of potato tubers (Li et al., 2009). Calcium silicate reduced downy mildew (*Peronospora manshurica*) in soybean (Nolla et al., 2006), and when applied with calcium chloride, white mold (*Sclerotinia sclerotiorum*) in dry beans (Paula Júnior et al., 2009). Foliar application of silicates might act as a physical barrier to pathogen infection, but root application of potassium metasilicate increased plant defenses and controlled downy mildew (*P. xanthii*) in cucumber (Liang et al., 2005). Silicon, along with *G. mosseae* and *Enterobacter* sp. UPMSSB7, suppressed white root rot caused by *Rigidoporus microporus*, similar to a fungicide treatment, while improving various growth parameters of rubber seedlings (Shabbir et al., 2021).

In addition to its nutritional importance, phosphite has a significant influence on disease suppression. Föster et al. (1998) reported a significant reduction in root and crown rot caused by *Phytophthora capsici* in pepper plants treated with phosphite. Shearer and Fairman (2007) observed that phosphite application resulted in a significant reduction in the mortality of *Banksia* spp. from *Phytophthora cinnamomi* infection. Silva et al. (2011) reported yield improvement and downy mildew reduction in soybean from potassium phosphite application. They also observed that potassium phosphite application supplemented with fungicidal treatments significantly controlled Asian soybean rust caused by *Phakospora pachyrhizi* and powdery mildew caused by *Microsphaera diffusa*. In another study, Zhao et al. (2013) found that applying phosphorus oxyanion solution to sweet oranges with huanglongbing, caused by *Candidatus Liberibacter* spp., reduced the severity of disease symptoms and significantly improved fruit yields.

6.7 Chitosan

Numerous studies have demonstrated the effect of chitosan on improved enzymatic activity, production of phytoalexins and pathogenesis-related proteins, lignin synthesis, and enhanced plant defenses, resulting in reduced disease incidence and increased postharvest quality (El Hadrami et al., 2010). Pichyangkura and Chadchawan (2015) thoroughly reviewed the plant protection offered by chitosan from several studies in vegetables, fruits, and ornamentals.

Published by Burleigh Dodds Science Publishing Limited, 2022.

Control of *F. oxysporum* f. sp. *radices-lycopersici* (Lafontaine and Benhanou, 1996) and *Xanthomonas gardneri* (Jail et al., 2014) in tomato, *Alternata*, *Fusarium avenaceus*, *F. culmorum*, *Epicoccum purpurascens*, and *S. sclerotiorum* in lemon balm (Szczeponek et al., 2006), *B. cinerea* and *P. viticola* in grapes when applied with copper sulfate (Aziz et al., 2006), *Colletotrichum capsici* in bell pepper during postharvest storage when applied with lemongrass oil (Ali et al., 2015), *Aspergillus flavus* (De Oliveira Pedro et al., 2013), *Aspergillus ochraceus* (Meng et al., 2020), and *Phytophthora infestans* (Huang et al., 2021) *in vitro*, and *F. oxysporum* in potato (Ren et al., 2021) are some of the examples of chitosan application in disease management.

7 Suppression of arthropod pests

7.1 AMF and other microorganisms

AMF can have an impact on herbivore damage by modifying the morphological and biochemical characters of the plant by augmenting defense mechanisms and imparting resistance. A recent review by Selvaraj and Thangavel (2021) discussed the role of AMF as a part of IPM against pest insects. Coppola et al. (2019) found that *Trichoderma atroviride* P1 in tomato contributed to the control of Egyptian cotton leafworm, *Spodoptera littoralis*, and the potato aphid, *Macrosiphum euphorbiae*, through direct and indirect defense mechanisms. An increased oxidative burst reaction and the production of volatile organic compounds that attract the parasitoid *Aphidius ervi* parasitizing aphids, and an enhanced production of protective enzymes against the leafworm, were found to be the defense mechanisms. A similar impact of increased defenses and attraction of *A. ervi* to parasitize *M. euphorbiae* was also seen in tomato from *T. harzianum* application, in an earlier study (Coppola et al., 2017). Some species of *Trichoderma* and their strains can also control pest insects through entomopathogenic activity. For example, *T. atroviride*, *Trichoderma citrinoviride*, and *T. harzianum* inhibited 73–85% of egg hatch in the bean weevil *Acanthoscelides obtectus*, and the polyphagous long-horned beetle *Xylotrechus arvicola* (Rodríguez-González et al., 2017). A *Trichoderma* sp. alone and in combination with *Bacillus thuringiensis* effectively controlled the brinjal shoot and fruit borer *Leucinodes orbonalis*, the cotton aphid *Aphis gossypii*, and the cotton leafhopper *Amrasca biguttula biguttula* (Nawaz et al., 2020). *Trichoderma longibrachiatum* controlled *L. orbonalis* and increased eggplant yields by 56% (Ghosh and Pal, 2016).

7.2 Plant growth-promoting rhizobacteria

As the inducers of plant defense mechanisms, PGPR elicit responses that influence herbivore infestations. While some PGPR may have entomopathogenic properties, most of them suppress pest populations through ISR (Disi et al., 2019). For example, Zehnder et al. (1997) reported that when cucumber plants were treated with *B. pumilus* INT-7, feeding damage by the cucumber beetle, *Diabrotica undecimpunctata howardi*, and the bacterial wilt caused by *Erwinia tracheiphila*, which the beetle transmits, were significantly reduced due to the increased synthesis of cucurbitacin. In cotton, treating plants with *Bacillus* spp. resulted in higher levels of gossypol and jasmonic acid, leading to reduced larval feeding of the beet armyworm *Spodoptera exigua* (Zebelo et al., 2016). In peppermint, treatment with *B. amyloliquefaciens* GB03 and *P. putida* SJ04 increased salicylic and jasmonic acid levels and essential oil production, similar to the response from the feeding of the sunflower looper *Rachplusia nu* (del Rosario Cappellari et al., 2020). Although this study did not measure the impact of PGPR on herbivores or herbivory, it demonstrated that PGPR elicit defense responses. Several studies also demonstrated the influence of PGPR on natural enemies. In arugula, treatment with *B. amyloliquefaciens* GBO3 and the presence of the fall armyworm *Spodoptera frugiperda*, and the diamondback moth *Plutella xylostella*, influenced the behavior of the predatory earwig *Doru lutiepes* (Bell et al., 2020). The presence of PGPR and herbivore damage had a synergism in attracting earwigs. In rice, reduced leaf folder damage and increased natural enemy activity were seen from the inoculation of *P. fluorescens* (Saravanakumar et al., 2008). Similarly, *P. fluorescens* WCS417r colonization of the thale cress *Arabidopsis thaliana* increased the activity of the parasitoid *Microplitis mediator* when the plants were infested with the cabbage moth *Mamestra brassicae* (Pangesti et al., 2015). PGPR suppressed the production of herbivore-induced terpene, methyl salicylate, and lilial, and enhanced plant growth and ISR.

7.3 Humic substances

A review by Joshi et al. (2015) presented several examples of insect control with vermicompost or vermiwash applications. Aphids, beetles, earworms, hornworms, mealybugs, and mites were controlled on cabbage, corn, cucumber, eggplant, mustard, pepper, and tomato in multiple studies. Humic acid also controlled the root-knot nematode *Meloidogyne incognita*, *in vitro* and *in situ*, and promoted banana growth (Seenivasan and Senthilnathan, 2018).

7.4 Algal and botanical extracts

Some algal and botanical extracts have both biostimulatory and insecticidal properties. It can be difficult to distinguish their roles in certain situations, but some examples will be discussed where biostimulants have an impact on arthropods. An earlier report by Stephenson (1966) indicated the indirect impact of a hydrolyzed seaweed product on aphids, red spider mites, and plant diseases, through increased plant resistance. Eckol, a phenolic extract of the brown seaweed *Ecklonia maxima*, increased various growth parameters, enzymatic activity, chlorophyll, carotenoid, and proline content, and reduced aphid infestations (Rengasamy et al., 2016). Increased myrosinase activity was responsible for the population reduction in the cabbage aphid *Brevicoryne brassicae*. González-Castro et al. (2019) found repellent and insecticidal properties of the extracts of the macroalgae *Caulerpa sertularioides*, *Laurencia johnstonii*, and *Sargassum horridum* against the Asian citrus psyllid *Diaphorina citri*.

Juglone, a quinoid compound extracted from walnut trees, has biostimulatory, algacidal, bactericidal, fungicidal, insecticidal, and herbicidal properties (Islam and Widhalm, 2020). Azadirachtin, which is commonly used as an insecticide and insect growth regulator, also has a biostimulatory effect (Dara, 2021).

7.5 Silicates and phosphites

Silicates applied as fertilizers benefit plants by strengthening the tissues and acting as a mechanical barrier, and also trigger immune responses. Numerous studies have demonstrated a significant reduction in arthropod populations and their damage from silicate applications (Laing et al., 2006; Moraes et al., 2019). Some of the mechanisms of resistance offered to plants by silicon include the reduced penetration of plant tissues, increased wearing of mouthparts, reduced digestibility and palatability of plant material, and damage to the insect's midgut epithelium (Alhousari and Greger, 2018). Additionally, soluble silicon in plants also attracts natural enemies and enhances the biocontrol of pests (Bakhat et al., 2018). For example, silicic acid application reduced damage by the cucurbit beetle *Diabrotica speciosa*, made the plants less preferable, and increased plant growth in potato without affecting their natural enemies (De Assis et al., 2012). Potassium silicate supplementation for chrysanthemum reduced 54% of mining by the American serpentine leafminer *Liromyza trifolii*, although it also reduced the plant height by 10% (Klittich and Parrella, 2014). Silicon application through irrigation significantly reduced the numbers of twospotted spider mites (*Tetranychus urticae*) and the levels of powdery mildew (*Podosphaera aphanis*) in strawberry (Liu et al., 2020a).

Similar to silicates, phosphites also have an impact on pest suppression. Foliar application of potassium phosphite in potato reduced the number of potato tuber moth (*Phthorimaea operculella*) larvae by more than 50%, without affecting parasitoid populations (Mulugeta et al., 2018). Similarly, foliar application of a phosphite fungicide containing sodium, potassium, and ammonium phosphites reduced the survival and prolonged the development of the Colorado potato beetle (*Leptinotarsa decemlineata*), and reduced the foliar damage while controlling diseases (Patterson and Alyokhin, 2014).

7.6 Chitosan

Chitosan has the potential to control multiple pests including arthropods. Derivatives of chitosan had significant insecticidal activity against *S. littoralis* (Rabea et al., 2005), *B. brassicae*, the oleander aphid *Aphis nerii*, and the wheat aphid *Schizaphid graminum* (Sahebzadeh et al., 2017) in laboratory assays. Colloidal chitosan from seafood waste showed significant antifeedant activity against *S. frugiperda* and reduced the feeding damage (Moorthy et al., 2021). Zhang and Tan (2003) reported insecticidal activity of chitosan against the cotton bollworm (*Helicoverpa armigera*), *P. xylostella*, *S. exigua*, the bird-cherry oat aphid (*Rhopalosiphum padi*), *A. gossypii*, the grain aphid (*Sitobion avenae*), the green peach aphid (*Myzus persicae*), the mealy plum aphid (*Hyalopterus pruni*), and the rose-grain aphid (*Metopolophium dirhodum*), with very high mortality in some species. Alfy et al. (2020) found that chitosan nanoparticles were effective against *S. littoralis*, the desert locust *Locusta migratoria*, and *M. incognita*. Fan et al. (2020) also reported that fluorinated derivatives of chitosan effectively killed the second-instar larvae and inhibited the egg hatch of *M. incognita*.

Chitosan is not only used for its direct and indirect effect on pest suppression. Its role as a nanocarrier of pesticides for safe and targeted delivery is also explored (Maluin and Hussein, 2020). For example, chitosan helped improve RNA interference efficacy against *S. frugiperda* (Gurusamy et al., 2020). Chitosan was also used for the controlled release of spinosad for increased environmental durability (Li et al., 2020). The monoterpenes carvacrol and linalool, nanoencapsulated in chitosan, increased the control efficacy against the pests *H. armigera* and *T. urticae* (Campos et al., 2018). In a different study, chitosan alone and as chitosan-reduced silver nanocrystals were detrimental to *H. armigera* (Murugan et al., 2021).

8 Tolerance to abiotic stresses

When plants undergo abiotic stresses such as salinity and drought, they are more vulnerable to pests and diseases. If biostimulants can improve plant growth under abiotic stresses, plants can use their resources to withstand biotic stresses. At the same time, plant stresses can also influence soil microbial populations (Liu et al., 2020b). For example, *G. mosseae* increased the salt tolerance of citrus seedlings with improved plant and root growth, photosynthesis, and ionic balance (Wu et al., 2010). *G. versiforme* stimulated plant growth and improved biomass of tangerine under water stress (Wu and

Xia, 2006). Improved nutrient absorption, higher accumulation of carbohydrates in roots and foliage, change in root plasticity, and higher antioxidative enzyme activity were among the factors that were found to impart salt and drought tolerance (Wu and Xia, 2006; He et al., 2007; Wu et al., 2010). Dual inoculation of *Rhizobium* sp. and *Enterococcus mundtii* increased seed germination, plant height, biomass, chlorophyll content, and nutrient uptake in mung bean growing in saline conditions (Kumawat et al., 2021).

Plant-derived protein hydrolysates improved plant growth, nutrient utilization, and yields of lettuce growing in pots under saline conditions (Lucini et al., 2015). The diameter of grapes was significantly improved, even under water-stressed conditions, when a protein hydrolysate was applied (Meggio et al., 2020). Although the differences were not significant, protein hydrolysates helped maintain the weight and number of fruit clusters under water stress.

Canellas et al. (2015) discussed the role of humic substances in regulating secondary metabolism and alleviating various stresses. Increasing the accumulation of phenolic compounds, improving osmoregulation, and regulating peroxidase activity are among the various influences of humic substances on plants. Similarly, macroalgae-based biostimulants also stimulate the production of plant defense mechanisms and other physiological responses that alleviate stress (Sharma et al., 2014). Silicate fertilizers increase the yield, quality, and shelf-life of grapes grown in calcareous gray desert soil (Zhang et al., 2017). Phosphites induce adaptive plant responses to various stresses and improve plant performance (Trejo-Téllez and Gómez-Merino, 2018). For example, under heat stress, phosphite induced the accumulation of stress proteins, reduced oxidative stress, and maintained photosynthetic efficiency in potato seedlings (Xi et al., 2020).

9 Multitrophic interactions and influencing factors

The effect of biostimulants can vary widely depending on various factors. In addition to neutral and positive effects, some biostimulant materials can negatively impact growth and yields (Nardi et al., 2016). AMF *Glomus caledonium*, *Glomus fasciculatum*, and *G. mosseae* increased the number of leaves per plant in the oxeye daisy *Leucanthemum vulgare* but influenced the levels of parasitism of the ragwort leafminer *Chromatomyia syngenesiae*, by *Diglyphus isaea* (Gange et al., 2003). Depending on the AMF species and their combinations, parasitism was higher, lower, or neutral in AMF-colonized plants. *Trichoderma* spp. that produce the volatile organic compound trichodiene, suppressed plant defenses and increased the emergence of *A. obtectus* and its damage to beans (Rodríguez-González et al., 2018). Application of glycine betaine did not improve strawberry fruit yields, but a product containing nutrients and humates improved yields by 11% (Dara, 2019d). Soy protein hydrolysate improved strawberry yields when applied alone but not in combination with a product containing *T. harzianum* and *T. virens* (Dara, 2020). A product containing soybean and corn oils stunted strawberry plant growth and reduced yields, probably from higher application rates (Dara, unpublished data). Similarly, an animal-derived protein hydrolysate reduced the weight of daughter plants in strawberry (Lisiecka et al., 2011). Seedling growth of cucumber was inhibited but the single-fruit weight and concentration of certain nutrients in shoots increased with *G. versiforme* (Wang et al., 2008). Overapplication of silicates reduced flower yields and caused growth abnormalities in ornamentals (Kamenidou et al., 2008).

Certain practices and other agricultural inputs might not be compatible with some biostimulants, especially if they have live microorganisms and require special storage, handling, and application. Both the biostimulant industry and the grower community should be aware of the potential negative impacts of biostimulants and work with the researchers to determine rates, frequencies, and application strategies that are appropriate for their crop situation.

10 Strategies of using biostimulants

Although numerous studies have demonstrated the potential of biostimulants in improving crop growth, health, and yields, it is always advisable to consider IPM or integrated crop management approaches that take advantage of multiple strategies and reduce the reliance on one or a few options. Arthropod pests and plant pathogens continuously evolve and develop resistance to various agricultural inputs, especially those that have pesticidal properties. Since some biostimulants also have a certain level of pesticidal properties, a continuous monitoring of potential negative interactions in the environment and an adjustment of the strategies as appropriate are recommended. In general, biostimulants can be used as seed or transplant treatments, or they can be applied throughout the crop cycle as seed treatments, to the soil, or as foliar sprays (Fig. 2). Based on several studies discussed in this chapter, biostimulants improved tomato seed germination, root and shoot growth, general health, yield, quality, and nutritional value, suppressed various pests and diseases, improved biocontrol activity, and reduced the need for synthetic fertilizers and the potential use of pesticides (Fig. 3). Using tomato as a model plant, the importance of biostimulants can be demonstrated in crop production and crop protection. In general, biostimulant supplements can enhance the current agronomic practices and build plant resilience to various biotic and abiotic stresses.

Building soil health should be one of the priorities for sustainable agriculture, and biostimulants play a major role in this area by adding organic matter, beneficial microorganisms, nutrients, and other essentials to the soil.

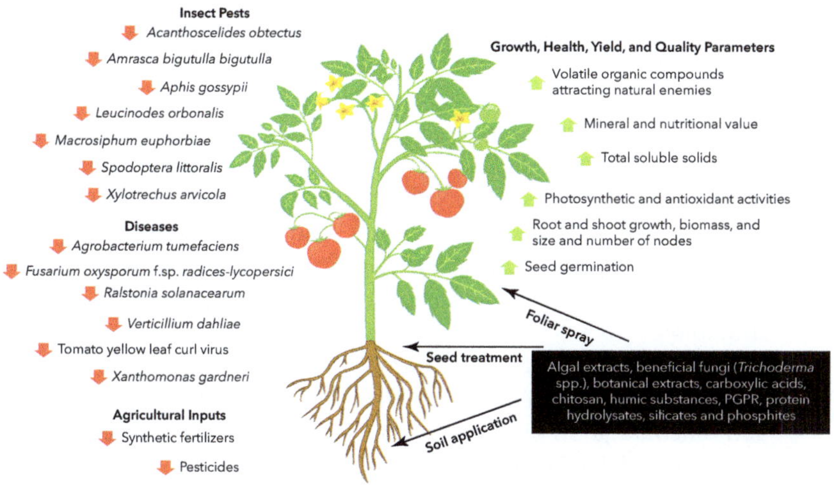

Figure 3 Tomato as a model crop for improving growth, health, yield, and parameters, and suppressing pests and diseases based on the various studies reviewed in this chapter.

It helps growers and crop care professionals to have a thorough knowledge of various biostimulants, their modes of action, interactions with other biotic and abiotic inputs and farming practices, and their indirect effect on pest and disease management. Biostimulants are part of a proactive strategy for maintaining plant health, while pesticides are curative measures to address specific issues. In other words, biostimulants are comparable to probiotics and vitamins, while pesticides are comparable to the medicines that humans take. Since biostimulants help improve nutrient uptake and trigger plant defenses against stressors, they contribute to the reduction of fertilizer and pesticide use and provide additional benefits beyond improved plant growth and yields. Modern-day agriculture can significantly benefit from incorporating biostimulant use as a good agricultural practice.

11 Challenges and future needs

Unlike the immediate effects of some fertilizer or pesticide inputs, the positive impact of biostimulants may not be as obvious and could easily be attributed to other crop production and protection practices. Additionally, the biological nature of several biostimulants also adds to the general misconception that biologicals are slow and less effective than synthetic agricultural inputs. The higher cost of some biostimulants and a lack of convincing applied research data are critical hurdles for the increased adaptation of biostimulants in many cropping systems. Based on the new IPM model and its emphasis on research, outreach, information management, and communication (Dara, 2019b), the future needs for integrating biostimulants as a critical part of crop production and IPM are as follows:

1. Develop collaborations among researchers, biostimulant manufacturers, and growers for conducting applied research and continuously generating data from various crop and field situations.
2. Understand various interactions among biostimulants, other agricultural inputs, agricultural practices, plants, arthropods, and microorganisms, and develop appropriate storage, handling, and use strategies.
3. Incorporate biostimulants as a part of cultural control of IPM.
4. Effectively disseminate the research findings through outreach activities and encourage improved communication among key players for building confidence and refining use strategies.
5. Invest in research to improve their formulations and their efficacy while optimizing their cost.

12 Conclusion

This chapter, probably the first of its kind, provided an overview of various biostimulant categories and how they can be used in improving plant growth, health, and yields, especially as a part of an IPM program (Fig. 2). Depending on the strain, source, form, formulation, application rates, and other factors, some of the biostimulants might be commercially available as biopesticides or biofertilizers. While biostimulatory and biopesticidal properties overlap for some biostimulant active ingredients, the main objective was to present the biostimulatory properties and the indirect effect of various active ingredients on arthropod pests, plant pathogens, and in some cases, plant-parasitic nematodes. With a growing interest in biostimulants and their potential, as well as the continued demand for sustainable food production, biostimulant use is expected to increase further in the near future.

13 Where to look for further information

13.1 Further reading

- Integrated pest and disease management in greenhouse crops (Springer) by Gullino et al. (2020).
- Biostimulants for sustainable crop production (Burleigh Dodds Science Publishing) by Youssef Rouphael et al. (2020).
- Biopesticides in organic farming: recent advances (Taylor and Francis) by Awasthi (2021).

Published by Burleigh Dodds Science Publishing Limited, 2022.

13.2 Key journals/conferences

- eJournal of Entomology and Biologicals (https://ucanr.edu/JEB).
- Frontiers in Sustainable Food Systems (https://www.frontiersin.org/journals/sustainable-food-systems).
- Journal of Economic Entomology (https://academic.oup.com/JEE).
- Entomological Society of America annual meetings (https://entsoc.org).

14 References

Aguín, O., Mansilla, P., Vilariño, A. and Sainz, M. J. (2004). Effects of mycorrhizal inoculation on root morphology and nursery production of three grapevine rootstocks. *Am. J. Enol. Vitic.* 55: 108–111.

Alfy, H., Ghareeb, R. Y., Soltan, E. and Farag, D. A. (2020). Impact of chitosan nanoparticles as insecticide and nematicide against *Spodoptera littoralis*, *Locusta migratoria*, and *Meloidogyne incognita*. *Plant Cell Biotehnol. Mol. Biol.* 21: 126–140.

Alhousari, F. and Greger, M. (2018). Silicon and mechanisms of plant resistance to insect pests. *Plants (Basel)* 7(2): 33.

Ali, A., Noh, N. M. and Mustafa, M. A. (2015). Antimicrobial activity of chitosan enriched with lemongrass oil against anthracnose of bell pepper. *Food Pack. Shelf Life* 3: 56–61.

Ali, M., Cheng, Z.-H., Hayat, S., Ahmad, H., Ghani, M. I. and Liu, T. (2019). Foliar spraying of aqueous garlic bulb extract stimulates growth and antioxidant enzyme activity in eggplant (*Solanum melongena* L.). *J. Integr. Agric.* 18(5): 1001–1013.

Amira, M., Lopez, D., Triki Mohamed, A., Khouaja, A., Chaar, H., Fumanal, B., Gousset-Dupont, A., Bonhomme, L., Label, P., Goupil, P., Ribeiro, S., Pujade-Renaud, V., Julien, J., Auguin, D. and Venisse, J. (2017). Beneficial effect of *Trichoderma harzianum* strain Ths97 in biocontrolling *Fusarium solani* causal agent of root rot disease in olive trees. *Biological Control* 110: 70–78.

Araujo, I. B., Peruch, L. A. M. and Stadnik, M. J. (2012). Effect of seaweed extract and silicate clay on Alternaria leaf spot and on the yield of green onion (*Allium fistulosum* L.). *Trop. Plant Pathol.* 37(5): 363–367.

Awasthi, L. P. (Ed) 2021. Biopesticides in organic farming: recent advances. CRC Press, 404 pages. Available at: https://www.google.com/books/edition/Biopesticides_in_Organic_Farming/1Y8kEAAAQBAJ?hl=en&gbpv=0&kptab=overview.

Aziz, A., Trotel-Aziz, P., Dhuicq, L., Jeandet, P., Couderchet, M. and Vernet, G. (2006). Chitosan oligomers and copper sulfate induce grapevine defense reactions and resistance to gray mold and downy mildew. *Phytopathology* 96(11): 1188–1194.

Backer, R., Rokem, J. S., Ilangumaran, G., Lamont, J., Praslickova, D., Picci, E., Subramanian, S. and Smith, D. L. (2018). Plant growth-promoting rhizobacteria: context, mechanisms of action, and roadmap to commercialization of biostimulants for sustainable agriculture. *Front. Plant Sci.* 9: 1473.

Baglieri, A., Cadili, V., Monterumici, C. M., Gennari, M., Tabasso, S., Montoneri, E., Nardi, S. and Negre, M. (2014). Fertilization of bean plants with tomato plants hydrolysates. Effect on biomass production, chlorophyll content and N assimilation. *Sci. Hort.* 176: 194–199.

Bakhat, H. F., Bibi, N., Zia, Z., Abbas, S., Hammd, H. M., Fahad, S., Ashraf, M. R., Shah, G. M., Rabbani, F. and Saeed, S. (2018). Silicon mitigates biotic stresses in crop plants: a review. *Crop Protec.* 104: 21-34.

Bakker, M. G., Manter, D. K., Sheflin, A. M., Weir, T. L. and Vivanco, J. M. (2012). Harnessing the rhizosphere microbiome through plant breeding and agricultural management. *Plant Soil* 360(1-2): 1-13.

Barone, V., Baglieri, A., Stevanato, P., Broccanello, C., Bertoldo, G., Bertaggia, M., Cagnin, M., Pizzeghello, D., Moliterni, V. M. C., Mandolino, G., Fornasier, F., Squartini, A., Nardi, S. and Concheri, G. (2018). Root morphological and molecular responses induced by microalgae extracts in sugar beet (*Beta vulgaris* L.). *J. Appl. Phycol.* 30: 1061-1071.

Battacharyya, B., Babgohari, M. Z., Rathor, P. and Prithiviraj, B. (2015). Seaweed extracts as biostimulants in horticulture. *Sci. Hort.* 196: 39-48.

Bell, K., Narajno-Guevara, N., dos Santos, R. C. D., Meadow, R. and Bento, J. M. S. (2020). Predatory earwigs are attracted by herbivore-induced plant volatiles linked with plant growth-promoting rhizobacteria. *Insects* 11(5): 271.

Bertsch, F., Ramírez, F. and Henríquez, C. 2009. Evaluación del fosfito como fuentefertilizante de fósforo vía radical y foliar. *Agron. Costarricense* 33: 249-265.

Cameron, D. D., Neal, A. L., van Wees, S. C. M. and Ton, J. (2013). Mycorrhiza-induced resistance: more than the sum of its parts? *Tren. Pl. Sci.* 18(10): 539-545.

Campos, E. V. R., Proença, P. L. F., Oliveira, J. L., Pereira, A. E. S., Ribeiro, L. N. M., Fernandes, F. O., Gonçalves, K. C., Polanczyk, R. A., Pasquoto-Stigliani, T., Lima, R., Melville, C. C., Della Vechia, J. F., Andrade, D. J. and Fraceto, L. F. (2018). Carvacrol and linalool co-loaded in β-cyclodextrin-grafted chitosan nanoparticles as sustainable biopesticide aiming pest control. *Sci. Rep.* 8(1): 7623.

Canellas, L. P., Olivares, F. L., Aguiar, N. O., Jones, D. L., Nebbioso, A., Mazzei, P. and Piccolo, A. (2015). Humic and fulvic acids as biostimulants in horticulture. *Sci. Hort.* 196: 15-27.

Cappelletti, M., Perazzolli, M., Nesler, A., Giovannini, O. and Pertot, I. (2017). The effect of hydrolysis and protein source on the efficacy of protein hydrolysates as plant resistant inducers against powdery mildew. *J. Bioprocess. Biotech.* 7: 5.

Chaoui, H., Edwards, C. A., Brickner, A., Lee, S. S. and Arancon, N. Q. (2002). Suppression of the plant diseases, *Pythium* (damping-off), *Rhizoctonia* (root rot) and *Verticillium* (wilt) by vermicomposts. *Brighton Crop Prot. Conf. Pests and Dis.* 2: 711-716.

Colla, G., Cardarelli, M., Bonini, P. and Rouphael, Y. (2017). Foliar applications of protein hydrolysate, plant and seaweed extracts increase yield but differentially modulate fruit quality of greenhouse tomato. *HortScience* 52: 1214-1220.

Colla, G., Nardi, S., Cardarelli, M., Ertani, A., Lucini, L., Canaguier, R. and Rouphale, Y. (2015). Protein hydrolysates as biostimulants in horticulture. *Sci. Hort.* 196: 28-38.

Coppola, M., Cascone, P., Chiusano, M. L., Colantuono, C., Lorito, M., Pennacchio, F., Rao, R., Woo, S. L., Guerrieri, E. and Digilio, M. C. (2017). *Trichoderma harzianum* enhances tomato indirect defense against aphids. *Insect Sci.* 24(6): 1025-1033.

Coppola, M., Cascone, P., Lelio, I. D., Woo, S. L., Lorito, M., Rao, R., Pennacchio, F., Guerrieri, E. and Digilio, M. C. (2019). *Trichoderma atroviride* P1 colonization of tomato plants enhances both direct and indirect defense barriers against insects. *Front. Physiol.* 10: 813.

Dara, S. K. (2017). Utilizing recycled food waste-based liquid compost: sustainable nutrient management option in strawberry. *Veg. West* 21: 16-17.

Dara, S. K. (2019a). Non-entomopathogenic roles of entomopathogenic fungi in promoting plant health and growth. *Insects* 10(9): 277.

Dara, S. K. (2019b). The new integrated pest management paradigm for the modern age. *J. IPM* 10(1): 1–9.

Dara, S. K. (2019c). Effect of microbial and botanical biostimulants with nutrients on tomato yield. *CAPCA Adv.* 22: 40–45.

Dara, S. K. (2019d). Evaluating the efficacy of anti-stress supplements on strawberry yield and quality. *UCANR eJ. Entomol. Biol.* Available at: https://ucanr.edu/blogs/blogcore/postdetail.cfm?postnum=31044.

Dara, S. K. (2020). Improving strawberry yields with biostimulants and nutrient supplements: a 2019–2020 study. *UCANR eJ. Entomol. Biol.* Available at: https://ucanr.edu/blogs/blogcore/postdetail.cfm?postnum=43631.

Dara, S. K. (2021). Impact of microbial, botanical, and organic acid-based biostimulants on strawberry health and yield. *CAPCA Adv.* 24: 30–35.

Dara, S. K. and Lewis, E. (2019). Evaluating biostimulant and nutrient inputs to improve tomato yields and crop health. *Prog. Crop Consult.* 4: 38–42.

Dara, S. K. and Peck, D. (2018). Evaluation of additive, soil amendment, and biostimulant products in Santa Maria strawberry. *CAPCA Adv.* 21: 44–50.

De Assis, F. A., Morae, J. C., Silveira, L. C. P., Francoso, J., Nascimento, A. M. and Antunes, C. S. (2012). Inducers of resistance in potato and its effects on defoliators and predatory insects. *Rev. Colomb. Entomol.* 38: 30–34.

De Oliveira Pedro, R., Takaki, M., Gorayeb, T. C., Del Bianchi, V. L., Thomeo, J. C., Tiera, M. J. and de Oliveira Tiera, V. A. (2013). Synthesis, characterization and antifungal activity of quaternary derivatives of chitosan on *Aspergillus flavus*. *Microbiol. Res.* 168(1): 50–55.

De Pascale, S., Rouphael, Y. and Colla, G. (2017). Plant biostimulants: innovative tool for enhancing plant nutrition in organic farming. *Eur. J. Hortic. Sci.* 82: 277–285.

Del Rosario Cappellari, L., Chiappero, J., Palermo, T. B., Giordano, W. and Banchio, E. (2020). Impact of soil rhizobacteria inoculation and leaf-chewing insect herbivory on *Mentha piperita* leaf secondary metabolites. *J. Chem. Ecol.* 46(7): 619–630.

Deniau, M. G., Bonafos, R., Chovelon, M., Parvaud, C.-E., Furet, A., Bertrand, C. and Marchand, P. A. (2019). Willow extract (*Salix* cortex), a basic substance of agronomical interests. *Int. J. Bio-Res. Stress Manag.* 10(4): 408–418.

Disi, J., Simmons, J. and Zebelo, S. (2019). Plant growth-promoting rhizobacteria-induced defense against insect herbivores. In: Maheshwari, D. and Dheeman, S. (Eds) *Field Crops: Sustainable Management by PGPR. Sustainable Development and Biodiversity* (vol. 123): 385–410, Springer, Cham.

Elarroussi, H., Elmernissi, N., Benhima, R., El Kadmiri, I. M., Bendaou, N., Smouni, A. and Wahby, I. (2016). Microalgae polysaccharides a promising plant growth biostimulant. *J. Algal Biomass Utln* 7: 55–63.

El Hadrami, A., Adam, L. R., El Hadrami, I. and Daayf, F. (2010). Chitosan in plant protection. *Mar. Drugs* 8(4): 968–987.

El Modafar, C., Elgadda, M., El Boutachfaiti, R., Abouraicha, E., Zehhar, N., Petit, E., El Aloui-Talibi, Z., Courtois, B. and Coutois, J. (2012). Induction of natural defence accompanied by salicyclic acid-dependant systemic acquired resistance in tomato seedlings in response to bioelicitors isolated from green algae. *Sci. Hort.* 138: 55–63.

Esitken, A., Karlidag, H., Ercisli, S. and Sahin, F. (2002). Effects of foliar application of *Bacillus subtilis* Osu-142 on the yield, growth and control of shot-hole disease (*Coryneum blight*) of apricot. *Gartenbauwiss* 67: 139–142.

Estrada-Ortiz, E., Trejo-Téllez, L. I., Gómez-Merino, F. C., Núñez-Escobar, R. and Sandoval-Villa, M. (2012). Phosphite on growth and fruit quality in strawberry. *Acta Hortic.* 947(947): 277–282.

Fan, Z., Qin, Y., Liu, S., Xing, R., Yu, H. and Li, P. (2020). Chitosan oligosaccharide fluorinated derivative control root-knot nematode (*meloidogyne incognita*) disease based on the multi-efficacy strategy. *Mar. Drugs* 18(5): 273.

Featonby-Smith, B. C. and van Staden, J. (1987). Effect of seaweed concentrate on yield and seed quality of *Arachis hypogaea*. *S. Afr. J. Bot.* 53(3): 190–193.

Fiorentino, N., Ventorino, V., Woo, S. L., Pepe, O., De Rosa, A., Gioia, L., Romano, I., Lombardi, N., Napolitano, M., Colla, G. and Rouphael, Y. (2018). *Trichoderma*-based biostimulants modulate rhizosphere microbial populations and improve N uptake efficiency, yield, and nutritional quality of leafy vegetables. *Front. Plant Sci.* 9: 743.

Fornes, F., Sánchez-Perales, M. and Guardiola, J. L. (2002). Effect of a seaweed extract on the productivity of 'de Nules' clementine mandarin and Navelina orange. *Botanica Marina* 45(5): 486–489.

Föster, H., Adaskaveg, J. E., Kim, D. H. and Stanghellini, M. E. (1998). Effect of phosphite on tomato and pepper plants and on susceptibility of pepper to Phytophthora root and crown rot in hydroponic culture. *Plant Dis.* 82(10): 1165–1170.

Gange, A. C., Brown, V. K. and Aplin, D. M. (2003). Multitrophic links between arbuscular mycorrhizal fungi and insect parasitoids. *Ecol. Lett.* 6(12): 1051–1055.

Gemin, L. G., Mógor, Á. F., Amatussi, J. D. O. and Mógor, G. (2019). Microalgae associated to humic acid as a novel biostimulant improving onion growth and yield. *Sci. Hort.* 256: 108560.

Ghosh, S. K. and Pal, S. (2016). Entomopathogenic potential of *Trichoderma longibrachiatum* and its comparative evaluation with Malathion against the insect pest *Leucinodes Orbonalis*. *Environ. Monit. Assess.* 188(1): 37.

Glick, B. R. (2012). Plant growth-promoting bacteria: mechanisms and applications. *Scintif* 2012: 963401.

Glinicki, R., Sas-Paszt, L. and Jadczuk-Tobjasz, E. (2010). The effect of plant stimulant/fertilizer Resistim on growth and development of strawberry plants. *J. Fruit Ornam Plant Res.* 18: 111–124.

Godlewska, K., Biesiada, A., Michalak, I. and Pcyga, P. (2019). The effect of plant-derived biostimulants on white head cabbage seedlings grown under controlled conditions. *Sustainability* 11(19): 5317.

Gómez-Merino, F. C. and Trejo-Téllez, L. I. (2015). Biostimulant activity of phosphite in horticulture. *Sci. Hort.* 196: 82–90.

González-Castro, A. L., Muñoz-Ochoa, M., Hernández-Carmona, G. and López-Vivas, J. M. (2019). Evaluation of seaweed extracts for the control of the Asian citrus psyllid *Diaphorina citri*. *J. Appl. Phycol.* 31(6): 3815–3821.

Gullino, M. L., Albajes, R., & Nicot, P. C. (Eds.). (2020). *Integrated Pest and Disease Management in Greenhouse Crops* (vol. 9). Springer Nature.

Gurusamy, D., Mogilicherla, K. and Palli, S. R. (2020). Chitosan nanoparticles help double-stranded RNA escape from endosomes and improve RNA interference in the fall armyworm, *Spodoptera frugiperda*. *Arch. Insect Biochem. Physiol.* 104(4): e21677.

He, Z. Q., He, C. X., Zhang, Z. B., Zou, Z. R. and Wang, H. S. (2007). Changes in antioxidative enzymes and cell membrane osmosis in tomato colonized by arbuscular mycorrhizae under NaCl stress. *Colloids Surf. B. Bioint.* 59: 128–133.

Hena, S., Gutierrez, L. and Croué, J. P. (2021). Removal of pharmaceutical and personal care products (PPCPs) from wastewater using microalgae: a review. *J. Hazard. Mater.* 403: 124041.

Hernandez, O. L., Calderín, A., Huelva, R., Martínez-Balmori, D., Guridi, F., Aguiar, N. O., Olivares, F. L. and Canellas, L. P. (2015). Humic substances from vermicompost enhance urban lettuce production. *Agron. Sustain. Dev.* 35(1): 225–232.

Huang, X. Q., You, Z. Y., Luo, Y., Yang, C. J., Ren, J., Liu, Y. L., Wei, G. J., Dong, P. and Ren, M. Z. (2021). Antifungal activity of chitosan against *Phytophthora infestans*, the pathogen of potato late blight. *Int. J. Biol. Macromol.* 166: 1365–1376.

Islam, A. K. M. M. and Widhalm, J. R. (2020). Agricultural uses of juglone: opportunities and challenges. *Agronomy* 10(10): 1500.

Jail, N. Gd, Luiz, C., Rocha Neto, A. Cd and Di Piero, R. M. (2014). High-density chitosan reduces the severity of bacterial spot and activates the defense mechanisms of tomato plants. *Trop. Plant Pathol.* 39(6): 434–441.

Janvier, C., Villenueuve, F., Alabouvette, C., Edel-Hermann, V., Mateille, T. and Steinberg, C. (2007). Soil health through soil disease suppression: which strategy from descriptors to indicators? *Soil Biol. Biochem.* 39(1): 1–23.

Jayaraj, J., Wan, A., Rahman, M. and Punja, Z. K. (2008). Seaweed extract reduces foliar fungal diseases on carrot. *Crop Prot.* 27(10): 1360–1366.

Ji, S., Liu, Z., Liu, B., Wang, Y. and Wang, J. (2020). The effect of *Trichoderma* biofertilizer on the quality of flowering Chinese cabbage and the soil environment. *Sci. Hort.* 262: 109069.

Jindo, K., Olivares, F. L., Malcher, D. J. D. P., Sánchez-Monedero, M. A., Kempenaar, C. and Canellas, L. P. (2020). From lab to field: role of humic substances under open-field and greenhouse conditions as biostimulant and biocontrol agent. *Front. Plant Sci.* 11: 426.

Joshi, R., Singh, J. and Vig, A. P. 2015. Vermicompost as an effective organic fertilizer and biocontrol agent: effect on growth, yield and quality of plants. *Rev. Environ. Sci. Biotechnol.* 14: 137–159.

Jung, S. C., Martinez-Medina, A., Lopez-Raez, J. A. and Pozo, M. J. (2012). Mycorrhiza-induced resistance and priming of plant defenses. *J. Chem. Ecol.* 38(6): 651–664.

Kamenidou, S., Cavins, T. J. and Marek, S. (2008). Silicon supplements affect horticultural traits of greenhouse-produced ornamental sunflowers. *Hort. Sci.* 43(1): 236–239.

Kamenidou, S., Cavins, T. J. and Marek, S. (2010). Silicon supplements affect floricultural quality traits and elemental nutrient concentrations of greenhouse produced gerbera. *Sci. Hort.* 123(3): 390–394.

Kiki, C., Rashid, A., Wang, Y., Li, Y., Zeng, Q., Yu, C. P. and Sun, Q. (2019). Dissipation of antibiotics by microalgae: kinetics, identification of transformation products and pathways. *J. Hazard. Mater.* 387: 121985.

Kim, M. J., Radhakrishnan, R., Kang, S. M., You, Y. H., Jeong, E. J., Kim, J. G. and Lee, I. J. (2017). Plant growth promoting effect of *Bacillus amyloliquefaciens* H-2-5 on crop plants and influence on physiological changes in soybean under soil salinity. *Physiol. Mol. Biol. Plants* 23(3): 571–580.

Klittich, D. and Parrella, M. (2014). Silicon supplementation: effects on chrysanthemum growth, leafminer populations, and parasitism. Proceedings of the IOBC/WPRS Working Group "Integrated Control in Protected Crops, Temperate Climate", Gent, Belgium, September 14–18, 2014 102: 117–119.

Kumawat, K. C., Sharma, P., Nagpal, S., Gupta, R. K., Sirari, A., Nair, R. M., Bindumadhava, H. and Singh, S. (2021). Dual microbial inoculation, a game changer?–bacterial biostimulants with multifunctional growth promoting traits to mitigate salinity stress in spring mungbean. *Front. Microbiol.* 11: 3491.

Lachhab, N., Sanzani, S. M., Adrian, M., Chiltz, A., Balacey, S., Boselli, M., Ippolito, A. and Poinssot, B. (2014). Soybean and casein hydrolysates induce grapevine immune responses and resistance against *Plasmopara viticola*. *Front. Plant Sci.* 5: 716.

Lafontaine, P. J. and Benhanou, N. (1996). Chitosan treatment: an emerging strategy for enhancing resistance of greenhouse tomato plants to infection by *Fusarium oxysporum* f. sp. *radices-lycopersici*. *Biocon. Sci. Technol.* 6(1): 111–124.

Laing, M. D., Gatarayiha, M. C. and Adandonon, A. (2006). Silicon use for pest control in agriculture: a review. *Proc. South Afr. Sugar Tech.* 80: 278–286.

Lal, R. (2016). Soil health and carbon management. *Food Ener. Sec.* 5(4): 212–222.

Larkin, R. P. (2008). Relative effects of biological amendments and crop rotations on soil microbial communities and soilborne diseases of potato. *Soil Biol. Biochem.* 40(6): 1341–1351.

Li, G. B., Wang, J. and Kong, X. P. (2020). Coprecipitation-based synchronous pesticide encapsulation with chitosan for controlled Spinosad release. *Carbohydr. Polym.* 249: 116865.

Li, H., Ding, X., Wang, C., Ke, H., Wu, Z., WAng, Y., Liu, H. and Guo, J. (2016). Control of tomato yellow leaf curl virus disease by *Enterobacter asburiae* BQ9 as a result of priming plant resistance in tomatoes. *Turk. J. Biol.* 40: 150–159.

Li, Y. C., Bi, Y., Ge, Y. H., Sun, X. J. and Wang, Y. (2009). Antifungal activity of sodium silicate on *Fusarium sulphureum* and its effect on dry rot of potato tubers. *J. Food Sci.* 74(5): M213–M218.

Liang, Y. C., Sun, W. C., Si, J. and Römheld, V. (2005). Effects of foliar- and root-applied silicon on the enhancement of induced resistance to powdery mildew in *Cucumis sativus*. *Plant Pathol.* 54(5): 678–685.

Lisiecka, J., Knaflewski, M., Spizewski, T., Fraszczak, B., Kaluzewicz, A. and Krazesinski, W. (2011). The effect of animal protein hydrolysate on quantity and quality of strawberry daughter plants cv. 'Elsanta'. *Acta Sci. Pol. Hort. Cult.* 10: 31–40.

Liu, B., Davies, K. and Hall, A. (2020a). Silicon builds resilience in strawberry plants against both strawberry powdery mildew *Podosphaera aphanis* and two-spotted spider mites *Tetranychus urticae*. *PLoS ONE* 15(12): e0241151.

Liu, H., Brettel, L. E., Qiu, Z. and Singh, B. K. (2020b). Microbiome-mediated stress resistance in plants. *Trends Pl. Sci.* 25(8): 733–743.

Lucini, L., Rouphael, Y., Cardarelli, M., Canaguier, R., Kumar, P. and Colla, G. (2015). The effect of a plant-derived biostimulant on metabolic profiling and crop performance of lettuce grown under saline conditions. *Sci. Hort.* 182: 124–133.

Maluin, F. N. and Hussein, M. Z. (2020). Chitosan-based agronanochemicals as a sustainable alternative in crop protection. *Molecules* 25(7): 1611.

Meggio, F., Trevisan, S., Manoli, A., Ruperti, B. and Quaggiotti, S. (2020). Systematic investigation of the effects of a novel protein hydrolysate on the growth, physiological parameters, fruit development and yield of grapevine (*Vitis vinifera* L., cv Sauvignon Blanc) under water stress conditions. *Agronomy* 10(11): 1785.

Meng, D., Garba, B., Ren, Y., Yao, M., Xia, X., Li, M.. and Wang, Y. (2020). Antifungal activity of chitosan against *Aspergillus ochraceus* and its possible mechanisms of action. *Int. J. Biol. Macromol.* 158: 1063–1070.

Molla, A. H., Manjurul Haque, M., Amdadul Haque, M. and Ilias, G. N. M. (2012). *Trichoderma*-enriched biofertilizer enhances production and nutritional quality of tomato (*Lycopersicon esculentum* Mill.) and minimizes NPK fertilizer use. *Agri. Res.* 1(3): 265–272.

Moorthy, A. V., Shanthi, M., Chinnaiah, C. and Senthil, K. (2021). Synthesis of colloidal chitosan from sea waste and its antifeedant effect against maize fall armyworm *Spodoptera frugiperda* (J. E. Smith). *J. Entomol. Zool. Stud.* 9(1): 418–421.

Moraes, J. C., Assis, F. A. and Assis, G. A. (2019). Use of silicon as resistance factor for plants against insect pests. In: Souza, B., Vázquez, L. and Marucci, R. (Eds) *Natural Enemies of Insect Pests in Neotropical Agroecosystems*: 497–508, Springer, Cham.

Mulugeta, T., Mulatu, B., Tekie, H., Yesuf, M., Andreasson, E. and Alexandersson, E. (2018). Phosphite alters the behaviroal response of potato tuber moth (*Phthorimaea operculella*) to field-grown potato. *Pest Manag. Sci.* 75(3): 616–621.

Murugan, K., Wang, L., Anitha, J., Dinesh, D., Amruthavalli, P., Vasanthakumaran, M., Paulpandi, M. and Hwang, J.-S. (2021). Insecticidal effect of chitosan reduced silver nanocrystals against filarial vector, *Culex quinquefasciatus* and cotton bollworm, *Helicoverpa armigera*. *Adv. Nano-Fert. Nano-Pest Agric.* 2021: 469–486.

Nadhirawaty, R. and Titah, H. S. (2019). Simultaneous bioaugmentation and biostimulation to remediate soil contaminated by ship dismantling in Bangkalan District, Indonesia. *J. Health Pollut.* 9(24): 191212.

Nandeeshkumar, P., Sudisha, J., Ramachandra, K. K., Prakash, H. S., Niranjana, S. R. and Shekar, S. H. (2008). Chitosan induced resistance to downy mildew in sunflower caused by *Plasmopara halstedii*. *Physiol. Mol. Plant Pathol.* 72(4–6): 188–194.

Nardi, S., Pizzeghello, D., Muscolo, A. and Vianello, A. (2002). Physiological effects of humic substances on higher plants. *Soil Biol. Biochem.* 34(11): 1527–1536.

Nardi, S., Pizzeghello, D., Schiavon, M. and Ertani, A. (2016). Plant biostimulants: physiological responses induced by protein hydrolyzed-based products and humic substances in plant metabolism. *Sci. Agric.* 73(1): 18–23.

Nawaz, A., Gogi, M. D., Naveed, M., Arshad, M., Sufyan, M., Binyameen, M., Islam, S. U., Waseem, M., Ayyub, M. B., Arif, M. J. and Ali, H. (2020). In vivo and in vitro assessment of *Trichoderma* species and *Bacillus thuringiensis* integration to mitigate insect pests of brinjal (*Solanum melongena* L.). *Egypt. J. Biol. Pest Control* 30(1): 60.

Nolla, A., Korndörfer, G. H. and Coelho, L. (2006). Efficiency of calcium silicate and carbonate in soybean disease control. *J. Plant Nutr.* 29(11): 2049–2061.

Olivares, F. L., Busato, J. G., de Paula, A. M., da Silva Lima, L., Aguiar, N. O. and Canellas, L. P. (2017). Plant growth promoting bacteria and humic substances: crop promotion and mechanisms of action. *Chem. Biol. Technol. Agric.* 4(1): 30.

Omara, A. E.-D., Hauka, F., Afify, A., El-Din, M. N. and Kassem, M. (2017). The role of some PGPR strains to biocontrol *Rhizoctonia solani* in soybean and enhancement the growth dynamics and seed yield. *Env. Biodiver. Soil Sec.* 1: 47–59.

Pangesti, N., Weldegergis, B. T., Langendorf, B., van Loon, J. J., Dicke, M. and Pineda, A. (2015). Rhizobacterial colonization of roots modulates plant volatile emission and enhances the attraction of a parasitoid wasp to host-infested plants. *Oecologia* 178(4): 1169–1180.

Pastor, V., Balmer, A., Gamir, J., Flors, V. and Mauch-Mani, B. (2014). Preparing to fight back: generation and storage of priming compounds. *Front. Plant Sci.* 5: 295.

Patterson, M. and Alyokhin, A. (2014). Survival and development of Colorado potato beetles on potatoes treated with phosphite. *Crop Protec.* 61: 38–42.

Paula Júnior, T. J., F. Vieira, R., Teixeira, H. and Eustáquio S. Carneiro, J. (2009). Foliar application of calcium chloride and calcium silicate decreases white mold intensity on dry beans. *Trop. Plant Pathol.* 34(3): 171–174.

Pedraza, R. O., Motok, J., Salazar, S. M., Ragout, A. L., Mentel, M. I., Tortora, M. L., Guerrero-Molina, M. F., Winik, B. C. and Díaz-Ricci, J. C. (2010). Growth-promotion of strawberry plants inoculated with *Azospirillum brasilense*. *World J. Microbiol. Biotechnol.* 26(2): 265–272.

Perazzolli, M., Dagostin, S., Ferrari, A., Elad, Y. and Pertot, I. (2008). Induction of systemic resistance against *Plasmopara viticola* in grapevine by *Trichoderma harzianum* T39 and benzothiadiazole. *Biological Control* 47(2): 228–234.

Perazzolli, M., Morett, M., Fontana, P., Ferrarini, A., Velasco, R., Moser, C., Delledonne, M. and Pertot, I. (2012). Downy mildew resistance induced by *Trichoderma harzianum* T39 in susceptible grapevines partially mimics transcriptional changes of resistant genotypes. *BMC Genom.* 13: 660.

Peterson, R. K. D., Higley, L. G. and Pedigo, L. P. (2018). Whatever happened to IPM? *Am. Entomol.* 64(3): 146–150.

Pichyangkura, R. and Chadchawan, S. (2015). Biostimulant activity of chitosan in horticulture. *Sci. Hort.* 196: 49–65.

Pindi, P. K. and Satyanarayana, S. D. V. (2012). Liquid microbial consortium-a potential tool for sustainable soil health. *J. Biofert Biopest* 3: 4.

Poveda, J., Hermosa, R., Monte, E. and Nicolás, C. (2019). *Trichoderma harzianum* favours the access of arbuscular mycorrhizal fungi to non-host Brassicaceae roots and increase plant productivity. *Sci. Rep.* 9(1): 11650.

Rabea, E. I., El Badawy, M. E., Rogge, T. M., Stevens, C. V., Höfte, M., Steurbaut, W. and Smagghe, G. (2005). Insecticidal and fungicidal activity of new synthesized chitosan derivatives. *Pest Manag. Sci.* 61(10): 951–960.

Raghavendra, V. B., Lokesh, S. and Prakash, H. S. (2007). Dravya, a product of seaweed extract (*Sargassum wightii*), induces resistance in cotton against *Xanthomonas campestris* pv. *malvacearum*. *Phytoparasitica* 35(5): 442–449.

Recep, K., Fikrettin, S., Erkol, D. and Cafer, E. (2009). Biological control of the potato dry rot caused by *Fusarium* species using PGPR strains. *Biological Control* 50(2): 194–198.

Ren, J., Tong, J., Li, P., Huang, X., Dong, P. and Ren, M. (2021). Chitosan is an effective inhibitor against potato dry rot caused by *Fusarium oxysporum*. *Physiol. Mol. Plant Pathol.* 113: 101601.

Rengasamy, K. R. R., Kulkarni, M. G., Pendota, S. C. and Van Staden, J. (2016). Enhancing growth, phytochemical constitutents and aphid resistance capacity in cabbage with foliar application of eckol–a biologically active phenolic molecule from brown seaweed. *New Biotechnol.* 33(2): 273–279.

Rickard, D. A. 2000. Review of phosphorus acid and its salts as fertilizer materials. *J. Plant Nutr.* 23: 161–180.

Rodríguez-González, Á., Casquero, P. A., Suárez-Villanueva, V., Carro-Huerga, G., Álvarez-García, S., Mayo-Prieto, S., Lorenzana, A., Cardoza, R. E. and Gutiérrez, S. (2018). Effect of trichodiene production by *Trichoderma harzianum* on *Acanthoscelides obtectus*. *J. Stored Prod. Res.* 77: 231–239.

Rodríguez-González, Á., Mayo, S., González-López, Ó., Reinoso, B., Gutierrez, S. and Caquero, P. A. (2017). Inhibitory activity of *Beuaveria bassiana* and *Trichoderma* spp. on the insect pests *Xylotrechus Arvicola* (Coleoptera: Cerambycidae) and

Acanthoscelides obtectus (Coleoptera: Chrisomelidae: Bruchinae). *Env. Mon. Assess.* 189: 12.

Ronga, D., Biazzi, E., Parati, K., Carminati, D., Carminati, E. and Tava, A. (2019). Microalgal biostimulants and biofertilisers in crop productions. *Agronomy* 9(4): 192.

Rose, M. T., Patti, A. F., Little, K. R., Brown, A. L., Jackson, W. R. and Cavagnaro, T. R. (2014). A meta-analysis and review of plant-growth response to humi substances: practical implications for agriculture. *Adv. Agron.* 124: 37–89.

Rouphael, Y., du Jardin, P., Brown, P., De Pascale, S. and Colla, G. (Eds) 2020. *Biostimulants for Sustainable Crop Production*. Burleigh Dodds Science Publishing. Available at: https://shop.bdspublishing.com/store/bds/detail/workgroup/3-190-84222.

Rouphael, Y., Franken, P., Schneider, C., Schwarz, D., Giovannetti, M., Agnolucci, M., De Pascale, S., Bonini, P. and Colla, G. (2015). Arbuscular mycorrhizal fungi act as biostimulants in horticultural crops. *Sci. Hort.* 196: 91–108.

Saberi, M., Sarpeleh, A., Askary, H. and Rafiei, F. (2013). Effects of wood vinegar and vermicompost combination in the control of *Verticillium dahliae* the causal agent of Verticillium wilt of greenhouse cucumber. *Appl. Entomol. Phytopathol.* 81: 0–0.

Sahebzadeh, N., Ghaffari-Moghaddam, M. and Sabagh, S. K. (2017). Toxicity of N-alkyl derivatives of chitosan obtained from adult of *Chrotogonus trachypterus* (Orthoptera, Acrididae) against the wheat, cabbage and oleander aphid (Hemiptera: Aphididae) species. *Jordan J. Biol. Sci.* 10: 49–55.

Salah, I. B., Aghrouss, S., Douira, A., Aissam, S., El Alaoui-Talibi, Z., Filali-Maltouf, A. and El Modafar, C. (2018). Seaweed polysaccharides as bio-elicitors of natural defenses in olive trees against Verticillium wilt of olive. *J. Plant Int.* 13: 248–255.

Saravanakumar, D., Lavanya, N., Muthumeena, B., Raguchander, T., Suresh, S. and Samiyappan, R. (2008). *Pseudomonas fluorescens* enhances resistance and natural enemy population in rice plants against leaffolder pest. *J. Appl. Entomol.* 132(6): 469–479.

Savvas, D. and Ntatsi, G. (2015). Biostimulant activity of silicon in horticulture. *Sci. Hort.* 196: 66–81.

Seenivasan, N. and Senthilnathan, S. (2018). Effect of humic acid on *Meloidogyne incognita* (Kofoid & White) Chitwood infecting banana (*Musa* spp.). *Int. J. Pest Manag.* 64(2): 110–118.

Selvaraj, A. and Thangavel, K. (2021). Arbuscular mycorrhizal fungi: potential plant protective agent against herbivorous insect and its importance in sustainable agriculture. In: Shrivastava, N., Mahajan, S. and Varma, A. (Eds) *Symbiotic Soil Microorganisms. Soil Biology* (vol. 60): 319–337, Springer, Cham.

Seyedbagheri, M.-M. (2010). Influence of humic products on soil health and potato production. *Potato Res.* 53(4): 341–349.

Shabbir, I., Samad, M. Y. A., Othman, R., Wong, M.-Y., Sulaiman, Z., Jaafar, N. M. and Bukhari, S. A. H. (2021). Evaluation of bioformulation of *Enterobacter* sp. UPMSSB7 and mycorrhizae with silicon for white root rot disease suppression and growth promotion of rubber seedlings inoculated with *Rigidoporus microporus*. *Biological Control* 152: 104467.

Shahid, M., Duma, C., Silvestre, J. and Pinelli, E. (2012). Effect of fulvic acids on lead-induced oxidative stress to metal sensitive *Vicia faba* L. plant. *Biol. Fertil. Soils* 48(6): 689–697.

Sharma, H. S. S., Fleming, C., Selby, C., Rao, J. R. and Martin, T. (2014). Plant biostimulants: a review on the processing of macroalgae and use of extracts for crop management to reduce abiotic and biotic stresses. *J. Appl. Phycol.* 26(1): 465–490.

Shearer, B. L. and Fairman, R. G. (2007). Application of phosphite in a high-volume foliar spray delays and reduces the rate of mortality of four *Banksia* species infected with *Phytophthora cinnamomic*. *Austral. Plant Pathol.* 36(4): 358–368.

Shukla, P. S., Borza, T., Critchley, A. T. and Prithiviraj, B. (2021). Seaweed-based compounds and products for sustainable protection against plant pathogens. *Mar. Drugs* 19(2): 59.

Silva, O. C., Santos, H. A. A., Dalla Pria, M. and May-De Mio, L. L. (2011). Potassium phosphite for control of downy mildew of soybean. *Crop Prot.* 30(6): 598–604.

Sousa, A. J. S., Silva, C. F. B., Sousa, J. S., Júnior, J. E. M., Freire, J. E. C., Sousa, B. L., Lobo, M. D. P., Monteiro-Moreira, A. C. O. and Grangeiro, T. B. (2019). A thermostable chitinase from the antagonistic *Chromobacterium violaceum* that inhibits the development of phytopathogenic fungi. *Enz. Microb. Technol.* 126: 50–61.

Spinelli, F., Fiori, G., Noferini, M., Srocatti, M. and Costa, G. (2009). Perspectives on the use of a seaweed extract to moderate the negative effects of alternate bearing in apple trees. *J. Hortic. Sci. Biotechnol.* 84(6): 131–137.

Stephenson, W. M. (1966). The effect of hydrolysed seaweed on certain plant pests and diseases. Proceedings of the Fifth Inter Seaweed Symposium, Halifax, Pergamon, August 25–28, 1965: 405–415.

Supraja, K. V., Behera, B. and Balasubramanian, P. (2020). Efficacy of microalgal extracts as biostimulants through seed treatment and foliar spray for tomato cultivation. *Ind. Crops Prod.* 151: 112453.

Szczeponek, A., Mazur, S. and Nawrocki, J. (2006). The usage of chitosan in protection of some peppermint and lemon balm pathogens. *Pol. Chitin Soc. Monogr.* 11: 193–200.

Tambascio, C., Covacevich, F., Lobato, M. C., de Lasa, C., Caldiz, D., Dosio, G. and Andreu, A. (2014). The application of K phosphites to seed tubers enhanced emergence, early growth and mycorrhizal colonization in potato (*Solanum tuberosum*). *Am. J. Plant Sci.* 5: 132–137.

Tanwar, A., Aggarwal, A. and Prakash, V. (2014). Effect of bioinoculants and superphosphate fertilizer on the growth and yield of broccoli (*Brassica oleracea* L. var. *italica* Plenck). *Nea Zea J. Crop Hort. Sci.* 42(4): 288–302.

Trejo-Téllez, L. I. and Gómez-Merino, F. C. (2018). Phosphite as an inductor of adaptive responses to stress and stimulator of better plant performance. In: Vats, S. (Ed) *Biotic and Abiotic Stress Tolerance in Plants*: 203–238, Springer, Singapore.

Uppal, A. K., El Hadrami, A., Adam, L. R., Tenuta, M. and Daayf, F. (2008). Biological control of potato *Verticillium* wilt under controlled and field conditions using selected bacterial antagonists and plant extracts. *Biological Control* 44(1): 90–100.

Van Hemelrijck, W., Hauke, K., Creemers, P., Mery, A. and Joubert, J.-M. (2013). Efficacy of a new oligosaccharide active against scab on apple. *Acta Hortic.* 1009: 45–52.

Vera, J., Castro, J., Gonzalez, A. and Moenne, A. (2011). Seaweed polysaccharides and derived oligosaccharides stimulate defense responses and protection against pathogens in plants. *Mar. Drugs* 9(12): 2514–2525.

Visconi, F., de Paz, J. M., Bonet, L., Jordà, M., Quiñones, A. and Intrigliolo, D. S. (2015). Effects of a commercial calcium protein hydrolysate on the salt tolerance of *Diospyros kaki* L. cv. "Rojo Brillante" grafted on *Diospyros lotus* L. *Sci. Hort.* 185: 129–138.

Wang, C., Li, X., Zhou, J., Wang, G. and Dong, Y. (2008). Effects of arbuscular mycorrhizal fungi on growth and yield of cucumber plants. *Commun. Soil Sci. Plant Anal.* 39(3–4): 499–509.

Wei, Z., Yang, X., Yin, S., Shen, Q., Ran, W. and Xu, Y. (2011). Efficacy of *Bacillus*-fortified organic fertilizer in controlling bacterial wilt of tomato in the field. *Appl. Soil Ecol.* 48(2): 152–159.

Wu, Q. S. and Xia, R. X. (2006). Arbuscular mycorrhizal fungi influence growth, osmotic adjustment and photosynthesis of citrus under well-watered and water stress conditions. *J. Plant Physiol.* 163(4): 417–425.

Wu, Q.-S., Zou, Y.-N. and He, X.-H. (2010). Contributions of arbuscular mycorrhizal fungi to growth, photosynthesis, root morphology and ionic balance of citrus seedlings under salt stress. *Acta Physiol. Plant.* 32(2): 297–304.

Xi, Y., Han, X., Zhang, Z., Joshi, J., Borza, T., Aqa, M. M., Zhang, B., Yuan, H. and Wang-Pruski, G. (2020). Exogenous phosphite application alleviates the adverse effects of heat stress and improves thermotolerance of potato (*Solanum tuberosum* L.) seedlings. *Ecotoxicol. Environ. Saf.* 190: 110048.

Xiong, J.-Q., Kurade, M. B. and Jeon, H.-H. (2017). Can microalgae remove pharmaceutical contaminants from water? *Tren Biotech.* 36: 30–44.

Yoshioka, Y., Ichikawa, H., Naznin, H. A., Kogure, A. and Hyakumachi, M. (2012). Systemic resistance induced in *Arabidopsis thaliana* by *Trichoderma asperellum* SKT-1, a microbial pesticide of seedborne diseases of rice. *Pest Manag. Sci.* 68(1): 60–66.

Zebelo, S., Song, Y., Kloepper, J. W. and Fadamiro, H. (2016). Rhizobacteria activates (+)-δ-cadinene synthase genes andinduces systemic resistance in cotton against beet armyworm (*Spodoptera exigua*). *Plant Cell Environ.* 39(4): 935–943.

Zehnder, G., Kloepper, J., Tuzun, S., Yao, C., Wei, G., Chambliss, O. and Shelby, R. (1997). Insect feeding on cucumber mediated by rhizobacteria-induced plant resistance. *Entomol. Exp. Appl.* 83(1): 81–85.

Zehnder, G. W., Murphy, J. F., Sikora, E. J. and Kloepper, J. W. (2001). Application of rhizobacteria for induced resistance. *Eur. J. Plant Pathol.* 107(1): 39–50.

Zhao, H., Sun, R., Albrecht, U., Padmanabhan, C., Wang, A., Coffey, M. D., Girke, T., Wang, Z., Close, T. J., Roose, M., Yokomi, R. K., Folimonova, S., Vidalakis, G., Rouse, R., Bowman, K. D. and Jin, H. (2013). Small RNA profiling reveals phosphorus deficiency as a contributing factor in symptom expression for citrus huanglongbing disease. *Mol. Plant* 6(2): 301–310.

Zhang, M., Liang, Y. and Chu, G. (2017). Applying silicate fertilizer increases both yield and quality of table grape (*Vitis vinifera* L.) grown on calcareous grey desert soil. *Sci. Hort.* 225: 757–763.

Zhang, M., Tan, T., Yuan, H. and Rui, C. (2003). Insecticidal and fungicidal activites of chitosan and oligo-chitosan. *J. Bioact. Compat. Poly* 18(5): 391–400.

Zhou, J., Feng, Z., Liu, S., Wei, F., Shi, Y., Zhao, L., Huang, W., Zhou, Y., Feng, H. and Zhu, H. (2021). CGTase, a novel antimicrobial protein from *Bacillus cereus* YUPP-10, suppresses *Verticillium dahliae* and mediates plant defence responses. *Mol. Plant Pathol.* 22(1): 130–144.

Published by Burleigh Dodds Science Publishing Limited, 2022.

Chapter 4

Improving application systems for bioprotectants in integrated pest management (IPM) programmes in horticulture

Clare Butler Ellis, Silsoe Spray Applications Unit Ltd, UK

1 Introduction

Spray application technique is an important component of pest management through the use of both chemical and biopesticide substances. While a good application technique can never make up for a poor product, an effective product can be made ineffective by inappropriate application.

The fundamental principles of good application are the same whether a biopesticide or a chemical pesticide is being used or whether it is being applied to a horticultural or an arable crop. However, there is a view that horticultural crops and biopesticides are more sensitive to application, and there is some justification for this - horticultural crops are often of higher value, might suffer more from cosmetic damage and many biopesticides are contact-acting only, so potentially much less robust. In addition, the majority of the published research into spray application of pesticides over the years has been aimed at cereals and broadacre crops, and as the biopesticide market is still relatively small, there has been much less industry-funded R&D. Gan-Mor and Matthews (2003)

http://dx.doi.org/10.19103/AS.2021.0095.04

concluded that there was a lack of investment in the development of effective formulations and delivery systems for biopesticides. More recently, a review of the application for biopesticides (Defra, 2011) concluded that the description of the principles of biopesticide application in a series of theoretical papers was good but there was little supporting experimental data.

Traditional field crop sprayers are hugely flexible, allowing sprays to be delivered relatively efficiently with a wide range of possible characteristics. There is an abundant choice of nozzles on the market around the world, allowing specific nozzles to be selected for each crop, pest and product combination, although it is probably more common for spray operators to use only one or two nozzles for the majority of applications. There are more specialist developments, particularly using air flows to manipulate droplet trajectories, that are used in horticultural applications. However, in many smaller horticultural enterprises, especially protected crops, the application equipment is much less sophisticated and relatively inflexible, being handheld or manually operated. Gan-Mor et al. (2014) reviewed research into equipment, techniques and biological control products from a very wide range of applications around the world but did not identify anything that would clearly bring benefits to the UK horticultural industry.

There is concern that biopesticides give variable performance in commercial practice and that performance might be improved by better grower practice, including optimising application technique. In the UK, the Amber project,[1] which began in 2016 (Chandler et al., 2020), set out to address some of these issues, and some of the findings from this work are now beginning to emerge (e.g. Butler Ellis et al., 2020). It is possible that the need to make direct contact between the biopesticide and the pest (because of the inability of the active component to be translocated) will require different application techniques from those used for chemical pesticides, but the fundamental requirements of good application practice are the same for both.

Detailed descriptions of the specific application techniques that are required for biopesticides would be immensely useful, but such information is not available, and probably never will be – the wide range of biopesticides, modes of action, pests and crops mean that there is likely to be a wide range of different application methods needed. However, understanding the basic principles of application together with the behaviour of the pest and the mode of action of the active substance will begin to allow those developing, testing and using all types of plant protection products to be able to make progress towards optimisation of application. While this chapter attempts to include as much information as possible relating specifically to the application

1 Amber (Application and Management of Biopesticides for Efficacy and Reliability) is funded by the UK Agriculture and Horticulture Development Board (project code CP158) using levy funds from UK growers.

of biopesticides, the majority of the data available are from other studies that explore the interaction between crop and application and are therefore applicable to all types of foliage-applied sprays.

2 Application technology in integrated pest management

While there is no recognised definition of how application technology should contribute to integrated pest management (IPM), it would be reasonable to suggest that the application method should result in the minimum amount of active substance used, whether a chemical or a biological pesticide. In addition, the application should minimise non-target exposures and environmental impacts. For sustainability, we need to achieve both of these in a practical, cost-effective manner.

Minimising the quantity of pesticide used requires ensuring the dose is delivered to the right location at the right time and in the right form. While many users believe that biopesticides are less forgiving of poor application than conventional chemistry, this also might be because we use chemical pesticides at relatively high doses to ensure their robustness. The 'right time, place and form' might be more crucial for biopesticides, but also could enable a significant reduction in doses if optimised for chemical pesticides too.

There is a wide range of crop structures in horticultural enterprises, with potentially different equipment required. In particular, tall row crops (such as tomatoes, cucumber and pepper in protected environments, and orchards, hops and vines outdoors) can be sprayed with a vertical boom arrangement, whereas shorter crops can be sprayed with a horizontal boom. The trajectory of the spray has a significant effect on the interception of spray by foliage, so we have to consider these types of crops separately, even though the underlying principles of application may be similar. Handheld pistol-type application equipment is commonly used for both types of crops and is obviously much more versatile and cost-effective for enterprises that cover a wide range of crops, but it has significant shortcomings, as will be discussed below.

3 Achieving the correct dose of active substance

The first step in ensuring the correct quantity of active substance on the target is the calibration of the application. This only guarantees that the required amount is delivered by the equipment/operator on average. There are two further components: quantity, or dose, retained on the plant and uniformity of distribution - that is, we need this same dose on every plant.

3.1 Calibration

Calibration is known to be an important part of preparation for spraying, but Butler Ellis et al. (2020) found that, for protected horticultural crops in the UK, it was in some cases overlooked. When handheld or manually moved equipment is used, the operator is responsible for the distribution of spray over the area and therefore the calibration of both the operator and the equipment is required. While this is possible to achieve, the operator is likely to behave differently during spraying a small test area for calibration compared with at the end of a long batch of spraying. Achieving consistency and a uniform dose is likely to be the biggest challenge, with good operator training and experience being very important.

Some biopesticide product labels indicate spraying a particular concentration of the product, rather than dose, without defining a water volume to be applied. This might suggest to users that calibration is not important, but where concentration is fixed, the delivered dose depends on the volume of water that is used, and this needs to be minimised if the product quantity is also to be minimised.

3.2 Uniformity of dose

Ensuring that every plant has the same dose is a challenging task. Boom sprayers are designed to give a distribution over the width of the boom that is as uniform as possible. Nozzles that meet the international standard ISO 16119-2 (International Standards Organisation, 2013) have uniformity requirements that try to ensure that, when operated at the correct height, pressure and spacing on a boom, have a coefficient of variation (CV) of spray delivery of no more than 7% or 9%. This is a theoretical ideal, and the actual CV becomes much higher in an outdoor environment, and when considering, for example, very small plants. For handheld applications where the operator is responsible for the distribution over the area to be treated, it is much more difficult to ensure uniformity of application, and this is a very important reason why booms are preferable to lances, and automated booms are preferable to manually moved ones. Foqué et al. (2012a) compared a handheld spray gun with a manually operated spray boom for the treatment of pot ivy plants and found that the spray boom gave the most uniform deposit over the treated area. Foqué et al. (2012b) found a similar improvement in uniformity when comparing a handheld spray gun with a vertical boom for bay laurel plants.

It is important not to neglect agitation in the spray tank, which is sometimes found to be lacking in spray equipment for protected crops (Butler Ellis et al., 2020). While chemical pesticides are formulated to ensure good mixing and stability, this is not true of biopesticides, and it is common for dispersions of spores in water to settle out without good agitation. This will inevitably compromise the dose delivered to the plant.

The viability of a biological entity after spraying can be important, particularly when it is relatively large. Studies have shown that the type of nozzle used can affect the shear forces and can result in damage to the entomopathogenic nematodes (Fife et al., 2005). Similar effects with smaller fungal spores have not been reported, however.

Good calibration and distribution mean that the operator knows and controls the dose delivered to the crop. This does not necessarily mean that the quantity of product deposited on the crop is optimum. It is rare for the required quantity of active substance *on the crop* for good control to be defined, but this is an important piece of information that could help optimise application if it were available. There is a number of different application factors that influence how much spray is deposited on a plant, including droplet size, velocity and trajectory, but one of the most important is the water volume.

3.3 Water volume

Water volume can have a significant impact on the quantity of spray deposited on the plant and, with field sprayers, it is commonly recommended, for maximum deposit, to use the minimum water volume. Butler Ellis et al. (2020) found that reducing volume increases the proportion of the active substance retained on pot chrysanthemum and basil. This is illustrated in Figs 1 and 2, with 'normalised volume' indicating the proportion of the spray liquid that was deposited on the plants. Thus if the concentration can be increased as the volume is reduced, the minimum possible volume will achieve the maximum amount of active substance on the crop. Brusselman et al. (2012) also found

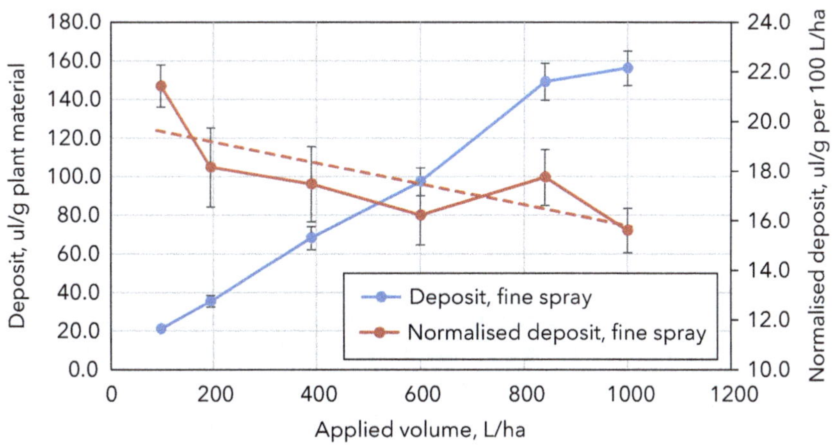

Figure 1 Quantity of spray liquid deposited on chrysanthemum pot plants for different applied volumes using an automated boom operating at different speeds. Source: From Butler Ellis et al., 2020.

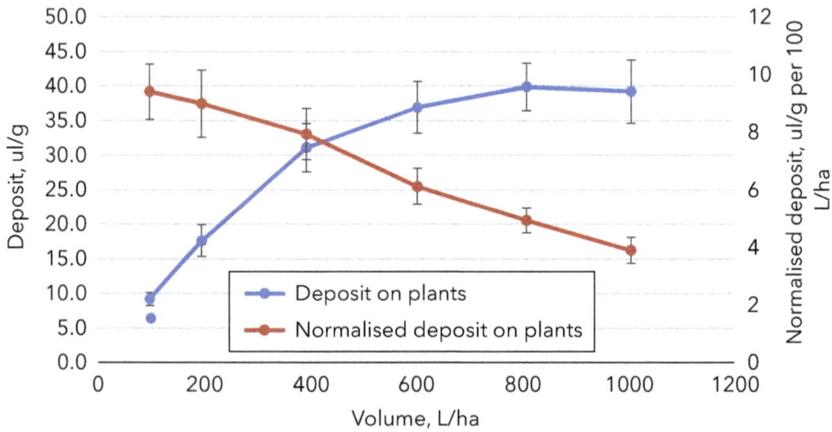

Figure 2 Quantity of spray deposited on basil pot plants for different applied volumes using an automated boom operating at different speeds. Source: From Butler Ellis et al., 2020.

that for nematode applications to vegetables, higher water volumes were associated with fewer nematodes on the plant.

Where concentration is specified and cannot be increased, there is a delivered spray volume that results in the maximum quantity of active substance on the crop. This is often referred to as 'just before run-off' in guidance, but there is no indication of what volume this refers to and it will, of course, be crop-dependent. It is very difficult for someone operating manual equipment to determine this maximum volume, and it is very subjective when judged by eye (and probably often overestimated). Figures 1 and 2 suggest that this occurs at around 600–1000 L/ha for basil and chrysanthemum, with the larger chrysanthemum plant being able to retain more spray liquid per gram of plant material than the smaller basil, and the peak for chrysanthemum occurring at a higher volume than for basil. However, many biopesticide product labels suggest much higher volumes than this, and therefore there is the potential for reducing waste by lowering volumes.

One of the reasons for high volumes in label recommendations is to include much larger crops with a vertical structure, such as tomato, cucumber and pepper. Because of the height of the crop, a different spraying technique is required, with a vertical boom providing the most uniform distribution, and when volumes are calculated per unit floor area, they can be very high. The optimum volume calculated per vertical crop area (sometimes called 'crop wall area') with a vertical boom may not be greatly different from that for short crops that could be sprayed with a horizontal boom.

A UK project (PC 136) funded by the then Horticultural Development Council between 1997 and 2000, but unpublished (final report available as of 16 September 2020 at https://nanopdf.com/download/pc-136-final_pdf), showed that a vertical boom with slightly upwards-angled nozzles gave the best distribution of spray over a vertical crop, and this appears to have become the standard for British growers with vertical booms. Some of the data from this project has been reassessed as part of the Amber project to inform application of biopesticides to vertical crops, and the effect of volume is shown in Fig. 3. The data is more variable than for smaller crops sprayed with horizontal booms, so it is difficult to identify an optimum volume for maximum spray deposit, but it appears to be around 1500 L/ha when calculated based on vertical crop area (rather than floor area). Only mean values were available from the historical data, but a new set of measurements in the Amber project shows the standard deviation in the error bars. When this data is normalised for volume, showing the potential deposit of active substance applied at constant dose (Fig. 4), this does not have the same clear relationship that we see with crops sprayed with a horizontal boom (e.g. Fig. 2). This suggests that deposits may be less sensitive to volume with such crop structures, although this could be due to more variability in the data, making it difficult to see trends.

There are difficulties in delivering low volumes with handheld equipment. The speed of movement of the operator is likely to be slow compared with an automated system, and this is often the limiting factor in reducing volume. It may be that where uniformity is poor, lower volumes are more likely to expose this weakness: that is, to ensure that some plants are not under-dosed, we have

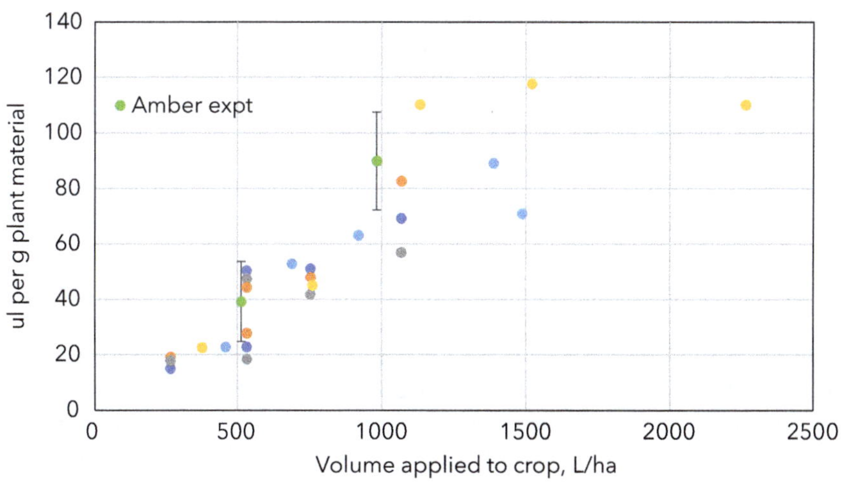

Figure 3 Spray liquid deposited on tomato leaves. Source: Data from Project PC136 re-calculated as volume applied to the vertical crop area (mean values only available) and new data from the Amber project.

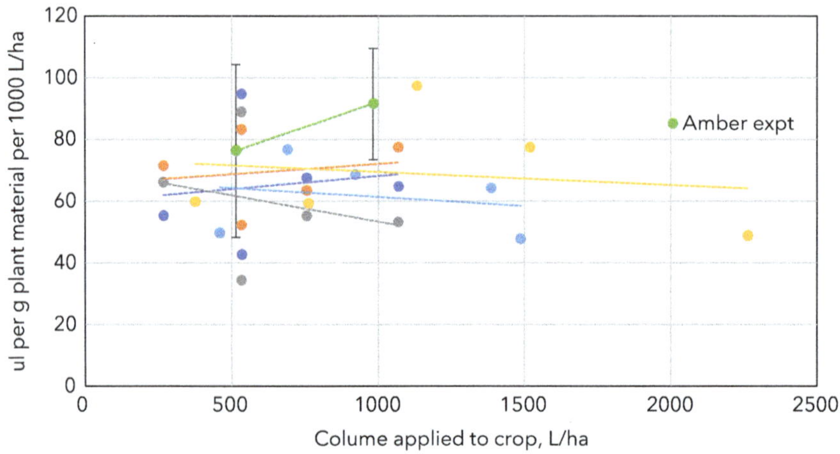

Figure 4 Normalised deposit of spray liquid on tomato leaves. Source: Data from Project PC136 re-calculated as volume applied to the crop (mean values only available) and new data from the Amber project.

to accept overdosing of others. A uniform application, which is most easily achieved by an automated boom, is an integral part of minimising product use.

It is worth remembering that ultra-low volume biopesticide applications can be successful, with the right formulation and application equipment (Bateman and Alves, 2000), although this work was predominantly aimed at insect control in developing countries, so pests, crops and climate were very different from typical Western European horticultural scenarios. However, this suggests that high volumes per se are not necessarily essential for biopesticides.

4 The right place: targeting the spray

In order to define what the 'right place' is to deliver the spray, some understanding of the pest or disease and the mode of action of the biopesticide is necessary. It is common to see a recommendation for a spray application that reaches the underside of leaves, but for conventional downward spraying techniques, it is almost impossible to achieve more than a tiny fraction of the spray on the leaf underside. Figure 5 shows the percentage of the deposited spray that is underneath chrysanthemum leaves for the same treatments shown in Fig. 1. Increasing the applied volume did not increase the quantity underneath the leaves. The fact that these products have worked in trials suggests either that the pest or disease will make contact with the active substance despite very little being deposited underneath the leaf or that the trials used a very different application technique from that used commercially.

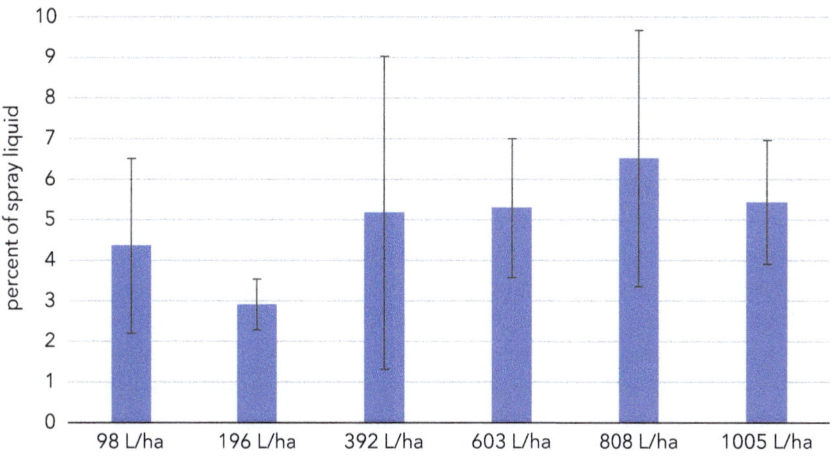

Figure 5 The quantity of spray liquid deposited on the underside of a leaf, as a percentage of the total deposited on the leaf of chrysanthemum pot plants.

Foqué and Nuyttens (2010) showed that using application technique to achieve significant improvements in the quantity deposited on the underside of leaves of ivy pot plants was challenging, although there were suggestions that air assistance (i.e. using an air flow to assist the transport of the spray) might be beneficial. Increasing the pressure with a spray gun (Foqué et al., 2012a) had no effect on the deposit on the underside of the leaf or penetration into the canopy.

Fungal spores that infect a plant through airborne transmission are very unlikely to deposit initially on the underside of a leaf and therefore a conventional downward spray is likely to be effective. For insects that are controlled through ingesting the active substance and which consume both sides of the leaf, a conventional spray onto the upper side of the leaves could be effective. Hall et al. (1995) developed a model of the dose acquisition of *Bacillus thuringiensis* (Bt), which included insect feeding and locomotory behaviour and a description of the spatial and size distributions of droplets on the leaf. Clearly, when the active substance cannot move, insect behaviour may be the dominating factor in determining efficacy, and this theoretical study demonstrates a methodology for evaluating the interaction between an insect and some application variables. However, a stationary insect that remains under the leaf and does not come into contact with other parts of the plant will always be difficult to control without specialist equipment.

Another common recommendation is to use a higher volume of spray when penetration into a dense canopy is required. There is no data to support this. Butler Ellis et al. (2012) showed that penetration of a cereal canopy is not increased by increasing the spray volume, and the same has been shown to be true of pot chrysanthemum (Butler Ellis et al., 2020). Using air flows to disturb the canopy is a recognised method of improving penetration of spray into the crop and potentially improving the distribution of active substance over the plant, but it is difficult to control, with too high an air velocity reflecting off the ground or bench, taking the fine spray with it and potentially giving a poorer distribution. Matching the air flow to the canopy is key to making air-assisted techniques work well but there is little, if any, information available to allow users to do this.

Penetration into the canopy can be manipulated to some extent by nozzle choice. More spray reached the soil underneath a pot chrysanthemum when a coarse spray was used compared with a fine spray (Butler Ellis et al., 2020). Deposition of spray on the lower stem of a cereal crop was increased by using larger, air-included droplets (Butler Ellis et al., 2012). This study also showed that a more vertical spray (80° compared with 110° fan angle) gave better penetration within the vertical structure of cereals. By contrast, with a dense broad-leaved crop, Robinson et al. (2000) concluded that angling sprays improve penetration when applying pesticides to potatoes.

Thus by matching the trajectory of the spray to the structure of the crop, it should be possible to increase penetration of a canopy. Foqué and Nuyttens (2010) suggested that both angling nozzles and air assistance improved penetration of ivy pot plants. A study of different application techniques on a tall plant (laurel bay) using a vertical boom (Foqué et al., 2012b) suggested that angling the spray, using air support or spraying in two consecutive passes with an opposite direction, did not result in a higher deposition or penetration capacity compared with the standard vertical spray boom.

The quantity deposited on the underside of leaves and the penetration into the crop appear to be better in some vertical crops, particularly where leaves have an angled orientation. The spray from one row will cover the top/front of the leaves, and the spray from the next row may penetrate through and provide some cover for the bottom/back of the leaves. This is shown in Fig. 6 with unpublished data from the Amber project.

More radical changes in application technique will be necessary to achieve large improvements in the penetration of a canopy or in deposits on the underside of leaves. Droplegs (Basil, 2002), which allow the spray to be released within the crop, might be thought to be a useful approach, but it is likely that their effectiveness will depend on the crop structure and the mode of action of the plant protection product. Bruyneel and Nuyttens (2010) found no benefit for applications to Brussels sprout plants, while Wraight and Ramos (2002) showed improved penetration of a potato crop and resultant improved control of the Colorado beetle. Other published evidence of the benefits of droplegs seems scarce.

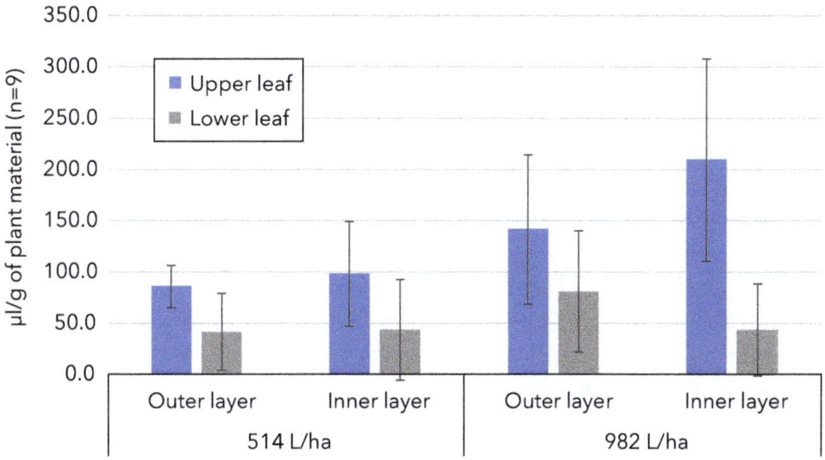

Figure 6 Spray liquid deposited on the upper side and the lower side of tomato leaves, at two volumes, applied to the crop. Source: Data from the Amber project.

5 The right time: ensuring good logistics

Ensuring that the application is made early enough - for example, before the plant is damaged or before the pest or disease gets out of control - is obviously important but application timing can be important for other reasons, particularly for biopesticides. Environmental conditions such as temperature, humidity and UV might influence the performance of the active substance; the diurnal behaviour of the pest might dictate the optimum timing in the day; or there might be a specific growth stage where the pest or disease is most vulnerable. Wraight and Ramos (2002) supported the assertion that timing is important by showing that applications from within the potato canopy provided greater control of Colorado potato beetle larvae than applications from above the canopy, when deployed against early instars, whereas applications deployed against late-instar larvae failed to provide useful control.

In addition, the persistence of biopesticides might be relatively short. Glare et al. (2012) suggest no more than a few days. Repeat applications are often required because of this, and therefore the total labour involved in using biopesticide products may be significantly higher than for chemicals.

Key to achieving optimum timing is good logistics, including a high work rate (in terms of the area treated per hour). With a conventional arable sprayer, the work rate is maximised by a high sprayer speed, wide booms, low volumes and fast loading and filling of the tank. This has led to the development of larger, faster and wider sprayers operating routinely at volumes of around 100 L/ha in the UK. The equipment for application in horticultural operations, particularly in protected crops where biopesticides are most commonly used, is very different.

Since for manually operated equipment the speed of the operator limits the work rate, increasing this might be expecting too much from a human being, particularly if the equipment is cumbersome or awkward to operate. Automating the movement of the sprayer down the rows whilst spraying would provide some advantages in speeding up the operation and making it consistent, so that a uniform dose is applied.

Reducing the applied water volume will reduce filling time, but as outlined above, reducing volume is also difficult when sprayers are moved manually. Most spray operators already use the smallest nozzles that are practical in order to cope with the relatively low speeds.

Matching tank size to the area to be treated will reduce the need to refill, but larger tanks will take longer to fill in the first place, so are of limited benefit.

Cleaning the spray equipment, and particularly the tank, will be important for biopesticides where a small amount of chemical contamination might affect the viability of any active biological entity, such as a fungal spore. Separate equipment for chemical and biological products might be advisable to minimise cross-contamination and reduce the time needed for cleaning.

Ease of operating is an important factor in achieving timely applications. This is a key reason why handheld lances and spray guns are still common;

Figure 7 Manually-operated mobile boom sprayer for a glasshouse growing ornamentals. Source: Photo courtesy of Bransford Webbs Plant Company.

they are simple, quick to set up and can cope with all types of protected crop structure, layout and crop. The Amber project found, anecdotally, that growers had variable success in moving from a handheld single nozzle system to a manually operated boom (Fig. 7). Manoeuvring between rows can be very difficult and where dimensions of beds vary between glasshouses or even sometimes within glasshouses, the design of such a boom sprayer can be challenging. An important step forward in application technology could be taken by designing the structure and the layout of glasshouses or polytunnels with the optimisation of application in mind, so that the best equipment available can be used.

Storage of biopesticides can also be an issue, since the shelf life of a biopesticide product might be shorter than conventional chemicals, and the storage requirements more challenging (e.g. a number of them need to be kept in a fridge). It might not be feasible to keep biopesticide products on-farm for long periods, so an efficient supply chain will be critical. Mixing biopesticides can be a slower process than mixing chemical pesticides, where formulation can be designed to facilitate rapid mixing, which can also compromise the timeliness of the application.

6 The right form: droplet size and coverage

Achieving the right dose at the right time and place is challenging enough, but there can also be benefits in ensuring the form of the deposit on the plants is optimum.

As well as influencing the quantity of active substance retained on the plant, the volume of water may be an important factor in achieving good coverage of the plant surface with spray liquid, in distributing the dose evenly and in maximising the efficacy of a biological entity, which might need a high level of humidity.

The size of the droplet in the spray will influence the quantity deposited on the plant and the coverage of the plant surface, with smaller droplets achieving higher deposits and greater coverage than larger ones. Figure 8 shows the relationship between volume and area covered for three different droplet sizes sprayed onto horizontally placed paper targets. This relationship will change for different surfaces and different spray liquid properties, and adjuvants and co-formulants can have a significant impact (Gent et al., 2003; Yu et al., 2009), but the basic principles remain. In a protected environment, there are no problems with spray drift and so fine sprays are commonly used.

Chapple et al. (2007) addressed the application of bioinsecticides from a theoretical perspective and identified some important concepts relating to how the atomisation of a spray liquid-containing microbial control agent might play out in practice. They identified a 'critical droplet diameter' as being

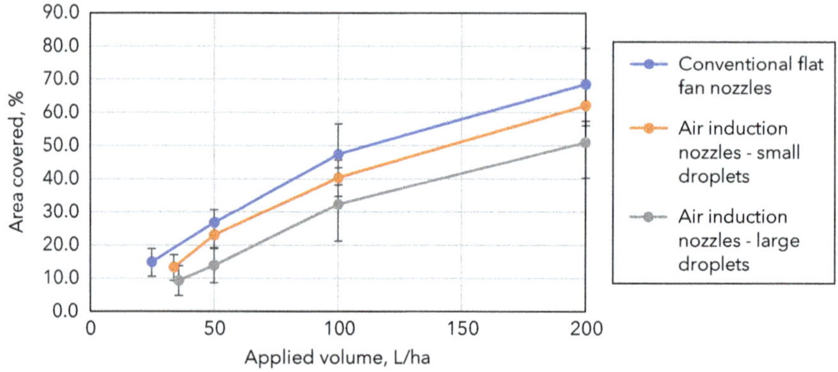

Figure 8 Area of a horizontal surface covered by spray with different applied volumes and different droplet sizes as produced by a conventional flat-fan nozzle and two different designs of air induction nozzle.

the size below which a droplet has a high probability of being 'empty' of the active substance. Their simplified analysis of a typical application calculated a critical droplet diameter of 149 μm, which suggests that a significant portion of the spray would contain no active substance. A similar analysis, for example, of a currently available product in the UK, Botanigard (Certis UK), used at the maximum concentration defined on the label, suggests a smaller critical diameter of around 50 μm. This is probably small enough to be irrelevant in most practical situations, but it is an important consideration when growers are encouraged to use high volumes (with resulting low concentrations) and fine sprays with a high percentage of small droplets.

7 Adjuvants

Adjuvants and co-formulants can influence the performance of agrochemicals in many ways (Knowles, 2012), but there is little information relating to their use for biopesticides. Irrespective of the active substance, the properties of the spray liquid will affect the quantity of spray retained on the plant, as well as the way it is distributed over the plant. Since many plants are relatively hydrophobic, a small amount of surfactant can be helpful to increase retention. Holloway et al. (2000) showed that surfactants had a small effect with wettable plants but became more important with unwettable plant surfaces. However, a surfactant is likely to also change the 'runoff' point which will reduce the optimum volume.

Surfactants can also improve spreading of the droplets on the plant surface. For chemical pesticides, surfactants are frequently built into the product, but this is much less common for biopesticides. Some biopesticide products recommend tank-mix adjuvants to achieve better retention and spreading, but it is by no means clear that this is always beneficial. Increasing droplet spread, when the active substance is a biological entity such as a spore, does not change the area of the plant that makes contact with the spore. In addition, the volume at which run-off occurs will be reduced and the rate of drying of the spray will be increased, so a high level of spreading might not be beneficial.

There is therefore scope to improve deposition and droplet spreading on the plant by the addition of adjuvants to biopesticides, which might be advantageous at lower volumes in situations where a high level of coverage is required, but this is not proven. However, the viability of the spore or other biological entity might be directly compromised, so it would be unwise to use such an additive without a recommendation from the biopesticide manufacturer who can establish compatibility.

Damage by UV is an issue that affects some biopesticides, and therefore adjuvants that can provide protection might be valuable. For example, Reddy et al. (2008) suggested Tinopal as a sunscreen for *Beauveria bassiana*.

8 Efficacy

The above discussion has assumed that there will be a correlation between measured characteristics of spray deposits and the performance of a biopesticide application, and this can only be ascertained with efficacy trials.

Conducting efficacy field trials for application is difficult for all types of plant protection products; differences between treatments can be small and variable, requiring a good trial design. Often, to test different equipment, larger plots are needed than for testing products. For a plant protection product used as part of IPM, it is particularly difficult when a holistic approach is needed. Data is therefore extremely limited.

The review by Defra (2011) identified a number of field trials testing application equipment and measuring levels of control, but motorised mist blowers and rotary atomisers were the most commonly tested application technique and few included the kind of equipment typically available in European horticulture.

Alternative approaches to conventional efficacy trials would be very valuable for establishing the importance of application in the performance of biopesticides. Controlled laboratory studies, where the biological variability can be reduced, would allow the underlying principles to be explored in a more cost-effective way. For example, a small-scale pilot experiment was successfully conducted as part of the Amber project (as yet unpublished) to investigate the effect of water volume on the control of spider mite in tomatoes. This technique needs further refinement but is a promising way forward for obtaining useful information where research funding is scarce.

9 Conclusion

Addressing basic good application practice is crucial in the application of biopesticides, and good, regular calibration is a very important part of this.

Moving towards automation should bring benefits, in terms of timeliness, uniformity and efficient delivery of active substance to the target.

Some information for users about the quantity of active substance on the plant that is needed to achieve control would allow more relevant research into application techniques to be undertaken. The mode of action of the active substance, the pest or disease and the crop structure are all important in identifying the optimum technique and ensuring that the correct dose reaches the right target location. Without this information, growers will continue to try to visibly cover every part of the plant with spray, and the result will be high levels of waste and potentially poorer control than the optimum.

Simple generic rules for the application of all biopesticides will probably never be possible to achieve and therefore guidance to users on a crop/pest or disease basis will be important.

Far more research on the interaction between performance and application is needed, for all types of plant protection products, to enable the principles of IPM to be implemented.

10 References

Basil, G. (2002). Dropleg, on-target application. *Pesticide Outlook* 13(2), 65–66.

Bateman, R. P. and Alves, P. T. (2000). Delivery systems for mycoinsecticides using oil-based formulations. *Aspects of Applied Biology* 57, 163–170.

Brusselman, E., Beck, B., Pollet, S., Temmerman, F., Spanoghe, P., Moens, M. and Nuyttens, D. (2012). Effect of the spray application technique on the deposition of entomopathogenic nematodes in vegetables. *Pest Management Science* 68(3), 444–453.

Bruyneel, L. and Nuyttens, D. (2010). Effect of spray application technology on the biological control of aphids in brussels sprouts. *Communication in Applied Biological Sciences* 75(1), 139–146.

Butler Ellis, M. C., Lane, A. G. and O'Sullivan, C. M. (2012). Distribution of spray applied to a cereal crop and the effect of application parameters on penetration. *Aspects of Applied Biology* 114, 347–354.

Butler Ellis, M. C., Lane, A. G., O'Sullivan, C. M., Chandler, D. and Prince, G. (2020). Application of biopesticides in the UK: lessons from the amber project and optimisation of applied volume. *Aspects of Applied Biology* 144, pp. 11–18.

Chandler, D., Prince, G., Talbot, D., Wedgwood, E., Gwynn, R., Jacobson, R., Butler Ellis, M. C. and Bennison, J. (2020). Helping growers get the best out of biopesticides: the AMBER project. *Aspects of Applied Biology* 144, 1–10.

Chapple, A. C., Downer, R. A. and Bateman, R. P. (2007). Theory and practice of microbial insecticide application. In: Lacey, L. A. and Kaya, H. K. (Eds), *Field Manual of*

Techniques in Invertebrate Pathology, Dordrecht: Springer. https://doi.org/10.1007/978-1-4020-5933-9_2.

Defra (2011). Review of literature and existing state of the art with respect to the application of microbial biopesticides in agriculture. Project PS2027. Available as of 16th September 2020 Available at: http://sciencesearch.defra.gov.uk/.

Fife, J. P., Ozkan, H. E., Grewal, P. S., Derksen, R. C. and Krause, C. R. (2005). Viability of a biological pest control agent through hydraulic nozzles. *Transactions of the ASAE* 48(1), 45–54.

Foqué, D. and Nuyttens, D. (2010). Effect of air support and spray angle on coarse droplet sprays in ivy pot plants. In: 2010 ASABE Annual International Meeting Sponsored by ASABE David L, Pittsburgh, PA, Lawrence Convention Center Pittsburgh, June 20–June 23, 2010, Paper number 1009014.

Foqué, D., Pieters, J. G. and Nuyttens, D. (2012a). Comparing spray gun and spray boom applications in two ivy crops with different crop densities. *HortScience* 47(1), 51–57.

Foqué, D., Pieters, J. G. and Nuyttens, D. (2012b). Spray deposition and distribution in a bay laurel crop as affected by nozzle type, air assistance and spray direction when using vertical spray booms. *Crop Protection* 41, 77–87.

Gan-Mor, S. and Matthews, G. A. (2003). Recent developments in sprayers for application of biopesticides–an overview. *Biosystems Engineering* 84(2), 119–125. ISSN 1537-5110. https://doi.org/10.1016/S1537-5110(02)00277-5.

Gan-Mor, S., Palevsky, E. and Matthews, G. A. (2014). Equipment for application of biopesticides. In: Sahayaraj, K. (Ed), *Basic and Applied Aspects of Biopesticides*, India: Springer, https://doi.org/10.1007/978-81-322-1877-7_19.

Gent, D. H.,Schwartz, H. F.and Nissen, S. J. (2003). Effect of commercial adjuvants on vegetable crop fungicide coverage, absorption, and efficacy. *Plant Disease* 87(5), 591–597.

Glare, T., Caradus, J., Gelernter, W., Jackson, T., Keyhani, N., Köhl, J., Marrone, P., Morin, L. and Stewart, A. (2012). Have biopesticides come of age? *Trends in Biotechnology* 30(5), 250–258, https://doi.org/10.1016/j.tibtech.2012.01.003.

Hall, F. R., Chapple, A. C. and Taylor, R. A. J. and Downer, R. A. (1995). Modelling the dose acquisition process of Bacillus thuringiensis: the influence of feeding pattern on survival. In: Hall, F. R. and Bany, J. (Eds), *Biorational Pest Control Agents: Formulation and Delivery*, ACS Symposium Series, vol 595, Washington, DC: American Chemical Society, pp, 68–78.

Holloway, P. J., Butler Ellis, M. C., Webb, D. A., Western, N. M., Tuck, C. R., Hayes, A. L. and Miller, P. C. H. (2000). Effects of some agricultural tank-mix adjuvants on the deposition efficiency of aqueous sprays on foliage, *Crop Protection* 19(1), 27–37.

International Standards Organisation (2013). ISO 16119-2:2013. *Agricultural and Forestry Machinery: Environmental Requirements for Sprayers. Part 2: Horizontal Boom Sprayers*, Geneva, Switzerland: ISO.

Knowles, A. (2012). *Chemistry and Technology of Agrochemical Formulations*, New York: Springer.

Reddy, N. P., Khan, P. A., Devi, K. U., Victor, J. S. and Sharma, H. C. (2008). Assessment of the suitability of Tinopal as an enhancing adjuvant in formulations of the insect pathogenic fungus Beauveria bassiana (Bals.) Vuillemin. *Pest Management Science* 64(9), 909–915. https://doi.org/10.1002/ps.1581. PMID: 18383501.

Robinson, T. H., Scott, T. A. J. and Marshall, D. J. (2000). The effect of nozzle angle and nozzle types on the deposition and biological performance of potato blight fungicides. *Aspects of Applied Biology* 57, 267–272.

Wraight, S. P. and Ramos, M. E. (2002). Application parameters affecting field efficacy of Beauveria bassiana foliar treatments against Colorado potato beetle Leptinotarsa decemlineata. *Biological Control* 23(2), 164–178.

Yu, Y., Zhu, H., Frantz, J. M., Reding, M. E., Chan, K. C. and Ozkan, H. E. (2009). Evaporation and coverage area of pesticide droplets on hairy and waxy leaves. *Biosystems Engineering* 104(3), 324–334, https://doi.org/10.1016/j.biosystemseng.2009.08.006.

Part 2

Using decision support systems in integrated pest management

Chapter 5

Advances in insect pest and disease monitoring and forecasting in horticulture

Irene Vänninen, Natural Resources Institute Finland (LUKE), Finland

1 Introduction

1.1 Knowledge domains of monitoring and forecasting research

Integrated pest management (IPM) is repeatedly characterized as a knowledge-intensive approach for plant protection. Thresholds for interventions and sampling for pest occurrence, the results of which inform whether a threshold has been reached, form the foundation and central organizing principles of all IPM activities (Castle and Naranjo, 2009). Information collected from the crop on occurrence or densities of pest(s) must be appraised in respect to the status of the crop system. The system's status concerns, among other things, crop phenology and its associated sensitivity to pests compared with pest density and prevailing and forecast weather conditions that will influence pest development.

Based on the information collected and its appraisal, a decision needs to be made: whether or not to intervene and how to intervene? Decisions, in turn, are appraised through a socioeconomic filter: crop value, cost of intervention, the farmer's level of risk taking and ecological, economic and social considerations of the sustainability of crop production. Farming objectives, the farmer's and society's values and attitudes and the farm's economic situation, all modulate

http://dx.doi.org/10.19103/AS.2021.0095.05

the selection of tools needed for information collecting and appraisal, as well as the decisions made based on that information.

To develop efficient tools for the implementation of monitoring and forecasting in practice, we need scientific research and knowledge and/or technological innovations from at least four different domains (Fig. 1): (1) biology and ecology of the pest and pest identity; (2) pest detection to estimate the pest's density or proportion of infested sampling units; (3) evaluation of the data on pest densities for decision-making with the help of pre-established reference thresholds that bring bioeconomics into the picture; and (4) forecasting of anticipated pest developments within and/or between the growing seasons. In the case of invasive pests, detecting and forecasting concern also their dispersal and establishment in new geographical regions. Forecasting can in some cases replace direct pest detection, but more often these two domains complement each other: forecasting results inform the grower about when and where to start monitoring.

At this point one thing must be made clear. Sampling and monitoring are not, strictly speaking, the same thing. Sampling is a one-time process of acquiring necessary information on the real-time pest status of a crop by using a sampling protocol (plan). Pests can be sampled for example by directly counting them on crop plants, by vacuuming, sweeping or beating to remove pests from plants for counting, or by destructive sampling by removing whole

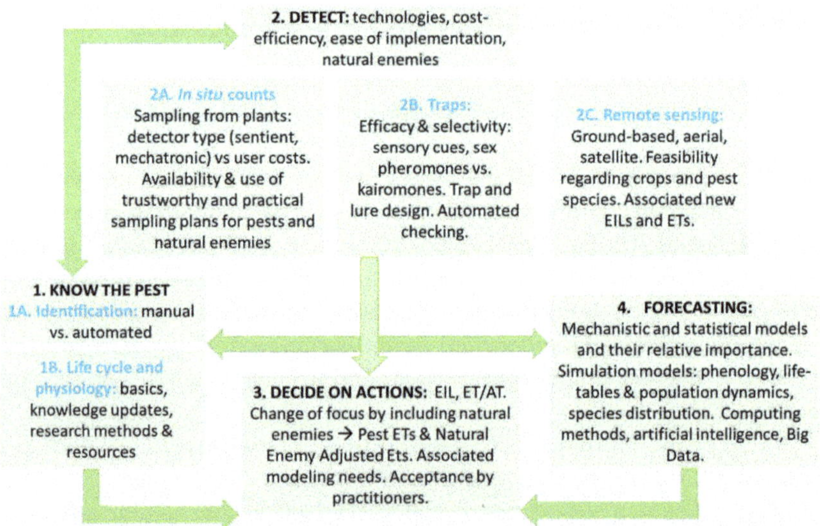

Figure 1 Four knowledge domains needed to develop tools for monitoring and forecasting insect pests. Identified challenges are listed after the domain name. EIL, economic injury level; ET, economic (damage) threshold; AT, action threshold. The arrows indicate the dependence of the domains on each other (see text for details).

Published by Burleigh Dodds Science Publishing Limited, 2022.

plants or plant parts for counting pests on them. Flying pests can be sampled using different types of traps: suction traps or traps that attract pests with colour, light or odour sources and that are equipped with a sticky surface or a vessel filled with liquid where insects are trapped. Depending on the sampling method, pest counts are either absolute or relative. Monitoring denotes the use of a sampling plan sequentially in time to monitor the change in the density of the target pest population – such repeated sampling can be called cascaded sampling. (Binns et al., 2000). Pest forecasting concerns the predicted occurrence and/or activity of biotic agent(s) that could cause economic damage to a crop, usually in the near future. Forecasting is customarily based on the knowledge concerning the impact of weather or climate on pest biology (Olatinwo and Hoogenboom, 2014) in order to predict the timing of pest attack or pest population development (Finch et al., 1996; Phelps et al., 1993)

The arrows in Fig. 1 depict how the four knowledge and technology domains associated with pest monitoring and forecasting interact with each other. Knowing the pest's biology, ecology and behaviour and being able to identify it are necessary for developing reliable detection methods and decision-making tools. Economic injury levels (EILs) and economic damage thresholds (ET) cannot be developed without understanding how the pest utilizes its plant resources in comparison with farmers' goals regarding those resources. Each method of detecting the pest requires associated knowledge on the relationship between pest counts and economic damage to the plant population. The nature of the economic damage imposes requirements on the nature of detection methods (e.g. how early the pest must be detected so as to be able to manage it with the existing tools for intervention). Pest identification issues are often intricately related to the detection method, since the latter influences the condition in which the pest is available for identification and how many other species interfere with the identification task.

The biological, ecological, phenological and sometimes even genetic fundamentals of the pest must be known so as to be able to develop forecasting models that accurately predict pest emergence and population dynamics. The forecast results, in turn, contribute to or wholly enable decision-making concerning interventions and their timing. Simulation models may reveal important new aspects of the pest's behavior, biology, ecology and/or bioeconomics that would be very laborious and resource-demanding to study empirically.

1.2 Mini reviews on 12 pest species to collect information on bottlenecks and challenges in the knowledge domains

For this chapter, the way of identifying the bottlenecks and challenges concerning monitoring and forecasting was as follows: a total of 12 key insect or mite

Published by Burleigh Dodds Science Publishing Limited, 2022.

pests of seven important fruit, berry and vegetable crops - apple, strawberry, cruciferous vegetables, onion, tomato + cucumber and carrot - were chosen for mini reviews on their status of monitoring and forecasting (Table 1). These pests included species that directly damage the end product and those that damage only non-saleable parts of the crop. Whiteflies were chosen as pests for a closer case study, as the author is most familiar with their biology and management. Through the 12 mini reviews, it was possible to obtain a picture of the status of research and implementation of monitoring and forecasting in horticultural crops that form important parts of the human diets all over the world, with emphasis on the boreal, temperate and Mediterranean climate zones.

Literature searches were made for the chosen pests on the status of the methods for their practical identification and detection and sampling programs, and also decision-making tools and forecasting of phenological development or population dynamics. The status of knowledge on the biology, ecology, etc., of the pests was reviewed only to the extent that was necessary to understand why there were bottlenecks in the above knowledge domains. Based on reviewing the status of monitoring and forecasting for the chosen pest species, the issues depicted in Fig. 1 were identified as crucial ones for the development of monitoring and forecasting systems for pests of horticultural crops. Other issues could have emerged had the sample of crops and pest species been larger.

1.3 Overview of bottlenecks and challenges in pest monitoring and forecasting

Sampling and detection require time, labour, and identification skills (Binns et al., 2000). There is a tendency for farmers and scouts to try to minimize the time needed for sampling based on in situ counts and checking traps (Agnello et al., 1994; Hamilton et al., 2004; Rincon et al., 2020). At the same time, farmers' tendency for risk aversion is influenced by the 'uncertainty associated with pest sampling and forecast information', EILs and ETs (Gent et al., 2011; Milner-Gulland and Shea, 2017). Uncertainties are caused by for example yearly trends in commodity values (Damos, 2014) and abiotic and biotic conditions that influence pests and natural enemies, efficacy of pesticides, yield levels and losses and the correlation between pest densities and economic losses (Johnson et al., 1992; te Beest et al., 2013). If exact information on these variables is not available at the time of sampling and forecasting pest population development, the inclusion of their variance, by way of a probability distribution, in calculations of ETs can be used, instead of explicitly known values. This results in probabilistic EILs (Peterson and Hunt, 2003), which instead of taking mean values for the parameters C, V, I and D in the EIL function (see Fig. 2), use the whole range of probabilities for values of these variables. For

Published by Burleigh Dodds Science Publishing Limited, 2022.

Table 1 Pests of five horticultural crops chosen for mini reviews that were used as the basis of identifying bottlenecks and stalled research concerning the pest species' monitoring and forecasting. EILs and ETs are available for all these pest species, either based on in situ counts or counts from traps, or both

Crop	Pests causing direct damage to the end product	Pests causing indirect damage to the crop
Apple	1. *Cydia pomonella* (codling moth)[1,2,5,6] 2. *Argyresthia conjugella* (apple fruit moth)[1,2,4,5,6]	3. *Panonychus ulmi* (European red mite)[4,5,6]
Strawberry	4. *Drosophila suzukii* (spotted wing drosophila) [2,3,4,5] 5. *Anthonomus rubi* (strawberry weevil)[2,4,5]	6. *Phytonemus pallidus* (strawberry mite)[4,5,6]
Crucifeours vegetables	7. *Delia radicum* (cabbage root fly)[2,3,4,5,6] 8. *Plutella xylostella* (diamondback moth) (direct pest in some crops)[1,4,5,6]	10. *P. xylostella* (indirect pest in some crops)[1,4,5,6]
Onions	11. *Delia antiqua* (onion fly) [3,5,6]	
Carrot	12. *Psila/Chamaepsila rosae* (carrot fly)[3,6]	
Tomato and cucumber (greenhouse)		13. Whiteflies (*Trialeurodes vaporariorum* greenhouse whitefly, *Bemisia tabaci* sweetpotato whitefly)[3,4,5,6]

[1] Sex pheromone available for attracting males to traps; [2]Other semiochemicals available or under product development for attracting both males and females, either alone or in combination of sex pheromones; [3]Visual traps commonly used for monitoring; [4]Sampling plans available for in situ counts from plants/soil for estimating abundance; [5]EILs and/or ETs available for decision-making at least in some crops; [6]Degree day or phenology models available for predicting emergence, flight peaks and/or population dynamics/trajectories, at least in some crops/regions. References will be given later in the text.

growers this means, for example, that they can choose from different levels of EILs depending on their risk aversion level. A risk-averse grower could choose a lower EIL that occurs only 25% of the time, due to uncertainties associated with the above variables (Higley and Peterson, 2009). Probabilistic EILs and ETs are rare, although they would address the very problem of uncertainty of EILs and ETs and farmers' risk aversion and its consequences for decision-making. Uncertainty of information tends to result in treating the crop even when there would be no need to treat it (Gent et al., 2011; McRoberts et al., 2011) or taking control actions before pest forecasts recommend taking action (Evans et al., 2017; Möhring et al., 2020). For farmers, the value of information about yield prospects and not suffering yield losses tends to be higher than that about pest density (Pannell, 1994). Farmers' risk aversion and uncertainty of information concerning pest numbers must be taken into account when developing and implementing sampling/monitoring and forecasting tools (Evans et al., 2017).

Pest insect identification is seldom a problem with traps based on female sex pheromones that only attract males to the traps (Witzgall et al., 2010).

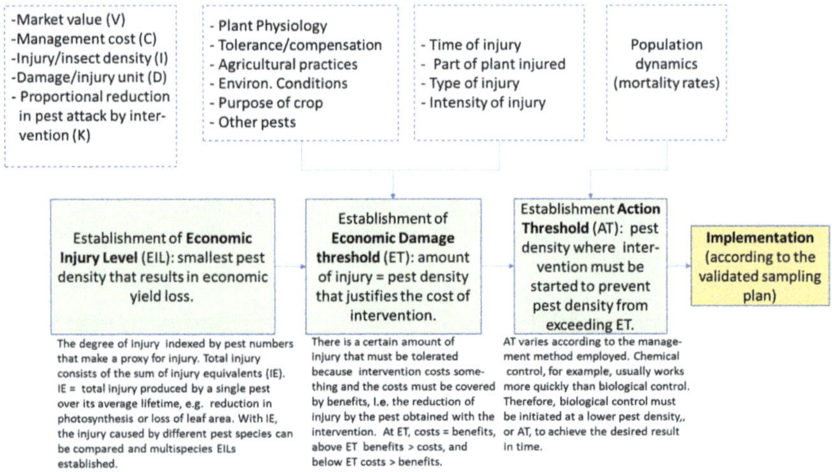

Figure 2 Phases of developing economic injury levels and economic thresholds for decision-making in pest management. Based on Radcliffe et al. (2009) and Higley and Pedigo (1996).

Sex pheromones are available for all three Lepidopteran pests mentioned in Table 1, but not all pest species use sex pheromones for attracting mates. The performance of sex pheromones is not always sufficient or it is too variable, as with *Plutella xylostella* (Evenden and Gries, 2010); their use could benefit from combining them with additional semiochemicals (Dai et al., 2008; Li et al., 2012). The high selectivity of sex pheromones can be problematic, too. With only male captures, it is difficult to make exact projections of crop damage, as it is ultimately the females that are responsible for the damage. The correlation between captured male counts and crop damage is not necessarily very good, thus the precision of trap catches in predicting damage needs improving for some pests (Adams, 2017). Therefore, it is desirable to also attract females to traps but without attracting too many non-targets.

Usually only a few traps based on sex pheromones are used per orchard or field, but even so, checking them still takes time and effort and requires walking to the traps to check them. This issue is addressed nowadays with traps equipped with cameras to enable automated checking of traps and even automated identification of trapped insects (see Section 2.1). Furthermore, the frequency of false negatives and overestimated positives (e.g. codling moth trapping) can be high with single, widely spaced traps, adding to the uncertainty of trapping results. (Adams, 2017). Males entering traps baited with sex pheromones do not always indicate damage potential, even if thresholds are exceeded. Knowing the pest's biology and ecology is important. The apple fruit moth (*Argyresthia conjugella*) lays eggs in apples only when there are insufficient rowan berries for the whole female population (Kobro et al., 2003; Tuovinen, 1987). It is therefore important to know not only the number of males coming to traps but also the number of rowan berries available for egg-laying with respect to the estimated size of the female population. Methods exist for this and should be combined with the use of traps (Edland, 1974; Tuovinen, 1987).

Food baits or plant-derived kairomones alone may not be as effective at attracting insects as sex pheromones (Landolt et al., 2007), and it can be difficult to find the right kinds of combinations of substances that would function equally well in all plant species backgrounds (Knudsen et al., 2017). The chemical ecology of herbivores has been a thriving domain of science only since the 1990s. There are still surprises to be revealed about how herbivores interact with their biotic environment, as exemplified by the recent advances in the chemical ecology of *Drosophila suzukii*.

Low sampling and trapping selectivity is an important bottleneck for reliable estimation of pest densities and can seriously hinder the uptake of sampling plans by practitioners. New, more efficient solutions for detecting and trapping pests lead to the need for developing new EILs, ETs/action thresholds (AT). The ET is theoretically dependent on the EIL (Pedigo et al., 1986), but it is very common for an ET to be established in isolation, as a practical action level arrived at through experience or direct field testing. This is because substantial research is required to derive an EIL (Fig. 2), so revisions of ETs are often based on experimentation (Hamilton et al., 2009a). Such experimental revisions may be possible for combinations of different pests that occur in the crop simultaneously, but once natural enemies are incorporated into EILs and ETs, mere empirical research can get very complicated. Modelling is needed to understand how natural enemies influence pest density in different scenarios of pest population dynamics.

Determination of EILs for crop plants had its peak period in the 1980s, and since then this research activity has been declining (Castle and Naranjo, 2009). The context for use of EILs and ETs is gradually changing due to removal of rapidly acting pesticides from the market for reasons of human and environmental health (Collier et al., 2020). Thus, EILs and ETs may gradually lose their importance as mere thresholds for curative intervention with pesticides. Binns et al. (2000) consider EIL too narrow a concept: in its current form, it does not take into account that a grower may have other values than profit that guide her/his decision-making.

> With increased reliance on biological control, cultural management and host plant resistance, pest control decision rules, and especially monitoring, will become even more important. Because it is often difficult to quantify the abundance of natural enemies, let alone predict their impact, monitoring [i.e. repeated sampling on time, particularly when there are multiple pest generations per growing season] is a useful tool for assessing the effectiveness of biological control.
>
> *(Binns et al., 1996)*

The incorporation of natural enemies in ETs brings along several challenges related to the uncertainty of factors that influence the EIL and ET and methods of threshold development. Growers in developed countries in particular tend to focus on economic returns and risk-averse management tactics, as explained above. Therefore, they may consider reliance on biological control too risky (Giles et al., 2017 and references therein) unless they have access to reliable, standardized impact data reports on biological control efficacy that are comparable across cropping systems, just as pesticide efficacy tests are reported (Giles et al., 2017). According to Macfadyen et al. (2015), when assessing the impact of natural enemies, the timing and consistency of mortality caused by natural enemies between seasons can be more important than the magnitude of mortality attributed to them. Impact assessments must be tailored according to the needs of farmers and the specific pest problems they face so that natural enemies can be incorporated as an integral part of decision-making, particularly upon reliance on conservation biological control (Macfadyen et al., 2015). But, as these authors state, studies on tangible quantified impacts of natural enemies associated with ETs are still scarce. Macfadyen et al. (2015) list the challenges and key research questions associated with assessing the impact of natural enemies and describe methods to assess the impacts. Whether obtaining such impact data is possible at all, and under what circumstances such data can be relied upon, is a matter of investing in empirical research and modelling of the impact of natural enemies, communicating the results and understanding farmers' perceptions and approaches to IPM when developing EILs and ETs. EILs and ETs incorporating natural enemies will require more thorough information on the role and biology of natural enemies in different crops as well as how to sample for them (Giles et al., 2017) (Fig. 3). This calls upon new biological, ecological and behavioural research on natural enemies and the factors that influence their performance. Modelling the interaction of pests and natural enemies also becomes more important than before in order to be able to incorporate their impact into EILs and ETs. To calculate economic impacts of natural enemies, modelling can be used as a starting point to create hypotheses that can be tested empirically in subsequent phases of research. One example is modelling the quantitative effects of entomopathogenic fungi in reducing yield losses of cereals by aphids (Saussure, 2019).

The development and use of new sampling protocols must be included in the skill repertoire of crop protection researchers, consultants and growers, and that also involves skills to use models for validation. Sampling methods for natural enemies must be developed and their robustness compared and established to satisfy the requirements of trustworthiness of sampling (Table 2). In this new situation, ensuring the practicality of sampling can become an issue, as it is challenging even now. The interaction between theory and practice will gain a heightened importance.

Validation of sampling plans that include natural enemies is even more important than validating sampling plans for pests alone, but Castle and Naranjo (2009) stated that validation of sampling plans is often not done despite the fact that methods have been available since the 1990s (Hamilton et al., 2009b; Hull and Beers, 1990; Naranjo and Hutchison, 1997). In 2009, roughly half of the ETs developed were not accompanied by a corresponding sampling plan for their effective implementation (Castle and Naranjo, 2009). This is unfortunate because validation is particularly important for sampling plans developed for applying a decision rule involving ETs, the corner stones of decision-making in IPM (Castle and Naranjo, 2009).

Success of insect pest management actions is largely determined by the efficacy of the management action, the appropriate life stage of the pest targeted and the appropriate timing of the intervention (Tonnang et al., 2017). Forecasting is undertaken to predict the timing of crucial events in pest lifecycles so that management actions can be applied when they are likely to be most effective (Collier et al., 2020). Forecasting is based on models, that is, representations of the construction and working of systems of interest. A model must be both realistic and simple. Two general types of modelling approaches are commonly used for pest forecasting in IPM: mechanistic and statistical. Mechanistic models can be theoretical (analytical), or they can be simulation models of a given specific system. Simulation models are meant for experimenting on how the output changes when input parameters and

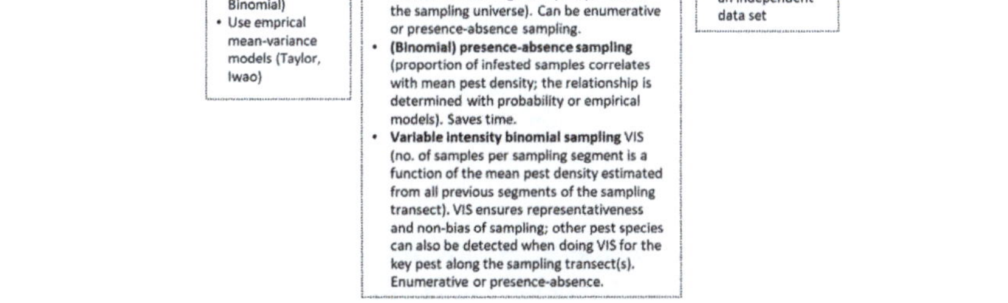

Figure 3 Phases of development of a formal validated sampling plan, with short descriptions of the empirical and probability models needed to obtain the required parameters in different phases and perform sampling. Including natural enemies in the sampling will add a new layer or several new layers to the picture, depending on how many natural enemy species must be taken into account. Based on Binns et al. (2000); Hoy (1991); Naranjo (2008); Pedigo and Buntin (1993).

Published by Burleigh Dodds Science Publishing Limited, 2022.

Table 2 Attributes of trustworthy sampling plans (Binns et al., 2000)

A trustworthy sampling plan has	What does it mean?
Good practicality	A simple enough procedure to appeal to users, not be misunderstood and must fit in with the agenda of the user regarding sampling time and the time it takes to collect the samples.
High representativeness	No unaccountable biases in the sampling protocol in selecting which instances of pest occurrence become sampled. This means several things (see bias)
Low bias	Bias is the difference between the expectation of an estimate and the true value. Bias is caused by many things, for example, by: (a) taking too few samples (can be a problem in sequential sampling plans if a minimum sample size is not given); (b) taking samples from just some parts of a field and leaving other parts unsampled; (c) selecting samples based on how easy it is to reach and process them; (d) selecting sample units on the basis of a characteristic which is correlated with the properties to be investigated by sampling; and (e) substituting omitted or rejected sample units by more readily observable units.
High reliability	No uncontrolled variables should influence the sampling protocol. They include the person collecting the data, weather, or possible diurnal behaviour of the pest (determines sampling time). Reliability and representativeness together ensure that the sample mean, m, is an unbiased estimate of the true pest density, μ.
Reasonable relevance	The estimate of pest abundance has a reasonable relationship to crop yield and loss.
Moderate to high precision depending on the goal and practicality of the sampling plan.	Refers to how close to its own expectation one single estimate is. Increases with increasing number of samples. A lower precision is often acceptable in sampling plans for decision-making in IPM. Moderate precision, is, however, required even then.
Good accuracy	A measure of how close an estimate is to what it is estimating. Can be estimated by the mean square error, mse (the greater the mse, the lower is the accuracy). Incorporates both bias and precision. Example: an estimate may have high precision, but low accuracy because of high bias.

configuration of input variables are changed in the system of interest. They should preferably involve stochasticity (i.e. at least one of the input or output variables is probabilistic) (Maria, 1997). In pest forecasting, the goal is to understand the role of different factors for the change in pest or natural enemy population dynamics (Prasad and Prabhakar, 2012).

The mechanistic approach is based on causality and tracks pest dynamics in a process model that simulates the development of the target organism and

Published by Burleigh Dodds Science Publishing Limited, 2022.

perhaps its host(s) and natural enemies. Mechanistic models are based on knowledge on how the input variables lead to the output(s), whereas statistical models give little or no insight into the specific processes leading to the output(s) (Baker et al., 2018). Mechanistic models enable the creation of new hypotheses concerning the functioning of the system of interest, whereas the purpose of statistical models is limited to prediction (Table 3). Statistical models are correlative. They range from simple regressions to much more complex models such as Bayesian models where probabilities represent all uncertainty contained in the input parameters and outputs, and finally to models based on artificial intelligence (AI) methods operating on very large data sets.

Mechanistic modelling involves two stages. First, the model is constructed and calibrated by using a subset of the available data. This is done with the help of simplified mathematical formulations of the assumed causal mechanism. Second, another subset of the data is used to validate and refine the model. Validating is needed to determine whether the range of possible input–output behaviours predicted by the model (the causality hypothesis) is consistent with experimental observations. After validation, the model can be used in applications where experiments are impossible or difficult to achieve (Baker et al., 2018). Magarey and Isard (2017) list problems and their causes and solutions associated with creating, parameterizing, validating and implementing mechanistic pest forecast models. Problems associated with the implementation phase can be rather challenging. Does the model represent stakeholder field observations? Are the model outputs useful to stakeholders, that is, do they give critical information for decision-making and is there time to act on the information in practice? Do stakeholders take the model into use at all due to time limitations or difficulties in understanding the model's outputs? These issues concern also the development of EILs and ETs.

Three subcategories of mechanistic models characterized by their input variables and output types are important for pest forecasting. First, phenology models are used to predict adult emergence time, peak adult flight and egg-laying activity, egg hatch or occurrence of damaging stages in order to optimize the timing of control measures. The measure of accumulated heat (degree days) provides a reference for the physiological age and development of organisms (Orlandini et al., 2017), but other weather factors such as rainfall or humidity can also be important, as shown for the emergence of the carrot fly, for example (Ovcharenko and Nikolaeva, 2020). Complete phenology models of insect and mite pests include mathematical functions for development time and rate, mortality, senescence, survival and reproduction of the species in question (Tonnang et al., 2017).

Phenology can be simulated in a deterministic or stochastic manner to yield the pest's life table parameters such as intrinsic rate of natural population increase (r_m), net reproduction rate (R0), finite rate of increase (λ), mean

Published by Burleigh Dodds Science Publishing Limited, 2022.

Table 3 Features that differentiate mechanistic and statistical forecast models

Feature	Mechanistic forecast models	Statistical forecast models
Basis of relationship between inputs and outputs	Causality (Baker et al., 2018).	Correlation (Baker et al., 2018).
Basic tools	Mathematical equations describing the phenomenon of interest; associated assumptions to simplify the mathematical expressions that describe the relation between inputs and output (Maria, 1997).	Tools such as regression and AI algorithms that filter out irrelevant or redundant information to discover relevant co-occurrences and dependencies in the data (Baker et al., 2018).
Focus is on	(1) Understanding the mechanisms of pest development and infestation progression via mimicking real-life events; (2) generation of novel hypotheses for causal mechanisms through observations (Baker et al., 2018). Deductive approach.	Prediction via isolating relevant inputs from the dataset for a given output. Only little or no insight into the specific processes leading to the outputs. The predictive power of machine learning models increases with the number of unique cases that can be observed and used to train the predictor (Baker et al., 2018). Inductive approach.
Dataset size	Small data sets are OK (Baker et al., 2018).	Large data sets on historical or current events (Baker et al., 2018).
Extrapolation of prediction beyond the observed conditions	Yes (Baker et al., 2018).	No. Outputs are dependent on the specific input data and can only make predictions that relate to patterns within the data supplied (Baker et al., 2018).
Generation of new hypotheses	Good possibilities (Orlandini et al., 2017).	Less or no possibilities (Orlandini et al., 2017).

generation time (G) and the doubling time (t) (Tonnang et al., 2017). Another goal is to simulate the impacts of different pest scenarios on crop production when both yield losses and the effect of changing climate on organisms must be taken into consideration (Donatelli et al., 2017). For this purpose, crop and pest behaviour models must be coupled, which is not always easy (Donatelli et al., 2017). The eventual goal is, in fact, to be able to create complete simulation models that include biological, ecological, economic and social processes and their interactions, all of which together influence crop production and guide on the ways of managing crop systems (see Tonnang et al., 2017 and their reference to Walters et al., 2016). Tonnang et al. (2017) list a range of research questions that must be answered to achieve such an ambitious modelling goal. In accordance with such goals, bioeconomic models are expected to gain more importance as tools for creating EILs and ETs (Tonnang et al., 2017; Zhang and Swinton, 2012). Management of simulations and updation of decision support systems that use the simulations present large obstacles for implementing simulation models for pest forecasting (Orlandini et al., 2017).

Improvements of phenology models are still needed to make the outputs more precise and to improve region or site specificity of outputs, as shown for root-feeding insect pests such as the carrot fly and the cabbage root fly (Bažok et al., 2012; Biron et al., 2002; Collier et al., 2020; Dreves, 2006; Villeneuve and Latour, 2017) and the codling moth (Damos and Soulopoulou, 2019). Producing phenological forecasting models can be complicated by genetic and behavioural differences between different pest (or natural enemy) populations, as exemplified by the winter morphs of *D. suzukii* (Amarasekare and Shearer, 2013; Shearer et al., 2016) and late and early emerging forms of *Delia radicum*. In some regions, these two genotypes occur (Biron et al., 2002), whereas in other regions, all flies emerge synchronously (Johansen and Meadow, 2006). Migrating species such as *P. xylostella*, combined with climate warming, bring new challenges to the development of phenological and other forecasting models (Zalucki et al., 2017; Zalucki and Furlong, 2008; Zhu et al., 2018).

Most of the earlier models lack a stochastic function for variability in development times among individuals within a population, resulting in prediction errors (Orlandini et al, 2017). Furthermore, for insect species that exhibit seasonality, diapause or aestivation in their life cycles should be accommodated in the models. This can be done with Monte Carlo simulation modelling (Orlandini et al., 2017). Mechanistic phenological models are often built considering the developmental rate function only. The models can be enriched and their accuracy increased by including elements of demographic models in them, such as the age distribution of individuals which exit from the overwintering phase, the age- and temperature-dependent profile of the fecundity rate function and the consideration of a temperature-dependent mortality rate function, as exemplified for *Cydia pomonella* by Pasquali et al. (2019).

Second, life table and population models aim at representing and understanding different or selected factors such as, for the diamondback moth, host plants (Jaleel et al., 2019), temperature (Ngowi et al., 2017), specific natural enemies (Tonnang et al., 2010, 2009) or a combination of factors (Li et al., 2016) that influence population development and abundance. A drawback for forecasting the time and size of population peaks is gaps in the ecological databases such as short-range dispersal, overwintering behaviour, colonization patterns and age-specific mortality including inter- and intraspecific competition (Prasad and Prabhakar, 2012). Such obstacles may in some cases be bypassed using statistical models based on neural networks and large data sets (Tonnang et al., 2010). There is also some imbalance as to modelling the life cycle and population dynamics of different kinds of pests. For example, soil-borne diseases have not been modelled to the same extent as airborne diseases. Primary infection by active inoculum accumulated within the rhizosphere of host plants is the main target for model development (Orlandini et al., 2017).

One important input for pest modelling is pest observations. Sharing pest observations with other growers would allow everyone to see a more comprehensive picture of pest activity, but this is rarely done and tools for doing it are lacking (Orlandini et al., 2017).

Third, species distribution models (SDM) use a variety of algorithms to estimate relationships between species locations and environmental conditions and predict and map habitat suitability (Franklin, 2010). SDMs for pest forecasting can be used at two scales: predicting the establishment risk of exotic pests in new geographic areas of distribution (Franklin, 2010) and predicting where to target pest management actions in site-specific pest management within crop fields or orchards (Méndez-Vázquez et al., 2019). At both scales, SDM output is a risk map. At the larger, geographic scale, SDM make use of georeferenced presence data of species, digital maps of environmental variables and correlative algorithms such as Bioclim, Genetic Algorhitm for Rule Production, Maxent, or General Linear Models. For predicting insect distribution at geographic scales, Tonnang et al. (2017) list the advantages and disadvantages of commonly used inductive (statistical) and deductive (causal) approaches to using SDMs. Modelling alone is not always enough to predict how exotic species behave in new areas of distribution. Surprising results can be obtained with closer research as exemplified by the better than expected overwintering ability of *D. suzukii* in regions with cold winters (Thistlewood et al., 2018).

At the field scale, the classic precision agriculture techniques of zoning are enriched with one or more spatially explicit ecological layers (species' niches) that are created on the basis of a pest's within-field distributional patterns (Méndez-Vázquez et al., 2019). In addition to the model itself, it must be parameterized using high-resolution environmental data sampled with such precision agriculture tools as wireless sensor networks for plants, soil and air measurements, drones and remote sensing.

The relative status of mechanistic and statistical forecast models is changing with the advancement of statistical modelling approaches based on AI. With the growing number of observations and variables, the need for mechanistic models has been reduced and the predictive power of statistical models has increased. This is because a very large number of unique cases that can be observed and used to train the mechanism-free predictor essentially represent the whole reality that is covered in the input–output relationship. In the omics research, high-throughput methods of data collection produce very large data sets, but according to Baker et al. (2018), the research community remains focused on producing a plethora of potential mechanistic models that explain small pieces of a much bigger picture. The use of big data for forecasting in pest management research is not commonplace as yet, either, as concluded by, for example, Fenu and Malloci (2021), who studied the kinds of AI algorithms used for forecasting the onset of disease at a pre-symptomatic stage. One important obstacle of course is how to collect large amounts of data needed for big data analyses.

2 Addressing key issues and challenges of pest monitoring and forecasting

2.1 Automated identification of insects from traps and plants

2.1.1 Fixed camera-based systems

Automated identification with up to 100% identification success was developed for *C. pomonella* as early as 2011; this required a modified trap equipped with mobile phone cameras of different resolutions and a commercial acquisition and data transfer system using wireless technology to transfer the images for analysis (Guarnieri et al., 2011). Some other, smaller, insects came to the trap also, but they did not interfere with identification. The technical and operating characteristics used by Guarnieri et al. (2011) are nowadays regarded as being fairly basic. When a pest-selective sampling method provides 'clean' samples without non-target species and when the orientation of individuals in the traps allows complete feature extraction, automated identification is relatively easy based on images taken of trapped individuals and image analysis developed with machine learning or AI algorithms. With deep learning algorithms even the 'wrong' orientation of insects in traps, background clutter and interference from uneven illumination in the images can be overcome (Wen et al., 2015). Developments since 2011 include independent power sources (such as solar-powered batteries) for the camera and for sending images via wireless channels plus, of course, more powerful data analysis techniques, such as artificial neural networks, that can be trained to identify patterns in images. The newest techniques allow dozens of species from different insect orders to be distinguished from each other (Cardim Ferreira Lima et al., 2020). For a recent review detailing the strengths and limitations of camera-based traps for pest insect detection, consult Preti et al., 2020.

Automated identification is unnecessary if insects can be trapped and their images sent to computer screens for identification by expert entomologists. High-quality images are required but that is not a problem nowadays, and for human eyes, the poor orientation of targets is not necessarily an obstacle for correct identification (unless very specific miniscule features must be seen to distinguish co-occurring species from each other). Automated traps equipped with specific or generic semiochemicals and cameras have been developed and applied successfully for distinguishing different fruit fly species in images sent to human experts (Shaked et al., 2018). Shaked et al. (2018) also made attempts to automate identification of individual fruit fly species to distinguish them from other species of fruit fly and from other insects, but that has turned out to be difficult so far. Traps with fixed cameras are, of course, expensive, but owing to the miniaturization of optics and electronics, material costs are becoming lower in addition to savings in travel and time of checking the traps.

2.1.2 Mobile camera-based systems

Cheap traps photographed with mobile cameras present another option to acquire pest images automatically. Roosjen et al. (2020) used cameras fixed to traps and also cameras attached to custom-made drones (unmanned aerial vehicles – UAV) to take images of *D. suzukii* flies on red sticky trap surfaces, under various illumination conditions in the field. Deep learning methodology was applied for feature detection. The results for images taken with static viewpoint cameras were satisfactory, but the positioning of cameras on the UAV produced lower-quality images and resulted in limited success in terms of automated identification. In the field, UAVs should be operating autonomously for them to be feasible as mobile 'photographers' of traps. It would not really make sense for a person to walk in an orchard manoeuvring UAVs within their sight, as the whole idea of automated trapping is to reduce human labour, making their presence unnecessary. In a presentation on the project (https://www.abim.ch/fileadmin/abim/documents/presentations2019/ABIM_2019_7_03_Johannes_Fahrentrapp.pdf), the authors concluded that trapping efficiency must be improved and that a landing platform for the UAV must be placed in front of the trap to take better images. Even so, they concluded that deep learning, in combination with high resolution, has good potential for the detection of small insects; it is the practical technical aspects of capturing insects and acquiring high-quality images that still need to be worked upon.

A commercial example of traps equipped with pheromones or other semiochemicals and fixed-position cameras are iScout-traps. Different types of iScout-traps are available for several insect species, for example, *D. suzukii*, *C. pomonella* and *P. xylostella*, and they include a vertically placed, coloured sticky trap with a camera in front of it (https://metos.at/iscout.com). Images are sent to a web portal for analysis.

2.1.3 Tools for identifying pests with challenging taxonomic features such as flies

It seems difficult to achieve automated identification of, for example, fly species that closely resemble each other, occur together in the field, and are attracted to the same traps and whose taxonomy involves observing, for example, minute details of genitalia and hairs on the legs and body. There are as yet no reports for automated identification attempts of *Delia* spp. or *Psila rosae*. Information on probabilities of host associations may help in the identification process (Savage et al., 2016), but even so, there is room for uncertainty. The visual traps used to detect the presence and relative abundance of *D. radicum* tend to capture a lot of non-target species that also include other *Delia* species not considered as pests. By combining visual stimuli with a feeding attractant and a new trap design, KLP+ traps by CSalomon have improved the selectivity of trapping *D. radicum*. The flies are retained by glue on the surface of a transparent plastic sheet surrounding the inside of a plastic cup (http://www.csalomontraps.com/4listbylatinname/pdffajonkentik/deliaradicum.pdf). Even so, the trap manufacturer lists a couple of other Delia species, flea beetles and Ceutorhynchus weevils that are attracted to the lure, too. Thus, whenever different *Delia* species occur together in a trap, identification is complicated and requires expert skills. The same problem of non-selective traps also concerns *P. rosae*, although the orange traps (e.g. https://www.andermattbiocontrol.com/sites/products/monitoring-systems/rebell-orange.html) are the most attractive for this species and the habitus of the carrot fly is more easily recognizable than that of different *Delia* species. Among carrot pests, the identity of psyllids in traps is also difficult to verify, as species living in wild relatives of carrot can end up in the traps, too, and can only be distinguished from the carrot psyllid based on differences in male genitalia (Nissinen, 2008). The costs of developing specific identification methods for regional pests make an obstacle for R&D.

Molecular methods have been suggested as a replacement for visual identification of *Delia* sp. from traps (Thöming et al., 2017). In some cases, however, the glue in sticky traps can compromise the possibilities of molecular identification, as shown for sandflies vectoring a human disease (Halada et al., 2018), whereas for some other species, the glue of sticky traps has been considered as an excellent preservative, and did not hinder the use of molecular markers for insect identification (Chen et al., 2014; Frey and Frey, 1995). The bottlenecks lie in getting sufficient amounts of DNA easily in field conditions and having to send the DNA samples to a laboratory for processing. Rapid identification tools based on extremely small amounts of DNA are available for laboratory use, as for example, for eggs, immatures and adults of a psyllid pest of potato (Sumner-Kalkun et al., 2020). DNA can be extracted from trapped individuals with Flinders Technology Associates FTA®cards, that is, chemically treated filter papers designed for the collection, preservation and shipment of biological samples for subsequent DNA and RNA analysis. Lemmetty and Vänninen (2014) used them successfully for extracting DNA from *Bemisia* adults on sticky traps. Up to now, FTA cards for DNA extraction have been shown to work also for 11 other genera of insects including beetles, leafhoppers, flies, psyllids and aphids that were either frozen or glued on sticky traps (Pusz-Bochenska et al., 2020). The method can be used also by growers and advisors in the field, but the cards must be sent to a laboratory for analysis. DNA barcodes can be used for identification, but this presupposes that there has been adequate *a priori* identification of barcoded specimens. The study by Savage et al. (2016) on taxonomy and ecology of *Delia* spp. shows how important the understanding of biological and ecological differences in respect to herbivore–host plant associations is in order to develop the accuracy of barcoding. Differences in the ecology of different populations can be reflected also in their DNA, use of host resources and pest status.

2.1.4 Automated identification based on optoacoustics

New approaches for automated identification are emerging from domains other than the visual sensory domain. By combining the acoustic sensory domain with advanced optical detection methods, new ways of identifying fly species are emerging. The wing beat frequency of fruit flies that enter tachometer traps equipped with attractants can be detected with optoelectronic devices based on LED (light-emitting diode) light probes and transformed into frequency profiles to differentiate between the species of interest expected to enter the trap (Potamitis et al., 2018, 2017, 2015). The fundamental frequencies of the wingbeats of the flies of interest must be known and used as a reference to measure if the energy of the bandwidth exceeds a threshold. If it does, it is a verified detection. Changes in wingbeat due to temperature differences can be accommodated. Species verification can be achieved either in situ or by transmitting the recordings and performing recognition on a server. The current system makes in situ decisions about the identity of flies entering a trap for only one species. But, based on the recordings that are stored inside the trap and transmitted further, the recognition scores are greatly improved and allow for better discrimination, even between fruit fly species, although at the cost of increased power consumption and decreased algorithmic complexity at the trap level. According to Rigakis et al. (2019), the wingbeat frequencies of *D. suzukii* are included in the research group's agenda for automated detection.

The acoustic domain is used also for differentiating whitefly species (Kanmiya, 2006, 1996; Kanmiya and Sonobe, 2002) and even sibling species of *Bemisia tabaci* from each other (Nakabayashi et al., 2017). Males of *Trialeurodes vaporariorum* and *B. tabaci* drum the leaf surface with their abdomen as mating behaviour in a species-specific manner (Kanmiya, 2006, 1996). These 'acoustic signatures' can be recorded with a sensitive microphone and compared with reference frequency profiles. Male acoustic signatures are likely to be used for species recognition during courtship behaviour. In the citrus whitefly, it is the females that drum the leaf surface and males orientate towards the females using vibrations as cues (Liao and Yang, 2015). Monitoring for the presence of cryptic species of *Bemisia* (that are indistinguishable on the basis of morphology), or finding mixed infestations of the greenhouse whitefly and *Bemisia*, would be possible with acoustic signatures. Of course, such a tool should be developed so that it does not require complex 'studios' to record the whitefly vibration sounds and so that ambient sounds do not interfere with the recording. The vibrations even provide avenues for controlling whiteflies (Yanagisawa et al., 2020)

2.2 Improved detection with more selective traps

Sex pheromones for monitoring and subsequently also for mass-trapping of Lepidopteran species have been available since the 1960s. Aggregation pheromones are currently also available for monitoring thrips (Kirkpatrick et al., 2017) and strawberry weevils (Cross et al., 2006b). Recently, compounds that are most likely to function as potent oviposition stimulants to female *D. suzukii* were found (Tait et al., 2020) and developed into what can be called an egg-sink that attracts females strongly even in the close presence of attractive berries on which to lay eggs (Rossi Stacconi et al., 2020a). Other functions of the compound mixture are being investigated, as the ingredients attract male flies, too. Depending on the attractive distance of the mixture to *D. suzukii*, and its selectivity, this finding can mean a breakthrough not only in monitoring the flies but also for their control through behavioural disruption.

To attract females selectively to traps, plant-derived kairomones, as attractants, have been studied intensively during the last 20 years. Such efforts have resulted in substantial advances in monitoring techniques for several horticultural pest insects (Table 4), but there is still work to do. The chemical ecology of *Delia radicum*, *D. antiqua* and *P. rosae* received a bout of research interest in the 1980s–1990s, but then there was a halt. Interestingly, since 2010, a French research group took the initiative to focus again on the chemical ecology of *D. radicum* (Kergunteuil et al., 2015, 2012; Lamy et al., 2018, 2017). They have produced new results on both attractive and repellent compounds for this species for the purpose of developing a push–pull strategy for managing the cabbage root flies. A similar bout of new research for better monitoring techniques is emerging for *D. antiqua* in Japan, Norway and the United States of America (Hoshizaki et al., 2020; Thöming et al., 2017; Willett et al., 2020).

Improving the selectivity of traps would have positive consequences for automated identification of pest species and for more efficient coverage of the sampling universe, resulting in more accurate decision-making. The following criteria for a good trap can be used to guide research on trap selectivity. Traps should: (a) specifically attract only the target pest insect, (b) be effective at capturing and retaining the majority of pest insects that come in contact with

Table 4 An overview of research and status of commercialization of semiochemicals for species included as case studies in this chapter

Species: Sex pheromone, other pheromones	Plant-derived semiochemicals (kairomones)
Cydia pomonella: Codlemone (main component (E,E)-8,10-dode-cadien-1-ol).	• Pear ester (ethyl (2E, 4Z)-2,4-decadienoate) (Light et al., 2001) used in combo traps with codlemone (Knight, 2010; Knight et al., 2019, 2014, 2005; Knight and Light, 2005a; Schmera and Guerin, 2012; Williams, 2016). Commercialized.
Argyresthia conjugella: Z-11-hexadecenyl acetate (Jaastad et al., 2002).	• Seven component kairomone mimic of rowan tree (Bengtsson et al., 2006; Knudsen and Tasin, 2015). Commercialized by Bio Innovate, SE.
Anthonomus rubi: Male-produced aggregation pheromone (grandlure I+ grandlure II + lavandulol in a 1:4:1 ratio); the mix is attractive to both sexes (Cross et al., 2006a; Innocenzi et al., 2001).	• A volatile from strawberry flowers, 1,4 dimethoxy-benzene (DMB), increases the attractiveness of the aggregation pheromone to both sexes of *A. rubi* (Baroffio et al., 2018; Mozūraitis et al., 2020). Commercialized.
Drosophila suzukii: A synthetic volatile blend based on six compounds found in the skin of egg-infested berries is attractive to both sexes. Commercialization in progress. (Kaiser et al., 2021; Rossi Stacconi et al., 2020a,b; Tait et al., 2020).	• Several types of odours of fermentation, microbial, fruit, and leaf origin attract the flies (Cloonan et al., 2018) and have been developed into commercial products (Cha et al., 2018; Frewin et al., 2017; Harmon et al., 2019; Kirkpatrick et al., 2017; Tonina et al., 2018; Zerulla et al., 2016).
Delia radicum: None known so far.	• Allyl-isothiocyanate, a feeding attractant (Finch and Skinner, 1982; Nottingham, 1988; Nottingham and Coaker, 1987, 1985) in a Coaster Lure Packet is used in a new trap design by CSolomon, resulting in increased selectivity of the trap for the flies http://www.csalomontraps.com/4listbylatinname/pdffajonken-tik/deliaradicum.pdf. • Oviposition stimulants have been found (Gouinguené et al., 2006; Kergunteuil et al., 2012). • Plants attractive to flies are characterized by high release rates of distinct terpenes (Kergunteuil et al., 2015).

Published by Burleigh Dodds Science Publishing Limited, 2022.

Plutella xylostella: (Z)-11-hexadecenal (Z 11-16:Ald), (Z)-11-hexa-decenyl acetate (Z11-16:Ac), (Z)-11-hexadecenol (Z11-16:OH) (Chow et al., 1977; Koshihara et al., 1978).	• Allyl isothiocyanate is a feeding attractant (Gupta and Thorsteinson, 1960) and an oviposition stimulant. • Isothiocyanate iberin and sulforaphane are potent oviposition stimulators (Renwick et al., 2006). • Plant-derived kairomones have been tested in combination with the sex pheromone (Dai et al., 2008; Li et al., 2012; Miluch et al., 2014; Reddy and Guerrero, 2000).
Delia antiqua: None known so far.	• n-propyl disulfide, n-propyl mercaptan are oviposition stimulants at short range (Ishikawa et al., 1978). • Methyl propyl and dipropyl trisulfides have attractant-oviposition stimulating activity (Matsumoto and Thorsteinson, 1968). • Decomposing onion bulbs are a potent attractant for adult flies, but a 2-phenylethanol + n-vakeric acid mixture captures flies 1.8–5.0 times more effectively (Ishikawa et al., 1984).
Psila rosae: None known so far.	• Both sexes respond to a number of volatile plant constituents from Apiaceae (Guerin and Visser, 1980). • Propeny-benzenes oftans-asarone and hexanal treated sticky traps increase fly catches synergistically (Guerin et al., 1983). • Oviposition stimulants have been found (Degen et al., 1999). • 2-phenylethanol and n-valeric acid embedded in silicon rubber bulbs worked as attractant in field tests (Ishikawa et al., 1987).
Whiteflies: A female sex pheromone identified by Yin and Maschwitz (1983) was developed into a (proto-type?) lure attached to yellow sticky trap by Chinese researchers (Wang et al., 2011)	• High nitrogen changes the profile of tomato volatile organic compounds (VOC) and attract *B. tabaci* (Islam et al., 2017). • Repellent terpenoid compounds have been identified (Bleeker et al., 2012, 2011). • Volatile Organic Compounds (VOCs) attractive to *T. vaporariorum* were identified from tomato and eggplant cultivars (Darshanee et al., 2017).

the trap, (c) provide early detection of the pest insect and (d) allow correlation of trap catch with subsequent crop infestation (Cloonan et al., 2018). When many attractive substances are available, as is the case for *D. suzukii*, odour cues from different sources may interfere with each other and reduce the pest's attraction to otherwise attractive odour combinations - a risk to be taken into account (Cloonan et al., 2019). On the other hand, the case with *Anthonomus rubi* shows that multi-pest traps are possible when the attractive compounds are specific enough for the different target pests, in which case they do not interfere with each other (Baroffio et al., 2018; Fountain et al., 2017).

Inspired by inconsistencies and high variability in trapping *C. pomonella* with sparsely placed sex pheromone traps, a new concept of understanding the interaction principles between traps and pest insects was developed recently (Adams, 2017; Miller et al., 2015). The new concept aims at being able to estimate absolute - instead of only relative - population densities in the landscape. This is done by specifying the attractive plume radius of the attractants and the proportion of target insects caught by the trap per areal unit. One outcome from the new concept is a recommendation for placing sex pheromone traps in line close to each other to reduce the variability of catches that are a problem when single traps placed far apart are used for trapping *C. pomonella*. Line trapping offers savings in time and cost when servicing aggregated versus distributed traps (Adams, 2017). Specific action thresholds were developed for the pear ester trap and compared with those of sex pheromone traps. Traps equipped with the kairomone pear ester improved female catches and predicted the egg hatching time of *C. pomonella* better than codlemone traps. The prediction improvement was based on cumulative degree-day totals required from Biofix until egg hatch: the degree-day totals had the lowest variability when the Biofix was based on the sustained catch of female moths in a pear ester-baited trap (Knight and Light, 2005b).

Trapping studies with *D. suzukii* show that lure attractiveness and selectivity can change during the season in relation to environmental temperatures and phenological, developmental and physiological stages of both the pest and its host plants. Different lures must be implemented in different periods and for different purposes, that is, for monitoring or mass trapping (Rodriguez-Saona et al., 2020; Tonina et al., 2018; Wong et al., 2018). The plant background can significantly influence trap catches, as shown for *D. suzukii* (Cha et al., 2018) and *A. conjugella* (Cha et al., 2018; Knudsen et al., 2017; Knudsen and Tasin, 2015). Thus, the same lure is not always appropriate in all contexts. The potential for trap catches to predict the density of larval populations depends on the crop species, pest generation, and density and geographical region, with variations due to climate and natural enemy complexes. Unique pheromone-based predictive models may be needed in different growing regions where the climate and the responses of the moths to pheromones vary. Furthermore, the presence and composition of natural enemies influence the relationship between trap catches of moths and the density of moth immature stages; therefore, a weighing factor is needed in the predictive model to take the effect of natural enemies into account (Miluch et al., 2013).

Depending on the target species, traps integrating both visual and olfactory cues can be superior tools for monitoring (Kirkpatrick et al., 2017). Trap design can have a crucial effect on trap catches as shown for *D. suzukii* (Kirkpatrick et al., 2017) and the multispecies trap for *A. rubi*, *Lygus* sp. (Fountain et al., 2017) and *D. radicum* (http://www.csalomontraps.com/4listbylatinname/pdffajonkentik/deliaradicum.pdf).

2.3 Remote sensing for reducing sampling time and overcoming sampling problems caused by patchy pest distribution

Most horticultural pest insects and mites are aggregated spatially. This is a nuisance when developing sampling plans: patchy distributions make accurate estimates of populations difficult, resulting in the implementation of the wrong management strategy. The seemingly more economical approach of taking fewer than the recommended number of samples has little value, because the representativeness of sampling suffers (see Table 2) and information is lost through apparent savings in human labour and time investments. A good example is *P. xylostella* in vegetables in Australia: stakeholders tend to favour fixed sample sizes, but even so, they take too few samples. This preferred sampling plan was shown to erode the criteria for reliable decision-making (Hamilton et al., 2006), although simpler and more time-saving sampling protocols are available (Hamilton et al., 2004) but seemingly do not match the criteria of stakeholders.

The effect of a patchy distribution also concerns phytophagous mites (Zahner and Baumgaertner, 1984), adult and immature whiteflies in, for example, tomato (Kim et al., 2001; Park et al., 2011b), eggs and consequently root-inhabiting maggots and pupae of cabbage root flies in cabbage fields (Bligaard, 1999; Finch et al., 1978, 1975), the carrot fly (Jens, 1983), eggs and onion maggots infesting the roots of seedlings and later the swollen plant part sitting on the soil (Whitfield et al., 1985) and larvae of *P. xylostella* that consume different types of *Brassica* vegetables (Chua and Lim, 1979). Automating the sampling or detection of patchily distributed pests or plants infested by them would be a desirable option to obtain sufficient numbers of observations with less time and labour. Such automation could concern either in situ counts of the pests themselves or measuring plant responses to the presence of the pests, that is, biotic stress.

Instead of farmers walking their crops, drones could do the sampling when this requires in situ counts or observations, particularly in very large cultivations.

Up until now, most research on the use of drones for crop protection purposes has focused on disease detection through plant phenotyping using multispectral, hyperspectral, RGB (red, green, blue), fluorescence and thermal images of canopies as well as landscape and habitat features (Gogoi et al., 2018; Zhang et al., 2019). In situ counts of insects from plants using cameras attached to drones are much more difficult. Illumination, angle of capture and shadows can significantly interfere with pest or symptom detection in images taken with drones. Furthermore, the visual and spectral cues used to detect pests may be the result of several different factors, including other insect species than the target ones that may be present in the crop. Therefore, verification by other methods of species detection is often necessary (Barbedo, 2019).

There have been, however, some proof-of-concept type experiments which suggest that drones could have a bigger role in sampling or monitoring horticultural pest insects, either for research purposes or for purposes of actual crop protection. Aerial thrips have been sampled with the help of sticky, customized Petri plates attached to a drone, to study trivial and long-distance dispersal of thrips above onion fields (Smith et al., 2015) and to monitor several insect species from the air above rice fields (Kim et al., 2018). Drones can also carry a sweep net and sample insects from vegetation in otherwise inaccessible places (Löcken et al., 2020) or fields that are too large to sample by walking (Kovanci et al., 2005). The same approach could be used to sample adult *P. xylostella* in *Brassica* vegetables, and even insect larvae can be sampled as shown by Löcken et al. (2020). Adult carrot flies and cabbage root flies move between crop fields and their surrounding habitats daily. Drone sampling may help to detect details of these flies' dispersal and local migration behaviour, for which studies on sampling aerial thrips with the help of drones pave the way for (Smith et al., 2015). In vineyards, drones are being developed for monitoring *Phylloxera* by taking hyperspectral, multispectral or RGB images of the foliage and correlating reflectance spectra with pest densities (Vanegas et al., 2018). Mini drones have even been developed as artificial predators for monitoring and killing moths in greenhouses (https://pats-drones.com/). For research purposes, drones equipped with UVA sensors could be used for detecting insects treated with fluorescent substances in ecological studies, for example, for finding out how insects move in their habitats or between habitats (Teickner et al., 2019). Fixed traps equipped with sensors measuring abiotic conditions in the monitoring area contribute to data collection on factors influencing pest occurrence and reproduction and may eventually be integrated with forecasting models for pest occurrence and population dynamics.

Infestation by phytophagous mites, aphids and whiteflies causes changes in the spectral reflectance of leaves and canopies. Detection of such changes has already been shown to be possible in soybeans infested by aphids (Marston et al., 2020) and whiteflies (Barros et al., 2021). The same methods are applied in greenhouses for pest detection through stress symptoms in plants, with the help of drones (https://www.greenhousemag.com/article/the-drones-are-close/). Pats drones have developed a phone-based app that uses GPS and mapping software to autonomously fly a drone to points in a field selected by the farmer. The images taken by the drone are interpreted by the app to provide an accurate green area index (GAI) and to count emerging plants. The quality of the image collected also is good enough to identify weeds and is claimed to be sufficiently accurate to capture insect damage on a single leaf (https://pats-drones.com/). In horticulture, low-growing crops such as cabbage, strawberry, onion and lettuce are the easiest candidates for monitoring using drone-based cameras to detect pests or their symptoms. Tall crops such as fruit trees, tomato and cucumber or berry bushes can be more challenging, due to their more complex vertical structure.

With remote sensing and AI, it has become possible to use plant, field and regional scale phenotypic information and integrate it into predictive and prescriptive management tools for monitoring, mapping and predicting outbreaks (Jung et al., 2021). When the whole field can be covered with remote sensing, implementing sampling plans that take the pest spatial distribution into account will decrease in importance. The remote sensing techniques include ground-based spectroradiometers, aerial photographic cameras, airborne digital multispectral and hyperspectral imaging systems, and moderate and high-resolution satellite imaging systems (Abd El-Ghany et al., 2020; Prabhakar and Thirupathi, 2018; Yang and Everitt, 2011). Radar-based technologies make a group of their own that is ground based but focuses on detecting insects in the air even in the night time so that it is possible to detect insect migrations in the dark (Abd El-Ghany et al., 2020; Prabhakar and Thirupathi, 2018).

The reviews by Abd El-Ghany et al. (2020) and Prabhakar and Thirupathi (2018) list the different vegetation indexes that are used to measure the stress level of plant canopies and what kind of stress types the indices can reveal. Glenn and Tabb (2019) compared different methods of determining the NDVI (normalized difference vegetation index) for apple trees. They concluded that NDVI is a useful tool when evaluating long-term crop changes such as pest damage, chronic water shortage and nutrient deficiencies that affect chlorophyll, whereas NDVI is not useful for acute stresses such as an irrigation pump failure or plugged irrigation lines that have an effect within days. Overall, remote sensing techniques for detecting pests or pest-caused changes in apple trees are not very advanced yet (Park et al., 2021), although the use of reflectance indices in leaves infested by different densities of mite pests was studied already in the 1990s (Penuelas et al., 1995), and a review of remote sensing and geospatial techniques for fruit tree management was published in 2010 (Panda et al., 2010).

Pests, diseases and their impacts on plants can also be detected based on the chemicals the infested plants emit into the air. E-noses are suitable for monitoring when they provide information on pest presence at a time when it is still possible to initiate, continue or intensify control actions. The use of e-noses is greatly enhanced by AI-based pattern recognition algorithms. There are, however, still challenges to overcome regarding sensor performance, sampling and detection in open areas, and scaling up measurements, as reviewed by Cui et al. (2018) from the point of view of IPM in vegetables and fruit trees.

An important advantage of visual and thermal remote sensing techniques, particularly those based on aircraft and satellites, is the improvement of spatial and temporal resolution compared with traditional methods for pest monitoring such as traps of different kinds or in situ counts. A disadvantage is that the remote-sensing tools tend to be expensive to use in small areas, particularly when time-series are the goal. Aerial images particularly are costly if repetitive imaging is required to study canopy status. Technical issues include distortions in images due to the relative motion of sensors and source. The glasshouse environment can be challenging for day-time spectral remote sensing because of inconsistent lighting, spectral scattering and shadows caused by glasshouse structures. Such problems were overcome when the images were taken after sunset with an active light source (Nguyen and Nansen, 2020). With this innovation, leafminers in the leaves of bok choy and spinach could be detected with >99% accuracy. Lastly, data processing requires specialized training for analysis of images (Prabhakar and Thirupathi, 2018).

Crop pests and diseases commonly occurring in continuous cropping pattern zones are best amenable to remote sensing, whereas crop pests/diseases that occur sporadically in time and space are less amenable to be monitored by remote sensing (Rao and Lakshmikantha, 2020). Because remote sensing of pests cannot be cost-effectively or technically applied to all pest species, the approach taken to evaluate its possibilities in detecting and forecasting pest insects in China is worth bringing up. Cock et al. (2016) compiled tables on symptoms, thresholds for action and options for pest management responses of main agricultural insect pests in important crops. Chinese cabbage represented horticultural crops in the study. The authors used the tables, among other things, to evaluate the scope for remote sensing of the pests in China and how the information generated or forecast would be used to improve pest management by existing agricultural extension services. Among the insect pests and diseases important in Chinese cabbage, three were concluded to benefit from the use of remote sensing: one viral and one fungal disease and the diamondback moth *P. xylostella*. The authors also list the currently existing ETs and intervention options for the pests of biggest importance and in so doing integrate with each other the need to adapt such thresholds with the new technology. For example, correlations must

Published by Burleigh Dodds Science Publishing Limited, 2022.

be produced for results of remote sensing based on vegetation indexes or hyperspectral profiles and associated pest densities to enable decision-making concerning intervention needs.

Results of remote sensing can be integrated into robotic platforms such as autonomously moving trolley-mounted robots or robotic platforms in greenhouses. The Ecoation robotic platform includes, among other things, an optochemical method for detecting the presence of whiteflies in tomato (https://www.ecoation.com/) based on changes in plant chemistry. Another example of the use of robots for the same purpose is the Greenpatrol-robot under development in Spain for use in tomato and cucumber crops in greenhouses (http://www.greenpatrol-robot.eu/Greenpatrol-robot). A user interface can be accessed by farmers to determine the robot's status and see a map of healthy and infected zones together with recommended actions. The robot is capable of identifying where pests are located and of returning to treat them. Under the farmer's instruction, the robot has the ability to spray the plant with pesticide. Scanning of the greenhouse by the robot is based on IPM strategy algorithms. There is no information, as yet, whether the system will at some point also include natural enemies as recommended actions.

2.4 Nano-inspired biosensors for plants

Various nano-inspired biosensors have been reported that range from detection of plant infections (fungal, viral and bacterial), abiotic stress, metabolic content, phytohormones, miRNAs, genetically modified (GM) plants to transcriptional and genetically encoded biosensors in a very short time span (Giraldo et al., 2019; Kumar and Arora, 2020). Combined with abiotic data at a microenvironmental level, the nanosensors would make a good reporting tool about the status of the crop in terms of plant health, as so called 'plant wearables'. A plant wearable can consist of an ultrathin and ultra-lightweight nanosensor, the aim being to attach flexible sensor devices directly on plant tissues such as leaves for continuous monitoring (Li et al., 2020). It is too early to tell whether they can be used for detecting horticultural pest insects and mites. They would probably complement, or replace, spectral imaging tools as biomarkers for detecting symptoms caused by various stressors, but it is not clear how closely they can fingerprint to give the stressors' identity. Sampling plans must be developed with a good understanding of the spatial distribution of the pest species and the criteria that must be fulfilled to achieve reliable information about pest occurrence in the field or greenhouse.

Owing to their small size, nanosensors could in principle be 'sown' on the crop in large numbers, or they could be taken to the crop and attached there according to a sampling plan that has a clear goal. If biotic stressors are detected only on a presence–absence basis in a few locations, there is a danger of the concept of the EIL and ET becoming obsolete. That threat notwithstanding, nanostructure-supported non-invasive detection tools combined with smartphones pave the way for fast and on-site diagnosis of plant diseases and long-term monitoring of plant health conditions. Such solutions are desirable particularly in resource-poor settings (Li et al., 2020). A potential application could be nanosensors attached to plants for detection of whitefly-induced biochemical changes in plants. The changes are systemic, and detection could be improved by placing the sensors in the top and middle layer of vertical plants to target leaves with L3 and L4 immatures and adults that induce the strongest local and systemic biochemical changes in the plant (Estrada-Hernández et al., 2009).

2.5 Natural enemy adjusted thresholds (NEETs)

Decision-making solutions (sampling plans and EILs and ETs) developed in the 1990s for managing phytophagous mites in apple have paved the way for new research that aims to include natural enemies in sampling plans and EILs (Nyrop, 1988; Park et al., 2000; Van Der Werf et al., 1994). Van Der Werf et al. (1994) developed a sequential sampling program for phytoseiid predators of mites in apple and continued the program development by modelling predator:prey ratios for phytoseiids and phytophagous mites in the same crop (Van Der Werf et al., 1994). Action thresholds for tetranychid mites in some ornamental plants were developed by incorporating phytoseiid mites as their control agents (Alatawi et al., 2005; Opit et al., 2003). Later on, Zhang and Swinton (2012, 2009) undertook pioneering work by developing natural enemy adjusted thresholds (NEETs) for soybean aphids in the United States of America. They introduced a new decision rule for judicious insecticide decisions using NEETs. The new threshold represents the pest population density at which insecticide control becomes optimal in spite of the opportunity cost of injury to natural enemies of the target pest. Bannerman et al. (2015), also working on soybean, in Michigan, compared the relative bias, precision and efficiency of sampling methods for natural enemies of the soybean aphid. Such studies are still rare though, and an additional challenge is that NEETs require modelling and computer skills. Tran and Koch (2017) determined the spatial patterns of predators of a pest aphid. Decision support systems are considered to be necessary for implementation of NEETs, as otherwise turning sampling results into decisions is too complicated. Automated sampling and identification is unlikely to become possible for all species of pest and their natural enemies; despite this, sampling plans should be as practical as possible in terms of the time and labour resources required.

It is the farmer who takes the risk of adjusting her plant protection strategy that may or may not involve natural enemies, and her decision concerning investment in biological control has repercussions concerning pesticide residues in food and the environment as well as the economics of the farm. Traditional cost–benefit analysis may not be good enough to estimate the socioeconomic costs and benefits of such investments comprehensively. How does the farmer, then, know whether it would be good for her or not to invest in using biological control? This question was recently addressed by Benjamin and Wesseler (2016), who used the maximum incremental social tolerable irreversible costs (MISTICs) as a tool for such evaluation. The output of the MISTIC modelling tool is an estimation of when the incremental reversible benefits of the IPM strategy that includes biocontrol outweigh possible irreversible costs of such a strategy by a minimum threshold. Only when the minimum threshold is exceeded should introduction of biocontrol as part of farm-level IPM be considered. At the same time, the output informs us how the individual farmer's decision influences society. The results obtained by Benjamin and Wesseler indicate that including biocontrol in potato, but not in maize, IPM is feasible. According to the authors, the reasons behind biocontrol being more feasible in potato could be lack of adequate pesticide control of diverse pests in potato, the sensitivity to residue-free products by society and the regulated use of IPM by authorities. Although theoretical, the modelling study by Benjamin and Wesseler highlights how the changed context that is likely to involve reduced efficacy and availability of pesticides changes the premises of making decisions concerning the adoption of alternative pest management strategies. Predicting the costs and benefits of biocontrol at the farm and societal level gives impetus to develop and target the development of NEETs at the crop level.

2.6 The value of information, sampling plans and economic/ action thresholds: the cases of **P. xylostella** *and* **T. absoluta**

Three papers from three different decades together highlight important issues that were brought up in the overview of bottlenecks and the challenges of monitoring and forecasting. The first is that when sampling for decision-making, there is a lower limit for the number of samples that must be taken to obtain reliable information for decision-making, whereas the maximum number of samples can be adjusted and resources thus saved. The second issue is that stakeholders tend to prefer fixed-size sampling with so low sample sizes that such sufficient information is not obtained. The third issue is farmers' risk aversion that tends to result in ignoring sampling results that recommend no action.

The issues culminate in the following questions: (1) which types of sampling plan give precise enough results for decision-making in practice and optimize the net value of sample information but would still be acceptable to users in terms of practicality; (2) how should ETs actually be implemented – as top-down recommendations that must be strictly followed or as learning tools that are developed together with the stakeholders through a bottom-up principle and that give room for the peculiarities of human risk aversion and associated decision-making style; and 3) how to support farmers in the implementation of sampling plans that may first seem complicated, but which in effect are quite practical and information laden at the same time? These questions concern sampling that requires in situ counts of pests in the crop, as they are considered more time consuming and laborious compared to using traps that concentrate pests in one or more locations.

'Sequential classification sampling plans' were strongly recommended by Binns et al. (2000). Sequential means that the sample size is not predetermined. Instead, the decision of whether to make a decision (intervene or not) is made during the sampling process, based on the cumulative counts of pests in samples collected so far. There must be some criterion with which to compare the cumulative count, of course, to be able to know if more samples must be taken or if sampling can be stopped and a decision taken. The person assessing the crop takes samples and checks whether the cumulative count of insects is below or above the stop boundary after every sample.

There can also be two stop boundaries, in fact; in this case, the sampling is called 'tripartite classification sampling'. With 'sequential tripartite classification sampling', the farmers get more information: depending on where they end up with the cumulative sample size, they must either intervene now or not intervene, they also receive a recommendation for when they should sample again on a later date (e.g. after 7 days). Because the sampling result now informs the farmers about future actions, so that the sampling plans are 'chained' in time, a tripartite sampling plan is also called a 'cascaded tripartite classification sampling plan'. If the sampling plan can advise the farmers reliably about when in the near future the next sampling bout should be undertaken, and the interval (e.g. being 7, 14 or 21 days) between current and future sampling plan depends on the current sampling result, the sampling plan is called 'adaptive frequency classification sampling' Nyrop et al. (1994). Cornell University produces tripartite classification sampling plans for apple growers for decision-making concerning the management of the European red spider mite (Anonymous, 2020). The plans are presented as charts and with instructions on how to undertake the sampling and how to compare the sampling results with the chart for decision-making.

Precision of a sampling plan refers to SEM/m, the ratio of the standard error to the sample mean (Green, 1970). A fixed, or predetermined, precision level of 0.25–0.30 is often considered sufficient for decision-making purposes in IPM. The total number of samples needed to be taken depends, then, on the degree of precision required. In sequential sampling, sample size is greatest when the population is estimated to be within a critical range of densities at which treatment may be necessary – it is not desirable to make a wrong decision.

Sample size is less when the population estimates are well above or below that range. The goal is to achieve acceptable sampling precision with minimal effort. In this way, situations of taking unnecessary samples can be avoided in contrast to fixed-size sampling plans, where the same number of samples is always taken (Binns et al., 2000; Pedigo and Buntin, 1993).

Because sequential tripartite classification or adaptive frequency classification sampling protocols give much more information to the grower than simpler sampling plans, in addition to that they save sampling resources, they were considered by their creators (Binns et al., 1996; Binns and Nyrop, 1992; Nyrop et al., 1994, 1989, p. 1989; van der Werf et al., 1997) to be much more valuable than traditional fixed-size sampling plans. But their reception among farmers has not been particularly encouraging as they can be considered too complicated by the practitioners. The same concerns variable intensity sampling plans, where sampling intensity (how many sampling units are taken) is reconsidered after every sampled segment of the sampling transect and which also fulfils the criterion of representatively sampling throughout the field (Pedigo and Buntin, 1993) (see also Fig. 3).

Monitoring the diamondback moth in broccoli fields in Australia at the beginning of the 2000s brought up the issue of sequential sampling again. Hamilton et al. (2004) developed a sequential sampling plan for the moth larvae. The plan used a dynamic action threshold that accounted for factors such as the prevalence of parasitism, crop growth stage and intended market destination. It was presented as a computer program. After a couple of years, it turned out that the sequential sampling plan was not used: it was considered too complex, and stakeholders preferred fixed-size sampling plans, but took usually only 10–20 samples per field, which according to Hamilton was clearly too few to produce reliable results. Furthermore, only a small part of broccoli fields was covered by such low sample sizes, thus sampling violated the criterion of representativeness. Sequential sampling plans demand that at least a minimum sample size always be taken from a field to satisfy the criterion of representativeness. To achieve even better representativeness, a variable sampling plan (see Fig. 2) can be applied.

In Hamilton's case of sampling for *P. xylostella*, he showed that to achieve a good enough level of precision using a fixed sampling size, at least 45 samples should be taken per field. This was in contrast to the industry standard of fixed sample size of 10–20 samples per field, too few to result in reliable decision-making. Hamilton et al. (2006) also showed that depending on the lower or higher action threshold (larvae per plant) used for decision-making, the probability of making type II error (not treating when there would be a need to treat) was higher for the lower AT, particularly if the sample size was too small (Hamilton et al., 2006). This meant that when taking only a few samples, risk aversive growers (who wanted to intervene at lower pest densities) were actually making the type II error more often than growers who accepted higher pest densities before intervening. Thus, the appropriate sample size is dependent upon both the AT and the level of type II error that one is prepared to accept. The question then becomes: what should be done to encourage the implementation of reliable and at the same time less resources-demanding sampling plans by the stakeholders to really help their decision-making?

Giles et al. (2017) offered an answer to the above question: 'These thresholds are more likely to be utilized by stakeholders when integrated into dynamic web-based IPM decision support systems that summarize pest management data [and push site-specific biological control management recommendations to decision-makers].' A very interesting approach to answer the same question was offered recently by Rincon et al., (2020), who also ended up recommending IT-based support systems – albeit in a less complicated form than that of Giles et al., to encourage farmers to take up more reliable sampling plans. However, their approach went deeper into the socioeconomic issues that may hinder implementation of sampling plans. They began with the issues concerning uncertainty and the value of information, and the time constraints of farmers and their risk aversion, but addressed also the issue of fixed ETs which in reality vary according to the variation in product price in particular.

Rincon et al. (2020) worked on a tomato greenhouse infested with *T. absoluta*, undertook intensive sampling in the crop to first determine explicitly the level of infestation, and then evaluated two sequential and two variable-intensity sampling (VIS) plans for the classification of pest density, by letting farmers do the sampling and measure the time needed for sampling. The re-sampling was undertaken both with computer simulations and by field trials. As a result they suggested a new approach to ETs: that they should be seen not as fixed recommendations but as *learning tools*. With the help of such a learning tool, the farmers can determine their own threshold, keeping in mind the reference, but at the same time adjusting their decision-making according to their experience, intuition and knowledge about market developments.

Eventually, Rincon et al. (2020) recommended variable intensity sampling, not sequential sampling. The value of the information produced by VIS was the highest, and the time needed for sampling was no more than that needed for binomial presence–absence sampling. The merits of the study by Rincon et al. (2020) are in their collaborative approach to developing the sampling plan and doing an intensive sampling first and in using it as a reference for the farmers, who tested the four different sampling plans. The farmers learned about the level of uncertainty of the sampling results directly and could themselves consult their risk-averse selves with regard to decision-making. However, they were equipped with objective knowledge about the pest situation. Rincon et al. (2020) concluded that implementation of the adaptive sampling plan requires simultaneous technological developments to make real-time calculations and deliver the information in an appropriate format. They foresee apps that will do the calculations for the sampler, as they must get instruction to continue or not continue sampling 'on the go'. In such situations, documentation based on voice recognition would be of great help as it frees the grower's or advisor's hands for actual work.

It must be noted that sampling based on counts in situ turned out to be adequate for a pest like *T. absoluta*, where the larval mines in leaves are easily seen and counted. VIS sampling that is based on counting specimens may not be the best option for pests such as whiteflies that are more difficult to see. VIS can be implemented also with presence–absence sampling, in which case the difficultly of counting very small and prolific specimens is avoided. Furthermore, automated counting methods like the app that recognizes either adult or immature stages of whiteflies (Anonymous, 2021) is exactly what is needed for situations like this, and such tools, once available for use in the field, can play a decisive role in the willingness of stakeholders to adopt a sampling plan.

2.7 Forecasting

Innovations regarding mechanistic models and advances in statistical models brought about by the use of AI and associated growing importance of Big Data and data quality are some of the key developments in forecasting for IPM. These developments concern not only horticulture but primary production in general. Several authors emphasize 'seeing the wood for the trees', that is, moving from reductionist mechanistic models towards more systemic or holistic approaches to modelling as a tool of pest forecasting in IPM. Harvey (2015) calls for Big Data for conducting meta-analyses and constructing powerful models for IPM in temperate horticulture. Orlandini et al. (2017) describe the need to develop agroclimatology-based mechanistic models in terms of how models are built, parameterized, validated and implemented to produce, as outputs, pest risk maps for long-term decision-making and preparedness, and pest forecasts for day-to-day decision-making. Tonnang et al. (2017), Baker et al. (2018) and Orlandini et al. (2017) all emphasize the creation of a modelling culture that should involve model developers, service providers (e.g. for weather data) and final users.

New approaches to data analysis such as Big Data algorithms emphasize collaboration between computer scientists and biologists. The literature is replete with reviews about Big Data and its possibilities, but what is really needed for pest forecasting are more empirical cases of the use of Big Data for improving forecasting at wider temporal and spatial scales than before. This need poses new requirements for multidisciplinary collaboration and for being able to identify what types of data are relevant and valuable now, even though they may not have been so before. This requires also that the importance of data collecting must be made understood to all stakeholders. Otherwise a large part of important data could remain only anecdotal and not become used (Harvey, 2015).

Orlandini et al. (2017) call for coupling of pest models with crop models, that is, a step towards a more systemic approach, since many current pest models do not have outputs that are easy to translate into pest impacts. They foresee also an increasing quantity of pest observations and a move toward the concept of big data and associated technologies for capturing, storing, managing and analysing data. Such a move is expected to result in a changed balance between mechanistic and statistical forecast models, as depicted by (Baker et al., 2018). A thorough understanding of mechanisms behind biological and ecological processes is still needed instead of mere working with 'black boxes' associated with statistical models. Developments in both modelling approaches are foreseen to better use them together, and thus get more out of the combination of deductive and inductive modelling approaches. Mechanistic models could be used by machine learning algorithms both as transient inputs and as a validating framework. (Baker et al., 2018).

At the same time, technical developments in remote sensing are also making mechanistic models more powerful by helping to get better data for model validation. One limitation of mechanistic weather-based phenology models has namely been that data sets of in situ temperature are very specific to the locality of weather stations, that is, the data have a low spatial and temporal resolution. Satellite-based remote sensors that continuously measure land surface temperatures over vast areas can nowadays be used for creating more accurate degree-day accumulation maps for large areas. One example is given by the studies of Marques da Silva et al. (2015) on *T. absoluta* in Portugal. Remotely measured land surface temperature data was combined with the threshold temperature and thermal constants of *T. absoluta* development. The outputs were risk maps depicting the number of generations produced by the pest in different parts of the country with respect to the spatial and temporal variation of degree-days sum. The authors found that the spatial resolution of such maps was better than those produced with in situ weather stations. Such resolution improvement can make a difference in site-specific management of the pest. The authors foresaw an early warning system that could geographically locate farmers associated with similar climatological patterns and could warn them when higher risk levels are reached.

At a smaller scale, precision agriculture technologies enable predicting the zoning of pest prone areas within fields for pests that have aggregated distributions. There are methods to partition fields in management zones, but zoning for pest management purposes requires spatially explicit ecological layers that are created on the basis of a pest's within-field distributional patterns.

Published by Burleigh Dodds Science Publishing Limited, 2022.

Such layers represent a species' ecological niche. The patterns, or zones, are generated via species distribution modelling (SDM) and high-resolution environmental data using tools such as drones and wireless sensor networks to map environmental conditions with high spatial and temporal resolutions. Méndez-Vázquez et al. (2019) delineated site-specific pest management zones in a lime orchard using SDM by first mapping pest-driving environmental features within the orchard via precision sampling tools. These measurements served as raw data for creating the spatially explicit ecological layer needed for predicting where selected pests would thrive. They worked with six virtual pest species (with known affinity to lime trees) and mapped, in a correlative manner, their known distributional ranges within the experimental orchard using a subset of real environmental predictors. Lastly, they evaluated the performance of the selected zoning models in terms of multivariate similarities between environmental preferences of pests and environmental characteristics of individual management zones. The authors concluded that the use of the ecological layer that was created for purposes of identifying pest management zones according to the environmental requirements of the pest species worked better than the classical zoning methods. The approach appears rather demanding in terms of measurements that need to be done but may be feasible to do in long-term habitats such as fruit orchards.

Tonnang et al. (2017) give an overview of advances in crop pest forecast modelling and present such approaches based on advanced mathematics, computer and physics theories. These approaches include artificial neural networks, cellular automata coupled with fuzzy logic, fractal, multi-fractal, percolation, synchronization and individual/agent-based approaches. Most of the new types of modelling tools presented by Tonnang et al. (2017) concern spatio-temporal dynamics of pest distributions and densities. For example, a cellular automata modelling was used to predict the risk of the invasion and natural spread of *T. absoluta* from Spain across Africa. The output revealed that *T. absoluta* could reach South Africa 10 years after being detected in Spain (Guimapi et al., 2016). The cellular automata model integrated NDVI, temperature, relative humidity and yield of tomato production. Artificial neural networks (ANN) are showing good promise in predicting pest dynamics more accurately. Examples include predictions for population densities of *P. xylostella* and its ichneumonid parasitoid *Diadegma semiclausum* (Tonnang et al., 2010) and forecasting paddy stem borer population occurrence (Yang et al., 2009). More recently, Yan et al. (2015) compared multiple regression (MR) and ANN for predicting monthly pest risks of *Thrips palmi* and *P. xylostella*. MR is the simplest and most widely used method for pest-risk prediction. The advantages of ANN are that the modelling can be conducted without prior knowledge: any relationship between given predictors and dependent variables can be learned by the neural networks, regardless of linearity or non-linearity. Non-normally distributed data, multicollinearity issues and data noise are also tolerable when training the network. But when using ANN, the importance of predictors to a given dependent variable cannot be explicitly identified by ANN as it performs like a black box.

Yan et al. (2015) list a number of characteristics of MR and ANN that have importance for the modelling process and output. These authors are among those that see benefits in combining the mechanistic and statistical modelling approaches. MR could be used to identify key variables that contribute to pest development, that is, MR can open up the black box that ANN cannot do and can reduce the model complexity and improve the training efficiency and/or accuracy (Yan et al., 2015). Likewise, Kumar et al. (2018) also found out that ANN produced more accurate outputs than classical regression in predicting incidence of two rice insect pest species and one natural enemy species using weather variables as inputs. ANN and fuzzy logic were used also to predict daily risk of the western flower thrips in roses grown in greenhouses using only four variables in comparison with earlier models that needed a large number of variables to produce desired output (Tay et al., 2020).

D. suzukii exemplifies several of the issues that Magarey and Isard (2017) list in their troubleshooting guide concerning mechanistic forecasting models. As *D. suzukii* is an invasive species, the biggest of problems or gaps concern details of its biology and ecology, but also a validation of models that predict the development of its life stages or fitness in the new regions of its distribution. Degree-day-based models and stage-structured models are not always validated to see how well they predict the inter-annual variation in the activity of the pest. Initial population development after invasion, survival at temperature extremes, and conditions necessary for the development of damaging populations are not known well enough to use these details as input for models. Existing models may rely on development measured in laboratory experiments (Kinjo et al., 2013; Tochen et al., 2016) instead of utilizing field detection data from traps or fruit samples. Data on winter survival and spring populations of *D. suzukii* in regions with cold winter conditions are limited (Dalton et al., 2011; Shearer et al., 2016; Stockton et al., 2018); thus, there is a need to further evaluate and predict fly activity using field-collected data on adult flies. The age structure of estimated populations suggests that trap and fruit infestation data are of limited value for validating models concerning the fly's development (Hamby et al., 2016; Tochen et al., 2016; Wiman et al., 2014). Despite these issues, *D. suzukii* activity was concluded to be predictable and environmental conditions (such as the annual number of days below 0°C, the number of winter and spring days above 10°C and the fly activity in the preceding year) can be used in temperate regions to provide regional risk warnings (Leach et al., 2019).

P. xylostella is a migrant species whose appearance in new regions comes as a nasty surprise. A climate niche model has been developed for the moth.

The model has been used to predict the global geographic distribution of *P. xylostella*, but the moth's seasonal abundance for various locations has not been used in construction of the model. The model was also used to 'predict' historical population dynamics in Hangzhou, China, over several years, and for analysis of recent outbreaks of the species in the British Isles (Zalucki and Furlong, 2008; Zhu et al., 2018).

Climate warming has created the need to update phenological models for many pests, including *Psila rosae*, *D. radicum* and *D. antiqua*, whose peak flight and start of egg-laying are predicted with phenological models. The likelihood of *D. radicum* producing a fourth generation in the autumn in the United Kingdom is foreseen if mean annual temperatures increase by 5 or 10°C, coupled with earlier hatching of the flies from overwintering pupae (Collier et al., 1991). The warming climate can influence the flight dynamics of pests such as the codling moth during the growing season (Roşu-Mareş et al., 2020), resulting in prolonged periods of control and increased damages to crops. Samietz et al. (2013) took advantage of the improved spatial and temporal resolutions of climate model projections and modelled the phenology and generations of the codling moth. Their results also indicate that the pest's impact on apples would increase and its management would become more difficult with climate warming. Using 49 climate indices and undertaking further analyses with climate projections, Bradshaw et al. (2019) predicted that in a 2–4°C warmer world, *B. tabaci* could pose a risk to outdoor UK crops, including vegetables, in July and August. Currently, the pest occurs outdoors at the latitudes of southern France.

3 Case study: whitefly sampling, monitoring and forecasting

3.1 Two forms of plant injury by whiteflies complicate EIL and ET development

The relationship between yield reduction and plant injury by whiteflies has been elusive for a long time, since whiteflies cause economic damage in two ways: directly by depleting plants of photosynthates and indirectly by secreting honeydew. Honeydew accumulates on leaves and fruits and causes two types of harm: firstly, by favouring the growth of saprophytic fungi that block access of light to leaves and secondly, by causing aesthetic and technical harm through contamination of fruits, resulting either in loss of grade A fruits or in the need for washing them prior to sale. Honeydew production by the adult and larval stages and the effect of temperature and nitrogen fertilization on its production are known (Blua and Toscano, 1994; Costa et al., 1999; Henneberry et al., 2001; Hong and Rumei, 1993). This could in principle be used as a proxy for estimating the variable feeding intensity of immatures. In practice, measuring honeydew production is difficult and must be limited for research purposes and the EILs and ETs must be based on the proportion of contaminated fruits.

In Fig. 4, the components and their correlations needed to understand the direct and indirect effects of whiteflies on yield quantity and quality are summarized. The light green and light blue components denote variables that have been measured traditionally to study whitefly economic damage impacts on plant yield: feeding and honeydew secretion. Johnson et al. (1992) showed that the yield of field grown tomato correlated negatively with cumulative immature whitefly-days (per 1 cm leaf disk of tomato). (The whitefly days indicate the prolonged pest pressure on the physiology of plants due to feeding by both adults and immature stages.) A 5% direct yield loss, but little loss to sooty mould, occurred after 69 cumulative immature whitefly days (with a maximum of 0.7 nymphs/cm^2 in weekly samplings per plot). On the other hand, a 5% yield loss in grade A fruit caused by sooty mould contamination alone would have been reached after 298 cumulative greenhouse whitefly days (peak density = 8.3 immatures/cm^2 tomato leaflet). At that point, a total loss of 26% tomato yield would have occurred due to the combination of feeding and sooty mould contamination. No clear initial plateau in yield could be discerned; instead, yield reduction was linear throughout the sampled levels of whitefly abundance. It appears that direct yield reduction was more important than contamination of fruits with honeydew.

3.2 Sampling, monitoring and identifying of whiteflies

In an ideal world, tomato or cucumber growers can choose a monitoring protocol that depends on the size of their crop and willingness to invest in monitoring the pest. In a relatively small greenhouse, say 5000 m^2, they can use yellow sticky traps (YST) for monitoring by placing them at a distance of at least 15–20 m^2 from each other (1 trap per 200–400 m^2, roughly) so that they do not produce spatially redundant information as shown by Kim et al. (2001) and Park et al. (2011b). Besides the usual YSTs, LED-enhanced or LED-based traps that attract whiteflies have been developed and equipped with fixed cameras that take an image at selected intervals and send it to a computer screen (Stukenberg et al., 2015; Stukenberg, 2018). Counting of adult whiteflies from YSTs is being made easier by automatic counting using machine-vision-based apps (https://play.google.com/store//apps/details?id=<aidev.cocis.makere re. org.whiteflycounter &hl=fi&gl=US) (https://www.koppert.com/natutec-scout/) (McCarthy et al., 2020). A statistical model, based on machine-vision data, for distinguishing *T. vaporariorum* and *B. tabaci* adults from each other in traps has also been developed (Moerkens et al., 2019). Even immature

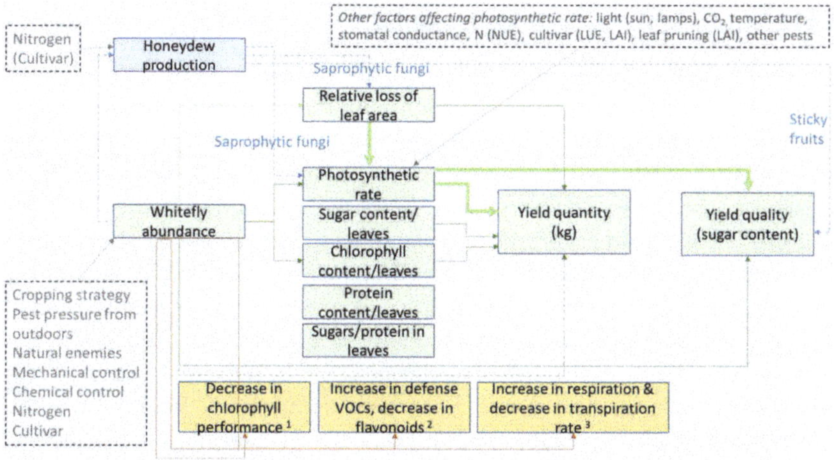

Figure 4 Factors needed to understand plant injury and economic damage caused to plants by *Trialeurodes vaporariorum*. Data for cucumber (green boxes) (Rumei and Liying, 1991) were used to indicate the relative strength of correlations between, plant growth indices, yield quantity and quality and abundance of *T. vaporariorum*. Thick green line: correlation significant at 0.01. Thin green line: correlation significant at 0.05. Dashed thin line: correlation almost significant. The correlations between honeydew production with growth indices are not based on empirical results, just sketched to show the relationships. The orange boxes give the most recent research targets for the impacts of whitefly on plants, and eventually yield. Their direct correlations with variables in the green and blue boxes are not known, but all correlate significantly with whitefly density. References: 1 (Cai et al., 2016; Chen et al., 2011) for tomato, 2 (Darshanee et al., 2017; Su et al., 2018) for tomato, 3 (Shannag and Freihat, 2009) for cucumber.

stages on leaves can nowadays be distinguished from each other and counted automatically, but such phenotyping is currently cost-effective and practical only for research purposes used in the laboratory (Anonymous, 2021).

From the traps, the grower monitors both the pest and a predatory bug *M. pygmaeus* and can conclude from their relative counts whether biocontrol by the predator is going well or not (Böckmann and Meyhöfer, 2017; Moerkens et al., 2020). Based on trap catches of the parasitoid *Encarsia formosa*, she can also conclude whether parasitism is going on well: six or more individuals per trap indicates the parasitoid is keeping the pest in control (Böckmann et al., 2014). Should the grower want to check the parasitization rate of puparia on the leaves, she can use the threshold of 80% black, parasitized pupae for concluding that the whitefly situation is in control (van Roermund et al., 1997). And if the grower is well updated, she knows that she must place Encarsia cards so that there is no more than an 8 m distance from one point to another. Otherwise there will be areas that parasitoids do not cover (Pérez et al., 2011).

3.3 Possibilities of physiological EILs

To succeed in biological control, the grower must have started control measures as soon as she observed the first whiteflies in the crop. The whiteflies reduce the chlorophyll content of leaves, the performance of the remaining chlorophyll gets poorer than normal and the stomatal conductance also decreases; these negative consequences take place in both tomato (Buntin et al., 1993) and cucumber (Rumei and Liying, 1991; Shannag and Freihat, 2009). Provided that there were no more than five to six adult whiteflies per plant in the beginning of the infestation (an EIL in greenhouse cucumber where the economic damage will develop into significant yield reductions in 2–3 months in the absence of control measures, as shown by Jeon et al. (2009)), such reductions in plant performance will not be reflected in the yield immediately. In cucumber, 20% of chlorophyll content of leaves can decrease before it results in yield decrease (Rumei and Liying, 1991); at this point, sooty moulds due to honeydew secretion are not present yet, so yield reduction is directly from the sucking action by whiteflies. So there is time to act after detecting the whiteflies but the length of time depends on the initial level of infestation and on stages of whiteflies of the initial infestation (Rumei and Liying, 1991).

The extent to which stomatal conductance and chlorophyll content and performance are reduced and transpiration is increased depends on how plants are fertilized, how much light they receive, how high fruit load they have and what other biotic and abiotic stressors are affecting them. The physiological direct yield response varies depending on environmental conditions, but this correlation remains understudied. Current portable phenotyping instruments such as fluorometers for measuring chlorophyll and secondary metabolites (Groher, 2019), chlorophyll meters (Chrysargyris et al., 2020; Vesali et al., 2017), porometers for measuring stomatal conductance (Buntin et al., 1993) and Fv/Fm-meters for measuring the maximum quantum efficiency of the plants' photosystem (Poudyal et al., 2019) can be used to determine physiological changes in crop plants. The data can be used for determining physiological EILs. Big Data approaches should help in revealing the correlations between such physiological responses and environmental factors.

3.4 Simultaneous mass trapping and monitoring with sticky traps: consequences to ETs?

In year-round tomato and cucumber crops in Finland at the beginning of the 2010s, nominal thresholds for initiating occasional treatments with selective pesticides were based on whitefly catches with sticky traps. Relatively few growers monitored whitefly dynamics with YSTs at that time. Following a collaborative project in 2010–12 (Vänninen et al., 2015), the use of sticky traps for monitoring increased and nominal thresholds developed with input from growers and advisors were taken up more widely. At the same time, however, large yellow sticky bands for mass trapping gained popularity, as an approach to improve whitefly management in year-round crops where biocontrol does not work as reliably as in summer crops. This has created a new situation: the small YSTs used for monitoring seem to catch fewer whiteflies nowadays, as the large yellow glue bands hung above every row of the crop seemingly compete with the small traps by attracting more whiteflies.

At the moment, there are no fixed thresholds and decision-making is based on the experience of growers and advisors, either in counting/observing adults from leaves in plant tops or monitoring whiteflies with YST or a combination of these approaches. Decision-making based on experience can work well but seems to be insufficient in years when whitefly pressure is high due to outdoor weather conditions that promote whitefly reproduction in wild plants near greenhouses; subsequently, the pest pressure in the autumn from outdoors is high when winter crops of tomato and cucumber are planted in the greenhouses and attract whiteflies from outdoors when the weather gets colder. The large yellow glue bands can also influence the spatial distribution of whiteflies in the crops. No formal studies on whitefly distribution in these cropping systems have been made so far but are now being conducted.

3.5 New approaches to whitefly monitoring

Spotting and locating whitefly hot spots is one goal of monitoring. This can be achieved using information on spatial distribution of whiteflies to cover the greenhouse area with appropriate placement of YSTs, as described above, or by plant sampling plans. Plant sampling is not a common procedure in greenhouse tomato and cucumber except for observing adults in tops of plants when working on the plants daily. Growers develop nominal thresholds based on experience of seeing adults on the top leaves and combine this information with YST counts, if the latter are used. Formal sampling plans for adult and immature counts have been developed for outdoor tomato crops. Sequential sampling plans with fixed precision for *B. tabaci* immatures were shown to reduce the number of needed leaf samples by 60–70% compared to conventional sampling plans based on fixed sample sizes (Gusmão et al., 2006, 2005) – a considerable reduction in time and labour for sampling. Better coverage of microclimate monitoring would help in recognizing risk zones for whitefly hot spots, as shown below.

For very large greenhouses where even YST monitoring is impractical, robotic platforms based on optochemical techniques are already available (https://www.ecoation.com). They scan plant rows for pests and fruit ripeness automatically but take in also human observations in digital form. The detection can be based on multispectral or hyperspectral imaging, but also on detection of volatile organic compounds emitted by infested plants – no information is available on this. A detection process that covers all plant rows makes the need for knowledge on spatial distribution of the pest obsolete and helps detect hotspots. The platform's AI system is so designed that it can correlate pest pressure and microclimatic conditions (temperature, humidity, light intensity, air currents) and therefore can help identify zones of high pest risk. The system also does future projections to forecast pest pressures within the coming week which helps to order the right amounts of biocontrol agents. Sophisticated AI-based statistical algorithms based on data that is accumulated continuously are doing the job that is too complex for humans alone (Ecoation, n.d.).

3.6 Simulation of whitefly population dynamics

Many mathematical simulation models have been created to map the responses of whiteflies to temperature. They include both linear and non-linear phenological models, with or without stochasticity, for development times or rates based on temperature (and host plant) (Chandi et al., 2021; Drost et al., 1998; Gamarra et al., 2020; Muñiz and Nombela, 2001; Nava-Camberos et al., 2001; Wang and Tsai, 1996). In most models air temperature is used, but leaf temperature has been shown to explain development time better (Park et al., 2011a). Life-table and population dynamic models for whiteflies include differential equations, matrix models, dimension-changeable matrix models and box car train models (Giessen et al., 1995; Hulspas-Jordaan and van Lenteren, 1989; van Roermund and van Lenteren, 1992; Yeow and Becker, 2018).

The interaction of whiteflies with their important natural enemy *Encarsia formosa* has been modelled by several researchers (e.g. Hulspas-Jordaan and van Lenteren, 1989; Yano et al., 1989). Rodríguez (2016) also included the whiteflies' interaction with powdery mildew that often occurs on tomato leaves together with whiteflies and can influence both the pest and its natural enemies. Rincon et al. (2015) produced an algorithm to simulate the effect of within-plant heterogeneity on whitefly-predator dynamics based on explicit prey spatial distributions. Such models can be used to scale-up functional responses of natural enemies. Giessen et al. (1995) explored the effects of antibiotic resistance of tomato plants on whitefly population development with a deterministic model. Moerkens et al. (2020) used simple statistical modelling based on whitefly and *Macrolophus* counts in YSTs to predict the success level of biocontrol. With the development of wireless sensors or moving robotic AI-platforms that measure microclimate on the go and accumulate data on pest densities at the same time, the predictive use of forecasting

Published by Burleigh Dodds Science Publishing Limited, 2022.

models is being integrated into whitefly management in decision support systems.

4 Conclusion

Pest monitoring in horticulture is developing technologically to reduce time and labour needed for sampling and to produce more accurate pest predictions. New ways of detecting pests based on selective e-traps, e-noses, cameras and acoustic signatures are already in use or emerging. Remote sensing of pests requires the development of new economic injury levels and ETs. The relative importance of mechanistic and statistical models is changing due to AI-technologies and Big Data. The use of Big Data will force researchers to collect, use and value data differently than before. The incorporation of natural enemies in ETs will take place gradually and require researchers to acquire modelling skills. Research for advancing monitoring and forecasting also must include the socioeconomic factors that determine whether new technologies will be implemented by farmers. Developing trustworthy sampling plans and forecasting models, and validating and implementing them in collaboration with stakeholders, remains important.

5 Future trends in research

The new, high-technological approaches to data collection and management appear very prominently in the research literature of pest management nowadays. The feasibility of these new technologies in annual and perennial horticultural crops of different sizes is an issue that needs to be considered as one of the research targets. Remote-sensing techniques require new ETs to be developed. Criteria can be developed in advance concerning the feasibility of new solutions for specific purposes as exemplified by the study of Cock et al. (2016) evaluating the usefulness of remote sensing for monitoring key agricultural pests. Feasibility evaluations should not, however, restrict empirical research too much, as practice often produces serendipitous new information and seeds for innovations that cannot be produced only by theory.

Remote sensing can indicate the occurrence of a pest in the crop, without having to pay attention to the pest's spatial distribution when collecting data. But can remote sensors tell also on the abundance of all pests or the abundance of natural enemies that influence the future trajectory of the pest population? Hardly. Phenological models will use increasingly site-specific data and become very accurate, but their output does not tell whether the pest really occurs at a specific site or how many pests there will be. Possibilities of using mobile phones for implementing accurate enough sequential sampling plans that are also user-friendly and not costly to use deserve research investments. The uncertainty of information is an issue that influences growers' willingness to follow ET recommendations. Does their willingness to rely on fixed ETs change if they obtain reliable evidence on how certain the ETs are or should ETs see as learning tools that help growers to adjust their behaviour regarding intervention decisions? Are ETs that take into account the level of risk aversion of growers more readily accepted than fixed one-option ETs? If ETs are probabilistic, does this increase or decrease growers' willingness to apply them? How to address the gap between the theory and practice of sampling and monitoring? Sampling plans are worth nothing if they serve only the CV of researchers and publication lists. Fitting them into the agenda and activities of stakeholders is the ultimate goal and must be included in research on sampling, monitoring and forecasting.

While digitalization is gaining a foothold in horticulture, the importance of biology, ecology and physiology of organisms is actually being accentuated with the advancement of new technologies. The Big Data collected must represent relevant life stages and events in the pests' life cycles. A similar requirement comes from the inclusion of natural enemies in ETs. Biological research is needed also to couple the knowledge concerning the pest, its natural enemies and the crop plant with each other for modelling purposes and for the timing of control actions such as push–pull technologies, use of trap crops or spatial and temporal performance of biocontrol agents. Sampling methods and decision thresholds for natural enemies will gradually become more important research targets. Studies on tangible quantified impacts of natural enemies associated with ETs are still scarce and must be incorporated in bioeconomic models. Improvements of phenology models of both pests and natural enemies are still needed to make the outputs more precise and to improve region- or site-specificity of outputs. Climate change is also contributing to this research need, as updated data are required on pest and natural enemy biology and interactions (Collier et al., 2020).

Sharing pest observations among growers and researchers with the help of current IT solutions should be encouraged by research to show its benefits for collecting Big Data and for being able to adjust sampling and management decisions with changing contexts of farming. Field-level precision farming decisions can then be combined with knowledge obtained from wider spatial and temporal scales that, depending on the pest species, can influence local pest conditions and decision-making. Using Big data successfully requires collaboration between theory (in the natural sciences) and practice (farmers, IPM researchers, advisors, AI specialists) so that we understand the type of data that is valuable and what are not.

Research on the application of automated identification and more selective trapping of insects will continue. New sensory domains (olfaction, acoustics) are already included in research agendas concerning insect and mite responses to their environment and the application of sensors to identify insects. For example, could wingbeat frequencies be used to distinguish species of *Delia* when they enter semi-selective traps? No information exists as yet. It seems that studies on the chemical ecology of several pest species are being revived now that research methods have become more sophisticated, our understanding of insect olfaction has increased and successful examples of kairomone traps are available as examples. New approaches must be compared with existing sampling methods in terms of efficacy and cost, to choose the best solution from the practical point of view.

6 Where to look for further information

6.1 Further reading

- A concise overview of sampling principles and practices in lecture form: Barbour, J. Sampling Insect Populations for Pest Management: http://pnwpestalert.net/uploads/meetings/BarbourIPMsampling.pdf.
- Binns, M. R., Nyrop, J. P., van der Werf, W. and Wopke, W. 2000. *Sampling and Monitoring in Crop Protection: The Theoretical Basis for Developing Practical Decision Guides*. CABI.
- Higley, L. G. and Pedigo, L. P. 1996. *Economic Thresholds for Integrated Pest Management*. University of Nebraska Press.
- Koul, O. and Cuperus, G. W. (Eds.). 2007. *Ecologically Based Integrated Pest Management*. CABI.
- Pedigo, L. P. and Buntin, G. D. 1993. *Handbook of Sampling Methods for Arthropods in Agriculture*. CRC Press.
- Remote sensing for insect pests – reviews and case studies (Potamitis et al., 2017; Roosjen et al., 2020; Suckling et al., 2020; Zhang et al., 2019).
- Review on the role of kairomones in IPM: (Murali-Baskaran et al., 2018).
- A review of sampling and monitoring methods for beneficial arthropods in agroecosystems (McCravy, 2018).
- Local, regional and global performance of models and effects on resource use of research: how to direct efforts and resources, also in the context of current IPM programs and monitoring and forecasting as one of its principles (Berlin et al., 2018).
- Big Data and IPM: Data types and collection (Zaza et al., 2018); Data processing (Pratheepa and Antony, 2018); Data valuation (Demirel and Kumral, 2021; Weersink et al., 2018).
- Integrated Pest Management Pheromones Market Size, Share & Trends Analysis Report (2020–2027): https://www.grandviewresearch.com/industry-analysis/ipm-pheromones.

6.2 Examples of companies offering tools for automated monitoring and identification of pests and platforms for documentation of scouting and sensor data

- https://metos.at/iscout/.
- https://www.trapview.com/v2/en/.
- https://www.ecoation.com/.
- https://arisbv.nl/en/vision-for-phenotyping/ornamental-crops-2/phenotyping-products/cirillo.
- https://www.botany.nl/en/cirillo.
- https://www.koppert.com/natutec-scout/.
- https://www.letsgrow.com.
- https://www.30mhz.com/products/platform/.

7 References

Abd El-Ghany, N. M., Abd El-Aziz, S. E. and Marei, S. S. 2020. A review: Application of remote sensing as a promising strategy for insect pests and diseases management. *Environ. Sci. Pollut. Res. Int.* 27(27), 33503–33515. https://doi.org/10.1007/s11356-020-09517-2.

Adams, C. G. 2017. Maximizing information yield from pheromone-baited monitoring traps: Estimating plume reach, trapping radius, and absolute density of Cydia pomonella (Lepidoptera: Tortricidae) in Michigan apple. *J. Econ. Entomol.* 110(2), 305–318.

Agnello, A. M., Kovach, J., Nyrop, J. P., Reissig, W. H., Breth, D. I. and Wilcox, W. F. 1994. Extension and evaluation of a simplified monitoring program in New York Apples. *Am. Entomol.* 40(1), 37–49. https://doi.org/10.1093/ae/40.1.37.

Alatawi, F. J., Opit, G. P., Margolies, D. C. and Nechols, J. R. 2005. Within-plant distribution of twospotted spider mites (Acari: Tetranychidae) on impatiens: Development of a presence–absence sampling plan. *J. Econ. Entomol.* 98(3), 1040–1047. https://doi.org/10.1603/0022-0493-98.3.1040.

Amarasekare, K. G. and Shearer, P. W. 2013. Laboratory bioassays to estimate the lethal and sublethal effects of various insecticides and fungicides on Deraeocoris brevis (Hemiptera: Miridae). *J. Econ. Entomol.* 106(2), 776–785. https://doi.org/10.1603/EC12432.

Anonymous 2020. Apple IPM sampling charts. Cornell University. Available at: https://nysipm.cornell.edu/sites/nysipm.cornell.edu/files/shared/documents/apple-IPM-sampling-charts.pdf.

Anonymous 2021. Botany en Aris richten joint venture PhenoVerius op voor insecttellingen op planten. *Groenten Nieuws.*

Baker, R. E., Peña, J. M., Jayamohan, J. and Jérusalem, A. 2018. Mechanistic models versus machine learning, a fight worth fighting for the biological community? *Biol. Lett.* 14(5), 29769297. https://doi.org/10.1098/rsbl.2017.0660.

Bannerman, J. A., Costamagna, A. C., McCornack, B. P. and Ragsdale, D. W. 2015. Comparison of relative bias, precision, and efficiency of sampling methods for natural enemies of soybean aphid (Hemiptera: Aphididae). *J. Econ. Entomol.* 108(3), 1381–1397. https://doi.org/10.1093/jee/tov009.

Barbedo, J. G. A. 2019. A review on the use of unmanned aerial vehicles and imaging sensors for monitoring and assessing plant stresses. *Drones*. 3(2), 40. https://doi.org/10.3390/drones3020040.

Baroffio, C. A., Sigsgaard, L., Ahrenfeldt, E. J., Borg-Karlson, A.-K., Bruun, S. A., Cross, J. V., Fountain, M. T., Hall, D., Mozuraitis, R., Ralle, B., Trandem, N. and Wibe, A. 2018. Combining plant volatiles and pheromones to catch two insect pests in the same trap: Examples from two berry crops. *Crop Prot.* 109, 1–8.

Barros, P. P. S., Schutze, I. X., Iost Filho, F. H., Yamamoto, P. T., Fiorio, P. R. and Demattê, J. A. M. 2021. Monitoring Bemisia tabaci (Gennadius) (Hemiptera: Aleyrodidae) infestation in soybean by proximal sensing. *Insects.* 12(1), 47. https://doi.org/10.3390/insects12010047.

Bažok, R., Ceranić-Sertić, M., Barčić, J. I., Borošić, J., Kozina, A., Kos, T., Lemić, D. and Čačija, M. 2012. Seasonal flight, optimal timing and efficacy of selected insecticides for cabbage maggot (Delia radicum L., Diptera: Anthomyiidae) control. *Insects.* 3(4), 1001–1027. https://doi.org/10.3390/insects3041001.

Bengtsson, M., Jaastad, G., Knudsen, G., Kobro, S., Bäckman, A.-C., Pettersson, E. and Witzgall, P. 2006. Plant volatiles mediate attraction to host and non-host plant in apple fruit moth, Argyresthia conjugella. *Entomol. Exp. Appl.* 118(1), 77–85. https://doi.org/10.1111/j.1570-7458.2006.00359.x.

Benjamin, E. O. and Wesseler, J. H. H. 2016. Social science perspectives on the bio-economy. A socioeconomic analysis of biocontrol in integrated pest management: A review of the effects of uncertainty, irreversibility and flexibility. *NJAS Wagening. J. Life. Sci.* 77, 53–60. https://doi.org/10.1016/j.njas.2016.03.002.

Berlin, A., Källström, H. N., Lindgren, A. and Olson, Å. 2018. Scientific evidence for sustainable plant disease protection strategies for the main arable crops in Sweden. A systematic map protocol. *Environ. Evid.* 7(1), 31. https://doi.org/10.1186/s13750-018-0141-3.

Binns, M. R. and Nyrop, J. P. 1992. Sampling insect populations for the purpose of IPM decision making. *Annu. Rev. Entomol.* 37(1), 427–453. https://doi.org/10.1146/annurev.en.37.010192.002235.

Binns, M. R., Nyrop, J. P. and Van Der Werf, W. 1996. Monitoring pest abundance by cascading density classification. *Am. Entomol.* 42(2), 113–121. https://doi.org/10.1093/ae/42.2.113.

Binns, M. R., Nyrop, J. P., van der Werf, W. and Werf, W. 2000. *Sampling and Monitoring in Crop Protection: The Theoretical Basis for Developing Practical Decision Guides*. Wallingford, UK; New York: CABI Publishing.

Biron, D. G., Coderre, D., Boivin, G., Brunel, E. and Nénon, J. P. 2002. Genetic variability and expression of phenological and morphological differences in populations of Delia radicum (Diptera: Anthomyiidae). *Can. Entomol.* 134(3), 311–327. https://doi.org/10.4039/Ent134311-3.

Bleeker, P. M., Diergaarde, P. J., Ament, K., Schütz, S., Johne, B., Dijkink, J., Hiemstra, H., de Gelder, R., de Both, M. T. J., Sabelis, M. W., Haring, M. A. and Schuurink, R. C. 2011. Tomato-produced 7-epizingiberene and R-curcumene act as repellents to whiteflies. *Phytochemistry* 72(1), 68–73. https://doi.org/10.1016/j.phytochem.2010.10.014.

Bleeker, P. M., Mirabella, R., Diergaarde, P. J., VanDoorn, A., Tissier, A., Kant, M. R., Prins, M., de Vos, M., Haring, M. A. and Schuurink, R. C. 2012. Improved herbivore resistance in cultivated tomato with the sesquiterpene biosynthetic pathway from a wild relative.

Proc. Natl. Acad. Sci. USA 109(49), 20124–20129. https://doi.org/10.1073/pnas.1208756109.

Bligaard, J. 1999. Damage thresholds for cabbage root fly [*Delia radicum* (L.)] in cauliflower assessed from pot experiments. *Acta Agric. Scand. B* 49(1), 57–64. https://doi.org/10.1080/09064719950135713.

Blua, M. J. and Toscano, N. C. 1994. Bemisia argentifolii (Homoptera: Aleyrodidae) development and honeydew production as a function of cotton nitrogen status. *Environ. Entomol.* 23(2), 316–321. https://doi.org/10.1093/ee/23.2.316.

Böckmann, E., Hommes, M. and Meyhöfer, R. 2014. Yellow traps reloaded: What is the benefit for decision making in practice? *J. Pest. Sci.* 88, 439–449. https://doi.org/10.1007/s10340-014-0601-7.

Böckmann, E. and Meyhöfer, R. 2017. Sticky trap monitoring of a pest–predator system in glasshouse tomato crops: Are available trap colours sufficient? *J. Appl. Entomol.* 141(5), 339–351. https://doi.org/10.1111/jen.12338.

Bradshaw, C. D., Hemming, D., Baker, R., Everatt, M., Eyre, D. and Korycinska, A. 2019. A novel approach for exploring climatic factors limiting current pest distributions: A case study of Bemisia tabaci in north-west Europe and assessment of potential future establishment in the United Kingdom under climate change. *PLoS ONE.* 14(8), e0221057. https://doi.org/10.1371/journal.pone.0221057.

Buntin, D. G., Gilbertz, D. A. and Oetting, R. D. 1993. Chlorophyll loss and gas exchange in tomato leaves after feeding injury by Bemisia tabaci (Homoptera: Aleyrodidae). *J. Econ. Entomol.* 86(2), 517–522. https://doi.org/10.1093/jee/86.2.517.

Cai, C., Xu, Y. and Cui, X. 2016. Physiological characteristics of different resistant tomato varieties in response to B-type Bemisia tabaci stress. *Chin. Agric. Sci.* 49, 2524–2533.

Cardim Ferreira Lima, M., Damascena de Almeida Leandro, M. E., Valero, C., Pereira Coronel, L. C. and Gonçalves Bazzo, C. O. 2020. Automatic detection and monitoring of insect pests—A review. *Agriculture* 10(5), 161. https://doi.org/10.3390/agriculture10050161.

Castle, S. and Naranjo, S. E. 2009. Sampling plans, selective insecticides and sustainability: The case for IPM as 'informed pest management'. *Pest Manag. Sci.* 65(12), 1321–1328. https://doi.org/10.1002/ps.1857.

Cha, D. H., Hesler, S. P., Wallingford, A. K., Zaman, F., Jentsch, P., Nyrop, J. and Loeb, G. M. 2018. Comparison of commercial lures and food baits for early detection of fruit infestation risk by Drosophila suzukii (Diptera: Drosophilidae). *J. Econ. Entomol.* 111(2), 645–652. https://doi.org/10.1093/jee/tox369.

Chandi, R. S., Kataria, S. K. and Fand, B. B. 2021. Effect of temperature on biological parameters of cotton whitefly, Bemisia tabaci (Gennadius) (Hemiptera: Aleyrodidae). *Int. J. Trop. Insect Sci.* 41(2), 1823–1833. https://doi.org/10.1007/s42690-020-00397-0.

Chen, M. S., Wheeler, S., Davis, H., Jeff Whitworth, R. J., Knutson, A., Giles, K. L., Royer, T. A. and Skinner, M. 2014. Molecular markers for species identification of Hessian fly males caught on sticky pheromone traps. *J. Econ. Entomol.* 107(3), 1110–1117. https://doi.org/10.1603/EC13384.

Chen, Y., Cui, X., Cai, C. and Cao, F. 2011. Effect of Bemisia tabaci (Gennadius) B-biotype on photosynthetic characteristics and activity of defense enzyme in different varieties of tomato. *Sci. Agric. Sin.* 44, 3547–3556.

Chow, Y. S., Lin, Y. M. and Hsu, C. L. 1977. Sex pheromone of the diamondback moth (Lepidoptera: Plutellidae). *Bull. Inst. Zool. Acad. Sin.* 16, 99–105.

Chrysargyris, A., Charalambous, S., Xylia, P., Litskas, V., Stavrinides, M. and Tzortzakis, N. 2020. Assessing the biostimulant effects of a novel plant-based formulation on tomato crop. *Sustainability*. 12(20), 8432. https://doi.org/10.3390/su12208432.

Chua, T. H. and Lim, B. H. 1979. Distribution pattern of diamondback moth, *plutella xylostella* (L.) (Lepidoptera: Plutellidae) on choy-sum plants. *Z. Angew. Entomol.* 88(1-5), 170-175. https://doi.org/10.1111/j.1439-0418.1979.tb02492.x.

Cloonan, K. R., Abraham, J., Angeli, S., Syed, Z. and Rodriguez-Saona, C. 2018. Advances in the chemical ecology of the spotted wing Drosophila (Drosophila suzukii) and its applications. *J. Chem. Ecol.* 44(10), 922-939. https://doi.org/10.1007/s10886-018-1000-y.

Cloonan, K. R., Hernández-Cumplido, J., De Sousa, A. L. V., Ramalho, D. G., Burrack, H. J., Della Rosa, L., Diepenbrock, L. M., Ballman, E., Drummond, F. A., Gut, L. J., Hesler, S., Isaacs, R., Leach, H., Loeb, G. M., Nielsen, A. L., Nitzsche, P., Park, K. R., Syed, Z., Van Timmeren, S., Wallingford, A. K., Walton, V. M. and Rodriguez-Saona, C. 2019. Laboratory and field evaluation of host-related foraging odor-cue combinations to attract Drosophila suzukii (Diptera: Drosophilidae). *J. Econ. Entomol.* 112(6), 2850-2860. https://doi.org/10.1093/jee/toz224.

Cock, M. J. W., Rui, T., Zhi, L., HuanHuan, W., McGillivray, L. A., Thomas, S. E., Cameron, K. H. and Feng, Z. 2016. The main agricultural insect and disease pests of China and implications for the use of remote sensing for their management. *CAB Rev*. 11, 14.

Collier, R., Mazzi, D., Folkedal Schjøll, A., Schorpp, Q., Thöming, G., Johansen, T. J., Meadow, R., Meyling, N. V., Cortesero, A. M., Vogler, U., Gaffney, M. T. and Hommes, M. 2020. The potential for decision support tools to improve the management of root-feeding fly pests of vegetables in western Europe. *Insects* 11(6), 369. https://doi.org/10.3390/insects11060369.

Collier, R. H., Finch, S., Phelps, K. and Thompson, A. R. 1991. Possible impact of global warming on cabbage root fly (Delia radicum) activity in the UK. *Ann. Appl. Biol.* 118(2), 261-271. https://doi.org/10.1111/j.1744-7348.1991.tb05627.x.

Costa, H. S., Toscano, N. C., Hendrix, D. L. and Henneberry, T. J. 1999. Patterns of honeydew droplet production by nymphal stages of Bemisia argentifolii (Homoptera: Aleyrodidae) and relative composition of honeydew sugars. *J. Entomol. Sci.* 34(3), 305-313. https://doi.org/10.18474/0749-8004-34.3.305.

Cross, J. V., Hall, D. R., Innocenzi, P. J., Hesketh, H., N Jay, C. and Burgess, C. M. 2006a. Exploiting the aggregation pheromone of strawberry blossom weevil Anthonomus rubi (Coleoptera: Curculionidae): Part 2. Pest monitoring and control. *Crop Prot.* 25(2), 155-166. https://doi.org/10.1016/j.cropro.2005.04.004.

Cross, J. V., Hesketh, H., Jay, C. N., Hall, D. R., Innocenzi, P. J., Farman, D. I. and Burgess, C. M. 2006b. Exploiting the aggregation pheromone of strawberry blossom weevil Anthonomus rubi Herbst (Coleoptera: Curculionidae): Part 1. Development of lure and trap. *Crop Prot.* 25(2), 144-154. https://doi.org/10.1016/j.cropro.2005.04.002.

Cui, S., Ling, P., Zhu, H. and Keener, H. M. 2018. Plant pest detection using an artificial nose system: A review. *Sensors (Basel)*. 18(2), 378. https://doi.org/10.3390/s18020378.

Dai, J., Deng, J. and Du, J. 2008. Development of bisexual attractants for diamondback moth, *Plutella xylostella* (Lepidoptera: Plutellidae) based on sex pheromone and host volatiles. *Appl. Entomol. Zool.* 43(4), 631-638. https://doi.org/10.1303/aez.2008.631.

Dalton, D. T., Walton, V. M., Shearer, P. W., Walsh, D. B., Caprile, J. and Isaacs, R. 2011. Laboratory survival of Drosophila suzukii under simulated winter conditions of the

Pacific Northwest and seasonal field trapping in five primary regions of small and stone fruit production in the United States. *Pest. Manag. Sci.* 67(11), 1368-1374. https://doi.org/10.1002/ps.2280.

Damos, P. 2014. Stochastic modeling of economic injury levels with respect to yearly trends in price commodity. *J. Insect Sci.* 14, 59. https://doi.org/10.1093/jis/14.1.59.

Damos, P. and Soulopoulou, P. 2019. Stage-specific probabilistic phenology model of Cydia pomonella (Lepidoptera: Tortricidae) using laboratory maximum likelihood parameter estimates. *J. Appl. Entomol.* 143(3), 250–261. https://doi.org/10.1111/jen.12589.

Darshanee, H. L. C., Ren, H., Ahmed, N., Zhang, Z. F., Liu, Y. H. and Liu, T. X. 2017. Volatile-mediated attraction of greenhouse whitefly Trialeurodes vaporariorum to tomato and eggplant. *Front. Plant Sci.* 8, 1285. https://doi.org/10.3389/fpls.2017.01285.

Degen, T., Buser, H.-R. and Städler, E. 1999. Patterns of oviposition stimulants for carrot fly in leaves of various host plants. *J. Chem. Ecol.* 25(1), 67–87. https://doi.org/10.1023/A:1020837116467.

Demirel, M. and Kumral, N. A. 2021. Artificial intelligence in integrated pest management. In: *Artificial Intelligence and IoT-Based Technologies for Sustainable Farming and Smart Agriculture*. Hershey, PA: IGI Global, 289–313.

Donatelli, M., Magarey, R. D., Bregaglio, S., Willocquet, L., Whish, J. P. M. and Savary, S. 2017. Modelling the impacts of pests and diseases on agricultural systems. *Agric. Syst.* 155, 213–224. https://doi.org/10.1016/j.agsy.2017.01.019.

Dreves, A. J. 2006. Phenology and monitoring of the cabbage maggot, *Delia radicum* (L.), in brassica root crops. PhD Thesis. Oregon State University, Corvallis, OR.

Drost, Y. C., van Lenteren, J. C. and van Roermund, H. J. W. 1998. Life-history parameters of different biotypes of Bemisia tabaci (Hemiptera: Aleyrodidae) in relation to temperature and host plant: A selective review. *Bull. Entomol. Res.* 88(3), 219–230. https://doi.org/10.1017/S0007485300025840.

Ecoation n.d. *Integrated Pest Management. How the Ecoation OKO Platform Uses Sensors and AI to Optimize IPM Programs and Reduce Treatment Cost in Greenhouses*. Available at: https://www.ecoation.com/_files/ugd/0a894f_a83293fa199c4f60a71254187a0b7f4c.pdf.

Edland, T. 1974. The rowan-berry moth (Argyresthia conjugella Zell.). Prognosis of attack on apple. Methods and results. A preliminary report. *Garneryrket.* 64, 524–532.

Estrada-Hernández, M. G., Valenzuela-Soto, J. H., Ibarra-Laclette, E. and Délano-Frier, J. P. 2009. Differential gene expression in whitefly Bemisia tabaci-infested tomato (*Solanum lycopersicum*) plants at progressing developmental stages of the insect's life cycle. *Physiol. Plant.* 137(1), 44–60. https://doi.org/10.1111/j.1399-3054.2009.01260.x.

Evans, K. J., Terhorst, A. and Kang, B. H. 2017. From data to decisions: Helping crop producers build their actionable knowledge. *Crit. Rev. Plant Sci.* 36(2), 71–88. https://doi.org/10.1080/07352689.2017.1336047.

Evenden, M. L. and Gries, R. 2010. Assessment of commercially available pheromone lures for monitoring diamondback moth (Lepidoptera: Plutellidae) in canola. *J. Econ. Entomol.* 103(3), 654–661. https://doi.org/10.1603/EC09339.

Fenu, G. and Malloci, F. M. 2021. Forecasting plant and crop disease: An explorative study on current algorithms. *Big Data Cogn. Comput.* 5(1), 2. https://doi.org/10.3390/bdcc5010002.

Finch, S., Collier, R. H. and Phelps, K. 1996. A review of work done to forecast pest insect attacks in UK horticultural crops. *Crop. Prot.* 15(4), 353–357. https://doi.org/10.1016/0261-2194(95)00135-2.

Finch, S. and Skinner, G. 1982. Trapping cabbage root flies in traps baited with plant extracts and with natural and synthetic isothiocyanates. *Entomol. Exp. Appl.* 31(2–3), 133–139. https://doi.org/10.1111/j.1570-7458.1982.tb03124.x.

Finch, S., Skinner, G. and Freeman, G. H. 1975. The distribution and analysis of cabbage root fly egg populations. *Ann. Appl. Biol.* 79(1), 1–18. https://doi.org/10.1111/j.1744-7348.1975.tb01516.x.

Finch, S., Skinner, G. and Freeman, G. H. 1978. Distribution and analysis of cabbage root fly pupal populations. *Ann. Appl. Biol.* 88(3), 351–356. https://doi.org/10.1111/j.1744-7348.1978.tb00724.x.

Fountain, M. T., Baroffio, C., Borg-Karlson, A.-K., Brain, P., Cross, J. V., Farman, D. I., Hall, D. R., Ralle, B., Rendina, P., Richoz, P., Sigsgaard, L., Storberget, S., Trandem, N. and Wibe, A. 2017. Design and deployment of semiochemical traps for capturing Anthonomus rubi Herbst (Coleoptera: Curculionidae) and Lygus rugulipennis Poppius (Hetereoptera: Miridae) in soft fruit crops. *Crop. Prot.* 99, 1–9. https://doi.org/10.1016/j.cropro.2017.05.001.

Franklin, J. 2010. *Mapping Species Distributions: Spatial Inference and Prediction*. Cambridge: Cambridge University Press.

Frewin, A. J., Renkema, J., Fraser, H. and Hallett, R. H. 2017. Evaluation of attractants for monitoring Drosophila suzukii (Diptera: Drosophilidae). *J. Econ. Entomol.* 110(3), 1156–1163. https://doi.org/10.1093/jee/tox081.

Frey, J. E. and Frey, B. 1995. Molecular identification of six species of scale insects (Quadraspidiotus sp.) by RAPD-PCR: Assessing the field-specificity of pheromone traps. *Mol. Ecol.* 4(6), 777–780. https://doi.org/10.1111/j.1365-294X.1995.tb00279.x.

Gamarra, H., Sporleder, M., Carhuapoma, P., Kroschel, J. and Kreuze, J. 2020. A temperature-dependent phenology model for the greenhouse whitefly Trialeurodes vaporariorum (Hemiptera: Aleyrodidae). *Virus. Res.* 289, 198107. https://doi.org/10.1016/j.virusres.2020.198107.

Gent, D. H., De Wolf, E. and Pethybridge, S. J. 2011. Perceptions of risk, risk aversion, and barriers to adoption of decision support systems and integrated pest management: An introduction. *Phytopathology* 101(6), 640–643. https://doi.org/10.1094/PHYTO-04-10-0124.

Giessen, W. A. V., Mollema, C. and Elsey, K. D. 1995. Design and use of a simulation model to evaluate germplasm for antibiotic resistance to the greenhouse whitefly (Trialeurodes vaporariorum) and the sweetpotato whitefly (Bemisia tabaci). *Entomol. Exp. Appl.* 76(3), 271–286. https://doi.org/10.1111/j.1570-7458.1995.tb01971.x.

Giles, K. L., McCornack, B. P., Royer, T. A. and Elliott, N. C. 2017. Incorporating biological control into IPM decision making. *Curr. Opin. Insect. Sci.* 20, 84–89. https://doi.org/10.1016/j.cois.2017.03.009.

Giraldo, J. P., Wu, H., Newkirk, G. M. and Kruss, S. 2019. Nanobiotechnology approaches for engineering smart plant sensors. *Nat. Nanotechnol.* 14(6), 541–553. https://doi.org/10.1038/s41565-019-0470-6.

Glenn, D. M. and Tabb, A. 2019. Evaluation of five methods to measure normalized difference vegetation index (NDVI) in apple and citrus. *Int. J. Fruit. Sci.* 19(2), 191–210. https://doi.org/10.1080/15538362.2018.1502720.

Gogoi, N. K., Deka, B. and Bora, L. C. 2018. Remote sensing and its use in detection and monitoring plant diseases: A review. *Agric. Rev.* 39, 307–313.

Gouinguené, S. P. D., Poiger, T. and Städler, E. 2006. Eggs of cabbage root fly stimulate conspecific oviposition: Evaluation of the activity and determination of an egg-associated compound. *Chemoecology* 16(2), 107–113. https://doi.org/10.1007/s00049-006-0335-y.

Green, R. H. 1970. On fixed precision level sequential sampling. *Res. Popul. Ecol.* 12(2), 249–251. https://doi.org/10.1007/BF02511568.

Groher, T. 2019. *Suitability of non-destructive sensors for monitoring physiological and biochemical responses of tomato leaves and fruits to abiotic stresses*. PhD Dissertation. Rheinische Friedrich-Wilhelms-Universität, Bonn, DE.

Guarnieri, A., Maini, S., Molari, G. and Rondelli, V. 2011. Automatic trap for moth detection in integrated pest management. *Bull. Insectol.* 64, 247–251.

Guerin, P. M., Städler, E. and Buser, H. R. 1983. Identification of host plant attractants for the carrot fly, *Psila rosae*. *J. Chem. Ecol.* 9(7), 843–861. https://doi.org/10.1007/BF00987809.

Guerin, P. M. and Visser, J. H. 1980. Electroantennogram responses of the carrot fly, Psila rosae, to volatile plant components. *Physiol. Entomol.* 5(2), 111–119.

Guimapi, R. Y. A., Mohamed, S. A., Okeyo, G. O., Ndjomatchoua, F. T., Ekesi, S. and Tonnang, H. E. Z. 2016. Modeling the risk of invasion and spread of *Tuta absoluta* in Africa. *Ecol. Complexity* 28, 77-93.

Gupta, P. D. and Thorsteinson, A. J. 1960. Food plant relationships of the diamond-back moth (plutella maculipennis (Curt.)). I. Gustation and olfaction in relation to botanical specificity of the larvae. *Entomol. Exp. Appl.* 3(3), 241–250.

Gusmão, M. R., Picanço, M. C., Guedes, R. N. C., Galvan, T. L. and Pereira, E. J. G. 2006. Economic injury level and sequential sampling plan for Bemisia tabaci in outdoor tomato. *J. Appl. Entomol.* 130(3), 160–166. https://doi.org/10.1111/j.1439-0418.2005.01032.x.

Gusmão, M. R., Picanço, M. C., Zanuncio, J. C., Silva, D. J. H. and Barrigossi, J. A. F. 2005. Standardised sampling plan for Bemisia tabaci (Homoptera: Aleyrodidae) in outdoor tomatoes. *Sci. Hortic.* 103(4), 403–412. https://doi.org/10.1016/j.scienta.2004.04.005.

Halada, P., Hlavackova, K., Risueño, J., Berriatua, E., Volf, P. and Dvorak, V. 2018. Effect of trapping method on species identification of phlebotomine sandflies by MALDI-TOF MS protein profiling. *Med. Vet. Entomol.* 32(3), 388–392. https://doi.org/10.1111/mve.12305.

Hamby, K. A., E. Bellamy, D., Chiu, J. C., Lee, J. C., Walton, V. M., Wiman, N. G., York, R. M. and Biondi, A. 2016. Biotic and abiotic factors impacting development, behavior, phenology, and reproductive biology of Drosophila suzukii. *J. Pest Sci.* 89(3), 605–619. https://doi.org/10.1007/s10340-016-0756-5.

Hamilton, A. J., Endersby, N. M., Schellhorn, N. A., Ridland, P. M., Rogers, P. M., Jevremov, D. and Baker, G. 2006. Evaluation of fixed sample-size plans for *Plutella xylostella* (Lepidoptera: Plutellidae) on broccoli crops in Australia. *J. Econ. Entomol.* 99(6), 2171–2176.

Hamilton, A. J., Schellhorn, N. A., Endersby, N. M., Ridland, P. M. and Ward, S. A. 2004. A dynamic binomial sequential sampling plan for *Plutella xylostella* (Lepidoptera: Plutellidae) on Broccoli and cauliflower in Australia. *J. Econ. Entomol.* 97(1), 127–135. https://doi.org/10.1093/jee/97.1.127.

Hamilton, A. J., Waters, E. K., Kim, H. J., Pak, W. S. and Furlong, M. J. 2009a. Sampling a weighted pest complex: Caterpillars in North Korean cabbage (Brassica oleracea var. capitata) crops. *Entomol. Exp. Appl.* 130(3), 282–289. https://doi.org/10.1111/j.1570-7458.2008.00815.x.

Hamilton, A. J., Waters, E. K., Kim, H. J., Pak, W. S. and Furlong, M. J. 2009b. Validation of fixed sample size plans for monitoring lepidopteran pests of Brassica oleracea crops in North Korea. *J. Econ. Entomol.* 102(3), 1336–1343. https://doi.org/10.1603/029.102.0361.

Harmon, D. S., Haseeb, M., Kanga, L. H. B. and Liburd, O. E. 2019. Evaluation of monitoring traps and lures for Drosophila suzukii (Diptera: Drosophilidae) in Berry plantings in Florida. *Insects.* 10(10), 313.

Harvey, C. D. 2015. Integrated pest management in temperate horticulture: Seeing the wood for the trees. *CAB Rev.* 10(28), 1–13.

Henneberry, T. J., Forlow, J. L. and De La Torre, T. 2001. *Honeydew Production by Sweetpotato Whitefly Adults and Nymphs* (Cotton: A College of Agriculture Report). The University of Arizona College of Agriculture and Life Sciences, Arizona.

Higley, L. G. and Pedigo, L. P. 1996. *Economic Thresholds for Integrated Pest Management*. Lincoln, NE: University of Nebraska Press.

Higley, L. G. and Peterson, R. K. D. 2009. Economic decision rules for IPM. In: *Integrated Pest Management. Concepts, Tactics, Strategies and Case Studies*. Cambridge: Cambridge University Press, 25–32.

Hong, L. and Rumei, X. 1993. Studies on honeydew excretion by greenhouse whitefly, Trialeurodes vaporariorum (Westw.) on its host plant, Cucumis sativus. *J. Appl. Entomol.* 115(1–5), 43–51. https://doi.org/10.1111/j.1439-0418.1993.tb00362.x.

Hoshizaki, S., Koshikawa, N., Toyoda, T. and Ishikawa, Y. 2020. Oviposition in the onion fly Delia antiqua (Diptera: Anthomyiidae) is socially facilitated by visual cues. *Bull. Entomol. Res.* 110(6), 677–683. https://doi.org/10.1017/S0007485320000152.

Hoy, C. W. 1991. Variable-intensity sampling for proportion of plants infested with pests. *J. Econ. Entomol.* 84(1), 148–157. https://doi.org/10.1093/jee/84.1.148.

Hull, L. A. and Beers, E. H. 1990. Validation of injury thresholds for European red mite (Acari: Tetranychidae) on 'yorking' and 'delicious' apple. *J. Econ. Entomol.* 83(5), 2026–2031. https://doi.org/10.1093/jee/83.5.2026.

Hulspas-Jordaan, P. M. and van Lenteren, J. C. 1989. The parasite-host relationship between Encarsia formosa (Hymenoptera: Aphelinidae) and Trialeurodes vaporariorum (Homoptera: Aleyrodidae). *XXX. Modelling Population Growth of Greenhouse Whitefly on Tomato.* Agric. Univ. Wageningen Papers 89-2, 1–54. Available at: https://library.wur.nl/WebQuery/wurpubs/fulltext/468011.

Innocenzi, P. J., Hall, D. R. and Cross, J. V. 2001. Components of male aggregation pheromone of strawberry blossom weevil, Anthonomus rubi Herbst. (Coleoptera : Curculionidae). *J. Chem. Ecol.* 27(6), 1203–1218. https://doi.org/10.1023/A:1010320130073.

Ishikawa, Y., Ikeshoji, T. and Matsumoto, Y. 1978. A propylthio moiety essential to the oviposition attractant and stimulant of the onion fly, Hylemya antiqua MEIGEN. *Appl. Entomol. Zool.* 13(2), 115–122. https://doi.org/10.1303/aez.13.115.

Ishikawa, Y., Matsumoto, Y., Tsutsumi, M. and Mitsui, Y. 1984. Mixture of 2-phenylethanol and n-valeric acid, a new attractant for the onion and seed-corn flies, Hylemya antiqua and H. Platura : Diptera: Anthomyiidae. *Appl. Entomol. Zool.* 19(4), 448–455. https://doi.org/10.1303/aez.19.448.

Ishikawa, Y., Matsumoto, Y., Tsutsumi, M., Mitsui, Y., Yamashita, K., Yoshida, M. and Shirai, E. 1987. Controlled release formulation of attractant for the onion and seed-corn flies, Hylemya antiqua and H. platura: Diptera: Anthomyiidae. *Appl. Entomol. Zool.* 22(3), 303–309. https://doi.org/10.1303/aez.22.303.

Islam, M. N., Hasanuzzaman, A. T. M., Zhang, Z. F., Zhang, Y. and Liu, T. X. 2017. High level of nitrogen makes tomato plants releasing less volatiles and attracting more Bemisia tabaci (Hemiptera: Aleyrodidae). *Front. Plant Sci.* 8, 466. https://doi.org/10.3389/fpls.2017.00466.

Jaastad, G., Bengtsson, M., Anderson, P., Kobro, S., Knudsen, G. and Witzgall, P. 2002. Sex pheromone of apple fruit moth Argyresthia conjugella (Lepidoptera: Argyresthiidae). *Agric. For. Entomol.* 4(3), 233–236. https://doi.org/10.1046/j.1461-9563.2002.00140.x.

Jaleel, W., Saeed, S., Saeed, Q., Naqqash, M. N., Sial, M. U., Aine, Q. U., Yanyuan, L., Rui, Z., He, Y. and Lu, L. 2019. Effects of three different cultivars of cruciferous plants on the age-stage, two-sex life table traits of *Plutella xylostella* (L.) (Lepidoptera: Plutellidae). *Entomol. Res.* 49(4), 151–157. https://doi.org/10.1111/1748-5967.12270.

Jens, K. 1983. Bestemmelse af æg-og Puppe antal af gulerodsfluen (Psila rosae F.) ved indsamling af jordprøver. *Tidsskr. Planteavl.* 87, 379–387.

Jeon, H. Y., Kim, H. H., Yang, C. Y., Kang, T. J. and Kim, D.-S. 2009. A tentative economic injury level for greenhouse whitefly on cucumber plants in the protective cultivation. *Korean J. Hortic. Sci. Technol.* 27, 81–85.

Johansen, T. J. and Meadow, R. 2006. Population differences in emergence of Brassica Root Flies (Diptera: Anthomyiidae). *Environ. Entomol.* 35(5), 1161–1165. https://doi.org/10.1093/ee/35.5.1161.

Johnson, M. W., Caprio, L. C., Coughlin, J. A., Tabashnik, B. E., Rosenheim, J. A. and Welter, S. C. 1992. Effect of Trialeurodes vaporariorum (Homoptera: Aleyrodidae) on yield of fresh market tomatoes. *J. Econ. Entomol.* 85(6), 2370–2376.

Jung, J., Maeda, M., Chang, A., Bhandari, M., Ashapure, A. and Landivar-Bowles, J. 2021. The potential of remote sensing and artificial intelligence as tools to improve the resilience of agriculture production systems. *Curr. Opin. Biotechnol.* 70, 15–22. https://doi.org/10.1016/j.copbio.2020.09.003.

Kaiser, C., Walton, V., Tait, G. and Stacconi, M. V. R. 2021. *U.S. Patent Application No. 17/259,535*.

Kanmiya, K. 1996. Discovery of male acoustic signals in the greenhouse whitefly, Trialeurodes vaporariorum (Westwood) (Homoptera: Aleyrodidae). *Appl. Entomol. Zool.* 31(2), 255–262. https://doi.org/10.1303/aez.31.255.

Kanmiya, K. 2006. Mating behaviour and vibratory signals in whiteflies (Hemiptera: Aleyrodidae). In: *Physiology, Behaviour, Ecology, and Evolution (Contemporary Topics in Entomology)*. Boca Raton, FL: CRC Press, 365–379.

Kanmiya, K. and Sonobe, R. 2002. Records of two citrus pest whiteflies in Japan with special reference to their mating sounds (Homoptera: Aleyrodidae). *Appl. Entomol. Zool.* 37(3), 487–495. https://doi.org/10.1303/aez.2002.487.

Kergunteuil, A., Dugravot, S., Danner, H., van Dam, N. M. and Cortesero, A. M. 2015. Characterizing volatiles and attractiveness of five brassicaceous plants with potential for a 'push-pull' strategy Toward the cabbage root fly, Delia radicum. *J. Chem. Ecol.* 41(4), 330–339. https://doi.org/10.1007/s10886-015-0575-9.

Kergunteuil, A., Dugravot, S., Mortreuil, A., Ralec, A. L. and Cortesero, A. M. 2012. Selecting volatiles to protect brassicaceous crops against the cabbage root fly, Delia radicum. *Entomol. Exp. Appl.* 144(1), 69–77. https://doi.org/10.1111/j.1570-7458.2012.01257.x.

Kim, H. G., Park, J.-S. and Lee, D.-H. 2018. Potential of unmanned aerial sampling for monitoring insect populations in rice fields. *Fla. Entomol.* 101(2), 330–334. https://doi.org/10.1653/024.101.0229.

Kim, J.-K., Park, J.-J., Park, H. and Cho, K. 2001. Unbiased estimation of greenhouse whitefly, Trialeurodes vaporariorum, mean density using yellow sticky trap in cherry tomato greenhouses. *Entomol. Exp. Appl.* 100(2), 235–243. https://doi.org/10.1046/j.1570-7458.2001.00868.x.

Kinjo, H., Kunimi, Y., Ban, T. and Nakai, M. 2013. Oviposition efficacy of Drosophila suzukii (Diptera: Drosophilidae) on different cultivars of blueberry. *J. Econ. Entomol.* 106(4), 1767–1771. https://doi.org/10.1603/EC12505.

Kirkpatrick, D. M., McGhee, P. S., Gut, L. J. and Miller, J. R. 2017. Improving monitoring tools for spotted wing drosophila, Drosophila suzukii. *Entomol. Exp. Appl.* 164(2), 87–93. https://doi.org/10.1111/eea.12602.

Knight, A. 2010. Improved monitoring of female codling moth (Lepidoptera: Tortricidae) with pear ester plus acetic acid in sex pheromone-treated orchards. *Environ. Entomol.* 39(4), 1283–1290. https://doi.org/10.1603/EN10034.

Knight, A. L., Hilton, R., Basoalto, E. and Stelinski, L. L. 2014. Use of glacial acetic acid to enhance bisexual monitoring of tortricid pests with kairomone lures in pome fruits. *Environ. Entomol.* 43(6), 1628–1640. https://doi.org/10.1603/EN14153.

Knight, A. L., Hilton, R. and Light, D. M. 2005. Monitoring codling moth (Lepidoptera: Tortricidae) in apple with blends of ethyl (E, Z)-2,4-decadienoate and codlemone. *Environ. Entomol.* 34(3), 598–603. https://doi.org/10.1603/0046-225X-34.3.598.

Knight, A. L. and Light, D. M. 2005a. Developing action thresholds for codling moth (Lepidoptera: Tortricidae) with pear ester-and codlemone-baited traps in apple orchards treated with sex pheromone mating disruption. *Can. Entomol.* 137(6), 739–747.

Knight, A. L. and Light, D. M. 2005b. Timing of egg hatch by early-season codling moth (Lepidoptera: Tortricidae) predicted by moth catch in pear ester-and codlemone-baited traps. *Can. Entomol.* 137(6), 728–738.

Knight, A. L., Mujica, V. and Larsson Herrera, S. 2019. Addition of terpenoids to pear ester plus acetic acid increases catches of codling moth (Lepidoptera: Tortricidae). *J. Appl. Entomol.* 143, 812–821.

Knudsen, G. K., Norli, H. R. and Tasin, M. 2017. The ratio between field attractive and background volatiles encodes host-plant recognition in a specialist moth. *Front. Plant Sci.* 8, 2206. https://doi.org/10.3389/fpls.2017.02206.

Knudsen, G. K. and Tasin, M. 2015. Spotting the invaders: A monitoring system based on plant volatiles to forecast apple fruit moth attacks in apple orchards. *Basic Appl. Ecol.* 16(4), 354–364.

Kobro, S., Søreide, L., Djønne, E., Rafoss, T., Jaastad, G. and Witzgall, P. 2003. Masting of rowan Sorbus aucuparia L. and consequences for the apple fruit moth Argyresthia conjugella Zeller. *Popul. Ecol.* 45(1), 25–30. https://doi.org/10.1007/s10144-003-0136-x.

Koshihara, T., Yamada, H., Tamaki, Y. and Ando, T. 1978. Field attractiveness of the synthetic sex-pheromone of the diamondback Moth, *Plutella xylostella* (L.). *Appl. Entomol. Zool.* 13(2), 138–141. https://doi.org/10.1303/aez.13.138.

Kovanci, O. B., Kovanci, B. and Gencer, N. S. 2005. Sampling and development of economic injury levels for Anthonomus rubi Herbst adults. *Crop Prot.* 24(12), 1035–1041. https://doi.org/10.1016/j.cropro.2005.02.007.

Kumar, S. T. P., Mazumdar, D., Kamei, D. and Kamei, A. 2018. Advantages of artificial neural network over regression method in prediction of pest incidence in rice crop. *Int. J. Agric. Stat. Sci.* 14, 357–363.

Kumar, V. and Arora, K. 2020. Trends in nano-inspired biosensors for plants. *Mater. Sci. Energy Technol.* 3, 255–273. https://doi.org/10.1016/j.mset.2019.10.004.

Lamy, F., Dugravot, S., Cortesero, A. M., Chaminade, V., Faloya, V. and Poinsot, D. 2018. One more step toward a push-pull strategy combining both a trap crop and plant volatile organic compounds against the cabbage root fly Delia radicum. *Environ. Sci. Pollut. Res. Int.* 25(30), 29868–29879. https://doi.org/10.1007/s11356-017-9483-6.

Lamy, F. C., Poinsot, D., Cortesero, A. M. and Dugravot, S. 2017. Artificially applied plant volatile organic compounds modify the behavior of a pest with no adverse effect on its natural enemies in the field. *J. Pest Sci.* 90(2), 611–621.

Landolt, P. J., Suckling, D. M. and Judd, G. J. R. 2007. Positive interaction of a feeding attractant and a host kairomone for trapping the codling moth, *Cydia pomonella* (L.). *J. Chem. Ecol.* 33(12), 2236–2244. https://doi.org/10.1007/s10886-007-9391-1.

Leach, H., Van Timmeren, S., Wetzel, W. and Isaacs, R. 2019. Predicting Within- and between-year variation in activity of the invasive spotted wing drosophila (Diptera: Drosophilidae) in a temperate region. *Environ. Entomol.* 48(5), 1223–1233. https://doi.org/10.1093/ee/nvz101.

Lemmetty, A. and Vänninen, I. 2014. Bemisia tabaci biotype Q determined for the first time on poinsettia crops in Finland and Sweden. *Ann. Zool. Fenn.* 51(6), 501–506. http://doi.org/10.5735/086.051.0605.

Li, P., Zhu, J. and Qin, Y. 2012. Enhanced attraction of *Plutella xylostella* (Lepidoptera: Plutellidae) to pheromone-baited traps with the addition of green leaf volatiles. *J. Econ. Entomol.* 105(4), 1149–1156. https://doi.org/10.1603/EC11109.

Li, Z., Yu, T., Paul, R., Fan, J., Yang, Y. and Wei, Q. 2020. Agricultural nanodiagnostics for plant diseases: Recent advances and challenges. *Nanoscale Adv.* 2(8), 3083–3094. https://doi.org/10.1039/C9NA00724E.

Li, Z., Zalucki, M. P., Yonow, T., Kriticos, D. J., Bao, H., Chen, H., Hu, Z., Feng, X. and Furlong, M. J. 2016. Population dynamics and management of diamondback moth (*Plutella xylostella*) in China: The relative contributions of climate, natural enemies and cropping patterns. *Bull. Entomol. Res.* 106(2), 197–214.

Liao, Y.-C. and Yang, M.-M. 2015. Acoustic communication of three closely related psyllid species: A case study in clarifying allied species using substrate-borne signals (Hemiptera: Psyllidae: Cacopsylla). *Ann. Entomol. Soc. Am.* 108(5), 902–911. https://doi.org/10.1093/aesa/sav071.

Light, D. M., Knight, A. L., Henrick, C. A., Rajapaska, D., Lingren, B., Dickens, J. C., Reynolds, K. M., Buttery, R. G., Merrill, G., Roitman, J. and Campbell, B. C. 2001. A pear-derived kairomone with pheromonal potency that attracts male and female codling moth, *Cydia pomonella* (L.). *Naturwissenschaften.* 88(8), 333–338. https://doi.org/10.1007/s001140100243.

Löcken, H., Fischer, O. W., Selz, J. and Boppré, M. 2020. 'Drone-Netting' for sampling live insects. *J. Insect Sci.* 20(5), 3. https://doi.org/10.1093/jisesa/ieaa086.

Macfadyen, S., Davies, A. P. and Zalucki, M. P. 2015. Assessing the impact of arthropod natural enemies on crop pests at the field scale. *Insect Sci.* 22(1), 20–34. https://doi.org/10.1111/1744-7917.12174.

Magarey, R. D. and Isard, S. A. 2017. A troubleshooting guide for mechanistic plant pest forecast models. *J. Integr. Pest Manag.* 8. https://doi.org/10.1093/jipm/pmw015.

Maria, A. 1997. Introduction to modeling and simulation. In: Andradóttir, S., Healy, K. J., Withers, D. H. and Nelson, B. L. (Eds) *Proceedings of the 29th Winter Simulation Conference*, 7–13. New York: ACM Press.

Marques da Silva, J. R., Damásio, C. V., Sousa, A. M. O., Bugalho, L., Pessanha, L. and Quaresma, P. 2015. Agriculture pest and disease risk maps considering MSG satellite data and land surface temperature. *Int. J. Appl. Earth Obs. Geoinf.* 38, 40–50. https://doi.org/10.1016/j.jag.2014.12.016.

Marston, Z. P. D., Cira, T. M., Hodgson, E. W., Knight, J. F., Macrae, I. V. and Koch, R. L. 2020. Detection of stress induced by soybean aphid (Hemiptera: Aphididae) using multispectral imagery from unmanned aerial vehicles. *J. Econ. Entomol.* 113(2), 779–786. https://doi.org/10.1093/jee/toz306.

Matsumoto, B. Y. and Thorsteinson, A. J. 1968. Effect of organic sulfur compounds on oviposition in onion maggot, Hylemya antiqua MEIGEN (Diptera: Anthomyiidae). *Appl. Entomol. Zool.* 3(1), 5–12. https://doi.org/10.1303/aez.3.5.

McCarthy, A., Long, D. and Grundy, P. 2020. Smartphone apps under development to aid pest monitoring. *GRDC Update*. Available at: https://grdc.com.au/__data/assets/pdf_file/0028/430858/GRDC-Update-Paper-Long-Derek-July-2020.pdf.

McCravy, K. W. 2018. A review of sampling and monitoring methods for beneficial arthropods in agroecosystems. *Insects.* 9(4), 170. https://doi.org/10.3390/insects9040170.

McRoberts, N., Hall, C., Madden, L. V. and Hughes, G. 2011. Perceptions of disease risk: From social construction of subjective judgments to rational decision making. *Phytopathology.* 101(6), 654–665. https://doi.org/10.1094/PHYTO-04-10-0126.

Méndez-Vázquez, L. J., Lira-Noriega, A., Lasa-Covarrubias, R. and Cerdeira-Estrada, S. 2019. Delineation of site-specific management zones for pest control purposes: Exploring precision agriculture and species distribution modeling approaches. *Comput. Electron. Agric.* 167, 105101. https://doi.org/10.1016/j.compag.2019.105101.

Miller, J. R., Adams, C. G., Weston, P. A. and Schenker, J. H. 2015. *Trapping of Small Organisms Moving Randomly: Principles and Applications to Pest Monitoring and Management*. Cham/Heidelberg/New York/Dordrecht/London: Springer.

Milner-Gulland, E. J. and Shea, K. 2017. Embracing uncertainty in applied ecology. *J. Appl. Ecol.* 54(6), 2063–2068. https://doi.org/10.1111/1365-2664.12887.

Miluch, C. E., Dosdall, L. M. and Evenden, M. L. 2013. The potential for pheromone-based monitoring to predict larval populations of diamondback moth, *Plutella xylostella* (L.), in canola (*Brassica napus* L.). *Crop Prot.* 45, 89–97. https://doi.org/10.1016/j.cropro.2012.11.023.

Miluch, C. E., Dosdall, L. M. and Evenden, M. L. 2014. Factors influencing male *Plutella xylostella* (Lepidoptera: Plutellidae) capture rates in sex pheromone-baited traps on Canola in Western Canada. *J. Econ. Entomol.* 107(6), 2067–2076. https://doi.org/10.1603/EC13371.

Moerkens, R., Brenard, N., Bosmans, L., Reybroeck, E., Janssen, D., Hemming, J. and Sluydts, V. 2019. Protocol for semi-automatic identification of whiteflies *Bemisia tabaci* and *Trialeurodes vaporariorum* on yellow sticky traps. *J. Appl. Entomol.* 143(6), 652–658.

Moerkens, R., Janssen, D., Brenard, N., Reybroeck, E., del Mar Tellez, M., Rodríguez, E., Bosmans, L., Leirs, H. and Sluydts, V. 2020. Simplified modelling enhances biocontrol decision making in tomato greenhouses for three important pest species. *J. Pest. Sci.* 94(2), 285–295 https://doi.org/10.1007/s10340-020-01256-0.

Möhring, N., Wuepper, D., Musa, T. and Finger, R. 2020. Why farmers deviate from recommended pesticide timing: The role of uncertainty and information. *Pest. Manag. Sci.* 76(8), 2787–2798. https://doi.org/10.1002/ps.5826.

Mozūraitis, R., Hall, D., Trandem, N., Ralle, B., Tunström, K., Sigsgaard, L., Baroffio, C., Fountain, M., Cross, J., Wibe, A. and Borg-Karlson, A. K. 2020. Composition of strawberry floral volatiles and their effects on behavior of strawberry blossom weevil, Anthonomus rubi. *J. Chem. Ecol.* 46(11–12), 1069–1081. https://doi.org/10.1007/s10886-020-01221-2.

Muñiz, M. and Nombela, G. 2001. Differential variation in development of the B-and Q-biotypes of Bemisia tabaci (Homoptera: Aleyrodidae) on sweet pepper at constant temperatures. *Environ. Entomol.* 30(4), 720–727.

Murali-Baskaran, R. K., Sharma, K. C., Kaushal, P., Kumar, J., Parthiban, P., Senthil-Nathan, S. and Mankin, R. W. 2018. Role of kairomone in biological control of crop pests: A review. *Physiol. Mol. Plant* 101, 3–15.

Nakabayashi, H., Mizutani, K., Ebihara, T., Wakatsuki, N., Uga, H., Kubota, K. and Ishii, M. 2017. Biotype identification of Bemisia tabaci by acoustical method. *J. Agric. Inform.* 8(3), 11–22. https://doi.org/10.17700/jai.2017.8.3.391.

Naranjo, S. E. 2008. Sampling arthropods. In: Kluwer, P.-Z. (Ed.) *Encyclopedia of Entomology* (vol. 3). Dordrecht/Boston/London: Academic Press Publishers, 1933–1945.

Naranjo, S. E. and Hutchison, W. D. 1997. Validation of arthropod sampling plans using a resampling approach: Software and analysis. *Am. Entomol.* 43(1), 48–57.

Nava-Camberos, U., Riley, D. G. and Harris, M. K. 2001. Temperature and host plant effects on development, survival, and fecundity of Bemisia argentifolii (Homoptera: Aleyrodidae). *Environ. Entomol.* 30(1), 55–63. https://doi.org/10.1603/0046-225X-30.1.55.

Ngowi, B. V., Tonnang, H. E. Z., Mwangi, E. M., Johansson, T., Ambale, J., Ndegwa, P. N. and Subramanian, S. 2017. Temperature-dependent phenology of *Plutella xylostella* (Lepidoptera: Plutellidae): Simulation and visualization of current and future distributions along the Eastern Afromontane. *PLoS ONE* 12(3), e0173590. https://doi.org/10.1371/journal.pone.0173590.

Nguyen, H. D. and Nansen, C. 2020. Hyperspectral remote sensing to detect leafminer-induced stress in bok choy and spinach according to fertilizer regime and timing. *Pest Manag. Sci.* 76(6), 2208–2216.

Nissinen, A. 2008. *Towards ecological control of carrot Psyllid (Trioza apicalis).* PhD Dissertation. University of Eastern Finland, Kuopio.

Nottingham, S. F. 1988. Host-plant finding for oviposition by adult cabbage root fly, Delia radicum. *J. Insect Physiol.* 34(3), 227–234.

Nottingham, S. F. and Coaker, T. H. 1985. The olfactory response of cabbage root fly Delia radicum to the host plant volatile allylisothiocyanate. *Entomol. Exp. Appl.* 39(3), 307–316. https://doi.org/10.1111/j.1570-7458.1985.tb00474.x.

Nottingham, S. F. and Coaker, T. H. 1987. Changes in flight track angles of cabbage root fly, Delia radicum, in diffuse clouds and discrete plumes of the host-plant volatile allylisothiocyanate. *Entomol. Exp. Appl.* 43(3), 275–278. https://doi.org/10.1111/j.1570-7458.1987.tb02222.x.

Nyrop, J. P. 1988. Sequential classification of prey/predator ratios with application to European red mite (Acari: Tetranychidae) and Typhlodromus pyri (Acari: Phytoseiidae) in New York apple orchards. *J. Econ. Entomol.* 81(1), 14–21. https://doi.org/10.1093/jee/81.1.14.

Nyrop, J. P., Agnello, A. M., Kovach, J. and Reissig, W. H. 1989. Binomial sequential classification sampling plans for European red mite (Acari: Tetranychidae) with special reference to performance criteria. *J. Econ. Entomol.* 82(2), 482–490. https://doi.org/10.1093/jee/82.2.482.

Nyrop, J. P., Binns, M. R., van der Werf, W. and Kovach, J. 1994. Cascading tripartite binomial classification plans to monitor European red mite (Acari,Tetranychidae) through a season; development and evaluation of a new methodology for pest monitoring. *Exp. Appl. Acarol.* 18(3), 123–153. https://doi.org/10.1007/BF02353682.

Olatinwo, R. and Hoogenboom, G. 2014. Chapter 4: Weather-based pest forecasting for efficient crop protection. In: Abrol, D. P. (Ed.) *Integrated Pest Management*. San Diego: Academic Press, 59–78. https://doi.org/10.1016/B978-0-12-398529-3.00005-1.

Opit, G. P., Margolies, D. C. and Nechols, J. R. 2003. Within-plant distribution of twospotted spider mite, Tetranychus urticae Koch (Acari: Tetranychidae), on ivy geranium: Development of a presence-absence sampling plan. *J. Econ. Entomol.* 96(2), 482–488. https://doi.org/10.1093/jee/96.2.482.

Orlandini, S., Magarey, R. D., Park, E. W., Sporleder, M. and Kroschel, J. 2017. Methods of Agroclimatology: Modeling approaches for pests and diseases. In: Hatfield, J. L., Sivakumar, M. V. K., Prueger, J. H. (Eds) *Agroclimatology: Linking Agriculture to Climate*. Agronomy Monographs 60. Madison, WI: ASA, CSSA, and SSSA, 453–488. https://doi.org/10.2134/agronmonogr60.2016.0027.

Ovcharenko, F. V. and Nikolaeva, Z. V. 2020. Особенности развития морковной МУХИ В условиях тверской области. [The features of carrot fly development under the conditions of Tver region]. *Вестник Алтайского Государственного Аграрного Университета*. 1, 31–37.

Panda, S. S., Hoogenboom, G. and Paz, J. O. 2010. Remote sensing and geospatial technological applications for site-specific management of fruit and nut crops: A review. *Remote Sens.* 2(8), 1973–1997. https://doi.org/10.3390/rs2081973.

Pannell, D. J. 1994. The value of information in herbicide decision making for weed control in Australian wheat crops. *J. Agric. Resour. Econ.* 19, 366–381.

Park, J., Park, K., Shin, K. and Cho, K. 2011a. Evaluation and comparison of effects of air and tomato leaf temperatures on the population dynamics of greenhouse whitefly (Trialeurodes vaporariorum) in cherry tomato grown in greenhouses. *Korean J. Hortic. Sci. Technol.* 29, 420–432.

Park, J.-J., Lee, J.-H., Shin, K.-I., Lee, S. E. and Cho, K. 2011b. Geostatistical analysis of the attractive distance of two different sizes of yellow sticky traps for greenhouse whitefly, Trialeurodes vaporariorum (Westwood) (Homoptera: Aleyrodidae), in cherry tomato greenhouses. *Aust. J. Entomol.* 50(2), 144–151. https://doi.org/10.1111/j.1440-6055.2010.00796.x.

Park, J.-J., Park, H., Kim, Y. H. and Cho, K. 2000. Application of sequential classification of prey/predator ratio test to Tetranychus urticae and Phytoseiulus persimilis system in greenhouse roses. *J. Asia-Pac. Entomol.* 3(2), 121–126. https://doi.org/10.1016/S1226-8615(08)60065-5.

Park, Y.-L., Cho, J. R., Choi, K.-H., Kim, H. R., Kim, J. W., Kim, S. J., Lee, D.-H., Park, C.-G. and Cho, Y. S. 2021. Advances, limitations, and future applications of aerospace and geospatial technologies for apple IPM. *Kor. Soc. Appl. Entomol.* 60, 135–143.

Pasquali, S., Soresina, C. and Gilioli, G. 2019. The effects of fecundity, mortality and distribution of the initial condition in phenological models. *Ecol. Modell.* 402, 45–58.

Pedigo, L. P. and Buntin, G. D. 1993. *Handbook of Sampling Methods for Arthropods in Agriculture*. Boca Raton, FL: CRC Press.

Pedigo, L. P., Hutchins, S. H. and Higley, L. G. 1986. Economic injury levels in theory and practice. *Annu. Rev. Entomol.* 31(1), 341–368.

Penuelas, J., Filella, I., Lloret, P., Munoz, F. and Vilajeliu, M. 1995. Reflectance assessment of mite effects on apple trees. *Int. J. Remote Sens.* 16(14), 2727–2733.

Pérez, D. E., Cantor, F., Rodríguez, D. and Cure, J. R. 2011. Dispersión de Encarsia formosa (Hymenoptera: Aphelinidae) parasitando *Trialeurodes vaporariorum* (Hemiptera: Aleyrodidae) en tomate bajo invernadero. *Rev. Colomb. Entomol.* 37, 210–216.

Peterson, R. K. D. and Hunt, T. E. 2003. The probabilistic economic injury level: Incorporating uncertainty into pest management decision-making. *J. Econ. Entomol.* 96(3), 536–542. https://doi.org/10.1603/0022-0493-96.3.536.

Phelps, K., Collier, R. H., Reader, R. J. and Finch, S. 1993. Monte Carlo simulation method for forecasting the timing of pest insect attacks. *Crop Prot.* 12(5), 335–342. https://doi.org/10.1016/0261-2194(93)90075-T.

Potamitis, I., Rigakis, I. and Fysarakis, K. 2015. Insect biometrics: Optoacoustic signal processing and its applications to remote monitoring of McPhail type traps. *PLoS ONE.* 10(11), e0140474. https://doi.org/10.1371/journal.pone.0140474.

Potamitis, I., Rigakis, I. and Tatlas, N. A. 2017. Automated surveillance of fruit flies. *Sensors (Basel).* 17(1), 110. https://doi.org/10.3390/s17010110.

Potamitis, I., Rigakis, I., Vidakis, N., Petousis, M. and Weber, M. 2018. Affordable bimodal optical sensors to spread the use of automated insect monitoring. *J. Sens.* 2018, 1–25.

Poudyal, D., Rosenqvist, E., Ottosen, C.-O., Poudyal, D., Rosenqvist, E. and Ottosen, C.-O. 2019. Phenotyping from lab to field – Tomato lines screened for heat stress using Fv/Fm maintain high fruit yield during thermal stress in the field. *Funct. Plant Biol.* 46, 44–55. https://doi.org/10.1071/FP17317.

Prabhakar, M. and Thirupathi, M. 2018. Remote sensing of pest and disease damage in different crops and cropping systems. In: *Agroforestry Opportunities for Enhancing Resilience to Climate Change in Rainfed Areas*. ICAR: Central Research Institute for Dryland Agriculture, Hyderabad, India, 170–179.

Prasad, Y. and Prabhakar, M. 2012. Pest monitoring and forecasting. In: Abrol, D. B. and Shankar, U. (Eds) *Integrated Pest Management: Principles and Practice*. Oxfordshire, UK: CABI Publishing, 41–57.

Pratheepa, M. and Antony, J. C. 2018. Outlook of various soft computing data preprocessing techniques to study the pest population dynamics in integrated pest management. In: *Soft Computing for Biological Systems*. Singapore: Springer, 187–200.

Preti, M., Verheggen, F. and Angeli, S. 2020. Insect pest monitoring with camera-equipped traps: Strengths and limitations. *J. Pest Sci.* 94(2), 203–217. https://doi.org/10.1007/s10340-020-01309-4.

Pusz-Bochenska, K., Perez-Lopez, E., Dumonceaux, T. J., Olivier, C. and Wist, T. J. 2020. A rapid, simple, laboratory and field-adaptable DNA extraction and diagnostic method suitable for insect-transmitted plant pathogen and insect identification. *Plant Health Prog.* 21(1), 63–68. https://doi.org/10.1094/PHP-09-19-0063-FI.

Radcliffe, E. B., Hutchison, W. D. and Cancelado, R. E. 2009. *Integrated Pest Management: Concepts, Tactics, Strategies and Case Studies*. Cambridge: Cambridge University Press.

Rao, P. N. and Lakshmikantha, B. P. 2020. Applications of geospatial technologies in plant health management. In: *Innovative Pest Management Approaches for the 21st Century*. Singapore: Springer Singapore Pte., 3–25.

Reddy, G. V. P. and Guerrero, A. 2000. Pheromone-based integrated pest management to control the diamondback moth *Plutella xylostella* in cabbage fields. *Pest Manag. Sci.* 56(10), 882–888. https://doi.org/10.1002/1526-4998(200010)56:10<882::AID-PS226>3.0.CO;2-T.

Renwick, J. A. A., Haribal, M., Gouinguené, S. and Städler, E. 2006. Isothiocyanates stimulating oviposition by the diamondback moth, *Plutella xylostella*. *J. Chem. Ecol.* 32(4), 755–766. https://doi.org/10.1007/s10886-006-9036-9.

Rigakis, I., Potamitis, I., Tatlas, N.-A., Livadaras, I. and Ntalampiras, S. 2019. A multispectral backscattered light recorder of insects' wingbeats. *Electronics.* 8(3), 277. https://doi.org/10.3390/electronics8030277.

Rincon, D. F., Hoy, C. W. and Cañas, L. A. 2015. Generating within-plant spatial distributions of an insect herbivore based on aggregation patterns and per-node infestation probabilities. *Environ. Entomol.* 44(2), 194–209. https://doi.org/10.1093/ee/nvu022.

Rincon, D. F., Rivera-Trujillo, H. F., Mojica-Ramos, L. and Borrero-Echeverry, F. 2020. *Revisiting Pest Sampling Plans in Light of Economic Uncertainty and Risk Aversion*. AgriXiv preprint. 22 p. https://doi.org/10.31220/osf.io/swxpe.

Rodríguez, H. A. B. 2016. *A spatially explicit individual-based model of the population dynamics of the greenhouse whitefly (Trialeurodes vaporariorum), Encarsia formosa and tomato powdery mildew (Oidium neolycopersici) in tomato crops*. PhD Dissertation. Georg-August-University Göttingen, Göttingen, Germany.

Rodriguez-Saona, C., Firbas, N., Hernández-Cumplido, J., Holdcraft, R., Michel, C., Palacios-Castro, S. and Silva, D. B. 2020. Interpreting temporal and spatial variation in spotted-wing drosophila (Diptera: Drosophilidae) trap captures in highbush blueberries. *J. Econ. Entomol.* 113(5), 2362–2371. https://doi.org/10.1093/jee/toaa153.

Roosjen, P. P., Kellenberger, B., Kooistra, L., Green, D. R. and Fahrentrapp, J. 2020. Deep learning for automated detection of Drosophila suzukii: Potential for UAV-based monitoring. *Pest. Manag. Sci.* 76(9), 2994–3002. https://doi.org/10.1002/ps.5845.

Rossi Stacconi, M., Tait, G., Nieri, R., Chave, R., Park, K., Blood, R. and Walton, V. M. 2020a. A new food-grade behavior disruptor as a management tool for spotted-wing drosophila. *Presented at the Website of North American Raspberry and Blackberry Growers*, 1.

Rossi Stacconi, M. V., Tait, G., Rendon, D., Grassi, A., Boyer, G., Nieri, R. and Walton, V. M. 2020b. Gumming up the works: Field tests of a new food-grade gum as behavioral disruptor for Drosophila suzukii (Diptera: Drosophilidae). *J. Econ. Entomol.* 113(4), 1872–1880. https://doi.org/10.1093/jee/toaa072.

Roşu-Mareş, S., Şofron, A. and Moldovan, C. 2020. Preliminary results on the changes in the flight dynamic of *Cydia pomonella* (L.) in North-Eastern Transylvania, under the influence of climate change. *Sci. Pap. Ser. B Hortic.* 64, 183–185.

Rumei, X. and Liying, Z. 1991. Functional relationship between pest abundance, plant physiology and yield as exemplified by the greenhouse whitefly, Trialeurodes vaporariorum Westw. (Hom., Aleyrodidae). *J. Appl. Entomol.* 112(1-5), 370-381. https://doi.org/10.1111/j.1439-0418.1991.tb01070.x.

Samietz, J., Stoeckli, S., Hirschi, M., Spirig, C., Höhn, H., Calanca, P. and Rotach, M. 2013. Impact of climate change on phenology and sustainable management of the codling moth (Cydia pomonella). *Impact Clim. Chang. Phenol. Sustain. Manag. Codling Moth Cydia Pomonella*. 91, 449-457.

Saussure, S. 2019. *Interactions Between the Cereal Aphid Pests Sitobion avenae and Rhopalosiphum Padi and Their Fungal Natural Enemies*. Ås: Norwegian University of Life Sciences.

Savage, J., Fortier, A.-M., Fournier, F. and Bellavance, V. 2016. Identification of Delia pest species (Diptera: Anthomyiidae) in cultivated crucifers and other vegetable crops in Canada. *Can. J. Arthropod. Identif.* 29, 1-40. https://doi.org/10.3752/cjai.2016.29.

Schmera, D. and Guerin, P. M. 2012. Plant volatile compounds shorten reaction time and enhance attraction of the codling moth (Cydia pomonella) to codlemone. *Pest. Manag. Sci.* 68(3), 454-461. https://doi.org/10.1002/ps.2292.

Shaked, B., Amore, A., Ioannou, C., Valdés, F., Alorda, B., Papanastasiou, S., Goldshtein, E., Shenderey, C., Leza, M., Pontikakos, C., Perdikis, D., Tsiligiridis, T., Tabilio, M. R., Sciarretta, A., Barceló, C., Athanassiou, C., Miranda, M. A., Alchanatis, V., Papadopoulos, N. and Nestel, D. 2018. Electronic traps for detection and population monitoring of adult fruit flies (Diptera: Tephritidae). *J. Appl. Entomol.* 142(1-2), 43-51. https://doi.org/10.1111/jen.12422.

Shannag, H. G. and Freihat, N. M. 2009. Gas exchange of cucumber, Cucumis sativus L., impaired by tobacco whitefly, Bemisia tabaci (Gennadius)(Hemiptera: Aleyrodidae). *Jordan J. Agric. Sci.* 5, 295-305.

Shearer, P. W., West, J. D., Walton, V. M., Brown, P. H., Svetec, N. and Chiu, J. C. 2016. Seasonal cues induce phenotypic plasticity of Drosophila suzukii to enhance winter survival. *BMC Ecol.* 16, 11. https://doi.org/10.1186/s12898-016-0070-3.

Smith, E. A., Fuchs, M., Shields, E. J. and Nault, B. A. 2015. Long-distance dispersal potential for onion thrips (Thysanoptera: Thripidae) and iris yellow spot virus (Bunyaviridae: Tospovirus) in an onion ecosystem. *Environ. Entomol.* 44(4), 921-930. https://doi.org/10.1093/ee/nvv072.

Stockton, D. G., Wallingford, A. K. and Loeb, G. M. 2018. Phenotypic plasticity promotes overwintering survival in a globally invasive crop pest, Drosophila suzukii. *Insects.* 9(3), 105. https://doi.org/10.3390/insects9030105.

Stukenberg, N., Gebauer, K. and Poehling, H.-M. 2015. Light emitting diode (LED)-based trapping of the greenhouse whitefly (*Trialeurodes vaporariorum*). *J. Appl. Entomol.* 139(4), 268-279. https://doi.org/10.1111/jen.12172.

Stukenberg, N. M. 2018. *LED based trapping of whiteflies and fungus gnats: From visual ecology to application*. Doctoral Thesis. Leibniz Universität, Hannover. http://doi.org/10.15488/4144.

Su, Q., Chen, G., Mescher, M. C., Peng, Z., Xie, W., Wang, S., Wu, Q., Liu, J., Li, C., Wang, W. and Zhang, Y. 2018. Whitefly aggregation on tomato is mediated by feeding-induced changes in plant metabolites that influence the behaviour and performance of conspecifics. *Funct. Ecol.* 32(5), 1180-1193. https://doi.org/10.1111/1365-2435.13055.

Suckling, D. M., Stanbury, M. E., Lennon, O., Colhoun, K. M., Chinellato, F. and El-Sayed, A. M. 2020. Kairomone and camera trapping New Zealand flower thrips, *Thrips obscuratus*. *Insects* 11(9), 622.

Sumner-Kalkun, J. C., Sjölund, M. J., Arnsdorf, Y. M., Carnegie, M., Highet, F., Ouvrard, D., Greenslade, A. F. C., Bell, J. R., Sigvald, R. and Kenyon, D. M. 2020. A diagnostic real-time PCR assay for the rapid identification of the tomato-potato psyllid, Bactericera cockerelli (Šulc, 1909) and development of a psyllid barcoding database. *PLoS One*. 15(3), e0230741. https://doi.org/10.1371/journal.pone.0230741.

Tait, G., Park, K., Nieri, R., Crava, M. C., Mermer, S., Clappa, E., Boyer, G., Dalton, D. T., Carlin, S., Brewer, L., Walton, V. M., Anfora, G. and Rossi-Stacconi, M. V. 2020. Reproductive site selection: Evidence of an oviposition cue in a highly adaptive Dipteran, Drosophila suzukii (Diptera: Drosophilidae). *Environ. Entomol.* 49(2), 355-363. https://doi.org/10.1093/ee/nvaa005.

Tay, A., Lafont, F. and Balmat, J.-F. 2020. Forecasting pest risk level in roses greenhouse: Adaptive neuro-fuzzy inference system vs artificial neural networks. *Inf. Process. Agric.* 8(3), 386–397. https://doi.org/10.1016/j.inpa.2020.10.005.

te Beest, D. E., Paveley, N. D., Shaw, M. W. and van den Bosch, F. 2013. Accounting for the economic risk caused by variation in disease severity in fungicide dose decisions, exemplified for Mycosphaerella graminicola on winter wheat. *Phytopathology*. 103(7), 666–672. https://doi.org/10.1094/PHYTO-05-12-0119-R.

Teickner, H., Lehmann, J. R. K., Guth, P., Meinking, F. and Ott, D. 2019. Recognize the little ones: UAS-based in-situ fluorescent tracer detection. *Drones*. 3(1), 20. https://doi.org/10.3390/drones3010020.

Thistlewood, H. M. A., Gill, P., Beers, E. H., Shearer, P. W., Walsh, D. B., Rozema, B. M., Acheampong, S., Castagnoli, S., Yee, W. L., Smytheman, P. and Whitener, A. B. 2018. Spatial analysis of seasonal dynamics and overwintering of *Drosophila suzukii* (Diptera: Drosophilidae) in the Okanagan-Columbia Basin, 2010–2014. *Environ. Entomol.* 47(2), 221–232.

Thöming, G., Schjoll, A. F. and Johansen, T. J. 2017. Developing tools for monitoring and forecasting of onion fly Delia antiqua in Norway. In: *Proceedings of the Working Group "Integrated Protection in Field Vegetables"*, Arenenberg, Switzerland, 2-6. IOBC WPRS Bull.

Tochen, S., Woltz, J. M., Dalton, D. T., Lee, J. C., Wiman, N. G. and Walton, V. M. 2016. Humidity affects populations of *Drosophila suzukii* (Diptera: Drosophilidae) in blueberry. *J. Appl. Entomol.* 140(1–2), 47–57. https://doi.org/10.1111/jen.12247.

Tonina, L., Grassi, A., Caruso, S., Mori, N., Gottardello, A., Anfora, G., Giomi, F., Vaccari, G. and Ioriatti, C. 2018. Comparison of attractants for monitoring *Drosophila suzukii* in sweet cherry orchards in Italy. *J. Appl. Entomol.* 142(1–2), 18–25. https://doi.org/10.1111/jen.12416.

Tonnang, H. E. Z., Hervé, B. D. B., Biber-Freudenberger, L., Salifu, D., Subramanian, S., Ngowi, V. B., Guimapi, R. Y. A., Anani, B., Kakmeni, F. M. M., Affognon, H., Niassy, S., Landmann, T., Ndjomatchoua, F. T., Pedro, S. A., Johansson, T., Tanga, C. M., Nana, P., Fiaboe, K. M., Mohamed, S. F., Maniania, N. K., Nedorezov, L. V., Ekesi, S. and Borgemeister, C. 2017. Advances in crop insect modelling methods–Towards a whole system approach. *Ecol. Modell.* 354, 88–103. https://doi.org/10.1016/j.ecolmodel.2017.03.015.

Tonnang, H. E. Z., Nedorezov, L. V., Ochanda, H., Owino, J. and Löhr, B. 2009. Assessing the impact of biological control of *Plutella xylostella* through the application of Lotka-Volterra model. *Ecol. Modell.* 220(1), 60–70. https://doi.org/10.1016/j.ecolmodel.2008.09.002.

Tonnang, H. E. Z., Nedorezov, L. V., Owino, J. O., Ochanda, H. and Löhr, B. 2010. Host-parasitoid population density prediction using artificial neural networks: Diamondback moth and its natural enemies. *Agric. For. Entomol.* 12, 233–242. https://doi.org/10.1111/j.1461-9563.2009.00466.x.

Tran, A. K. and Koch, R. L. 2017. Spatial patterns and sequential sampling plans for predators of Aphis glycines (Hemiptera: Aphididae) in Minnesota Soybean. *Environ. Entomol.* 46(3), 663–673. https://doi.org/10.1093/ee/nvx040.

Tuovinen, T. 1987. *Pihlajanmarjakoin ennustemenetelmä (Tiedote No. 12/1987), MTT Tiedote*. Jokioinen: Maatalouden tutkimuskeskus.

van der Werf, W., Nyrop, J. P., Binns, M. R. and Kovach, J. 1997. Adaptive frequency classification: A new methodology for pest monitoring and its application to European red mite (Panonychus ulmi, Acari: Tetranychidae). *Exp. Appl. Acarol.* 21(6/7), 431–462. https://doi.org/10.1023/A:1018431912984.

Van Der Werf, W., Nyrop, J. P. and Hardman, J. M. 1994. Sampling predator/prey ratios to predict cumulative pest density in the mite-predatory mite system *Panonychus ulmi-Typhlodromus pyri* in apples. Asp. *Appl. Biol.* 37, 41–51.

Vanegas, F., Bratanov, D., Powell, K., Weiss, J. and Gonzalez, F. 2018. A novel methodology for improving plant pest surveillance in vineyards and crops using UAV-based hyperspectral and spatial data. *Sensors (Basel).* 18(1), 260. https://doi.org/10.3390/s18010260.

van Roermund, H. J. W. and van Lenteren, J. C. 1992. *The Parasite-Host Relationship Between Encarsia formosa (Hymenoptera: Aphelinidae) and Trialeurodes vaporariorum (Homoptera: Aleyrodidae) XXXV. Life-History Parameters of the Greenhouse Whitefly Parasitoid Encarsia formosa as a Function of Host Stage and Temperature*. Wageningen Agr. Univ. Papers 92-3, 1–147. Available at: https://library.wur.nl/WebQuery/wurpubs/fulltext/281747.

van Roermund, H. J. W., van Lenteren, J. C. and Rabbinge, R. 1997. Biological control of greenhouse whitefly with the parasitoid *Encarsia formosaon* tomato: An individual-based simulation approach. *Biol. Control.* 9(1), 25–47. https://doi.org/10.1006/bcon.1997.0512.

Vänninen, I., Pereira-Querol, M. and Engeström, Y. 2015. Generating transformative agency among horticultural producers: An activity-theoretical approach to transforming integrated pest management. *Agric. Syst.* 139, 38–49.

Vesali, F., Omid, M., Mobli, H. and Kaleita, A. 2017. Feasibility of using smart phones to estimate chlorophyll content in corn plants. *Photosynthetica* 55(4), 603–610. https://doi.org/10.1007/s11099-016-0677-9.

Villeneuve, F. and Latour, F. 2017. A few biological specificities of the carrot root fly (Psila rosae Fabre) for a more accurate forecasting in carrot (Daucus carota L.) production. *Acta Hortic.* 1153, 193–200. https://doi.org/10.17660/ActaHortic.2017.1153.28.

Walters, J. P., Archer, D. W., Sassenrath, G. F., Hendrickson, J. R., Hanson, J. D., Halloran, J. M., Vadas, P. and Alarcon, V. J. 2016. Exploring agricultural production systems and their fundamental components with system dynamics modelling. *Ecol. Modell.* 333, 51–65. https://doi.org/10.1016/j.ecolmodel.2016.04.015.

Wang, K. and Tsai, J. H. 1996. Temperature effect on development and reproduction of silverleaf whitefly (Homoptera: Aleyrodidae). *Ann. Entomol. Soc. Am.* 89(3), 375–384. https://doi.org/10.1093/aesa/89.3.375.

Wang, Y.-H., Hu, X.-M., Lin, X.-H., Geng, W.-H. and An, D.-R. 2011. Pest control efficacy of yellow board and sex pheromone on Trialeurodes vaporariorum. 西北農業學報 *Northwest Agric. J.* 20, 194–198.

Weersink, A., Fraser, E., Pannell, D., Duncan, E. and Rotz, S. 2018. Opportunities and challenges for big data in agricultural and environmental analysis. *Annu. Rev. Resour. Econ.* 10(1), 19–37. https://doi.org/10.1146/annurev-resource-100516-053654.

Wen, C., Wu, D., Hu, H. and Pan, W. 2015. Pose estimation-dependent identification method for field moth images using deep learning architecture. *Biosyst. Eng.* 136, 117–128. https://doi.org/10.1016/j.biosystemseng.2015.06.002.

Whitfield, G. H., Carruthers, R. I., Lampert, E. P. and Haynes, D. L. 1985. Spatial and temporal distribution of plant damage caused by the onion maggot (Diptera: Anthomyiidae). *Environ. Entomol.* 14(3), 262–266. https://doi.org/10.1093/ee/14.3.262.

Willett, D. S., Filgueiras, C. C., Nyrop, J. P. and Nault, B. A. 2020. Field monitoring of onion maggot (Delia antiqua) fly through improved trapping. *J. Appl. Entomol.* 144(5), 382–387. https://doi.org/10.1111/jen.12740.

Williams, D. 2016. *Development of mass-trapping methods for Codling Moth females in disrupted orchards* (Final Report, Project Number: MT12000). Horticulture Innovation Australia Limited, Sydney.

Wiman, N. G., Walton, V. M., Dalton, D. T., Anfora, G., Burrack, H. J., Chiu, J. C., Daane, K. M., Grassi, A., Miller, B., Tochen, S., Wang, X. and Ioriatti, C. 2014. Integrating temperature-dependent life table data into a matrix projection model for Drosophila suzukii population estimation. *PLoS One.* 9(9), e106909. https://doi.org/10.1371/journal.pone.0106909.

Witzgall, P., Kirsch, P. and Cork, A. 2010. Sex pheromones and their impact on pest management. *J. Chem. Ecol.* 36(1), 80–100. https://doi.org/10.1007/s10886-009-9737-y.

Wong, J. S., Wallingford, A. K., Loeb, G. M. and Lee, J. C. 2018. Physiological status of Drosophila suzukii (Diptera: Drosophilidae) affects their response to attractive odours. *J. Appl. Entomol.* 142(5), 473–482. https://doi.org/10.1111/jen.12497.

Yan, Y., Feng, C. C., Wan, M. P. H. and Chang, K. T. T. 2015. Multiple regression and artificial neural network for the prediction of crop pest risks. In: International Conference on Information Systems for Crisis Response and Management in Mediterranean Countries. Second International Conference ISCRAM-med 2015, Tunis, Tunisia, October 28–30, 2015. Proceedings. Cham: Springer, 73–84.

Yanagisawa, R., Suwa, R., Takanashi, T. and Tatsuta, H. 2020. Substrate-borne vibrations reduced the density of tobacco whitefly Bemisia tabaci (Hemiptera: Aleyrodidae) infestations on tomato, Solanum Lycopersicum: An experimental assessment. *Appl. Entomol. Zool.* 56(2), 157–163. https://doi.org/10.1007/s13355-020-00711-9.

Yang, C. and Everitt, J. H. 2011. Remote sensing for detecting and mapping whitefly (*Bemisia tabaci*) infestations. In: Thompson, W. M. O. (Ed.) *The Whitefly, Bemisia tabaci (Homoptera: Aleyrodidae) Interaction with Geminivirus-Infected Host Plants: Bemisia tabaci, Host Plants and Geminiviruses*. Dordrecht: Springer Netherlands, 357–381. https://doi.org/10.1007/978-94-007-1524-0_13.

Yang, L., Peng, L., Zhang, L.-M., Zhang, L.-L. and Yang, S. 2009. A prediction model for population occurrence of paddy stem borer (*Scirpophaga incertulas*), based on back propagation artificial neural network and principal components analysis. *Comput. Electron. Agric.* 68(2), 200–206. https://doi.org/10.1016/j.compag.2009.06.003.

Yano, E., van Lenteren, J. C., Rabbinge, R. van Vianen, A. and Dorsman, R. 1989. The parasite-host relationship between *Encarsia formosa* (Hymenoptera: Aphelinidae) and *Trialeurodes vaporariorum* (Homoptera: Aleyrodidae). *XXXII. Simulation Studies of the Population Growth of Greenhouse Whitefly on Egg Plant, Cucumber, Sweet Pepper and Gerbera*. Wageningen Agric. Univ. Papers 89-2, 75–99. Available at: https://library.wur.nl/WebQuery/wurpubs/fulltext/218439.

Yeow, K. and Becker, M. 2018. JIS: Pest population prognosis with escalator boxcar train. *Presented at the 2018 IEEE International Conference on Industrial Engineering and Engineering Management (IEEM)*, Bangkok, TH, 381–385. https://doi.org/10.1109/IEEM.2018.8607724.

Yin, L. T. and Maschwitz, U. 1983. Sexual pheromone in the Green House Whitefly Trialeurodes vaporariorum Westw. *Z. Angew. Entomol.* 95(1–5), 439–446. https://doi.org/10.1111/j.1439-0418.1983.tb02665.x.

Zahner, P. and Baumgaertner, J. 1984. Sampling statistics for Panonychus ulmi (Koch) (Acarina, Tetranychidae) and Tetranychus urticae Koch (Acarina, Tetranychidae) feeding on apple trees. *Res. Popul. Ecol.* 26(1), 97–112. https://doi.org/10.1007/BF02515510.

Zalucki, M. P. and Furlong, M. J. 2008. Predicting outbreaks of a migratory pest: An analysis of DBM distribution and abundance. In: *The Management of the Diamondback Moth and Other Crucifer Insect Pests. Presented at the Fifth International Workshop.* Beijing, China: China Agricultural Science and Technology Press, 122–131.

Zalucki, M. P., Li, Z. Y., Yonow, T., Kriticos, D., Furlong, M. J. and Feng, X. 2017. Modelling the population dynamics & management of diamondback moth: The role of climate, natural enemies & cropping patterns. *Mysore J. Agric. Sci.* 51, 9–17.

Zaza, C., Bimonte, S., Faccilongo, N., La Sala, P., Contò, F. and Gallo, C. 2018. A new decision-support system for the historical analysis of integrated pest management activities on olive crops based on climatic data. *Comput. Electron. Agric.* 148, 237–249. https://doi.org/10.1016/j.compag.2018.03.015.

Zerulla, F. N., Capezzone, F. and Zebitz, C. P. W. 2016. Field performance of different lures for Drosophila suzukii. *Presented at the 17th International Conference on Organic Fruit-Growing*, Hohenheim, DE.

Zhang, J., Huang, Y., Pu, R., Gonzalez-Moreno, P., Yuan, L., Wu, K. and Huang, W. 2019. Monitoring plant diseases and pests through remote sensing technology: A review. *Comput. Electron. Agric.* 165, 104943. https://doi.org/10.1016/j.compag.2019.104943.

Zhang, W. and Swinton, S. M. 2009. Incorporating natural enemies in an economic threshold for dynamically optimal pest management. *Ecol. Modell.* 220(9–10), 1315–1324. https://doi.org/10.1016/j.ecolmodel.2009.01.027.

Zhang, W. and Swinton, S. M. 2012. Optimal control of soybean aphid in the presence of natural enemies and the implied value of their ecosystem services. *J. Environ. Manage.* 96(1), 7–16. https://doi.org/10.1016/j.jenvman.2011.10.008.

Zhu, L., Li, Z., Zhang, S., Xu, B., Zhang, Y., Zalucki, M. P., Wu, Q. and Yin, X. 2018. Population dynamics of the diamondback moth, *Plutella xylostella* (L.), in northern China: The effects of migration, cropping patterns and climate. *Pest. Manag. Sci.* 74(8), 1845–1853. https://doi.org/10.1002/ps.4885.

Chapter 6

Advances in proximal sensors to detect crop health status in horticultural crops

Catello Pane, CREA – Research Centre for Vegetable and Ornamental Crops, Italy

1 Introduction

Digital technologies can assist farmers in managing plant diseases of horticultural crops in various ways. They can increase the ability to monitor the incidence of disease within both extensive and intensive types of cultivation, provide important information on the spatial/temporal distribution of disease symptoms and provide objective data to feed into disease forecasting models (Gao et al., 2020). The potential of these technologies lies in the increasing accuracy and reliability of sensor and other technologies, which increasingly match and surpass farmers' traditional decision-making processes (Nikolaidis, 2008). Proximal sensing systems acquire information by placing electronic instruments in contact with or in close proximity to the surface of the 'sensed object' (Zubler and Yoon, 2020). They are a subset of the so-called remote sensing technologies, which involve rapid, non-invasive analysis of plants without making physical contact with them. Proximal technologies include handheld, fixed, robotic or tractor-mounted devices. Other technologies are based on satellite, airborne or unmanned aerial vehicle (UAV) platforms (Oerke, 2020).

Global production of fresh and processed vegetables is increasing because they are such an excellent source of fibre, vitamins, antioxidants and other health-promoting compounds. They are cultivated both in the ground (open

http://dx.doi.org/10.19103/AS.2021.0095.06

fields and/or greenhouses/polytunnels) and increasingly in innovative soil-less systems, including hydroponics, aquaponics and indoor growing chambers with artificial lighting. Vegetable crops are subject to various fungal diseases exacerbated by the intensive exploitation of soils, continuous cropping and plant production in warm, humid environments that are conducive to the development of diseases. Pathogenic diseases have a serious economic impact on the sector, causing significant reductions in yield, together with high levels of food waste by reducing the aesthetic appearance, nutritional quality and shelf-life of horticultural products. Appropriate and effective phytopathological management is crucial for the competitiveness of the industry and for consumer safety.

The interest in more sustainable production systems, where plants are grown with reduced use of synthetic pesticides in favour of eco-friendly alternatives, is currently the main driver of innovation in plant protection strategies. Scientific research is continuously developing new tools for plant disease control and for the improvement of soil health, which avoid the use of synthetic chemistry. This includes new environmentally benign plant protection molecules and new pathogen-resistant and/or tolerant varieties (Juroszek et al., 2020).

Digital technologies have stimulated a growing interest in the detection of the health status of horticultural crops as part of decision support systems designed to increase the precision of disease prevention and treatment (Lázaro et al., 2020). By extending the sensing capacity of the grower both in space (i.e. monitoring the phytopathological status of field crops on a large scale) and in time (i.e. early identification of outbreaks and the capacity for short- and medium-term forecasting to drive targeted interventions), electronic devices contribute to improving the efficient use of resources and the success of disease management strategies.

Sensors act in two main ways to identify diseases:

- sensing significant changes that occur when plants are attacked by a pathogen; and
- detecting the occurrence of climatic conditions favourable to pathogenesis at the micro-environmental level.

When pathogens attack host plants, they produce a number of effects, ranging from early physiological changes, the appearance of specific and/or general symptoms of disease, loss of vigour, a reduced growth rate, and, in some cases, plant death. All of these effects can be captured by, for example, optoelectronic and olfactory devices that may allow the early and non-destructive diagnosis of plant disease. This can inform the most suitable intervention strategy (i.e. type and rate of the active ingredient to be used, and targeted application in the areas affected). In the second approach, specific micro-environmental conditions linked mainly to temperature, humidity and free water on the foliage are necessary to complete the first phases of pathogenesis, such as propagule germination, hyphal growth and host penetration. Continuous monitoring of these conditions can help to forecast future infection risks and plan targeted treatments, which can be more effective than traditional scheduled management using fungicides.

Innovative digital technologies that can be deployed include:

- sensors for the detection of micro-environmental parameters in the field;
- electronic noses; and
- imaging and optical sensors.

These sensors can then be linked among them to equip a network of sensors functional to the internet of things and artificial intelligence, to forecasting and decision-making models (Nawaz et al., 2020). These technologies are changing approaches to phytosanitary control, as well to the genetic improvement of vegetable crops for pathogen resistance characteristics, through their incorporation into high-tech phenotyping platforms (Ilakiya et al., 2020).

Key issues and challenges that are being faced in this subject, however, still remains a concern, in particular: (i) exchanging knowledge between the research world and farmers about how sensors work, what they measure and which information coming from them is expendable in plant disease management; (ii) make digital technologies and sensors easier available for growers, reducing costs, dimensions and application difficulties; (iii) gain high effective accuracy of the monitoring methods, increasing the value of the information and reduce false alerts; (iv) increase cost-efficient farm internet connections, software use, digitalization and professional background of operators.

2 Optoelectronic devices for detecting disease in vegetable plants

Optoelectronic proximal sensors work by acquiring reflected and/or emitted energy from plants, taking into account both the spatial distribution of the signals (imaging) and their site-specific higher resolution value (non-imaging). They provide a non-destructive contactless diagnosis of impaired plant status due to any adverse interaction with pathogens.

When a leaf surface is hit by an external radiation source, part of that radiation (called reflectance) is reflected back towards the external environment (Fig. 1). Reflectance is the measure of the ability of an object to reflect part of the incident radiation. It is a dimensionless quantity expressed as the ratio between the intensity of the reflected radiation and the intensity of the incident radiation. The set of electromagnetic waves of different lengths that are

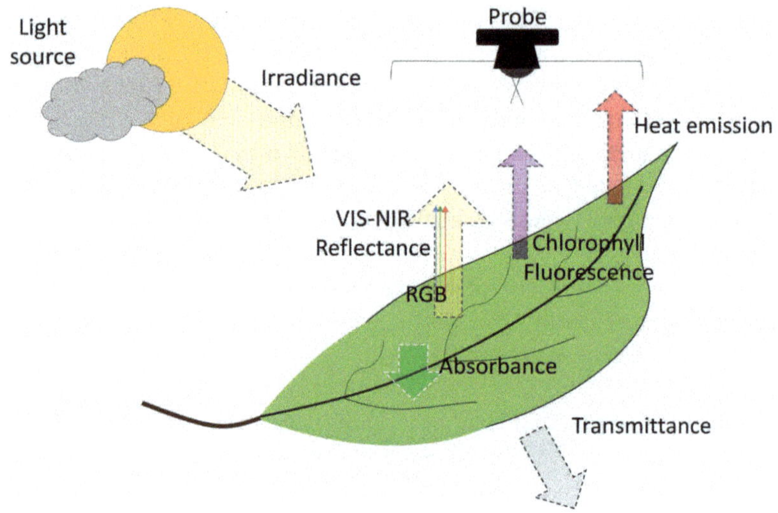

Figure 1 Scheme of how light irradiance can be partitioned through a targeted leaf among VIS-NIR spectrum reflectance, including RGB, chlorophyll fluorescence, heat emission, absorbance and transmittance and which light/energy net-losses can be recorded by the optoelectronic probes.

reflected represents the reflectance spectrum. Information from this spectrum is captured by spectral sensors. The electromagnetic spectrum can be split into different spectral regions including visible and near-mid-long infrared spectra. The detectable limits of the spectrum are defined by the type, technological setting and analytical capacity of currently available optoelectronic sensors. Imaging involves processing numerical data related to images acquired using optical sensors and the subsequent visualization of the results. These are associated with variations in physicochemical characteristics such as colour, thermographic temperature and chlorophyll fluorescence, which represent symptoms of pathogen attack.

Hyperspectral sensors are able to cover wide regions of the electromagnetic spectrum, which can include visible and near-infrared wavelengths in the range 400–3000 nm, according to the power of single or combined instruments, for example, those also using non-imaging technologies, such as spectroradiometers and cameras. Imaging hyperspectral sensors are able to capture the spectroscopic signature of a plant's status (i.e. the trace reflectance value for each measurable wavelength), including health and disease symptoms (Lowe et al., 2017). Prabhakar et al. (2013), for example, radiometry to identify leaves with different levels of yellow mosaic virus symptoms. The hyperspectral probe acquires information continuously in the designated spectrum range, with a pre-set resolution level given by the minimum step between each measured wavelength.

Hyperspectral technology involves the acquisition of a set of wavelengths *per* measuring point in non-imaging sensors. A hyperspectral image obtained by a camera with a well-defined width and height is a hypercubic dataset in which each pixel has the corresponding series of reflectance values. In order to reduce data redundancy, data processing simplifies the broad numeric array into the most significant spectral wavelengths (i.e. a few bands or narrow spectral intervals and/or their combinations). Hyperspectral analyses use vegetative indices related to a plethora of physiological (i.e. photosynthesis, water relations, plant vitality), morphological-structural (i.e. leaf area index (LAI), green biomass, growth rate, soil coverage) and biochemical traits (i.e. degree of pigmentation related to the content of chlorophyll, anthocyanins, carotenoids) (Haboudane et al., 2004; Xue and Su, 2017). Vegetative indices can be used to identify diseases and/or to highlight correlations between causal factors (pathogen) and physiological damage (symptoms) (Fig. 2). Zhao et al. (2016), for example, related the spatial distribution of angular leaf spot disease symptoms of cucumber through hyperspectral mapping of chlorophyll (*chl*) in leaves using the vegetative index *Chl-HSI*. An alternative is to use *de novo* hyperspectral algorithms extracted from reflectance measurements and machine-learning models comparing diseased and healthy samples, which can also be used for reliable detection and identification of phytopathologies in plants (Mahlein et al., 2013).

Many studies have shown the use of reflectance-derived information for the (early) detection of both direct and indirect symptoms caused by

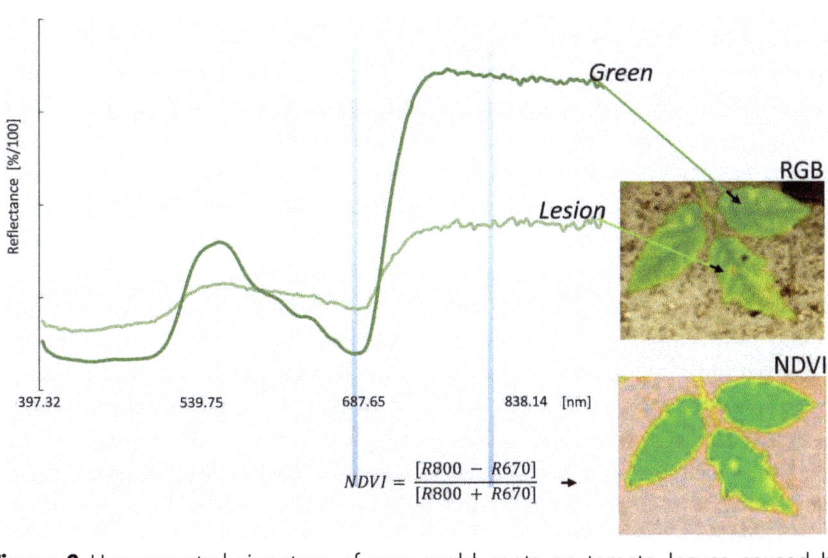

Figure 2 Hyperspectral signature of grey mold spots on tomato leaves caused by *Botrytis cinerea*, in comparison with healty leaf spectrum as, also, highlighted by RGB and NDVI images.

pathogen attacks, for example, those produced by *Corynespora cassicola* and *Xanthomonas perforans* on tomatoes (Abdulridha et al., 2020a) and *Rhizoctonia solani* on sugarbeet (Barreto et al., 2020). Bienkowski et al. (2019) used data from potato foliage reflectance spectra (Blue Wave Spectrometer 350–1150 nm), combined with supervised training of back propagation neural networks, to identify pre-symptomatic samples affected by a set of different diseases. Gold et al. (2020) detected early infections of late blight and early blight in potatoes just 1 day after inoculation by using a contact leaf spectroradiometer (350–2500 nm) and identified normalized difference spectral indices using partial least squares discriminant analysis. Hyperspectral technology was able to identify the *Arabidopsis thaliana/Sclerotinia sclerotiorum* pathosystem only 3 h after artificial inoculation of plants (Liang et al., 2019).

Hyperspectral fingerprinting of plant disease symptoms has been carried out in laboratory settings as a basis for possible in-field sensor applications. As an example, machine learning narrowed a 380–1023 nm dataset down to just five wavelengths with the potential as a non-invasive method to identify early blight and late blight diseases on tomato leaves (Xie et al., 2015). The study of reflectance data from the canopy of *Sclerotinia*-infected celery indicated the red, near-infrared, and blue wavelengths as the most significant raw spectral bands for disease prediction (Huang and Apan, 2006). Investigation of signatures for each stage of powdery mildew infection of squash, scanned by benchtop hyperspectral imaging system (range of 380–1030 nm, 281 channels, resolution of 2.1 nm) identified the most significant bands to monitor the development of the disease (Abdulridha et al., 2020b). Manganiello et al. (2021) identified high-performing hyperspectral vegetation indices in the 400–1000 nm spectral range, very useful to track the biocontrol activity of *Trichoderma* spp. against soil-borne diseases in baby-leaf vegetables (Fig. 3). A short-wavelength infrared (SWIR) spectral imaging system (950–1650 nm) has been used to extract key wavelengths identifying *Burkholderia cepacia* infection of onions (Wang et al., 2012b). Lu et al. (2018) used a line scanning spectrograph to investigate the reflectance spectra and absolute reflectance difference spectra in the wavelength range of 500–1000 nm; these were used to create multispectral images at 560575 nm and 720 nm with the potential to detect tomato yellow leaf curl virus infection in the field. High-resolution probes used in fundamental research can generate new knowledge useful in developing cheaper, more practical spectral-based sensors, such as multispectral sensors, for automated diagnosis and/or monitoring of plant diseases in the field.

Multispectral sensors are able to work in well-defined spectral bands (a typical limit is 25) to simplify subsequent data processing. They consist of a hybrid camera system that can be directly associated via software to plant disease vegetation indices and reveal changes in plant physiology attributable to the pathogen. A five-channel camera working on the RGB and red-edge (RE) bands

Figure 3 Hyperspectral snapshot camera during an image acquiring stage on vegetable cropping system to tentatively capture disease symptoms.

was able to separate late blight infected potato plants from healthy ones using the vegetation indices SR Clgreen, RI, TCARI, TCARI/OSAVI-2, ClRed-Edge, and RE NDVI (Fernández et al., 2020). Radiometric evaluations using single-waveband reflectance values and selected vegetative indices have been used to assess the severity of *Cercospora beticola* necrosis in sugar beet (Steddom et al., 2005). A multispectral imaging system based on two synthetic narrowband indices has been developed for disease diagnosis of *Pyrenopeziza brassicae* on *Brassica napus* (Veys et al., 2019). A linear regression model showed that NIR and RE bands can be used to identify grey mold leaf infection in tomatoes caused by *Botrytis cinerea* as early as 9 h after infection (Fahrentrapp et al., 2019). With the simplification of probe numbers and complexity, multispectral sensors are often used in phenotyping platforms (Svensgaard et al., 2014), on tractor-based crop scanners for on-the-go observations (Bourgeon et al., 2016) and in the development of more accurate, low-cost and portable multispectral optical devices for disease detection (Kitić et al., 2019; Habibullah et al., 2020; Wang et al., 2020).

Lower-cost optical sensors for identifying vegetable diseases include a digital camera, which uses red-green-blue (RGB) channels analysed by software to extract useful information about changes in plant appearance. The cost-effectiveness and compactness of RGB cameras encourage their use in many applications in disease monitoring. They can be seen as simplified multispectral cameras able to detect information from three wavelengths in the spectrum visible range, red (700.47 nm), green (546.09 nm) and blue (435.79 nm) signals. This technology can identify disease symptoms based on changes in colour of affected plant organs (Padmavathi and Thangadurai, 2016). In the case of multispectral scanners, a number of RGB vegetation indices have been developed to predict plant traits to indicate possible disease (McKinnon and Hoff, 2017; Lussem et al., 2018). Artificial intelligence algorithms can be used to identify visual symptoms of the disease from the analysis of coloured images (Camargo and Smith, 2009). Image processing and pattern recognition methods are used to identify colour variations crucial for disease symptom recognition (Zhang et al., 2015). These rely on the representativeness of sample collections showing all possible shades related to possible disease symptoms. Johannes et al. (2017) have developed an image-processing algorithm based on detecting hot spots in wheat that allows recognition of diseases by different devices. Mohanty et al. (2016) have developed accurate image classifiers for smartphone-assisted disease diagnosis.

Thermal sensors are used to detect infrared energy emitted from an object in the long infrared (IR) region with wavelengths ranging from 8000 nm to 14 000 nm. Infrared energy is converted to thermographic temperature points and displayed as an image showing temperature scattering (pixel by pixel) known as a thermogram. Patterns of heat on a plant surface can be related to cooling/warming dynamics. Thermal variations in a plant are mainly driven by changes in the surrounding environment (i.e. solar radiation, air temperature, relative humidity). Thermal variations in the plant are determined by stomatal conductance regulating O_2, CO_2 and vapour exchanges (transpiration and respiration rate), as well as relative water content, physiological sap flow, cellular and tissue vitality.

Thermal temperature shifts are dependent on environmental conditions, biotic damage type and how and when the transpiration rate is affected. For example, Oerke et al., (2006) observed different thermal imaging patterns in cucumber leaves infected with downy mildew and exposed to different conditions of humidity and air temperature. In the early stages of *Pseudoperonospora cubensis* infection, areas of swelling on cucumber leaves were visible in thermograms as cooled areas; as the disease progressed, these areas increased in temperature (Lindenthal et al., 2005). In this way, thermograms can be used to identify disease symptoms. Thermal imaging is non-specific but potentially effective in early recognition of the disease, depending on the degree of resolution (the lowest differential of temperature detectable) of the camera. Wang et al. (2012a) used an IR camera, able to detect in the range 7500–13 000 nm, for non-destructive detection of pre-symptomatic leaf symptoms of Fusarium wilt in cucumber. The camera was able to detect symptoms 5 days after inoculation with a thermographic temperature difference of about 3.4°C. Tomato cucumber mosaic virus infections were observable 3 days after infection by a thermal imaging detection system equipped with a 0.06°C sensitive camera (Zhu et al., 2018). Thermal imaging can show the spatial distribution of symptoms, giving additional information on the dynamic of pathogenesis, for example, as shown in pre-symptomatic detection of potato late blight (Prashar and Jones, 2014). Thermal analysis has been used to estimate the virulence of different species causing leaf blight (belonging to the genus *Alternaria*) on oilseed rape leaves (Baranowski et al., 2015).

As in the case of hyperspectral indices, there are thermal long-wave infrared stress indices relating to air, canopy or leaf temperatures in dry or wet plants associated with general physiological parameters, which can be used to identify disease symptoms (Pineda et al., 2021). A crop water-stress index, calculated by computing temperature values using discriminant and logistic regression analysis, can identify *Rhizoctonia*-diseased lettuce plants (Sandmann et al., 2018). Rispail and Rubiales (2015) found a significant correlation between the surface leaf temperature of pea varieties and their susceptibility to *Fusarium. oxysporum f. sp. pisi*. Thermal imaging has also been proposed to assess seed viability by observing variations in heat flow and can be used to separate viable and non-viable seeds (Men et al., 2017).

Chlorophyll fluorescence imaging and non-imaging sensors capture fluorescence emissions from photochemical processes related to photosynthesis. Plants use photosynthesis to produce organic carbon biomass. The process, through photosystems I and II, entraps light energy into ATP molecules. Under light stimuli, leaves dissipate excess energy by emitting fluorescence in the red and far-red spectral range with two peaks at 685 nm and 740 nm. The majority of fluorescence is emitted by photosystem II. Fluorescence measurements carried out under different types of incident light modulation provide indicators of plant photochemical status that can be used to discriminate between healthy and infected plants (Rolfe and Scholes, 2010; Pérez-Bueno et al., 2016).

Advances in technology, such as fluorometers, have simplified the detection and visualization of chlorophyll fluorescence. Hand-held camera devices collecting fluorescence data can be used to monitor physiological changes in photosynthetic efficiency, identifying damage caused by plant disease affecting photosynthetic active leaf surfaces by limiting water and nutrient use (Gorbe and Calatayut, 2012). Disease symptoms that are produced, for example, *Bremia lactucae* in lettuce (Bauriegel et al., 2014), *Ralstonia solanacearum* in tomatoes (Kim et al., 2019) and *Uromyces appendiculatus*, *Phaeoisariopsis griseola* and *Colletotrichum lindemuthianum* in beans (Bassanezi et al., 2002), can be highlighted with pixel-based analyses of Fv/Fm images. Chlorophyll fluorescence imaging has been used for the early detection of tracheofusariosis (caused by *F. oxysporum*) in cucumber (Zhou et al., 2020). Fusarium infections in tomatoes have been identified by analysing the quantum yield of photosystem II (Wagner et al., 2006). Pineda et al. (2017) used artificial neural networks, logistic regression analyses and support vector machines trained with a set of imaging data, based on the fluorescence ratio F520/F680, to distinguish healthy fruit from those affected by soft rot and leaves affected by powdery mildew. Fluorescence quenching analysis allowed the early diagnosis of *Pseudomonas syringae* pv. *Phaseolicola* infections in bean (Rodríguez-Moreno et al., 2008).

3 Sensors for the detection of micro-environmental parameters related to disease outbreaks

This category includes all those sensors that allow the detection of microclimatic parameters in the cultivation environment. Sensors measure parameters such as temperature, humidity, wind speed, and free water which may promote pathogen spread. They can also measure parameters such as solar radiation, light intensity, water potential, wind direction and speed, and amount of rain which affect plant vigour and thus susceptibility to pathogenic infection. There is a wide range of devices, most recently using wireless technology with continuous data acquisition and remote data management (rather than a direct connection to a server). The sensors can be applied both in the environment in which the crop is developing, for monitoring soil, air and water conditions (extra-canopy), and at the plant level (intra-canopy) based on the concept of a 'plant sentinel' using single plants to estimate conditions for the surrounding crop (Fig. 4).

The information acquired by the network of sensors can be processed by algorithms based on well-calibrated forecasting models to identify threshold values related to the risk of infection. A forecast of the risk of disease, for example, enables a farmer to adjust the climate within a greenhouse, preventing thermal fluctuations and/or water levels/humidity that might promote disease. Using environmental data as inputs, these systems estimate the risk of disease occurring and alert growers when conditions might become favourable to pathogenic attacks. Models need to be calibrated and validated according to both the crop and the pathogen to ensure accurate prediction of disease risk. For example, downy mildew (*Phytophthora infestans*), a key disease in many vegetables, relies on the activity of oomycetes and are themselves highly

Figure 4 Compact climatic station for continuous monitoring of environmental parameters close to tomato plants.

dependent on environmental factors, which can be monitored and modelled. The sporulation of *Bremia lactucae* on lettuce, for example, occurs at the optimum temperature of 15°C and increases markedly with relative humidity values higher than 90%. Infection of rocket with *Hyaloperonospora parasitica* has been shown to occur at temperatures between 10°C and 26°C and relative humidity of 85%.

Electronic noses are another type of device that has been proposed for rapid and non-destructive detection of plant disease. Infected plants release a distinctive set of volatile organic compounds (VOCs) as a reaction to biotic stressors as part of plant airborne defence signalling or pathogen-associated molecular patterns (Heil and Ton, 2008). The so-called electronic noses (eNose) are devices capable of acquiring real-time information about the chemical and physical nature of gas emitted by plants including VOCs, effectively smelling the scent of plant diseases (Wilson, 2013). The eNose works by detecting variations in electrical conductivity related to the presence of gas and/or other volatile compounds in the atmosphere. This kind of device allows non-destructive, continuously acquired and early (though non-specific) diagnosis of plant diseases (Wilson, 2020; Cellini et al., 2017). eNose technologies have been proposed for the vegetable sector, for example, to identify damage by pests (Laothawornkitkul et al., 2008), bacterial infections in potatoes (Biondi et al., 2014), and grey mould and powdery mildew in tomato (Sun et al., 2016; Ghaffari et al., 2012). However, despite their potential, eNose devices need further development before they can be widely used in the horticultural sector (Sharifi et al., 2018).

4 Case study: digital and mechatronic applications on baby leaf vegetable quality chain

Baby leaf vegetables, including wild rocket, baby lettuce and baby spinach, and more other minor herbaceous species, are developed under intensive greenhouse systems in order to supply a minimally processed salad chain. The attention of consumers for ready-to-eat fresh vegetable foods leads to the preference of products from reduced or no pesticide-based cultivations. However, these conditions entail high soil-environment exploitation and contribute to making crops more susceptible to soil-borne pathogens, increasing phytopathological risks and requirements for effective disease management practices alternative to synthetic fungicides. Targeted use of resources and productive factors such as the plant protection means is focused by this case study[1] aimed at implementing advanced digital and mechatronic applications for quality horticultural supply chains, such as those of greenhouse baby leaf vegetables. In particular, research integrated digital supporting systems (i.e. hyperspectral and thermal imaging, non-imaging spectral probe and disease risk alert devices) in order to enhance performances of sustainable disease management tools (Manganiello et al., 2021) and identify digital-based variables associated to levels of disease on the specific crops (Santonicola et al., 2019) and weed infestation (Pallottino et al., 2020).

Greenhouse climatic station, equipped with temperature, humidity, foliar wetting (Fig. 5), photosynthetically active radiation, and pyranometre sensors, served to continuously monitor crop and elaborate threshold values conducive to the development of major diseases. On the other hand, spectral and thermal optoelectronic probes are applied in different growing conditions in order to find, by imaging and non-imaging elaborations, new vegetation indices and/or by artificial intelligence models able to discriminate symptomatic from healthy plants (Fig. 6).

5 Conclusion

In the next decades, horticultural production will face challenges such as climate change, loss of soil fertility and the need to feed an increasing world population. The increasing threat of disease could dramatically undermine the resilience of production systems that are currently transitioning towards more agroecological methods of cultivation driven by the growing focus on environmental protection and the production of healthy and safe food. The concept of 'producing more with less' which underpins these systems will

1 The current research is carried out in the framework of the project AgriDigit-Agrofiliere (Integrated digital technologies for the sustainable strengthening of agro-food production and transformation, financed by the Italian Ministry of Agriculture, Food and Forestry Policies (Mipaaf), grant DM 36503.7305.2018 of 20/12/2018).

Figure 5 Electronic leaf to assess foliar wetting on wild rocket cultivation.

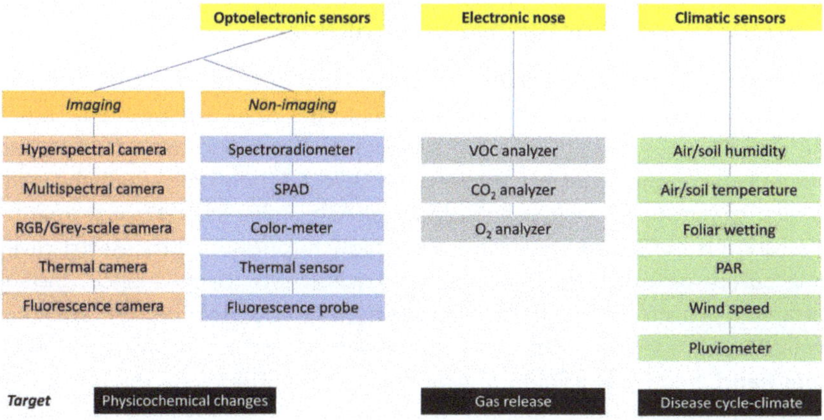

Figure 6 Classification of sensors applicable for the disease management support in vegetable cropping systems.

involve enhancing the efficacy of disease control methods by targeting their application more efficiently using digital and other technologies.

The proximal sensors discussed in this chapter can be integrated into small hand-held instruments, mounted in fixed locations, embedded in operating machines moving through the crop or carried by light drones flying just above crops (Pirna and Lache, 2010). Production systems involving proximal sensors assume the digitalization of increasing amounts of the primary production chain, giving operators the ability to continuously assess and improve methods of cultivation. A key area for development is to increase accuracy in predicting disease risk. It will be essential to develop a holistic approach. Concepts such as 'internet of things', 'artificial intelligence', 'forecasting' and 'decision-making', 'modelling', 'robotic' and 'mechatronic' will need to be integrated with concepts such as agroecology, biological control and low input practices to achieve sustainable improvements in production. Machine vision technologies based on analysis of observed parameters using vegetation indices will, for example, allow the targeted application of bio-protectants and biopesticides to replace reliance on fungicides (Ampatzidis et al., 2017).

In time, proximal sensing may facilitate the transition to a fully 'smart' horticulture. Machine learning is an important element in this transition. It can be applied to different aspects of plant disease management: identification/detection, classification, quantification and prediction through big data mining and the possibility of self-calibration with sensors as dynamic, self-regulating systems (Yang and Guo, 2017). Deep learning of large data sets using pattern recognition approaches will improve these various aspects of disease management (Ferentinos, 2018) though there are challenges linked to plant-disease management (Barbedo, 2018).

6 Where to look for further information

Specifically, research in proximal sensing applied at vegetable cropping systems could evolve by producing digital support to the crop disease management in a wider range of host/pathogen systems. The optimum would be to have substantial and significant information on how sensors work regarding the whole disease spectrum of each vegetable crop. As parallel step, the automatized acquisition of measurements and their high throughput processing giving on accurate outputs would be pursued with more perseverance. Indeed, research on this topic moves on balance between discovering new electronic technologies, such as sensors, hardware components/software, and, on the other hand, finding out field applicative solutions. Therefore, the continuous knowledge exchanging among all the involved scientists (engineers, computer scientists, physicians, agronomists, biologists, etc.) is crucial to reach successful, affordable and reliable results. However, in order to plan future research aimed at making available knowledge, technology and usefulness in horticultural crop health status detection, following a stepwise approach that take in account the system complexity is necessary. In perspective, innovation work in proximal sensoring for vegetable crops will focus on the reduced size and costs of the sensors, easy to apply under the cultivation conditions (i.e. wireless, resistant to adverse working conditions, durable, energetically autonomous, etc.) and to connect with the on farm or on cloud control unit. On the other hand, high-performing software for big data management needs to be coupled with high-significant system biology models and/or algorithm to increase the accuracy of the previsions or of the interpretations. All of these technological aspects have to be defined and developed in the course of specific lab to field-experimentations paying very much attention to check the true link between the acquired data and the observed biological phenomena, which, in this specific case, are involved in disease occurrence and/or symptom expression. Because big data management is very time-consuming stage of the research, and sometime the bottleneck is difficult to overcome, the scientific solidity of the dataset has greatest importance in the flow-work. The most interesting discovering are possible when different experts from the universities, research centres, developing start-up worldwide, etc., are able to dialogue with each other constructively. This scientific field takes advantage of real multidisciplinarity.

7 References

Abdulridha, J., Ampatzidis, Y., Kakarla, S. C. and Roberts, P. (2020a). Detection of target spot and bacterial spot diseases in tomato using UAV-based and bench-top-based hyperspectral imaging techniques, *Precis. Agric.* 21, 955–978.

Abdulridha, J., Ampatzidis, Y., Roberts, P. and Kakarla, S. C. (2020b). Detecting powdery mildew disease in squash at different stages using UAV-based hyperspectral imaging and artificial intelligence, *Biosyst. Eng.* 197, 135–148.

Ampatzidis, Y., De Bellis, L. and Luvisi, A. (2017). iPathology: robotic applications and management of plants and plant diseases, *Sustainability* 9(6), 1010. https://doi.org/10.3390/su9061010.

Baranowski, P., Jedryczka, M., Mazurek, W., Babula-Skowronska, D., Siedliska, A. and Kaczmarek, J. (2015). Hyperspectral and thermal imaging of oilseed rape (*Brassica napus*) response to fungal species of the genus *Alternaria*, *PLoS ONE* 10(3), e0122913.

Barbedo, J. G. A. (2018). Factors influencing the use of deep learning for plant disease recognition, *Biosyst. Eng.* 172, 84–91.

Barreto, A., Paulus, S., Varrelmann, M. and Mahlein, A. K. (2020). Hyperspectral imaging of symptoms induced by *Rhizoctonia solani* in sugar beet: comparison of input data and different machine learning algorithms, *J. Plant Dis. Prot.* 127(4), 441–451.

Bassanezi, R. B., Amorim, L., Bergamin Filho, A. B. and Berger, R. D. (2002). Gas exchange and emission of chlorophyll fluorescence during the monocycle of rust, angular leaf spot and anthracnose on bean leaves as a function of their trophic characteristics, *J. Phytopathol.* 150(1), 37–47.

Bauriegel, E., Brabandt, H., Gärber, U. and Herppich, W. B. (2014). Chlorophyll fluorescence imaging to facilitate breeding of *Bremia lactucae*-resistant lettuce cultivars, *Comput. Electron. Agr.* 105, 74–82.

Bienkowski, D., Aitkenhead, M. J., Lees, A. K., Gallagher, C. and Neilson, R. (2019). Detection and differentiation between potato (*Solanum tuberosum*) diseases using calibration models trained with non-imaging spectrometry data, *Comput. Electron. Agric.* 167, 105056.

Biondi, E., Blasioli, S., Galeone, A., Spinelli, F., Cellini, A., Lucchese, C. and Braschi, I. (2014). Detection of potato brown rot and ring rot by electronic nose: from laboratory to real scale, *Talanta* 129, 422–430.

Bourgeon, M. A., Paoli, J. N., Jones, G., Villette, S. and Gée, C. (2016). Field radiometric calibration of a multispectral on-the-go sensor dedicated to the characterization of vineyard foliage, *Comput. Electron. Agric.* 123, 184–194.

Camargo, A. and Smith, J. S. (2009). An image-processing based algorithm to automatically identify plant disease visual symptoms, *Biosyst. Eng.* 102(1), 9–21.

Cellini, A., Blasioli, S., Biondi, E., Bertaccini, A., Braschi, I. and Spinelli, F. (2017). Potential applications and limitations of electronic nose devices for plant disease diagnosis, *Sensors (Basel)* 17(11), 2596.

Fahrentrapp, J., Ria, F., Geilhausen, M. and Panassiti, B. (2019). Detection of gray mold leaf infections prior to visual symptom appearance using a five-band multispectral sensor, *Front. Plant Sci.* 10, 628.

Ferentinos, K. P. (2018). Deep learning models for plant disease detection and diagnosis, *Comput. Electron. Agr.* 145, 311–318.

Fernández, C. I., Leblon, B., Haddadi, A., Wang, J. and Wang, K. (2020). Potato late blight detection at the leaf and canopy level using hyperspectral data, *Can. J. Remote Sens.* 46(4), 390–413.

Gao, Z., Luo, Z., Zhang, W., Lv, Z. and Xu, Y. (2020). Deep learning application in plant stress imaging: a review, *AgriEngineering* 2(3), 430–446.

Ghaffari, R., Laothawornkitkul, J., Iliescu, D., Hines, E., Leeson, M., Napier, R., More, J. P., Paul, N. D., Hewitt, C. N. and Taylor, J. E. (2012). Plant pest and disease diagnosis: electronic nose and support vector machine approach, *J. Plant Dis. Prot.* 119(5–6), 200–207.

Gold, K. M., Townsend, P. A., Chlus, A., Herrmann, I., Couture, J. J., Larson, E. R. and Gevens, A. J. (2020). Hyperspectral measurements enable pre-symptomatic detection and differentiation of contrasting physiological effects of late blight and early blight in potato, *Remote Sens.* 12(2), 286.

Gorbe, E. and Calatayut, A. (2012). Applications of chlorophyll fluorescence imaging technique in horticultural research: a review, *Sci. Hort.* 138, 24–35.

Habibullah, M., Mohebian, M. R., Soolanayakanahally, R., Bahar, A. N., Vail, S., Wahid, K. A. and Dinh, A. (2020). Low-cost multispectral sensor array for determining leaf nitrogen status, *Nitrogen* 1(1), 67–80.

Haboudane, D., Miller, J. R., Pattey, E., Zarco-Tejada, P. J. and Strachan, I. B. (2004). Hyperspectral vegetation indices and novel algorithms for predicting green LAI of crop canopies: modeling and validation in the context of precision agriculture, *Remote Sens. Environ.* 90, 337–352.

Heil, M. and Ton, J. (2008). Long-distance signalling in plant defence, *Trends Plant Sci.* 13(6), 264–272.

Huang, J. F. and Apan, A. (2006). Detection of *Sclerotinia* rot disease on celery using hyperspectral data and partial least squares regression, *J. Spat. Sci.* 51(2), 129–142.

Ilakiya, T., Parameswari, E., Davamani, V., Swetha, D. and Prakash, E. (2020). High-throughput crop phenotyping in vegetable crops, *Pharm. Innov. J.* 9(8), 184–191.

Johannes, A., Picon, A., Alvarez-Gila, A., Echazarra, J., Rodriguez-Vaamonde, S., Navajas, A. D. and Ortiz-Barredo, A. (2017). Automatic plant disease diagnosis using mobile capture devices, applied on a wheat use case, *Comput. Electron. Agr.* 138, 200–209.

Juroszek, P., Racca, P., Link, S., Farhumand, J. and Kleinhenz, B. (2020). Overview on the review articles published during the past 30 years relating to the potential climate change effects on plant pathogens and crop disease risks, *Plant Pathol.* 69(2), 179–193.

Kim, J. H., Bhandari, S. R., Chae, S. Y., Cho, M. C. and Lee, J. G. (2019). Application of maximum quantum yield, a parameter of chlorophyll fluorescence, for early determination of bacterial wilt in tomato seedlings, *Hortic. Environ. Biotechnol.* 60(6), 821–829.

Kitić, G., Tagarakis, A., Cselyuszka, N., Panić, M., Birgermajer, S., Sakulski, D. and Matović, J. (2019). A new low-cost portable multispectral optical device for precise plant status assessment, *Comput. Electron. Agric.* 162, 300–308.

Laothawornkitkul, J., Moore, J. P., Taylor, J. E., Possell, M., Gibson, T. D., Hewitt, C. N. and Paul, N. D. (2008). Discrimination of plant volatile signatures by an electronic nose: a potential technology for plant pest and disease monitoring, *Environ. Sci. Technol.* 42(22), 8433–8439.

Lázaro, E., Makowski, D., Martínez-Minaya, J. and Vicent, A. (2020). Comparison of frequentist and bayesian meta-analysis models for assessing the efficacy of decision support systems in reducing fungal disease incidence, *Agronomy* 10(4), 560.

Liang, J., Li, X., Zhu, P., Xu, N. and He, Y. (2019). Hyperspectral reflectance imaging combined with multivariate analysis for diagnosis of *Sclerotinia* stem rot on *Arabidopsis thaliana* leaves, *Appl. Sci.* 9(10), 2092.

Lindenthal, M., Steiner, U., Dehne, H. W. and Oerke, E. C. (2005). Effect of downy mildew development on transpiration of cucumber leaves visualized by digital infrared thermography, *Phytopathology* 95(3), 233–240.

Lowe, A., Harrison, N. and French, A. P. (2017). Hyperspectral image analysis techniques for the detection and classification of the early onset of plant disease and stress, *Plant Methods* 13, 80.

Lu, J., Zhou, M., Gao, Y. and Jiang, H. (2018). Using hyperspectral imaging to discriminate yellow leaf curl disease in tomato leaves, *Precis. Agric.* 19, 1–16.

Lussem, U., Bolten, A., Gnyp, M. L., Jasper, J. and Bareth, G. (2018). Evaluation of RGB-based vegetation indices from UAV imagery to estimate forage yield in grassland, *Int. Arch. Photogramm. Remote Sens. Spatial Inf. Sci.* XLII-3, 1215–1219.

Mahlein, A.-K., Rumpf, T., Welke, P., Dehne, H.-W., Plümer, L., Steiner, U. and Oerke, E.-C. (2013). Development of spectral indices for detecting and identifying plant diseases, *Remote Sens. Environ.* 128, 21–30.

Manganiello, G., Nicastro, N., Caputo, M., Zaccardelli, M., Cardi, T. and Pane, C. (2021). Functional hyperspectral imaging by high-related vegetation indices to track the wide-spectrum *Trichoderma* biocontrol activity against soil-borne diseases of baby-leaf vegetables, *Front. Plant Sci.* 12, 630059.

McKinnon, T. and Hoff, P. (2017). Comparing RGB-Based Vegetation Indices with NDVI for Drone Based Agricultural Sensing. AgribotixLLC AGBX021-17, pp. 1–8.

Men, S., Yan, L., Liu, J., Qian, H. and Luo, Q. (2017). A classification method for seed viability assessment with infrared thermography, *Sensors (Basel)* 17(4), 845.

Mohanty, S. P., Hughes, D. P. and Salathé, M. (2016). Using deep learning for image-based plant disease detection, *Front. Plant Sci.* 7, 1419.

Nawaz, M. A., Khan, T., Rasool, R. M., Kausar, M., Usman, A., Bukht, T. F. N., Ahmad, R. and Ahmad, J. (2020). Plant disease detection using internet of thing (IoT), *Int. J. Adv. Comput. Sci. Appl.* 11, 505–509.

Nikolaidis, I. (2008). Sensor networks and the law of accelerating returns, *IEEE Network* 4, 2–3.

Oerke, E. C. (2020). Remote sensing of diseases, *Annu. Rev. Phytopathol.* 58, 225–252.

Oerke, E. C., Steiner, U., Dehne, H. W. and Lindenthal, M. (2006). Thermal imaging of cucumber leaves affected by downy mildew and environmental conditions, *J. Exp. Bot.* 57(9), 2121–2132.

Padmavathi, K. and Thangadurai, K. (2016). Implementation of RGB and grayscale images in plant leaves disease detection–comparative study, *Indian J. Sci. Technol.* 9(6), 1–6.

Pallottino, F., Pane, C., Figorilli, S., Pentangelo, A., Antonucci, F. and Costa, C. (2020). Greenhouse application of light-drone imaging technology for assessing weeds severity occurring on baby-leaf red lettuce beds approaching fresh-cutting, *Span. J. Agr. Res.* 18, e0207.

Pérez-Bueno, M. L., Pineda, M., Cabeza, F. M. and Barón, M. (2016). Multicolor fluorescence imaging as a candidate for disease detection in plant phenotyping, *Front. Plant Sci.* 7, 1790.

Pineda, M., Barón, M. and Pérez-Bueno, M.-L. (2021). Thermal imaging for plant stress detection and phenotyping, *Remote Sens.* 13, 68.

Pineda, M., Pérez-Bueno, M. L., Paredes, V. and Barón, M. (2017). Use of multicolour fluorescence imaging for diagnosis of bacterial and fungal infection on zucchini by implementing machine learning, *Funct. Plant Biol.* 44(6), 563–572.

Pirna, C. and Lache, S. (2010). A new mechatronic approach for underlieing decisional processes in precision agriculture, *Bull. Transil.* 3, 27–32.

Prabhakar, M., Prasad, Y. G., Desai, S., Thirupathi, M., Gopika, K., Rao, G. R. and Venkateswarlu, B. (2013). Hyperspectral remote sensing of yellow mosaic severity and associated pigment losses in *Vigna mungo* using multinomial logistic regression models, *Crop Prot.* 45, 132–140.

Prashar, A. and Jones, H. G. (2014). Infra-red thermography as a high-throughput tool for field phenotyping, *Agronomy* 4(3), 397–417.

Rispail, N. and Rubiales, D. (2015). Rapid and efficient estimation of pea resistance to the soil-borne pathogen *Fusarium oxysporum* by infrared imaging, *Sensors (Basel)* 15(2), 3988–4000.

Rodríguez-Moreno, L., Pineda, M., Soukupová, J., Macho, A. P., Beuzoń, C. R., Barón, M. and Ramos, C. (2008). Early detection of bean infection by *Pseudomonas syringae* in asymptomatic leaf areas using chlorophyll fluorescence imaging, *Photosynth. Res.* 96(1), 27–35.

Rolfe, S. A. and Scholes, J. D. (2010). Chlorophyll fluorescence imaging of plant-pathogen interactions, *Protoplasma* 247(3–4), 163–175.

Sandmann, M., Grosch, R. and Graefe, J. (2018). The use of features from fluorescence, thermography, and NDVI imaging to detect biotic stress in lettuce, *Plant Dis.* 102(6), 1101–1107.

Santonicola, L., Villecco, D., Pentangelo, A. and Pane, C. (2019). Detecting downy mildew symptoms on wild rocket leaves by hyperspectral imaging. Proceedings of the XXV National Congress of Italian Phytopathological Society, Milan, Italy, 16–18 September 2019, 131.

Sharifi, R., Lee, S. M. and Ryu, C. M. (2018). Microbe-induced plant volatiles, *New Phytol.* 220(3), 684–691.

Steddom, K., Bredehoeft, M. W., Khan, M. and Rush, C. M. (2005). Comparison of visual and multispectral radiometric disease evaluations of Cercospora leaf spot of sugar beet, *Plant Dis.* 89(2), 153–158.

Sun, Y., Wang, J. and Cheng, S. (2016). Predicting. attacked time of tomato seedling by e-nose based on kernel principal component analysis. Proceedings of the ASABE Annual International Meeting, Orlando, FL, USA, 17–20 July 2016.

Svensgaard, J., Roitsch, T. and Christensen, S. (2014). Development of a mobile multispectral imaging platform for precise field phenotyping, *Agronomy* 4(3), 322–336.

Veys, C., Chatziavgerinos, F., Al Suwaidi, A., Hibbert, J., Hansen, M., Bernotas, G., Smith, M., Yin, H., Rolfe, S. and Grieve, B. (2019). Multispectral imaging for presymptomatic analysis of light leaf spot in oilseed rape, *Plant Methods* 15, 4.

Wagner, A., Michalek, W. and Jamiolkowska, A. (2006). Chlorophyll fluorescence measurements as indicators of fusariosis severity in tomato plants, *Agron. Res.* 4, 461–464.

Wang, L., Duan, Y., Zhang, L., Wang, J., Li, Y. and Jin, J. (2020). LeafScope: a portable high-resolution multispectral imager for *in vivo* imaging soybean leaf, *Sensors (Basel)* 20(8), 2194.

Wang, M., Ling, N., Dong, X., Zhu, Y., Shen, Q. and Guo, S. (2012a). Thermographic visualization of leaf response in cucumber plants infected with the soil-borne pathogen *Fusarium oxysporum* f. sp. *cucumerinum*, *Plant Physiol. Biochem.* 61, 153–161.

Wang, W., Li, C., Tollner, E. W., Gitaitis, R. D. and Rains, G. C. (2012b). Shortwave infrared hyperspectral imaging for detecting sour skin (*Burkholderia cepacia*)-infected onions, *J. Food Eng.* 109(1), 38–48.

Wilson, A. D. (2013). Diverse applications of electronic-nose technologies in agriculture and forestry, *Sensors (Basel)* 13(2), 2295–2348.

Wilson, A. D. (2020). Non-invasive early disease diagnosis by electronic-nose and related VOC-detection devices, *Biosensors* 10(7), 73.

Xie, C., Shao, Y., Li, X. and He, Y. (2015). Detection of early blight and late blight diseases on tomato leaves using hyperspectral imaging, *Sci. Rep.* 5, 16564.

Xue, J. and Su, B. (2017). Significant remote sensing vegetation indices: a review of developments and applications, *J. Sens.* 2017, 1–17.

Yang, X. and Guo, T. (2017). Machine learning in plant disease research, *Eur. J. Bio. Med Res.* 3(1), 6–9.

Zhang, S. W., Shang, Y. J. and Wang, L. (2015). Plant disease recognition based on plant leaf image, *J. Anim. Plant Sci.* 25, 42–44.

Zhao, Y. R., Li, X., Yu, K. Q., Cheng, F. and He, Y. (2016). Hyperspectral imaging for determining pigment contents in cucumber leaves in response to angular leaf spot disease. *Sci. Rep.*, 6, 27790.

Zhou, C., Mao, J., Zhao, H., Rao, Z. and Zhang, B. (2020). Monitoring and predicting Fusarium wilt disease in cucumbers based on quantitative analysis of kinetic imaging of chlorophyll fluorescence, *Appl. Opt.* 59(29), 9118–9125.

Zhu, W., Chen, H., Ciechanowska, I. and Spaner, D. (2018). Application of infrared thermal imaging for the rapid diagnosis of crop disease, *IFAC Papers OnLine* 51(17), 424–430.

Zubler, A. V. and Yoon, J. Y. (2020). Proximal methods for plant stress detection using optical sensors and machine learning, *Biosensors* 10(12), 193.

Chapter 7

Advances in decision support systems (DSSs) for integrated pest management in horticultural crops

Mark W. Ramsden, ADAS, UK; and Aoife O'Driscoll, NIAB, UK

1 Introduction

Pressure to meet growing global demands for high-quality horticultural crops, under tightening financial margins, has created intensively managed production systems. Injurious pathogens, invertebrates and plants (collectively 'pests') can pose a risk to both the quantity and quality of crops and require detailed management to maximise sustainable production and profit. Growers are faced with the challenge of managing a dynamic suite of concurrent pest populations, whilst minimising environmental impacts (Tilman et al., 2011). Broad-spectrum pesticides are powerful tools across agriculture; however, their widespread repeat application has had well-documented repercussions on non-target species, pests and their natural enemies in crops (Pimentel, 2005).

Various techniques can be employed to manage pests in order to avoid economic injury to crops, often requiring a detailed understanding of the interactions between crop and pests, and how they are affected by biotic and abiotic conditions. Integrated pest management (IPM) systems aim to reduce reliance on plant protection products (PPPs) through the implementation of combinations of pest-control methods to prevent and/or reduce infestations.

http://dx.doi.org/10.19103/AS.2021.0095.07

Despite IPM being a priority area for development, food production is still heavily reliant on pesticides, due to the challenges of incorporating many of the IPM measures within existing management practices and a lack (perceived or otherwise) of viable alternatives (Damos, 2015). Successful IPM systems depend on effective decision-making associated with the selection and implementation of strategic, tactical and operational techniques (Rossi et al., 2012; Barzman et al., 2015), defined in Table 1.

Together, strategic, tactical and operational IPM techniques form an overall IPM Plan. Decision support systems (DSSs) are interactive systems (usually software based) that help users identify and solve problems and make decisions as part of developing an IPM strategy. A good DSSs will enable the user to make effective strategic, tactical and/or operational decisions with confidence, and in full knowledge of the associated risks and assumptions.

Strategic techniques require many farm-specific considerations to be accounted for. Strategic DSSs therefore tend to be generic guides to improve the overall IPM uptake, such as LEAF's 'Simple Sustainable Integrated Pest Management' Guide and scoring system in the UK (LEAF, 2020). The majority of DSSs available currently is associated with tactical techniques, supporting decisions on when to monitor crops for pest activity and when action may be required to prevent economic injury, such as forecasting systems, monitoring systems and thresholds (Damos, 2015; Jones et al., 2010; Collier et al., 2020). Operational decisions usually relate to the implementation of management practices in response to variable pest pressure at the level of the plant or the whole crop and are manual (e.g. spot treatment of early signs of pest infestations) or automated (e.g. variable-rate spray technology) (Rossi et al., 2012). Some tactical decisions may also be fully automated, where specific actions are defined and are implemented in response to set conditions – such as climate controls in a greenhouse – though these still require initial decision making when set up and further monitoring thereafter.

The primary purpose of DSSs is to enable users to make more effective decisions by presenting the likelihood of a given scenario (Rose et al., 2016). The use of in-crop monitoring, forecasting models and thresholds of economic injury are good basic tools for inclusion in an overall IPM plan (Ramsden et al., 2017), and many DSSs build on and/or combine these tools to enable users to optimise their decision-making. They are never a replacement for grower experience or industry support but should add to other sources of information to improve decision-making. To achieve this, DSSs should provide accurate, transparent information about the pest and the DSSs user should be able to relate this to a specific decision or action. DSSs may also improve users' understanding of the pest(s) and enable them to collect monitoring data that promote self-assessment of the effectiveness of their IPM strategy, tactics and/or operations.

The aim of this chapter is to highlight the current availability of strategic, tactical and operational DSSs in horticulture and provide a summary of what

Table 1 Examples of the types of IPM Decision Support Systems available, and their typical inputs and outputs

DSSs type	Definition	Example	Typical input(s)	Typical output(s)
Strategic	*Application of many precautionary, supportive and preventative measures*	IPM Self-assessment systems (e.g. LEAF IPM Guide and scoring system)	Detail on crop and non-crop management prior to, during and after the cropping period	Scoring of overall IPM strategy and/or specific elements, with indication of potential areas for improvement
Tactical	*Continuous monitoring and associated decisions and rules for application of pesticides and/or bioprotectants*	Weather-based forecast (e.g. cabbage root fly forecasting) information available through outputs in the AHDB Pest Bulletin	Daily/hourly air/soil temperature, rainfall and humidity	Estimated risk of given pest arriving and/or requiring action
		Observation-based risk assessment (e.g. aphid and diamond-back moth) monitoring available through the Rothamsted Insect Survey	Report presence/absence of pest at field scale at given time	Risk map or table of pest occurrence over time
		Action threshold (e.g. economic thresholds for pollen beetle or phoma leaf spot in Brassica crops)	Abundance of pest per unit area/portion of crop	Indication on whether or not action is required, and if so, what action may be appropriate
		Combination (e.g. VM Plasmopara) to support treatment decisions for downy mildew on grapevines, available through VitiMeteo	Climate, presence of pests and crop phenology data	Predicts infection events but also phenology and leaf area development. Spray intervals are determined by crop growth periods
Operational	*Checking the success of measures taken*	Manual operational decisions such as placement of biocontrol agents close to new pest infestations (e.g. Olivo.net) to support treatment decisions for olive fruit fly	Observation of pest activity in crop at fine scale	Provides spatial and temporal patterns of the pest population and allows for precision treatments against the pest.
		Automated operational decisions (e.g. variable sprayer technology)	Image-based data input	Automated application of treatment where pest(s) are predicted/detected

users should look for in DSSs. We will then discuss the future role of DSSs and supporting platforms, and the benefits these will have for the horticultural sector.

2 The availability of integrated pest management decision support systems in horticulture

Decision support systems, particularly tactical systems, are intrinsically linked to crop monitoring as they require crop observation data either as an input into the system, or as a follow-up to DSSs outputs to assess actual pest pressure. Monitoring is a core part of IPM and is time consuming, with associated economic costs. Many DSSs enable users to target monitoring in crops or parts of crops that are most vulnerable to pests, and where effective management of infestations would result in economic gains. Such DSSs will provide users with the confidence to know whether targeted crop monitoring is required, information on how this should be done and details of economic thresholds (Fig. 1; Ramsden et al., 2017). This two-step decision process allows for application of any IPM techniques for reducing pest risks, whether they are preventative or curative.

2.1 Tactical decision support systems

Tactical systems are mostly associated with a decision on the application of treatment (chemical pesticide or alternative). These systems enable users to assess (i) the likelihood of crop injury due to a given pest and/or (ii) the economic damage that is likely to occur if no action is taken. Both the likelihood of crop injury and economic damage can be quantified with respect to biotic and/or abiotic conditions, enabling an overall quantification of risk that can guide the user on how best to proceed. The likelihood of crop injury can be assessed

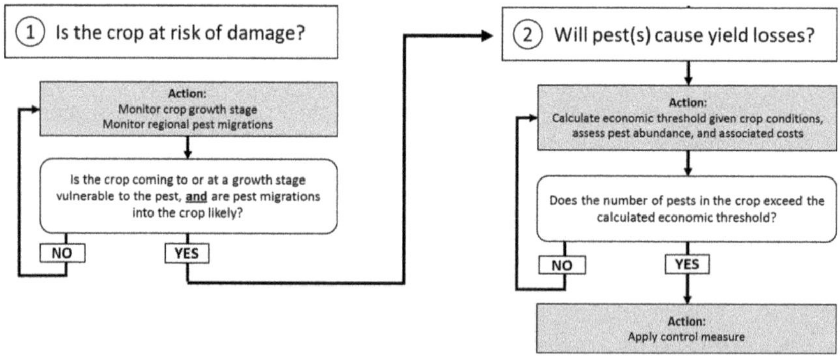

Figure 1 Decision flow for management of horticultural pests (adapted from Ramsden et al. 2017).

based on whether the crop is at a vulnerable growth stage, whether the pest is likely to be present and the ability of the crop to tolerate the pest. Tactical DSSs can be broken down into forecasting systems or observation systems. Forecasting systems model likelihood based on pest phenology and behaviour associated with climatic data. The output ranges from a basic estimation of when the pest is first likely to arrive, to suggested treatment periods that combine the arrival of the pest with knowledge of the life cycle to estimate when it is best to take action and so on (Collier et al., 2020). Observation systems, in contrast, present data collected from monitoring stations over space/time and provide evidence of pest activity, from which growers and advisors can infer the likelihood of infestation in their crops. Observation systems are more open to interpretation since they offer no prediction of future pest infestations. Forecast systems and/or observation-based systems can be combined with knowledge of crop variety, growth stage and crop condition to give an overall assessment of the likelihood of crop injury. This knowledge is usually achieved through inspection of the crop by the grower or the agronomist. If a DSSs predicts a low likelihood of pest activity and/or the crop is not at a vulnerable growth stage, no further action is necessary beyond routine crop monitoring. Where these systems suggest that activity is likely, more detailed monitoring of the crop is required to assess pest levels and the potential for economic injury, using thresholds (Ramsden et al., 2017).

Economic thresholds are a valuable DSSs within IPM for assessing whether control measures, such as pesticides, are necessary. They are usually defined in terms of the numbers of a pest per unit area, per plant, or per part of plant, above which action should be taken (Ramsden et al., 2017). While many thresholds are available for pests across horticulture crops, they are rarely used and there are significant pests for which thresholds are lacking. In tandem with this, even when reference thresholds are available, pesticides are still applied and there are several reasons for this. For example, many thresholds are based on outdated, unsubstantiated or inaccessible research, and so are considered unreliable by growers and agronomists, while assessment procedures for others are impractical within current production systems (Ramsden et al., 2017). Given these concerns, and in the context of fluctuations in the value of crops and the cost of PPPs, growers understandably err on the side of caution.

2.2 Strategic and operational decision support systems

A fully integrated pest management plan requires a holistic approach across crops, which balances the requirements for individual pest management against the wider farm management systems (Lechenet et al., 2017). Developing such an approach is unavoidably complex, often requiring diversification of cropping systems with wider rotations and use of different varieties, inputs and off-crop management – all of which are highly knowledge intensive. The context-specific nature of pest management relies on local expertise and experience, meaning that the best support for strategic decision making is usually found through engagement with networks of peers and technical experts rather than based on complex strategic DSSs (Lechenet et al., 2017).

Operational DSSs enable users to respond to the success of the strategic and/or tactical control measures in the crop, as part of ongoing monitoring efforts. Generally, operational decisions involve fine-scale inspection and action where early signs of pest infestation are identified. As a result, most operational decisions are based on presence/absence and are often clear-cut decisions or automated actions, for example, variable application technology automating pesticide application based on image-based detection of pest presence (Mohanty et al., 2016). Use of automated technology can have further benefits where systems are able to detect pest presence prior to symptoms being visible to growers. Since many of these systems require no further input from the user after being set up, they effectively become decision-*making* systems rather than decision support systems.

3 What to look for in decision support systems for pest management

DSSs are developed as a result of various initiatives and ambitions. The target end users are normally growers and/or agricultural advisors, and most DSSs are presented in very simple formats to encourage wider use. Simplification has a benefit in making DSSs outputs readily accessible. However, it can create barriers to use where outcomes are perceived as untrustworthy due to a lack of transparency or sophistication. Conversely, if the system is too complex, its users may be put off by the time taken to provide the required inputs and interpret the outputs. DSSs are most effective where they are used in tandem with supplementary information from peers and technical experts. Ideally, the system will have been developed with direct input from representative target users. This ensures that it is fit for the purpose and creates a network of informed users who trust and support the final system (Jakku and Thorburn, 2010). Designing sophisticated tools with simple user interfaces can also miss opportunities to support growers with specific challenges, such as supporting adherence to legislative requirements (Rose et al., 2016).

Most modern DSSs provide quantifiable information about the likelihood and/or scale of pest infestations, based on validated data appropriate to the target user. In addition, DSSs provide information about the pest and its ecology, directly or through supporting materials, that further supports the user in understanding and managing pests. This can occur either by improving their ability to interpret, adapt and apply the DSSs outputs to specific cropping situations, or by providing insight into combinations of techniques which may improve overall pest management.

3.1 Decision support systems: inputs, outputs and validation

Input requirements for DSSs vary from relatively simple information, such as the timing of management activities (e.g. drilling date), to a complex integration of climate data, agronomic factors and pest observations. DSSs outputs may be straightforward. Economic thresholds, for example, provide benchmarks and/or ranges of values against which observations can be compared and decisions made. Interpretation of outputs may, however, require a great deal of additional information and understanding, especially where DSSs are bound by assumptions about some of the factors (e.g. soil type, crop variety/health, availability of treatments etc.) that may not always apply to an individual case. Whatever the inputs and outputs, the associated assumptions and a summary description of the algorithm behind the outputs should ideally be available. Advanced systems may, however, combine multiple algorithms, requiring data on factors not available to the system or approximations derived from available data. These systems are inevitably difficult to summarise, but publication of their core principles encourages user engagement and improves uptake. It also allows for adaptation of a system for use in other regions with similar crops/pests (Wallhead and Zhu, 2017). When a new DSSs is launched, it important that there is clear communication of its aims, limitations and benefits to target users in order to maximise effective application by end users (Wilson et al., 2009).

In order to convince growers and advisors of the value of a DSSs, validation testing must demonstrate a reliable outcome, both in terms of improved pest management and economic returns (Jorgensen et al., 2020). Ideally, validation takes place alongside DSSs development, together with direct engagement with potential users (Jakku and Thorburn, 2010). This establishes a relationship between the developer and the relevant stakeholders, involving both parties in the initial concept and requirements. Technical development of the DSSs algorithms and associated identification of assumptions and limitations is then undertaken by the developers, and the system then validated on farms within the initial user group. Further testing in the wider community ensures that the DSSs is working correctly, and then the user interface can be developed (Oliver et al., 2012). Research-led DSSs development is often the result of short-term projects and tends to focus on the development of technical aspects of the algorithm behind the system. These projects tend to engage well with stakeholders early on, but ultimately leave the resultant DSSs largely unsupported, often without any form of user interface. This can lead to good systems being lost due to atrophy and a lack of peer-to-peer validation within the farming community (Gent et al., 2009). Research-led development also focuses on the biological factors associated with pest distribution, impact and management, rather than short-term economic implications of pest infestations. This can result in conflicts between long-term sustainable management and the need for immediate solutions to a pest infestation (Wilson et al., 2009). For example, where a DSSs is built on an assumed understanding that weed dispersal is predominantly driven by human activity, users who do not agree or are uncomfortable with this assumption may disregard the outputs because it does not promptly address their specific problem. Instead, the user may resort to an alternative control option – usually a synthetic pesticide – which provides short-term gains but does not contribute to long-term sustainable management strategies (Wilson et al., 2009).

Development of a DSSs that focuses too heavily on the user experience and/or short-term benefits can also limit the long-term viability and use of the system. While it is critical that DSSs development is user-centric (Parker and Sinclair, 2010), ongoing updates associated with rapid advances in technology can render the associated equipment and software obsolete in a short period of time. Even if past versions are still functional, they may no longer be supported by the company from which they were purchased (Kanatas et al., 2020). The repeated and/or frequent changes in the DSSs user-interface and/or content can be off-putting or distracting to the user (Kanatas et al., 2020).

3.2 Decision support systems as tools for knowledge exchange

A combination of biotic and abiotic conditions is responsible for crop performance, including damage by pests. The impact that pests can have on a crop relate to the type of injury they cause, and the other stresses the plant is under at that time (Boote et al., 1983; Ramsden et al., 2017). The collection and analysis of data during the development of the algorithms behind DSSs can provide a great deal of insight into the ecology of pests and their interaction with the crop. This knowledge can have value in addition to the DSSs outputs, enabling a more adaptive approach both to DSSs interpretation and to wider integration with other IPM techniques. This is especially important where DSSs outputs may be complex or counter-intuitive at face value and require a programme of knowledge exchange (Bruce, 2016). Knowledge exchange activities can therefore have a wider impact than the DSSs itself, equipping growers with a wide range of skills for crop protection that can remain long after an individual system has been discarded.

3.3 Decision support systems as tools for self-assessment

Pest management decisions are among some of the most difficult for growers and agronomists to make. Pest infestations resulting in financial losses can severely damage an agronomist's reputation, while there is little chance of repercussions against decisions to use pesticides when the risk of infestation is low or uncertain. Although the need to reduce pesticide applications is acknowledged widely, few tools exist to enable quantitative on-farm assessment of the effectiveness of different IPM techniques. The only route to self-improvement is largely trial and error, rather than detailed review of specific techniques.

IPM strategies are widely employed by many growers (Creissen et al., 2019) and their success is often quantified as a reduction in the Treatment Frequency Index (TFI), the number of recommended doses applied to a crop area (Lechenet et al., 2017). This is an effective tool for assessing the impact of IPM strategies at a regional level. The TFI is less informative for individual growers and advisors for evaluating and improving their own IPM strategies. Growers are understandably reluctant to replace practices they have developed over many years with alternative strategies that may incur large setup costs, and which may require long-term trade-offs between yield/quality and consistency/sustainability. Radical changes associated with transition towards advanced IPM can discourage growers, even though evidence shows that it can be more economically and environmentally sustainable than 'conventional' pest control strategies in the long term (Lechenet et al., 2017). Increasingly, DSSs are being employed to demonstrate risk relative to critical values. Such benchmarking enables growers to interpret outputs within the context of the wider risk in their region, or against past cropping experiences, and adapt their management accordingly. This network approach, usually consisting of small sub-groups of peers supported by facilitators and/or technical experts, enables adaptable interpretation and development of crop management within the context of wider farm management considerations (Clarke et al., 2019). Pest observation networks are the clearest current example of self-assessment facilitated by a DSSs. The observation network primarily acts as a data source for relative pest abundance across a region, indicating where infestations have been reported and where detailed monitoring is advised. This data also enables growers to benchmark their own observations against their peers, and then look to understand the drivers of such differences. These examples can then be used as case studies to promote wider uptake of observation networks. In France, the DEPHY network has achieved significant reductions in pesticide applications by using approaches such as this.

3.4 Case studies

The usefulness of pest and disease models to forecast pest risk and gather biological data has been well documented in the scientific literature. Relatively few of these model approaches have been translated into practical forecasting tools and related DSSs which growers can use in a commercial setting. DSSs available on three economically important diseases and pests are summarized below; powdery mildew on strawberry (*Podosphaera aphanis*); downy mildew on onions (*Peronospora destructor)* and cabbage root fly (*Delia radicum)*.

3.4.1 Powdery mildew (P. aphanis) on strawberry

Strawberry powdery mildew caused by the ascomycete fungus *P. aphanis* (Wallr.) (Peries, 1962) can infect all aerial parts of the strawberry plant and is a major constraint in strawberry production. Powdery mildew can be particularly difficult to manage in polytunnel-grown crops if the environment within the tunnel is not appropriately controlled. Hot and humid conditions prevail under the tunnels when they are closed and such conditions are conducive to the infection and spread of strawberry powdery mildew (Palmer, 2007, Carisse et al., 2013a,b). *P. aphanis* has many asexual generations of the fungus in a season and it is these asexually produced conidiospores which give rise to the seasonal epidemics of the disease. *P. aphanis* can develop at temperatures ranging between 5°C and 30°C and at a relative humidity as low as 12% (Peries, 1962; Amsalem et al., 2006). The speed with which it completes its life cycle is dependent on the influence of temperature and humidity on growth from conidiospore germination through to mycelium development and finally to spore production. After infection, the latency period (from spore germination to spore production) ranges from 14 days down to as low as 7 days under favourable conditions (Carisse et al., 2013b). One of the difficulties associated with predicting disease development for strawberry powdery mildew simply by using weather data is the wide range of temperature and relative humidity combinations under which the fungus can germinate and develop. Current weather-based models consider the interaction among these weather variables and several decision-support tools have been developed for the management of strawberry powdery mildew which are commercially available.

In the United States, the UC Davis Powdery Mildew Risk Assessment Model, developed at first for grape powdery mildew, was adapted for strawberry powdery mildew (Miller et al., 2003; Landry et al., 2010). It is based on a risk index between 0 and 100 points. In general, points are added to the risk index when four consecutive hours of temperature between 18°C and 27°C are observed during a day, otherwise, points are deducted. The index value is then coupled with the concentration of airborne conidia and both are considered within the model to provide a recommendation for suitable time intervals between pesticide applications as well as advice on which product should be applied.

In the United Kingdom, a rule-based prediction system for strawberry powdery mildew is available to growers, which calculates when *P. aphanis* inoculum present at a site poses the highest risk of developing into an epidemic. This is determined by calculating the percent of completion of a disease cycle. This is based on the number of hours with favorable conditions for germination, growth and sporulation (Dodgson et al., 2008). The system records the accumulated number of hours (up to 144) of disease-conducive conditions (temperature range 15–25°C and RH > 60%) needed for the development of the pathogen and identifies high-risk days when sporulation may occur, thus allowing growers to spray at the optimal time. Twice-weekly monitoring of disease severity in field trials using this software in 2014 and 2015 showed that strawberry crops that followed the predicted spray programme had between two and four less fungicide applications than the commercial programme, as well as less severe disease. A more user-friendly web-based system was introduced in 2017 which has improved the efficient use of fungicides with different modes of action and reduced fungicide applications. During the 2019 season, the system was validated on eight commercial sites throughout the United Kingdom and is now available for growers to purchase.[1]

SafeBerry is a DSSs that has been developed in Italy (Eccel et al., 2010). Like the UC Davis model, SafeBerry calculates a potential risk index which combines a basic risk index calculated using fixed traits such as varietal susceptibility, presence of overhead irrigation, plant density and a daily risk index calculated from variable traits such as phenological stage of the crop, disease incidence and time of disease onset (Carisse and Fall, 2021). The system determines temperature suitability for disease development based on measured and forecasted temperatures and a recommendation on whether to spray is given, along with a suggestion of appropriate fungicides (Eccel et al., 2010).

In France, the Centre Technique au Service de la Filière Fruits et Legumes (Ctifl) has developed a predictive model for strawberry powdery mildew (Bardet and Vibert, 2011). Five stages of the *P. aphanis* life cycle (dispersion, infection, mycelium growth, sporulation and disease progression) are modelled as functions of different environmental parameters obtained from weather data, including temperature, relative humidity and rain. A risk index is then calculated for a 4-day mobile period where dispersion and infection are likely to occur (Bardet and Vibert, 2011). The model is available through the web platform Inoki as a paid service (Ctifl, 2017).

3.4.2 Downy mildew (*P. destructor*) in onions

Onion downy mildew caused by *P. destructor* is an obligate oomycete fungal pathogen that can cause heavy yield losses in both salad and bulb onion crops. Once established in a crop, the disease can become readily airborne and

1 https://www.ceresagritech.org/projects/strawberry-decision-support/

inoculum (as conidia) is dispersed locally and over longer distances to infect other onion crops. The biology of this pathogen has been studied since the 1930s (Cook, 1932; Yarwood, 1937, 1943), with studies in the 1980s further investigating weather variables and environmental changes which promote *P. destructor* infection and disease spread. The production of spores of *P. destructor* occurs only at nights during favourable conditions (Yarwood, 1937; Leach et al., 1982). When humidity is high (<95%) from 08:00 am onwards, sporulation may occur at temperatures as low as 10°C, whereas the optimal temperature rises to 18°C when the threshold humidity does not begin until after 03:00 am (Hildebrand and Sutton, 1984a). Mature sporangia are mostly dispersed by wind and there is no active release mechanism *per se;* however, *P. destructor* sporangia can be vigorously discharged into the air in response to a reduction in relative humidity (Leach et al., 1982). After infection, a latency period of 13 days is observed at warmer temperatures (25°C/17°C for day/night) and 15–17 days at cooler temperatures (18°C for day/night, Cook, 1932; Hildebrand and Sutton, 1984b).

These findings led to the development of the DOWNCAST model in the 1980s (Jesperson and Sutton, 1987). DOWNCAST is a rule-based predictive model which is still in use in several regions around the world, such as the eastern regions of Canada. DOWNCAST predicts sporulation and infection events of *P. destructor*; however, it cannot predict the quantity of sporangia produced (Gilles and Kennedy, 2004). Tests of the model in Canada suggested that it gave positive predictions of sporulation for 38 out of the 45 nights where sporulation was observed (Jesperson and Sutton, 1987). Observations of *P. destructor* on onion trap plants in Australia also suggested that the model could predict infection events (Fitzgerald and O'Brien, 1994). However, in field tests in the Netherlands, DOWNCAST gave positive predictions of sporulation for only 11 out of the 24 nights observed (de Visser, 1998). This suggested that although the model could be a useful tool in drier, more continental climatic conditions, it was not consistently accurate in the maritime climate of mid-to-northern Europe where weather conditions are often highly variable. Other models such as ONIMIL, Zwipero and MILIONCAST were later developed to further improve on predicting the risk of *P. destructor* sporulation or infection or both (Battilani et al., 1996; Friedrich et al., 2003; Gilles et al., 2004).

MILIONCAST (acronym for MILdew on onION foreCAST) is a risk-based model which was developed based on data from controlled-environment studies investigating the effect of temperature and humidity on downy mildew sporulation (Gilles et al., 2004). Downy mildew infection risk factors are a culmination of sporulation risk, infection risk and latent period development risk. Visual symptoms in the form of lesions are expected to appear when a threshold for each of these risk factors exceeds a set score. MILIONCAST gave more correct predictions of sporulation than the DOWNCAST model, however the prediction was still only an indicator of the potential of *P. destructor* to sporulate during a night. The actual number of spores produced during a night is dependent on the level of infection and may vary among cultivars and different strains of *P. destructor* (Van der Heyden et al., 2020). Further developments have now made it possible to predict sporulation risk from temperature and air humidity readings at more frequent intervals (Gilles et al., 2004).

This model is now available as a commercial risk prediction service via CropMonitor (www.cropmonitor.co.uk). The scores are used to predict the risks of sporulation, initial infection and latent period development, but not the risk of symptom development. When both the sporulation and infection risk thresholds have been exceeded, it is presumed that infection will occur. Once the latent period development threshold has been exceeded, it is assumed that symptoms will become visible in the crop. This is an important factor for growers to consider when using the model, because the spray regime should be targeted at the initiation of infection and not at the initiation of symptom development. The model is likely to be most useful for timing the application of systemic fungicides targeted at preventing early infections, and less so for contact fungicides which are thought to have curative activity but in practice can be less effective. In future, it is hoped that the model will be used in conjunction with a lateral flow diagnostic (LFD) test, developed in partnership with Mologic Ltd. The LFD test detects increasing concentrations of *P. destructor* spores in collected air samples, thereby suggesting an increased risk of disease development in crops dependent on environmental conditions (Kennedy and Wakeham, 2008). Work is ongoing to provide onion growers more robust data on the correlation between spore numbers and disease development to demonstrate the value of using both tools for effective decision making in a commercial setting.

3.4.3 Cabbage root fly (*D. radicum*)

The cabbage root fly (*D. radicum*) is a major pest of cruciferous crops in temperate regions of the world. After overwintering as pupae in the soil, adult flies start to emerge in Spring (Finch and Collier, 1983); eggs are usually laid in the soil close to the base of the plant and newly hatched larvae emerge about 6 days later (Finch et al., 1975). Damage is due to feeding by the fly larvae which move through the soil to burrow into the roots. Larvae feed for about 3 weeks on the roots and stems of plants and then pupate close to the roots. In leafy brassicas, slight damage to the root system may not be a problem but heavy damage may cause the plant to wilt and even die (Finch, 1993). Cabbage root fly can occur as different 'diapause' or overwintering phenotypes in some countries (Collier and Finch, 1983, 1988). Different populations have varying temperature requirements for diapause completion/fly emergence, which is a challenge for accurate prediction of activity (Hawkes, 1972; Finch and Collier, 1987). The first generation of adult flies usually emerge in April–May and a second generation in July–August. In warmer regions, a third generation of flies may emerge in late August to September (Eckenrode and Chapman, 1972). Once early emerging flies resume their development at the end of diapause, they require time to feed, mate and mature their eggs. This time to development is equivalent to about 80 degree days (D°) above 6°C (Eckenrode and Chapman, 1972; Finch and Collier, 1983). The second biotype known as 'late-emerging flies', requires the presence of an additional phase in the course of diapause, which corresponds to 280D° above a base temperature of 7°C (Roberts, 1994). In some areas, both early emerging and late-emerging *D. radicum* populations are present (Finch et al., 1986; Johansen and Meadow, 2005, 2006), and overlapping generations of these two biotypes may lead to continual pest pressure during the growing season.

Several degree-day models are available which predict the timing of the various stages in the cabbage root fly life cycle and serve as a basis for decision making by growers in countries across Europe (Collier et al., 2020). Within Europe, two forecasting systems are based on simulation models. These are the UK models (Collier et al., 1991; Phelps et al., 1993) and the SWAT models from Germany (Hommes and Gebelein, 1996). The models simulate development of a field population of *D. radicum*, starting with the overwintering stage (diapause as pupa) and following development through to summer dormancy (aestivation) for the first, second and, where appropriate, third generation until autumn. The models are based on a series of rate equations which describe the relationship between the rate of development of *D. radicum* and temperature. All models require current local weather data or standard measurements of air and soil temperatures taken at agrometeorological stations.

In the UK, the Agriculture and Horticulture Development Board (AHDB) Pest Bulletin (https://www.syngenta.co.uk/ahdb-pest-bulletin) is an online service with forecast outputs and on-site monitoring data for *D. radicum*, as well as other root-feeding larvae of the Diptera species. The forecast indicates when flies of each generation will emerge from pupae and when they will lay eggs. The forecast outputs are based on weather data collected by Plantsystems Limited and funded by Syngenta. While the forecasts predict the timing of fly activity, they do not predict the number of insects present in the crop, as this depends on the size of the local fly population. For control purposes, it is considered that the pest has become active when 10% of the cabbage root flies have emerged/laid eggs, and that the peak of activity has occurred when 50% have emerged and/or laid eggs.

In Germany, the SWAT model has not been formally validated to date due to a lack of appropriate digitally available monitoring data. Work is underway to achieve this, and a new software version of the SWAT model, which is compatible with modern operating systems, is under development (Collier et al., 2020). Forecast information is disseminated through ISIP (Information System for Integrated Plant Production) (https://isip.de/isip/servlet/isip-de) which provides advisors with relevant information on request.

In Norway, a degree-day model for *D. radicum* has been developed based on spring emergence and the oviposition period (Brodal et al., 2007; Meadow et al., 2020). Damage thresholds are also being used for the *D. radicum/D. floralis* complex in Norway; these thresholds are related to the plants' developmental stage and indicate the number of eggs per plant that can be tolerated before a reduction in growth and yield is expected. VIPS (Varsling Innen PlanteSkadegjørere) is a web-based DSSs designed to assist Norwegian growers in the management of pests, diseases and weeds of cereals, vegetables and fruit crops. For cabbage root fly, it produces warnings based on weekly observations of oviposition related to damage thresholds for the different development stages (www.vips-landbruk.no). VIPS was established in 2001 as a collaborative project between Bioforsk and Norwegian Agricultural Extension Service (NAES) and consists of a network of more than 80 automatic weather stations across crop production areas, weather forecasts from the Norwegian Meteorological Institute and biological/field observations collected by NAES (Collier et al., 2020).

A forecasting model based on degree days for *D. radicum* is also available in Denmark based on local soil temperatures for individual postal code areas, using a forecasting model provided by PlanteInfo (www.planteinfo.dk), a DSSs used by Danish growers and agricultural advisers. The current local emergence of *D. radicum* in Denmark is communicated online via the advisory service HortAdvice[2] for growers and advisors to observe when risks of oviposition by first- and second-generation flies are highest. In Switzerland, information is provided via the 'Gemüsebau Info', an electronic bulletin coordinated by the competence centre for agricultural research, and Agroscope, covering all growing regions and the three official languages. In France, information is given through the 'Bulletins de Santé du Végétal' coordinated by the regional agricultural chambers and is available online.[3]

Forecasts are validated using insect monitoring data, usually yellow water traps, yellow sticky traps or by sampling plants for eggs and larvae (Finch and Skinner, 1974; Finch, 1992). Tools for more efficient monitoring, like automatic counting of pests and predators on yellow sticky traps using smartphone images, are under active development and will encourage growers to monitor

2 SEGES. Available online: https://www.landbrugsinfo.dk/Planteavl/Plantevaern/Skadedyr/Skadedyrsarter/Sider/PL_Kaalflue_varsling.aspx.
3 Bulletins de Santé du Végétal (BSV). Available online: http://draaf.bretagne.agriculture.gouv.fr/Bulletins-de-Sante-du-Vegetal-BSV.

in a more standardised way in future (Rustia et al., 2018; Moerkens et al., 2020, Pattnaik et al., 2020).

3.5 Gaps and limitations

The output from DSSs needs to be reliable, practical, and cost-effective. Systems are usually designed to be easy to use, but many require substantial input from the user that may be difficult and/or time-consuming to collect. Oversimplification of a system may, however, necessitate a lot of assumptions, restricting its ability to make trustworthy predictions. Above all, DSSs must be demonstrably cost-effective. Their role in pest management is undermined where the time taken or the cost of associated equipment makes it uneconomical to undertake assessments at the frequency required, or where outputs do not consistently support effective management of the pest. The distribution of most pests in crops is patchy rather than uniform, and monitoring may be required in multiple locations across a field. In farms with several crops in production at any given time, proposed monitoring requirements can quickly become unachievable even when several assessments can be done concurrently. DSSs can help address this, by enabling effective targeted assessments. Too often, DSSs are developed independent of a route to integration within the context of other crop management requirements. The challenge remains to develop DSSs that can be reliable for individual producers, accessible across relevant sectors and regions, and readily included alongside existing management practices.

4 Integrated pest management decision support system platforms

The World Wide Web (WWW) has provided virtual space for sharing knowledge, removing barriers to publishing, accessing and exploiting data (Damos, 2015). DSSs, once developed in isolation, are increasingly being combined to provide a suite of DSSs accessed via online dashboards. Further expansion into smart phone applications can provide DSSs outputs directly to growers in the field, promising rapid alert systems for key crop pests along with associated decision support. Incorporation of graphical information-based systems can provide real-time, targeted support, and the increasing role of machine learning will require rapid sharing of data inputs from users via online systems (Mahlein et al., 2012; Hayashi et al., 2019). These systems can, however, become so complicated in their design and functionality that the core principles are hidden from the user. As such, they either become decision-*making* systems, since the user is no longer able to question the outputs, or become redundant as the user no longer trusts them. Online systems can also automate collation of input data, such as climate data, through open access sources, user-provided sources (e.g. growers' own weather stations) or additional pay-per-use services. Where online platforms are sufficiently independent and robust, they can also provide a source of acceptable criteria for justifying management decisions, whether to regulatory bodies, wholesalers, food processers or consumers (Rossi et al., 2012).

As demonstrated in these case studies, there are often multiple DSSs available for a given pest, and different DSSs can have different input requirements and present outputs in different formats. A key benefit of multi-DSSs platforms is the ability to automatically integrate inputs (e.g. climate data) and standardise DSSs output formats, enabling the relative assessment of different pest risks during the cropping season via a range of multimedia tools and supportive materials (text, graphs, maps, photos and video)(Rossi et al., 2012).

4.1 Case study: VIPS – an Open-Source technology platform for integrated pest management decision support

VIPS is an Open Source technology platform developed for IPM decision support in Norway (Fig. 2). The structure of the system makes all available IPM tools for pests and diseases within a cropping system open for further development and validation and provides a single point of access for users looking to use DSSs. VIPS therefore brings growers, advisors, researchers, software developers and other IPM stakeholders together within one system. This also facilitates rapid release of new tools, without any delays or reprogramming of models when research and development are completed. At present, 24 models are running in VIPS, covering all pest and disease models used in Norway (vips-landbruk .no). VIPS also includes a decision support tool for management of weeds in cereals, based on the Danish Plant Protection Online.

4.1.1 Data input

The structure behind VIPS makes it possible to connect it to any online source of input data, such as weather stations, external forecast systems and automated insect traps. It can also replace missing weather data where necessary and use locally sourced information from observation networks where required by models within the platform.

4.1.2 Data output

Model output is typically presented with colour codes in red, yellow or green to show the risk of infection/infestation (high to low). Each model is also explained and presented in more detail, providing any assumptions or conditions that should be considered when interpreting the outputs.

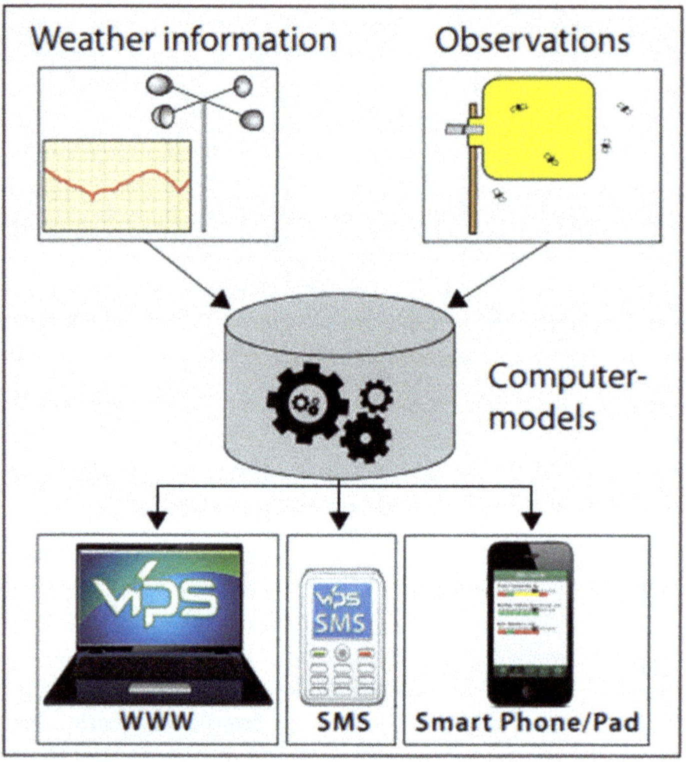

Figure 2 Flow chart showing data input and output in the VIPS system (adapted from Skog et al., 2015).

4.1.3 Mobile enabled

VIPS mobile is a version of the VIPS forecasting service adapted for tablets and smart phones, providing updated model forecasts to the user in the field. The user can configure the application to personal preferences (location, crop and pest) (Fig. 2). A text message service is also available, where users can receive forecasts from selected models and locations.

A current EU-funded project, IPM Decisions (H2020 project No. 817617) is developing a Europe-wide Platform for IPM DSSs, building on VIPS services across arable and horticultural sectors.

5 Conclusion

Beginning in the 1950s, the 'Green Revolution' ushered in an era of synthetic chemical pesticides. As the demand for produce increased, so did the reliance on pesticides to boost production on a grander scale. As the repercussions of extensive use of these pesticides on human health, non-target species and the wider environment became apparent, the development of DSSs that supported the avoidance and the targeted use of pesticides progressed rapidly. The development of DSSs was not reflected in their uptake, and the wider barriers to implementation are now becoming clearer. While numerous DSSs have been developed, relatively few are used routinely (Rossi et al., 2012). Uptake is thought to be constrained by the following (Shtienberg, 2013):

i regional fragmentation of DSSs development and user communities, limiting widespread awareness and access to what is available,
ii lack of quantification of the economic, environmental and consumer benefits of using DSSs – often limited by insufficient access to suitably large datasets spanning a wide range of sites and seasons,
iii inadequate testing of DSSs for their accuracy and predictive value, and/or limited data available to undertake testing,
iv short-term funding curtailing long-term updation and user support,
v DSSs addressing single pests, while growers need to account for multiple pests, and
vi DSSs which are insufficiently risk-averse to meet grower needs.

These constraints continue to limit the adoption of DSSs for horticultural crops and should be considered during the development of new systems. In addition to addressing the behavioural constraints to uptake, DSSs developers should also consider the socio-economic drivers such as farm size and type, and user experience. For example, younger growers, well-educated growers and those on relatively larger farms are most likely to use DSSs (Rose et al., 2016). Since their use is influenced by their farming styles and practices, DSSs should ideally be designed with direct input from the target user groups (Kanatas et al., 2020).

Systems are becoming increasingly user-centric, built on engagement with growers and advisors from inception that ensures clarity in the assumptions, calculations and interpretation of outputs. Advanced DSSs are, however, becoming so advanced in the ways in which they incorporate multiple factors and/or report on multiple pests concurrently, that application of the outputs into practical situations increasingly requires a great deal of trust in the providers. This trust is built on long-term interactions with providers, facilitated through comparison of multiple system outputs on the same pest assessment, or through shared reporting of outcomes within grower/advisor groups and networks.

Above all else, DSSs should only ever be used to *support* pest management decisions and will never replace location and crop-specific experience based on the user's own crop monitoring and technical knowledge. DSSs should help growers make effective decisions amid the increasing complexity of crop protection. In an industry with growing regulation and supervision of the use of pesticides and other inputs both by government and processors, DSSs can play a vital role in effective decision making. Increased reliance on pesticides has gone hand in hand with a delegation of responsibility in making hard decisions for crop protection. The pressures applied to growers and their agronomists to improve yield and quality have in the past made it all but impossible to reduce high levels of inputs, made easy by the accessibility of relatively cheap and easy-to-use pesticides. As agricultural priorities shift towards long-term economic and environmental sustainability, the responsibility for pest management is shifting back onto growers, along with the associated risks.

IPM requires more than a combination of independent management decisions. A holistic approach is necessary, based on a rounded understanding of crop physiology, nutrition requirements and pest biology. This requires a knowledge base far wider than a few individuals can be expected to retain. Instead, local stakeholders need to harmonise their vision and strategies to engender long-term sustainability, with the input of relevant experts where necessary (Swiergiel et al., 2019). DSSs should be used as a part of this process, but only as a conduit for improving the understanding of key decision points and to help draw conclusions in combination with monitoring data from the targeted crop area. In the future, we are going to see more systematic approaches to decision support, incorporating a range of stakeholders accessing DSSs through shared platforms. These will engrain much greater trust in systems, providing assurance that effective pest management can be maintained (Bailey et al., 2009; Geiger et al., 2010; Lechenet et al., 2014). It will also support innovation in the development of new DSSs – not just for use in IPM systems, but across the whole of the value chain in horticultural production systems.

6 Future trends in research

- Fundamental research will continue, since this ensures that current and new systems evolve alongside our best understanding of crop performance.
- Improved development of user interfaces will also continue.
- Greater emphasis will be made on the integration of systems, to enable wider access to, and to develop trust in, the DSSs. This will go alongside greater ongoing support for systems, to maintain and grow their use among farming networks.
- Economic thresholds remain the most sought-after, but poorly trusted DSSs within horticulture. Improved development of new thresholds is necessary to account for (i) crop and varietal variation and/or the influence of agronomic factors in resistance/tolerance to pests; (ii) improve quantification of pests' potential to cause damage to crops, (iii) the best

traits to assess in order to estimate damage and (iv) fluctuations in the availability and costs of control measures and crop value (Ramsden et al., 2017).

7 Where to look for further information

7.1 Further reading

- The University of California Statewide IPM program, http://ipm.ucanr.edu/.
- VIPS, Varsling Innen PlanteSkadegjørere. www.vips-landbruk.no (accessed on 11 Feb 2021).
- SEGES, https://www.landbrugsinfo.dk/Planteavl/Plantevaern/Skadedyr/Skadedyrsarter/ Sider/PL_Kaalflue_varsling.aspx (accessed on 11 Feb 2021).
- Syngenta, AHDB Pest Bulletin. https://www.syngenta.co.uk/ahdb-pest-bulletin (accessed on 11 Feb 2021).
- Information System for Integrated Plant Production (ISIP). https://isip.de/isip/servlet/isip-de (accessed on 11 Feb 2021).
- Horta srl: a permanent platform for enhancing results from research in the agroalimentary sector through https://www.horta-srl.it/.
- IOBC Working groups – https://www.iobc-wprs.org/expert_groups/wg.html.

7.2 Key journals and conferences

- Association of Applied Biologists Conference series, https://www.aab.org.uk/conferences.
- Journal of Crop Protection, https://www.journals.elsevier.com/crop-protection.

7.3 Major international research projects

- AFARCLOUD – Aggregate Farming in the Cloud is a H2020 EU-funded project No. 783221; http://www.afarcloud.eu/.
- ENDURE – European Network for the Durable Exploitation of crop protection strategies is an FP6-FOOD EU funded-project No. 607851; http://www.endure-network.eu/.
- IPM Decisions is a H2020 EU-funded project No. 817617; https://www.ipmdecisions.net/.
- IPMWorks is a H2020 EU-funded project No. 101000339; https://ipmworks.net/.
- MODEM_IVM – A web-based system for real-time Monitoring and Decision Making for Integrated Vineyard Management is an FP7-SME EU-funded project No. 262059; http://www.modem-ivm.eu/.

8 Acknowledgements

Production of this chapter was supported in part by the IPM Decisions project, which has received funding from the European Research Council (ERC) under the European Union's Horizon 2020 research and innovation programme (grant agreement 817617).

9 References

Amsalem, L., Freeman, S., Rav-David, D., Nitzani, Y., Sztejnberg, A., Pertot, I. and Elad, Y. 2006. Effect of climatic factors on powdery mildew caused by *Sphaerotheca macularis* f. sp. *fragariae* on strawberry. *European Journal of Plant Pathology* 114(3), 283–292.

Bailey, A. S., Beraglia, M., Fraser, I. M., Sharma, A. and Douarin, E. 2009. Integrated pest management portfolios in UK arable farming: results of a grower survey. *Pest Management Science* 62(9), 1030–1039.

Bardet, A. and Vibert, J. 2011. Strawberry powdery mildew. *Infos-Ctifl* 276, 38–44.

Barzman, M., Bàrberi, P., Birch, A. N. E., Boonekampe, P., Dachbrodt-Saaydeh, S., Graf, B., Hommel, B., Jensen, J. E., Kiss, J., Kudsk, P., Lamichhane, J. R., Messean, A., Moonen, A., Ratnadass, A., Ricci, P., Sarah, J. and Sattin, M. 2015. Eight principles of integrated pest management. *Agronomy for Sustainable Development* 35(4), 1199–1215. https://doi.org/10.1007/s13593-015-0327-9.

Battilani, P., Rossi, V., Racca, P. and Giosue, S. 1996. ONIMIL, a forecaster for primary infection of downy mildew of onion 1. *EPPO Bulletin* 26(3–4), 567–576.

Boote, K. J., Jones, J. W., Mishoe, J. W. and Berger, R. D. 1983. Coupling pests to crop growth simulators to predict yield reductions. *Phytopathology* 73(11), 1581–1587.

Brodal, G., Folkedal, A., Hole, H., Brevig, C. and Rafoss, T. 2007. VIPS-Warning and prognoses of pests and diseases in Norway. In: *Proceedings of the NJF 23rd Congress* (pp. 26–29).

Bruce, T. J. A. 2016. The CROPROTECT project and wider opportunities to improve farm productivity through web-based knowledge exchange. *Food and Energy Security* 5(2), 89–96.

Carisse, O. and Fall, M. L. 2021. Decision trees to forecast risks of strawberry powdery mildew Caused by *Podosphaera aphanis*. *Agriculture* 11(1), 29.

Carisse, O., Morissette-Thomas, V. and Van der Heyden, H. 2013a. Lagged association between powdery mildew leaf severity, airborne inoculum, weather, and crop losses in strawberry. *Phytopathology* 103(8), 811–821.

Carisse, O., Lefebvre, A., Van der Heyden, H., Roberge, L. and Brodeur, L. 2013b. Analysis of incidence-severity relationships for strawberry powdery mildew as influenced by cultivar, cultivar type, and production systems. *Plant Disease* 97(3), 354–362.

Clarke, C., Brightman, D., Roques, S., Kendall, S., Kindred, D. R. and Sylvester-Bradley, R. 2019. Operating Grower Innovation Groups (FIGs) for testing yield enhancing ideas using on-farm experimentation. Conference Paper: 12th European Conference on Precision Agriculture. https://doi.org/10.3920/978-90-8686-888-9_110.

Collier, R., Mazzi, D., Folkedal Schjoll, A., Schorpp, Q., Thoming, G., Johansen, T. J., Meadow, R., Meyling, N. V., Cortesero, A. M., Vogler, U., Gaffney, M. T. and Hommes, M. 2020. The potential for decision support tools to improve the management of root-feeding fly pests of vegetables in western Europe. *Insects* 11(6), 369–385.

Collier, R. H. and Finch, S. 1983. Effects of intensity and duration of low temperatures in regulating diapause development of the cabbage root fly (*Delia radicum*). *Entomologia Experimentalis et Applicata* 34(2), 193–200.

Collier, R. H. and Finch, S. 1988. Thermal requirements for cabbage root fly, *Delia radicum*, development. In: Cavalloro, R. and Pelerents, C. (Eds) *Progress on Pest Management in Field Vegetables*, 21–25. Available at: https://www.taylorfrancis.com/chapters/edit/10.1201/9781003079347-5/thermal-requirements-cabbage-root-fly-delia-radicum-development-collier-finch.

Collier, R. H., Finch, S. and Phelps, K. 1991. A simulation model for forecasting the timing of attacks of Delia radicum on cruciferous crops 1. *EPPO Bulletin* 21(3), 419–424.

Cook, H. T. 1932. Studies on the downy mildew of onion and the causal organism, *Peronospora Destructor*. 143.

Creissen, H., Jones, P., Tranter, R., Girling, R., Jess, J., Burnett, F., Gaffney, M., Thorne, F. and Kildea, S. 2019. Measuring the unmeasurable? A method to quantify adoption of Integrated Pest Management practices in temperate arable farming systems. *Pest Management Science* 72(12), 3144–3152.

Ctifl 2017. Fraise - Oïdium - v. (Centre technique au service de la filière fruits et légumes) 7, 1. Available at: http://www.ctifl.fr/inoki/Pages/Presentation/Modele.aspx?id=8.

Damos, P. 2015. Modular structure of web-based decision support systems for integrated pest management: a review. *Agronomy for Sustainable Development* 35(4), 1347–1372.

de Visser, C. L. M. 1998. Development of a downy mildew advisory model based on downcast. *European Journal of Plant Pathology* 104(9), 933–943.

Dodgson, J., Hall, A. and Parker, S. 2008. Rule based prediction system to optimize fungicide applications for control of *Podosphaera aphanis*. In: *VI International Strawberry Symposium* (pp. 355–358).

Eccel, E., Fratton, S., Ghielmi, L., Tizianel, A., Shtienberg, D. and Pertot, I. 2010. Application of a non-linear temperature forecast post-processing technique for the optimization of powdery mildew protection on strawberry. *Italian Journal of Agrometeorology-Rivista Italiana di Agrometeorologia* 15(2), 5–14.

Eckenrode, C. J. and Chapman, R. K. 1972. Seasonal adult cabbage maggot populations in the field in relation to thermal-unit accumulations. *Annals of the Entomological Society of America* 65(1), 151–156.

Finch, S. 1992. Improving the selectivity of water traps for monitoring populations of the cabbage root fly. *Annals of Applied Biology* 120(1), 1–7.

Finch, S. 1993. Integrated pest management of the cabbage root fly and the carrot fly. *Crop Protection* 12(6), 423–430.

Finch, S. and Collier, R. H. 1983. Emergence of flies from overwintering populations of cabbage root fly pupae. *Ecological Entomology* 8(1), 29–36.

Finch, S., Collier, R. H. and Skinner, G. 1986. Local population differences in emergence of cabbage root flies from south-west Lancashire: implications for pest forecasting and population divergence. *Ecological Entomology* 11(2), 139–145.

Finch, S. and Collier, R. H. 1987, June. Monitoring and forecasting attacks of cabbage root fly and carrot fly in vegetable crops. Acta Horticulturae International Symposium on Diptera Pests in Vegetable Crops 219, 39–46.

Finch, S. and Skinner, G. 1974. Some factors affecting the efficiency of water-traps for capturing cabbage root flies. *Annals of Applied Biology* 77(3), 213–226.

Finch, S., Skinner, G. and Freeman, G. H. 1975. The distribution and analysis of cabbage root fly egg populations. *Annals of Applied Biology* 79(1), 1-18.

Fitzgerald, G. S. and O'Brien, R. G. 1994. Validation of 'Downcast' in the prediction of sporulation-infection periods of *Peronospora destructor* in the Lockyer Valley. *Australian Journal of Experimental Agriculture* 34(4), 537-539.

Friedrich, S., Leinhos, G. M. E. and Löpmeier, F.-J. 2003. Development of ZWIPERO, a model forecasting sporulation and infection periods of onion downy mildew based on meteorological data. *European Journal of Plant Pathology* 109(1), 35-45.

Gent, D. H., De Wolf, E. and Pethybridge, S. J. 2009. Perceptions of risk, risk aversion, and barriers to adoption of decision support systems and integrated pest management: an introduction. *Phytopathology* 101(6), 640-643.

Geiger, F., Bengtsson, J., Berendse, F., Weisser, W. W., Emmerson, M., Morales, M. B., Ceryngier, P., Liira, J., Tscharntke, T., Winqvist, C., Eggers, S., Bommarco, R., Part, T., Bretagnolle, V., Plantegenest, M., Clement, L. W., Dennis, C., Palmer, C., Onate, J. J., Guerrero, I., Hawro, V., Aavik, T., Thies, C., Flohre, A., Hanke, S., Fischer, C., Goedhart, P. W. and Inchausti, P. 2010. Persistent negative effects of pesticides on biodiversity and biological control potential on European farmland. *Basic and Applied Ecology* 11(2), 97-105.

Gilles, T. and Kennedy, R. 2004. Evaluation of Milioncast, a forecaster for onion downy mildew, with historical data. In: Spencer-Philips, P. and Jeger, M. (Eds) *Advances in Downy Mildew Research* (Vol. 2, pp. 81-89). Springer, Dordrecht. https://doi.org/10.1007/978-1-4020-2658-4_4.

Gilles, T., Phelps, K., Clarkson, J. P. and Kennedy, R. 2004. Development of MILIONCAST, an improved model for predicting downy mildew sporulation on onions. *Plant Disease* 88(7), 695-702.

Hawkes, C. 1972. The diurnal periodicity and cycle of behaviour of the adult cabbage root fly (*Erioischia brassicae*). *Annals of Applied Biology* 70(2), 109-118.

Hayashi, M., Tamai, K., Owashi, Y. and Miura, K. 2019. Automated machine learning for identification of pest aphid species (Hemiptera: Aphididae). *Applied Entomology and Zoology* 54(4), 487-490. https://doi.org/10.1007/s13355-019-00642-0.

Hildebrand, P. D. and Sutton, J. C. 1984a. Effects of weather variables on spore survival and infection of onion leaves by *Peronospora destructor*. *Canadian Journal of Plant Pathology* 6(2), 119-126.

Hildebrand, P. D. and Sutton, J. C. 1984b. Relationships of temperature, moisture, and inoculum density to the infection cycle of *Peronospora destructor*. *Canadian Journal of Plant Pathology* 6(2), 127-134.

Hommes, M. and Gebelein, D. 1996. Simulation models for the cabbage root fly and the carrot fly Integrated Control in Field Vegetable Crops. IOBC WPRS Bulletin. Working Group "Integrated Protection in Field Vegetables". In: Finch, S. and Brunel, E. (Eds) Proceedings of the Meeting at Guitté, (France). ISBN 92-9067-085-1, 19(11), 60-65.

Jakku, E. and Thorburn, P. J. 2010. A conceptual framework for guiding the participatory development of agricultural decision support systems. *Agricultural Systems* 103(9), 675-682.

Jesperson, G. D. and Sutton, J. C. 1987. Evaluation of a forecaster for downy mildew of onion (*Allium cepa* L.). *Crop Protection* 6(2), 95-103.

Johansen, T. J. and Meadow, R. 2005. Emergence patterns of Norwegian brassica root fly populations. *Comptes Redus de la Réunion* 28(4), 25-29.

Johansen, T. J. and Meadow, R. 2006. Population differences in emergence of brassica root flies (Diptera: Anthomyiidae). *Environmental Entomology* 35(5), 1161–1165.

Jones, V. P., Brunner, J. F., Grove, G. G., Petit, B., Tangren, G. V. and Jones, W. E. 2010. A web-based decision support system to enhance IPM programs in Washington tree fruit. *Pest Management Science* 66(6), 587–595.

Jorgensen, L. N., Matzen, N., Ficke, A., Nielsen, G. C., Jalli, M., Ronis, A., Andersson, B. and Djurle, A. 2020. Validation of risk models for control of leaf blotch diseases in wheat in the Nordic and Baltic countries. *European Journal of Plant Pathology* 157(3), 599–613.

Kanatas, P., Travlos, I. S., Gazoulis, I., Tataridas, A., Tsekoura, A. and Antonopoulos, N. 2020. Benefits and limitations of decision support systems (DSS) with a special emphasis on weeds. *Agronomy* 10(4), 548–558.

Kennedy, R. and Wakeham, A. J. 2008. Development and use of detection systems for the sporangia of *Peronospora destructor*. *Aspects of Applied Biology* 89, 55–56.

Landry, C., Carisse, O. and Bouchard, J. 2010. Developing response strategies against powdery mildew of strawberry based on risk estimation and fungicide effectiveness. Project QP0018CO Final Report. Available at: https://nbwildblue.ca/wp-content/uploads/Public/ResearchReports/Report_1001_QP0018CO_EN.pdf.

Leach, C. M., Hildebrand, P. D. and Sutton, J. C. 1982. Sporangium discharge by *Peronospora destructor*: influence of humidity, red-infrared radiation, and vibration. *Phytopathology* 72(8), 1052–1056.

LEAF Publication 2020. Simple sustainable integrated pest management. Available at: https://s3-eu-west-1.amazonaws.com/leaf-website/Simply-Sustainable-Integrated-Pest-Management-FINAL.,pdf.

Lechenet, M., Bretagnolle, V., Bockstaller, C., Boissinot, F., Petit, M. S., Petit, S. and Munier- Jolain, N. M. 2014. Reconciling pesticide reduction with economic and environmental sustainability in arable farming. *PLoS ONE* 9(6), e97922.

Lechenet, M., Dessaint, F., Py, G., Makowski, D. and Munier-Jolain, N. 2017. Reducing pesticide use while preserving crop productivity and profitability on arable farms. *Nature Plants* 3(3), 17008. https://doi.org/10.1038/nplants.2017.8.

Mahlein, A. K., Oerke, E. C., Steiner, U.. and Dehne, H. 2012. Recent advances in sensing plant diseases for precision crop protection. *European Journal of Plant Pathology* 133(1), 197–209. https://doi.org/10.1007/s10658-011-9878-z.

Meadow, R., Thoming, G. and Schjøll, A. F. 2020. IPM tools for control of cabbage root fly (*Delia radicum*) in Brassica vegetables. *IOBC-WPRS Bulletin* 153, 60–63.

Moerkens, R., Janssen, D., Brenard, N., Reybroeck, E., del Mar Tellez, M., Rodríguez, E., Bosmans, L., Leirs, H. and Sluydts, V. 2020. Simplified modelling enhances biocontrol decision making in tomato greenhouses for three important pest species. *Journal of Pest Science* 94, 1–11.

Mohanty, S. P., Hughes, D. P. and Salathe, M. 2016. Using deep learning for image-based plant disease detection. *Frontiers in Plant Science* 7, 1419. https://doi.org/10.3389/fpls.2016.01419.

Miller, T. C., Gubler, W. D., Geng, S. and Rizzo, D. M. 2003. Effects of temperature and water vapor pressure on conidial germination and lesion expansion of *Sphaerotheca macularis* f. sp. *fragariae*. *Plant Disease* 87(5), 484–492.

Oliver, D. M., Fish, R. D., Winder, M., Hodgson, C. J., Heathwaite, A. L. and Chadwick, D. R. 2012. Valuing local knowledge as a source of expert data: grower engagement and the design of decision support systems. *Environmental Modelling and Software* 36, 76–85.

Palmer, S. A. 2007. Strawberry powdery mildew: epidemiology and the effect of host nutrition on disease (Doctoral dissertation). Available at: https://digital.library.adelaide.edu.au/dspace/handle/2440/57974.

Parker, C. and Sinclair, M. 2010. User-centred design does make a difference. The case of decision support systems in crop production. *Behaviour and Information Technology* 20(6), 449–460.

Pattnaik, G., Shrivastava, V. K. and Parvathi, K. 2020. Transfer learning-based framework for classification of pest in tomato plants. *Applied Artificial Intelligence* 34(13), 981–993.

Peries, O. S. 1962. Studies on strawberry mildew, caused by *Sphaerotheca macularis* (Wallr. ex Fries) Jaczewski* I. Biology of the fungus. *Annals of Applied Biology* 50(2), 211–224.

Phelps, K., Collier, R. H., Reader, R. J. and Finch, S. 1993. Monte Carlo simulation method for forecasting the timing of pest insect attacks. *Crop Protection* 12(5), 335–342.

Pimentel, D. 2005. Environmental and economic costs of the application of pesticides primarily in the United States. *Environment, Development and Sustainability* 7(2), 229–252. https://doi.org/10.1007/s10668-005-7314-2.

Ramsden, M. W., Kendall, S. L., Ellis, S. A. and Berry, P. M. 2017. A review of economic thresholds for invertebrate pests in UK arable crops. *Crop Protection* 96, 30–43.

Roberts, L. G. 1994. Temperature and period dependence in the post-diapause development of the cabbage root fly. Delia radicum (Doctoral thesis). Available at: https://pearl.plymouth.ac.uk/bitstream/handle/10026.1/2523/LORNE%20GEORGE%20ROBERTS.PDF?sequence=1.

Rose, D. C., Sutherland, W. J., Parker, C., Lobley, M., Winter, M., Morris, C., Twining, S., Ffoulkes, C., Amano, T. and Dicks, L. V. 2016. Decision support tools for agriculture: Towards effective design and delivery. *Agricultural Systems* 149, 165–174.

Rossi, V., Caffi, T. and Salinari, F. 2012. Helping growers face the increasing complexity of decision-making for crop protection. *Phytopathologia Mediterranea* 51(3), 457–479.

Rustia, D. J. A., Lin, C. E., Chung, J. Y. and Lin, T. T. 2018. A real-time multi-class insect pest identification method using cascaded convolutional neural networks. In: *9th International Symposium on Machinery and Mechatronics for Agricultural and Biosystems Engineering (ISMAB)* (p. 67).

Shtienberg, D. 2013. Will decision support systems be widely used for the management of plant disease? *Annual Review of Phytopathology* 51, 1–16.

Skog, T.-E., Cucak, M., Gasi, F., Grahić, J., Zovko, M., Eikemo, H., Hole, H., Schjøll, A. F., Netland, J., Trandem, N., Rafoss, T., Meadow, R.. and Nordskog, B. 2015. VIPS–an open source technology platform for prognosis and decision support. Poster: Plant Health for Sustainable Agriculture, Ljubljana, Slovenia.

Swiergiel, W., Manduric, S., Ramert, B., Porcel, M. and Tasin, M. 2019. Development of sustainable plant protection programs through multi-actor Co-innovation: an 8-year case study in Swedish apple production. *Journal of Cleaner Production* 234, 1178–1191.

Tilman, D., Balzar, C., Hill, J. and Befort, B. L. 2011. Global food demand and the sustainable intensification of agriculture. *Proceedings of the National Academy of Sciences of the United States of America* 108(50), 20260–20264.

Van der Heyden, H., Dutilleul, P., Charron, J. B., Bilodeau, G. J. and Carisse, O. 2020. Factors influencing the occurrence of onion downy mildew (*Peronospora destructor*) epidemics: trends from 31 years of observational data. *Agronomy* 10(5), 738.

Wallhead, M. and Zhu, H. 2017. Decision support systems for plant disease and insect management in commercial nurseries in the midwest: A perspective review. *Journal of Environmental Horticulture* 35(2), 84-92.

Wilson, R. S., Hooker, N., Tucker, M., Lejeune, J. and Doohan, D. 2009. Targeting the grower decision making process: A pathway to increased adoption of integrated weed management. *Crop Protection* 28(9), 756-764.

Yarwood, C. E. 1937. The relation of light to the diurnal cycle of sporulation of certain downy mildews. *Journal of Agricultural Research* 54(54), 365-373.

Yarwood, C. E. 1943. Onion downy mildew. *Hilgardia* 14(11), 595-691.

Part 3

Improving integrated pest management techniques and implementation

Chapter 8

The use of agronomic practices in integrated pest management programmes in horticulture

Aude Alaphilippe, Claude Bussi, Marion Casagrande, Tarek Dardouri and Sylvaine Simon, INRAE UERI Gotheron, France; Pierre-Eric Lauri, INRAE UMR ABSys, France; Amélie Lefèvre, INRAE Agroecological Vegetable Systems Experimental Facility, France; and Mireille Navarrete, INRAE UR Ecodeveloppement, France

1 Introduction

As defined in the IOBC General Technical Guideline (Wijnands et al., 2018a), integrated pest management (IPM) requires preventive (indirect) measures on pest, disease and weed status to decrease pest populations and disease inoculum before intervening with direct plant protection measures. Prevention and/or elimination of pests, diseases (including plant physiological disorders) and weeds can be supported by the design and management of the agroecosystem at two scales. First, at the field scale, the challenge is to build a coherent cropping system by coordinating a set of decisions such as the choice of appropriate resistant/tolerant cultivars, pruning, fertilisation and water supply, crop sequences over time, and field design, to prevent the development of pest infestations and disease infections. Second, at the farm scale, the challenge is to spatially allocate cropping systems to field plots and to introduce natural biodiversity to design a farmscape favourable to pest natural enemies, while taking into account economic, human and organisational constraints. IPM is a part of the integrated

http://dx.doi.org/10.19103/AS.2021.0095.09

production concept based on agroecology and a systems approach, which not only allows for the development of more sustainable systems adapted to local agro-ecological conditions but also holds the key to optimising resource use (internal and external) and overall farm performance (Wijnands et al., 2018b). IPM promotes the design of coherent and resilient systems that reduce both pest attacks and damage to cultivated plants. It is based on methods and practices that foster plant tolerance to pests and diseases, hereafter referred to as 'pests' for brevity, and the control of pests by their natural enemies. This chapter specifically focuses on agronomic IPM practices that contribute directly or indirectly to plant health. Since plant susceptibility to pests also involves the plant nutrient status, we also include in this chapter related to plant physiological disorders, which are due to non-pathogenic factors and decrease the crop quality, causing important economic losses. Finally, practices directly targeting pests that are extensively presented in other parts of the book, such as Biological Control (Part 1), Decision Support (Part 2) and Physical Control (Part 3), are excluded from this chapter.

Although the systemic use of agronomic practices to manage plant health is appropriate for most production systems, this chapter focuses on fruit and vegetable production. Fruit trees are perennial crops, for which each tactical decision regarding annual management has consequences for several years, due not only to possible carry-over effects but also to paybacks of a given growing season or stage in the orchard lifetime. Orchards are commonly planted in rows with grassy ground cover between the rows, except in dry countries where the inter-rows are often weed-free bare-ground. This spatial arrangement, as well as its permanence, affords opportunities for pest-suppressive design in time and space. In this chapter, we focus on IPM practices in orchards, seen as a relevant case study to address the complexity of the agroecosystem. IPM practices adoption could scale-out only if consistent with the agri-food systems requirements (specifications about fruit physical appearance, shelf life, etc.) and vice versa if agri-food systems take into account such practices. Thus we also discuss the interest to include our proposal in the more general framework of agri-food systems, based on an example from vegetable production.

The IPM practices considered in this chapter are presented in Fig. 1 and are related to:

a the *crop and its annual management* (i.e. tactical decisions regarding such annual crop management) (Section 2);
b the enhancement of *plant diversity in the cropping system at the field scale* (i.e. strategic decisions such as planting additional cultivars, introducing cover crops or companion plants, intercropping for dilution and barrier effects, designing agroforestry systems) (Section 3); and
c their *combination* in coherent IPM agroecosystems that upscale from the field and farm scale (Section 4).

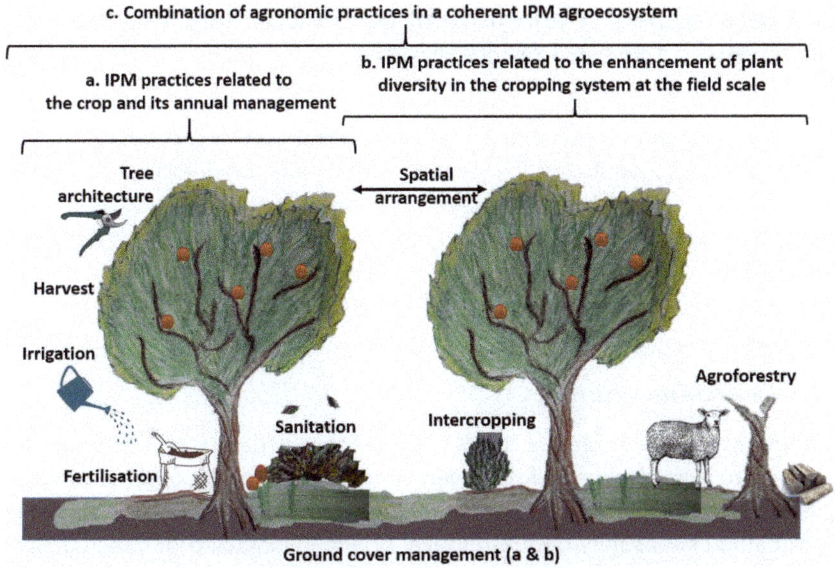

Figure 1 Representation of the agronomic practices related to (a) the crop and its annual management; (b) enhancement of plant diversity in the cropping system at the field scale and (c) their combination.

All practices detailed in this chapter cover several functions and/or services. We consider here as practices both the technique and its application by a farmer, which is context-dependent (Casagrande et al., 2017). The primary function of most of these practices (Fig. 1a) is to support fruit production, such as pruning, whereas their application may either disrupt pests and diseases (e.g. removal of inoculum) or stimulate them (e.g. extensive pruning increases shoot growth and therefore the risks of aphid infestation) as a secondary function. Other practices primarily contribute to enhancing biodiversity (Fig. 1b), such as intercropping or agroforestry, and they support pest control as a secondary function. The set of agronomic practices presented in Fig. 1 aims at improving the agroecosystem and its resilience through biological and ecological processes and the provision of ecosystem services while minimising synthetic external inputs. As such, these IPM practices could be considered as agroecological practices (Wezel et al., 2014).

In this chapter, we first describe and illustrate various agronomic IPM practices (Section 2) and plant diversity-based practices (Section 3) with detailed examples from fruit production, mainly on pome and stone fruit. We then provide some insights on the design of cropping systems that combine the above-mentioned IPM practices at field, farm and agri-food system scales (Section 4). We also show that designing such systems calls for additional research and new approaches.

2 Integrated pest management practices related to the crop and its annual management

In this section, we develop different aspects of crop management - including agronomic practices - that should be considered in order to decrease plant and orchard susceptibility to various pests and physiological disorders: tree training, mineral fertilisation, water supply and soil management. The description follows the evolutionary timeline of a fruit tree growing season as summarised in Fig. 2 and includes fruit harvest as well as sanitation practices as stand-alone solutions or combined with other cropping practices.

2.1 Tree architecture

Plant genotypes have various degrees of genetically determined resistance or tolerance to pests. However, for a given genotype, manipulation of architecture constitutes an efficient, although partial, lever to control pest propagation. Such manipulations may be the removal of already infected or infested parts, the reduction of contact between susceptible plant parts and pests, or the creation of unfavourable conditions for pests' development. These means are considered as elements of a 'pest avoidance strategy' (initially developed on disease by Ando et al., 2007) that the grower may implement on fruit trees. Both spatial and temporal levels are strongly intertwined and should be considered simultaneously.

At the whole tree scale, large trees are generally more infested/infected due to various causes: attractiveness/more effective interception of pests, microclimate and lower efficacy of either the foraging of natural enemies or of spraying (e.g. codling moth, *Cydia pomonella* L.; Kuehrt et al., 2006; rosy apple aphid, *Dysaphis plantaginea* Pass.; Stoeckli et al., 2008a).

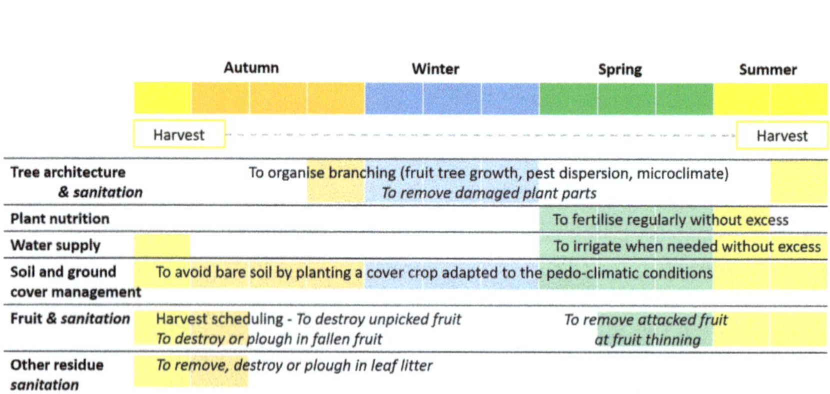

Figure 2 Evolutionary timeline of the agronomic practices detailed in this chapter.

At the shoot and whole tree scales, long shoots are usually more infested by aphids, and for a longer period during the season, than short shoots (e.g. green aphid, *Aphis* spp., on apple; Stoeckli et al., 2008b). Therefore, winter pruning or excessive fertilisation that increases the proportion of long shoots on trees, and the duration of growth of these shoots, also increases aphid infestation (e.g. green aphid, *Myzus persicae* Sulzer on peach; Grechi et al., 2008). These results are embodied in the 'plant vigour hypothesis' (Price, 1991) which states that an increase in shoot vigour corresponds to an increase in the cohort of young leaves that are softer and richer in sap with high nutritive content (Price, 1991; Mody et al., 2009). This phenomenon may be counterbalanced by the fact that too many shoots growing at the same time on trees may also increase competition between them and thus reduce the nutritive richness of individual shoots and their attractiveness towards aphids ('sink competition hypothesis'; Larson and Whitham, 1997). A first lesson is, therefore, to be able to define a good pruning strategy, not only to control tree size at the desired height and volume but also to avoid excessive pruning that could entail vigorous regrowth, which would be conducive to pest infestation and may put fruiting at a disadvantage. Various training systems have been proposed in the past to combine high fruiting quality and efficient pest management (Lauri, 2008; Lauri et al., 2009; Tustin et al., 2011) through good control of the homogeneity of shoot vigour (Lauri et al., 2016a).

One of these systems, implemented in apple orchards, known as centrifugal training, is based on the selective removal of young fruiting spurs, called artificial extinction, to enhance canopy porosity to light and consequently to better control fruit load and quality over consecutive years (Willaume et al., 2004; Lauri and Corelli-Grappadelli, 2014). This system has been compared across 3 consecutive years to a central axis reference system without artificial extinction, to ascertain whether these pruning strategies would affect pest development on apple trees. These tree training and pruning treatments were managed according to organic growing guidelines. Results generally showed that pests, namely rosy apple aphid, red spider mite (*Panonychus ulmi* (Koch)), green aphid and disease, namely scab (*Venturia inaequalis* (Cooke) G. Winter), were less prevalent on centrifugal trained trees than on the reference trees. Three main hypotheses could explain the positive effects of centrifugal training: the removal of pests or inoculum by way of artificial extinction; the resulting increase in light and air penetration within the canopy; and the greater distances between growing shoots (Simon et al., 2012). However, the fact that scab increased on centrifugal trained trees, following a period of heavy rain before harvesting one year, indicated that the effects of tree manipulation were only partial.

Another study on these training systems, with and without artificial extinction, was focused on the infestation dynamics of rosy apple aphid (Simon et al., 2012). Results showed that decreasing spur density on fruiting branches also increases the degree of branching of the remaining spurs and reduces the proportion of infested spurs. This suggests that a high degree of branching of current-year shoots is likely to hamper pest dynamics within the tree canopy through an effect on aphid dispersion within the tree (Simon et al., 2012).

2.2 Plant nutrition and fertilisation

The plant vigour hypothesis, which holds that most of the phloem feeders prefer fast-growing plants, suggests that a higher nitrogen supply could increase insect – in particular aphid – abundance by fostering plant growth (Rousselin et al., 2018). Moreover, plant fertilisation modifies the nutrient status of the tree, that is the plant part concentration and composition in carbon (C) and nitrogen (N), which might, in turn, affect pests.

Regarding N concentrations in the tree, significant differences were found between plant parts: amino acids – especially arginine – being the main form of storage (Gomez et al., 2020). These concentrations vary in function of the fertilisation levels. Decreasing N inputs might decrease the plant susceptibility to aphids, such as for the infestation of peach trees by *M. persicae* (Jordan et al., 2020). However, target leaf N concentrations for optimisation of aphid resistance are still to be investigated, considering that the effect of N leaf content on aphid performance is controversial (Matis et al., 2008; Sauge et al., 2010; Mace and Mills, 2015).

While possibly mitigating the susceptibility to aphids, mineral fertilisation of the orchard is also likely to modify biochemical composition of the fruit, which is decisive for the development of physiological fruit disorders such as apple bitter pit or apricot pitburn (Macheix et al., 1990). Pitburn, a post-harvest physiological disorder of the fruit, is frequently detected in apricot (Amiot et al., 1997). It consists of a brown discolouration around the apricot stone, due to the oxidation of phenolic compounds. Pitburn appears because of stressors that disrupt compartmentalisation inside cells and consequently result in leakage of phenolic compounds from the vacuole into the cytoplasm (Macheix et al., 1990). Some factors favour this disorder in orchards, such as water stress and high temperatures (Kotze and Bothma, 1989). Mineral fertilisation is also likely to contribute to pitburn as for other fruit physiological disorders; for instance, bitter bit in apples is known to be closely related to plant mineral fertilisation (Lewis et al., 1977).

Greater pitburn incidence when N fertilisation was increased suggested that moderate N applications were less detrimental to fruit quality than applications of large amounts of N (Bussi and Amiot, 1998). Hence, a significant positive correlation was detected between pitburn incidence and the fruit N content. Increasing N fertilisation also decreased fruit calcium (Ca) concentration (Bussi et al., 2003). Possible explanations for this include, first, that Ca oxalate may be formed in the leaves in response to nitrate assimilation (Van Egmond, 1979), resulting in high demand for Ca in these plant parts, which would make Ca less available to the fruit. A second reason for the observed Ca depletion may be the higher growth rate obtained, with N inducing a high leaf/fruit ratio, and then a dilution and limitation of the Ca able to reach the fruit (Perring and Preston, 1974). Alongside the Ca depletion, pitburn increases, thus supporting the hypothesis that there is a significant Ca effect on fruit resistance to pitburn (Fig. 3).

Kotze and Bothma (1989) also detected a higher incidence of pitburn when irrigation for fruit trees failed, and the induced deficit of Ca might contribute to explaining this result. Limited water supply to the trees is currently responsible for Ca depletion in perennial plants because Ca translocation largely results from the water transpiration flow through the aerial part of the tree (Appel and Stipes, 1984). Ca has already been shown to delay senescence in fruit by protecting cell membranes from disorganisation (Mc Guire and Kelman, 1984). It may likewise protect apricot from disorders of brown discolouration by reducing damage inside the cell, and thus avoiding the oxidation of phenols by oxidative enzymes (Macheix et al., 1990).

Consequently, optimised N fertilisation, as well as water supply, is likely to mitigate apricot pitburn through their effect on Ca uptake and allocation into the tree.

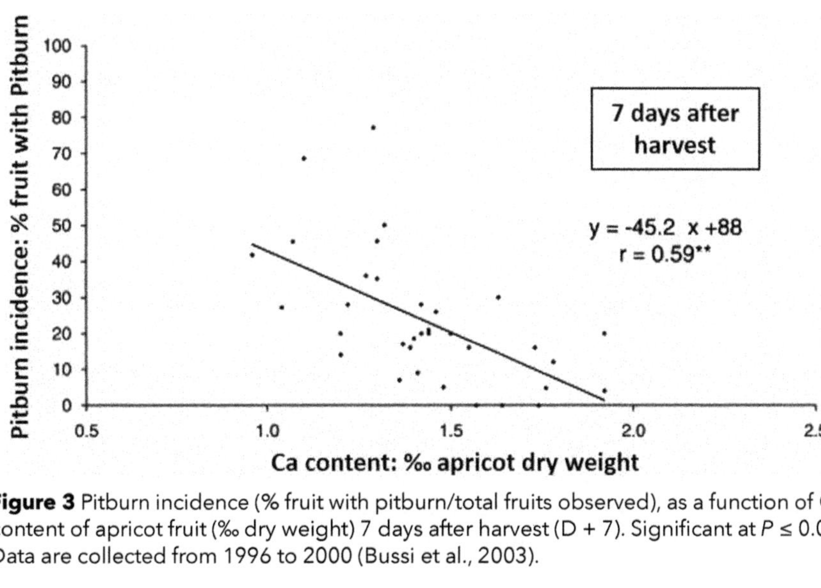

Figure 3 Pitburn incidence (% fruit with pitburn/total fruits observed), as a function of Ca content of apricot fruit (‰ dry weight) 7 days after harvest (D + 7). Significant at $P \leq 0.01$. Data are collected from 1996 to 2000 (Bussi et al., 2003).

2.3 Water supply

Several examples illustrate that the availabilities of both nitrogen and water resources are among the main abiotic factors modifying plant physiology and growth, with an impact on the performance of diseases and herbivorous insects such as aphids. However, the extent to which the modification of plant growth is responsible for such an impact remains unclear.

Additional water is applied in orchards for its direct impact on fruit quality (Naor et al., 2001). Regulated deficit irrigation (RDI) consists of appropriate and mild water restrictions during fruit growth, to improve fruit quality attributes (Girona et al., 2005), particularly increased total soluble solids in the fruit. In peach under limited water supply, Gibert et al. (2005) also detected a decrease in fruit water loss due to the formation of a thicker fruit cuticle, suggesting that microcracking of the fruit epidermis might be reduced. Given that fruit cracks are assumed to be the preferential pathway of pathogen invasion (Gibert et al., 2007), water restrictions in the peach orchard logically reduced the susceptibility of the peach to disease such as brown rot (Li et al., 1989a).

Brown rot, mainly caused by *Monilinia fructicola* (G. Wint.) in California and *Monilinia laxa* (Aderh. and Ruhl.) in Europe (Mari et al., 2004), is a major stone fruit disease in the south of France (Mercier et al., 2003), which can also affect apple (Barkai-Golan, 2001). Weather conditions conducive to brown rot lead to considerable economic losses in the fruit sector (Palou et al., 2003). According to Mercier et al. (2008), brown rot incidence in peach orchards was reduced significantly under RDI compared with conventional irrigation, from around 3 weeks before fruit maturity in the orchard until 1 week after fruit maturity in post-harvest conditions (Fig. 4). Because of the increased total soluble solids in the fruit and the limited incidence of brown rot due to water deprivation, nutritional quality and storage suitability of the fruit produced was considered higher under mild water restrictions than under conventional irrigation in a peach orchard (Mercier et al., 2008).

Water restriction in orchards generally decreases fruit diameters, therefore possibly limiting monetary returns from the fruit crop (Lampinen et al., 1995). In this respect, Mercier et al. (2009) investigated irrigation with alternating periods when water was withheld, and then subsequent re-watering, to possibly facilitate the enhancement of fruit total soluble solids and fruit resistance to pathogens, without limiting peach diameter at maturity. Nevertheless, compared to optimal and regular irrigation schedules when the same total amount of water was applied, water loss by the peach increased when re-irrigation was performed after water was withheld, inducing a higher brown rot incidence (Mercier et al., 2009). Miller et al. (1997) had already observed that re-applying water following deficit irrigation in kiwifruit during the final stage of fruit growth induced

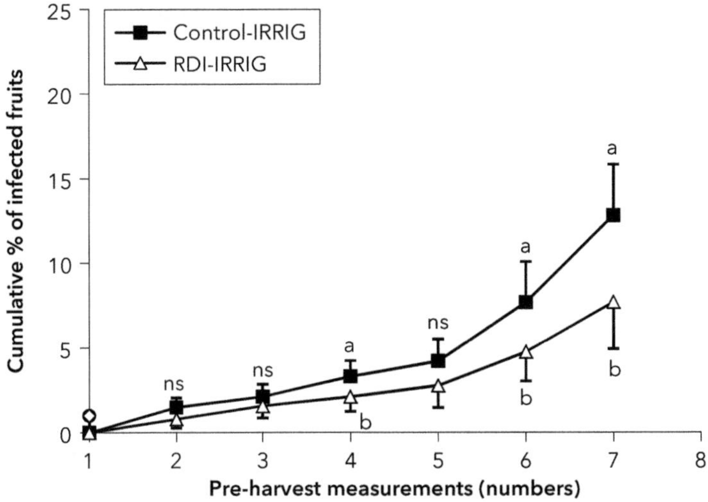

Figure 4 Time-course of cumulative brown rot incidence on peach for the combined 3 years (2003, 2004 and 2005) in relation to irrigation treatments. Control-IRRIG, conventional irrigation; RDI-IRRIG, water deprivation according to regulated deficit irrigation (RDI). Brown rot incidence was assessed twice a week, from around 3 weeks before fruit maturity until 1 day before harvest (seven measurements). Vertical bars denote the positive or negative standard error of the averaged cumulative percentage of infected fruits in each treatment. Different letters indicate statistical difference at $P \leq 0.05$ using the Fisher's test (LSD); ns, not significant (Mercier et al., 2008).

problems of fruit cracking and *Botrytis cinerea* (Pers.) invasion. Sekse (1995) with cherry, and Peet (1992) with tomato, also reported that cuticular fractures on the fruit surface worsened under irregular water supply. In peach, re-irrigation after water deprivation should be limited in order to avoid the deleterious effect of an increasing brown rot incidence (Li et al., 1989b). Increased peach susceptibility to brown rot when water is withheld and subsequently applied at the end of fruit growth confirms that such an irrigation regime should not be recommended. On the contrary, fruit growers have to maintain regular irrigation during the last period of fruit growth to avoid problems with increasing pathogen infection, as observed under an irregular irrigation regime. In this respect, the challenge facing fruit farmers is to prevent any excess in fruit diameter shrinkage or swelling under a regime where water is withheld then re-irrigated as also shown in apple (Ebel et al., 1995).

2.4 Soil and ground cover management

Whereas during the mid-1970s and early 1980s all the grass in fruit orchards was removed with herbicides in order to improve harvest yields, it is now common to have grass in the inter-rows between fruit trees, to decrease the risk of erosion (Hipps et al., 1990) and to ease the traffic while limiting soil compaction. So as for a long time now, the standard management system in stone fruit orchards under localised irrigation has been mown grass inter-rows (i.e. grass alleys) while maintaining herbicide use on the tree-rows (i.e. 2 m wide strips centred along the tree-line) in order to destroy the grass and thus to limit water competition between weeds and trees. Concerning those tree-rows, an important challenge still exists for fruit growers to reduce herbicide use, in compliance with IPM and the regulatory ban of several active ingredients. Alternative cultural practices to herbicide-weeded strips on the tree-rows in modern orchards have been investigated to limit underground water contamination by herbicides and negative effects on the environment and human health.

Tillage-based methods on the tree-rows are commonly used by growers to remove weeds competing with the trees for water and other nutrients. However, tillage-based soil management for weed control has some drawbacks: it can interfere with the development of superficial roots; the physical, chemical and biological properties of the soil can be disturbed (Stagnari et al., 2009), and erosion and runoff may potentially be increased (Duran Zuazo et al., 2008; Gomez et al., 2009). Another alternative to herbicide use in orchards consists of sowing cover crops that compete with weeds but not with the fruit trees. Moreover, cover crops and green manures may be essential agroecological practices for environment-friendly orchards. They can enhance soil fertility, reduce risks of water contamination, avoid erosion and runoff and also contribute to decreasing pest pressure (Wezel et al., 2014; Vukicevich et al., 2016).

In peach orchards, Parker and Meyer (1996) found that the effect of ground management on root patterns depends on the cover crop species: nimblewill grass (*Muhlenbergia schreberi*) increased root density and improved root distribution when compared with other species such as brome (*Bromus mollis*) and bahia grass (*Paspalum notatum*). Root density, earthworm density, water infiltration rate, nitrogen content and water availability were shown to increase under a white clover cover sown on the peach tree-rows as compared to bare soil, with mechanically weeded rows (Parveaud et al., 2012). These results demonstrated the agronomic advantages of nitrogen-fixing plants used as cover crops in peach orchards. Additionally, the positive effect of ground cover on the tree-row in peach orchards on tree growth and pest populations (Parker and Meyer, 1996), and on biotic and abiotic interactions (Meagher and Meyer, 1990), were also pointed out.

Concerning the effect of ground cover on susceptibility to brown rot in peach, experimental data suggested that white clover on the tree-row compared with conventional soil management was likely to limit post-harvest brown rot incidence, particularly after heavy rainfall (Gomez and Mercier, 2008). Covering soil surfaces with white clover could possibly reduce the inoculum released from fallen fruit. It was also assumed that this effect could be caused by slower water infiltration into the soil profile due to the retention of water by a cover crop. Because peach roots take up water from deeper soil layers than do white clover, water availability to tree roots was limited (Bussi et al., 2016). This would subsequently limit the detrimental consequences of excessive water absorption, especially after rainstorms. Large variations in the water supply to fruit trees had already been shown to favour the increase of cuticular fractures on the fruit surface of apples (Irving and Drost, 1987) and the increase of fruit cracking (Sekse, 1995; Peet, 1992). It was also shown that heavy rainfall during the last period of fruit growth stimulated fruit peel cracking on pomegranate (Galindo et al., 2014). As compared to bare soil on peach tree-rows, a clover ground cover mitigated the water availability in deep soil layers after heavy rainfall as corroborated by soil water potential measurements (Ψ soil) (Bussi et al., 2016). A clover ground cover could therefore reduce susceptibility to pathogens compared to current bare soil on peach tree-rows, especially in years with showery weather, for example with rainfall higher than 50 mm during the last period of fruit growth, as in the years 2011 and 2013 (Fig. 5). The frequency of sporadic summer rainstorms in Mediterranean areas is expected

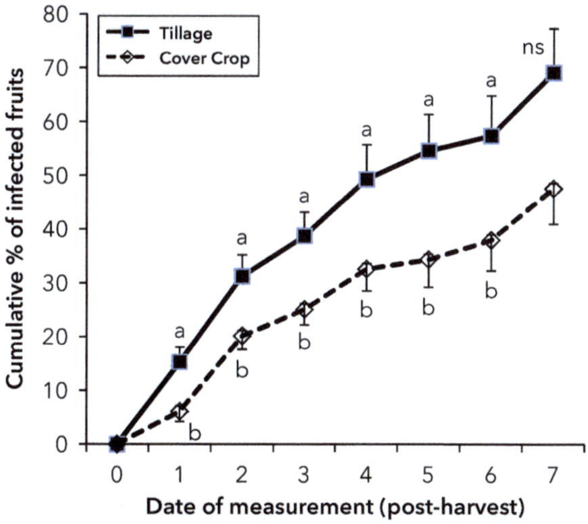

Figure 5 Time-course of cumulative brown rot incidence on peach for the combined 2 years (2011 and 2013) depending on soil management in the tree row: tillage or ground cover crop (White clover). Brown rot incidence was assessed every day in post-harvest conditions for 7 days. Vertical bars denote a positive or negative standard error of the averaged cumulative percentage of infected fruits in each treatment. Different letters indicate statistical difference at $P \leq 0.05$ using a Fisher's test (LSD); ns, not significant (Bussi et al., 2016).

to increase in the near future (Lobell et al., 2011), making the implementation of cultural practices such as soil ground cover feasible to limit brown rot incidence in peach orchards.

2.5 Harvest, damaged fruit and plant residue management

Agronomic practices related to the management of fruit trees (tree architecture, fertilisation, irrigation, soil management, etc.), presented above, contribute to reducing orchards' susceptibility to major pests, diseases and physiological disorders such as aphids, brown rot or pitburn. Sanitation practices, consisting of removing or destroying the infested and infected parts of cultivated plants are an important way to reduce the pest population or disease inoculum. In apple orchards, the main sanitation practices are related to the management of leaf litter and unpicked or fallen fruit.

In the case of apple scab, one of the main actions to decrease the pathogen inoculum consists of managing the leaf litter. Various options exist, such as urea or lime sulphur applications (Holb, 2006), but the most efficient option consists of the removal of leaf litter from orchard inter-rows, combined with the ploughing-in of leaves within the rows. This method permits a 95% decrease in aerial ascospore density and a two-thirds decrease in leaf scab severity in the following year (Gomez et al., 2007). Another interesting sanitation practice is the pruning of infected spurs, which is effective in reducing the powdery mildew *Podosphaera leucotricha* (Ellis & Everh.) incidence on apple (Holb and Kunz, 2016). This is carried out as part of winter pruning and does not require any specific action. Other approaches to crop residue removal may reduce the development rate of diseases, as in the case of anthracnose in guava orchards, caused by the fungus *Gloeosporium psidii* Delacr. (Fischer et al., 2017).

Concerning insect infestation of fruit, a good harvest schedule can decrease both pest and disease incidence. For instance, a 2-day, compared to a 3-day, harvest schedule significantly decreases the numbers of *Drosophila suzukii* (Mats.) in raspberry. Moreover, removing fallen fruit by burying or destroying it decreases pest populations (Leach et al., 2018).

2.6 Conclusion

Controlling pests or physiological disorders in the orchard by implementing suitable agronomic practices is not always sufficient since the effects are partial and depend on complex biological and ecological processes. However, in perennial crops such as fruit, such partial effects are useful to limit the increase in pest populations and pathogen inoculum over the years, thereby increasing the orchard's resilience and capacity to face climatic hazards. Figure 6 sums up the main IPM solutions and the associated processes described in this section.

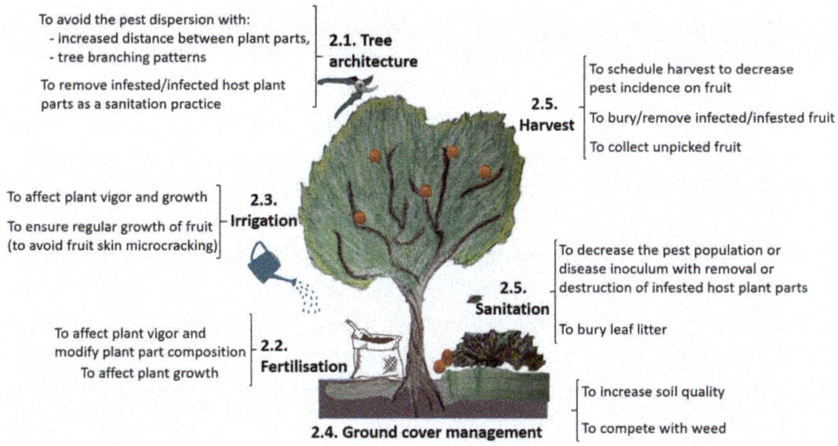

Figure 6 Representation of the agronomic practices related to annual management of crops, presented in Section 2.

3 Integrated pest management practices related to the enhancement of plant diversity in the cropping system at field scale

The implementation of orchard practices as alternatives to chemicals to control major fruit pests helps to reduce the number of insecticide and fungicide applications while maintaining an economically acceptable percentage of marketable fruit. This decrease in pesticide use may in turn limit their detrimental impact on pests' natural enemies. As mentioned in the IOBC General Technical Guideline (Wijnands et al., 2018a), it is necessary to both (1) preserve and favour natural enemies by 'soft' practices in the orchard and (2) introduce ecological infrastructures inside and outside the orchard to enhance a supportive conservation-oriented biological control of key pests by their antagonists. This section is dedicated to increasing plant diversity in the orchard using companion plants and intercropping (Section 3.1), as well as diversification with other cash crop(s) (Section 3.2) in order to disrupt pests and/or enhance pest suppression. Such diversification influences the spatial arrangement of the orchard and requires rethinking of both the crop design at a field scale and the pluri-annual crop management.

3.1 Intercropping with companion plants

Studies carried out on various crops have shown the effectiveness of some plant associations in significantly reducing agricultural pest populations. In contrast to the intensification of cropping systems and monoculture, the supervised introduction of some plants within the crop to be protected can provide us with various services that can disadvantage pests and keep their populations below a damage threshold (Ben Issa et al., 2017). The companion plants (CPs) or crops to be introduced into cropping systems can play different roles to regulate pests. Their effect can be direct, either physical or chemical, or indirect and mediated by natural enemies of pests or other organisms. Although little documented, the physically constraining effect of hedgerows on the dispersion of pests such as oriental moths has been reported in apple orchards (Garcia-Salazar et al., 2007). Around orchards, hedgerows can serve as physical barriers against the inflow of pests and diseases. CPs may also have an efficient mode of action related to their volatile organic compounds (VOCs) that can have various effects on the pest (e.g. disturbance, repulsion, attraction and masking of host plant VOCs). Finally, plant diversification in the orchard and its surroundings favours natural enemies through the provision of habitat and food resources, to the benefit of conservation biocontrol.

One of the proposed approaches in pest management is the application of the push-pull strategy (Pickett et al., 2014). This strategy, based on a combination of attractive and repellent actions, attracts disturbed pests away from the crop to trap plants that are more attractive than the crop to be protected (Malézieux et al., 2009; Mutiga et al., 2010; Tang et al., 2012). Attractive plants (placed at the edge of the plot) and repellent or disturbing plants to be associated with the plants to be protected through intercropping are used for this purpose (Mutiga et al., 2010; Ben Issa et al., 2017).

The presence of repellent plants in association with host plants makes the environment unfavourable to the installation of pest populations, particularly through repulsive and/or disruptive activity of their VOCs (Heil and Silva Bueno, 2007; Sun et al., 2015). These VOCs could also mask host odours and therefore impede host plant location. However, there is a lack of experimental validation of these mixed CP-crop systems, particularly due to the difficulty of conducting system experiments on the processes involved, at the appropriate scale to display significant effects (Deletre et al., 2016; Finch and Collier, 2012). Traditional knowledge (Shrivastava et al., 2010) enriched by scientific research has helped to identify plant species producing VOCs and/or VOC mixtures that repel and disrupt pest behaviour (Parolin et al., 2012a; Ben Issa et al., 2017) and are likely to serve as repellents in orchards against fruit tree aphids. Most of these volatile molecules are produced by aromatic plants (Tang et al., 2012) of the Lamiaceae family (e.g. mint (*Mentha piperita*), thyme (*Thymus vulgaris*), rosemary (*Rosmarinus officinalis*), lavender (*Lavandula latifolia*) and basil (*Ocimum basilicum*)) (Ikeura et al., 2012; Wubie et al., 2014; Ben Issa et al., 2017; Pascual-Villalobos et al., 2017). The Asteraceae family also includes many species with repellent properties such as *Tagetes* spp., white yarrow (*Achillea millefolium*) and feverfew (*Tanacetum parthenium*) (Kahn et al., 2017; Dardouri et al., 2019a,b, 2021). Plants such as garlic (*Allium sativum*), leek (*A. porum*) and chives (*A. schoenoprasum*) belonging to the Amaryllidaceae family also disturb the olfactory behaviour of pests (Thibout and Auger, 1997; Amarawardana et al., 2007). Other species belonging to various botanical families also have repellent effects (Dancewicz and Gabryś, 2008).

CPs can moreover act indirectly by attracting and promoting the presence and development of pest natural enemies by providing them with shelter and food (Song et al., 2013; Song et al., 2014), thus fostering conservation biocontrol.

However, to successfully use CPs as a control strategy, in addition to the choice of species, other parameters that influence VOCs composition and emission, such as the choice of genotypes to be introduced (Fig. 7) (Dardouri et al., 2019b), planting density and cultural practices, need to be considered.

Finally, to facilitate producers' adoption of this strategy in commercial orchards, spatial arrangements must be proposed to maintain access to the trees and permit mechanisation in the orchard.

Another important role of CPs in crops is conservation biocontrol that can be enhanced by an increase in the abundance, fecundity, longevity and/or foraging efficacy of natural enemies, i.e. predatory and parasitoid organisms (Landis et al., 2000). The idea is to benefit through time and space from a 'task force' of natural enemies likely to control a range of pests at any moment of their lifecycle. This is permitted by the provision of habitat (e.g. shelter, reproduction or diapause area and unsprayed area) and improvements in the quantity and quality of food, including floral resources (pollen, nectar) as well as alternative

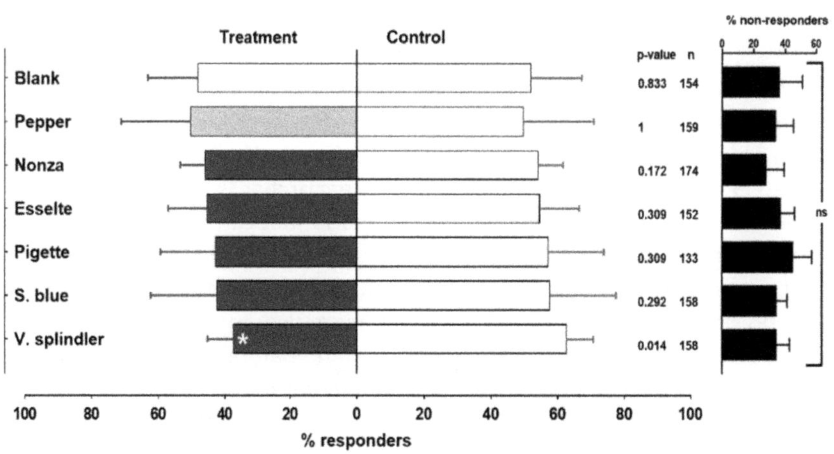

Figure 7 Green peach aphid (*Myzus persicae*) responders' orientation in the presence of a rosemary clone in one side of the olfactometer (grey bars, left part of the graph) and control (no plant) in the other one (white bars, right part). Of the five rosemary clones tested, only the Voltz Splindler clone was found to be repellent (Dardouri et al., 2019b).

prey (Landis et al., 2000). Most natural enemies need habitat(s) complementary to the crop, for instance for mating, diapausing or escaping disruptive practices in the crop, or finding alternative food in the case of a food shortage in the crop, as well as specific food to complete their life cycle (e.g. nectar for parasitoids or syrphid adults) (Wäckers et al., 2007). Those requirements in terms of habitat and food are hardly fulfilled in monoculture crops. The adjunction of plant assemblages (either woody or herbaceous, i.e. hedgerows, ground cover or strips) to the main crop is a way to compensate for such a lack of resources in the crop. Since they are perennial and multi-layered habitats (trees, grass), orchards are complex systems likely to contribute to, and to host, such plant diversification to the benefit of natural enemies and pest control, provided disruptive practices are not implemented (McKerchar et al., 2020).

Yet even though these principles are common knowledge in the scientific community, applications in crops are a subject of debate. While a general increase in arthropod diversity follows an increase in plant diversity in most studies on orchards, the decrease in pest abundance is not always reported and the benefits for pest management are seldom considered (Simon et al., 2010). This is partly related to complex underlying processes when the agroecosystem is considered (e.g. fostering pests by increased plant resources or negative interactions among natural enemies such as intra-guild predation), and to the difficulty of experimenting at the relevant functional scale with respect to the dispersion potential of particular organisms and/or landscape effects (Tscharnke et al., 2005, 2007). Moreover, interactions between pests and organisms other than predators can occur: for example, alternative sources of aphids on faba beans may decrease the abundance of the rosy apple aphid on apple through the diversion of commensalist ants (Pålsson et al., 2020).

Despite partial and sometimes inconsistent effects (Brown, 2001), early (e.g. Wyss et al., 1995) and more recent studies in apple orchards have shown the benefit of interplanting apple with tailored flower strips to control the rosy apple aphid in apple orchards (Sigsgaard and Kramer Jacobsen, 2017; Cahenzli et al., 2019). One documented case study in pear orchards proposes guidelines to select a psyllid-suppressive plant assemblage that can be specified for any layer (grass, shrubs and trees). These guidelines include three dimensions to consider: (i) a ban on plant species hosting pear pests or diseases; (ii) the selection of species hosting a rich natural enemy complex targeting psyllids; and (iii) the design of an assemblage provisioning natural enemies with habitat and/or resources all year long, including in winter (Simon et al., 2010). Such plant assemblages include plant species with specific functional traits, such as early season and late-flowering species for floral resources, evergreen leaves and hollow stems for winter shelter, alternative prey hosting, etc. Most of the studies displaying pest control benefits generated by CPs (Simon et al., 2010) consider pests such as mites, psyllids and aphids that can be tolerated at high or moderate population levels in the orchard before damage occurs. Moths that can cause serious fruit damage at low population levels are more challenging targets for conservation biocontrol (but see Irvin et al., 2006). The planting or sowing of CP assemblages to foster conservation biocontrol, therefore, deserves further consideration in on-farm and on-station experiments and surveys. They serve to gather, integrate and disseminate knowledge not only on biological processes but also on the multi-functionality of CPs and their contribution to other ecosystem services such as pollination or soil nitrogen availability, in addition to pest suppression (Demestihas et al., 2017). Finally, economic, operational and design aspects of CP implementation in the orchard are crucial aspects to be studied for the development of these multispecies systems.

3.2 Diversification with cash crops

In contradiction to the previous section, diversification with other cash crops implies that the associated plants also generate a valuable product. Irrespective of the association types, the purpose is multi-fold to: increase crop diversity at a field scale, increase the land-equivalent ratio, benefit from synergistic effects, and so on. Decreasing pest damage is one aspect. This diversification can include a single crop species and associate several cultivars within the orchard. In this case, similar processes of dilution-barrier effects, that is, less favourable conditions for pests due to a lower proportion of susceptible host plant and a spatial arrangement less favourable to dispersion, occur with the introduction of genetic diversity. A cultivar mixture in an organic apple orchard limited scab and powdery mildew infection and rosy apple aphid infestation (Parisi et al., 2013) compared to pure stands. This diversification implies a decrease in the fruit tree density, as in extensive orchards, which is likely to limit pest dispersion and development. In other cases, this diversification is termed agroforestry, which is the intentional combination in a given plot of trees and associated animals (sometimes and not developed here) and plants of various biological types: herbaceous or shrub. Based on analyses of traditional agroecosystems that are well developed in tropical forests (e.g. traditional village-agroforests in West Java (Michon et al., 1983)), agroforestry was conceptualised to designate 'interfaces and synergies between agricultural and forestry components' (Van Noordwijk et al., 2019). Cocoa-based agroforestry systems are probably among the most studied fruit tree-based systems in the tropics from the point of view of agronomy and biodiversity (Nijmeijer et al., 2019), and pests and diseases (Andres et al., 2016).

The idea of food-producing woody crops such as fruit trees has aroused increasing interest in recent years, and the concepts of fruit tree-based agroforestry systems (Lauri et al., 2016b, 2019) and multifunctional woody polyculture (Lovell et al., 2018) have been proposed. These systems combine the structural and temporal aspects of agroforestry (e.g. vertical stratification and horizontal design of tree species; timing of planting and sowing), and the functional aspects underpinned by ecological principles (e.g. functional interactions between plants through facilitation and competition, and between plants and animals) (Lauri and Simon, 2019; Lauri et al., 2020). The architecture and physiology of individual fruit tree species have been studied extensively in past decades, in relation to genetics and breeding, and to agronomic practices (e.g. apple (Lauri and Laurens, 2005)), with a focus on fruit texture and resistance to both biotic and abiotic stresses, taking advantage of significant advances in genetics, genomics and bioinformatics (Laurens et al., 2018). However, we still have little knowledge about how fruit trees behave in more complex systems where above- and belowground interactions between plants may be exacerbated (Lauri, 2021). Recent research in a walnut–apple agroforestry system shows that shade adaptation traits in apple increase with a decrease in light intensity, and that it is only at the level of ca. 65% reduction in light (i.e. 600 ± 138 MJ m^{-2}) that floral initiation is significantly impeded (Pitchers et al., 2021). However, apple genetic diversity, for both rootstock and scion, should be explored further with regard to the capacity of apple to grow in multispecies systems where, for example, stratification of root systems is a way to optimise belowground water and nutrient use.

3.3 Conclusion

All the practices (Fig. 8) that increase biodiversity at a field scale can provide benefits for pest control, even though these aspects are not always considered or evaluated. Further basic and applied research is necessary to understand the

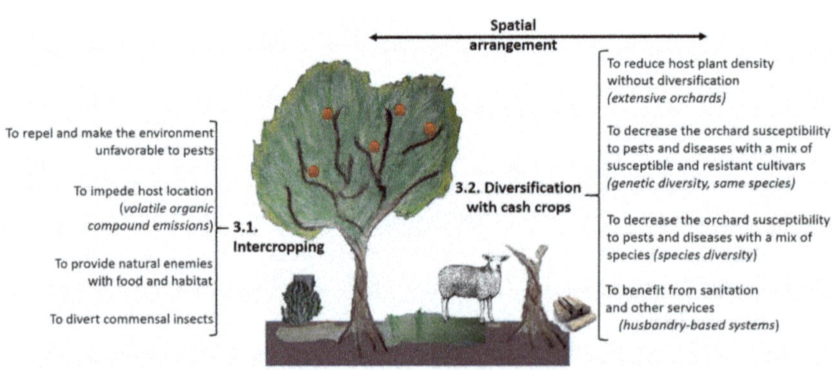

Figure 8 The agronomic practices related to the diversification of the system and its (re) design, presented in Section 3.

complex underlying processes and the magnitude of such effects on pests, and to develop further applications in crops.

4 Designing cropping systems that combine agronomic integrated pest management practices at field, farm and agri-food system scales

The challenge in designing innovative IPM fruit and vegetable production systems is to combine a range of practices and plant materials in a coherent system that should be operable for the farmer. The objective is to combine the benefits or even develop synergies among the identified partial effects of a range of IPM practices in order to control some pests and/or to increase the resilience of the fruit and vegetable production systems when biotic or abiotic stresses occur.

To assess interactions, overlap and compatibility between different levers of action against pests, the systems approach that considers the whole cropping system with on-station or on-farm trials offers an interesting basis. As Lechenet et al. (2017) pointed out the system experiment enables one to assess both the overall effectiveness of the strategy and the consistency between components of the cropping system. System experiments are also useful to interact with stakeholders. In Section 4.1, we present some case studies of combinations of practices and how they can contribute to drastically reducing pesticide use. In Section 4.2, we present some types of innovative designs in agroforestry. Finally, in Section 4.3, we discuss the involvement of a range of stakeholders in a participatory approach for the design of such complex systems.

4.1 Combining agronomic integrated pest management practices at a field scale

4.1.1 The BioREco system experiment (2005–2015): Performance of conventional, low-input and organic apple-tree orchard systems

The 'BioREco' system experiment investigated the possibilities of decreasing pesticide use in apple orchards (Simon et al., 2017, 2019). Three systems were tested: conventional as a reference, low-input and organic, and each system was planted with the same three cultivars, differing in disease susceptibility. Compared to the conventional system that was similar to the local orchards (Labeyrie et al., 2018), pesticide use at full production (2009–2015) was decreased by 38-45% on average in systems combining disease low-susceptible cultivars, alternative methods to the use of synthetic pesticides (e.g. mechanical weeding, biocontrol, sanitation and mating disruption) and an accurate evaluation of the risk of damage. In those systems, the yield was either similar (low-input) or lower (organic farming systems) than the reference. The multicriteria evaluation that considered agronomic, environmental, technico-economic and feasibility aspects showed the benefits and the limits of the experimental systems. For instance, the use of copper penalised the environmental score for the organic system compared with the low-input system. Conversely, the low-input system was costlier than the conventional one, with no upgrading in long supply chains to compensate for such extra cost. Finally, the low-input system was feasible at the expense of periodic monitoring and scouting activities, and reactivity to apply treatments, as soon as the risk for the orchard is confirmed. This work also enabled interactions between stakeholders in the food system concerning the experimental approach, the orchards, the multicriteria evaluation, and the results and knowledge necessary to design innovative IPM and organic orchards.

4.1.2 The Ecopeach system experiment (2013–2018): Evaluation of peach tree crop systems under low-input practices. Environmental, agronomic and economic assessments

Peach tree crops are high consumers of agricultural inputs, particularly for plant protection, with an average TFI higher than 20 in the south of France. The aim of the Ecopeach project (2013–2018) (Plenet et al., 2019) was to assess an innovative peach orchard system leading to a drastic reduction in agricultural inputs while maintaining economic and agronomic sustainability of the orchard, as compared to conventional orchards. Ecopeach is a network of experiments conducted by eight partners at seven sites scattered across the south of France. The experiments combine many levers of action, ranging from structural planting choices to the association of various alternative methods in the annual orchard management. Data analysis includes 29 systems × years (Plenet et al., 2019). Experimental results showed that phytosanitary inputs decreased (by more than 50%), mainly through: (i) limitation of agronomic inputs (irrigation and fertilisation) likely to reduce the susceptibility of peach to pathogens while maintaining suitable profitability in the fruit production and (ii) more risky spraying strategies that actually ensured a satisfying protection as compared to conventional spraying strategy. Compared to the conventional system, production costs were reduced under the low-input system, which partly compensated for the decrease in revenue caused by slightly lower marketable yields. Fruit quality was improved under low-input practices although economic returns did diminish, but only moderately. A small increase in the price paid to the producers (€0.15/kg of fruit) using low-input systems would be sufficient to offset the average decrease in marketable yield, and thus to promote the transition towards horticultural systems based on low-input practices, especially pesticide reduction.

4.2 Redesign and crop diversification: the need to upscale from the field to the farm scale

Agroforestry (see Section 3.2) and the advocacy for ecosystem services as major levers of production (e.g. Malézieux, 2012) have opened doors to the development of innovative orchard ecosystems from scratch, through participatory approaches and exploratory design (Simon et al., 2019; Penvern et al., 2018). For example, a diversified and multifruit production 'pest suppressive' area including cultivated and companion plant species has been co-designed with various actors in the objective of producing fruit without any pesticides. The main processes targeted are barrier-dilution effects, as well as plant-mediated (bottom-up) processes, such as trap cultivars, pest and disease low-susceptibility cultivars and natural enemy-mediated (top-down) processes, aside from agronomic (e.g. resource sharing) and operational considerations. The design considers both field and supra-field scales. Changes in practices due to the spatial design (e.g. ergonomics) and in the management of the agroecosystem, made 'on the way' in a dynamic 'observe and adjust' approach are necessary to manage the orchard with due consideration for all biotic interactions. The trajectory and performance of this system are now being assessed through multicriteria evaluation, including the provision of fruit and other products, ecosystem services, organisational aspects for labour and the valorisation of all products and sub-products. Such innovative systems do not encompass the field scale only. They are also implemented at the farm scale through various means: connection of agroecological infrastructures to the farm and landscape, design of production and production-supporting areas (e.g. hot spots of biodiversity), labour organisation in multi-production systems (specialist vs. generalist skills). Eventually, connections with the agri-food scale are also developed through the marketing of diversified, irregular and small amounts of each product. The challenge is to achieve trade-off in the design and management choices to account for all those scales and dimensions.

4.3 Dealing with complexity: involving stakeholders of the agri-food system in the design of multi-specific cropping systems

The previous sections illustrated some case studies of innovative cropping systems that were co-designed to combine a set of farming practices at field scale (Section 4.1) and/or to organise natural and cultivated biodiversity at both field and farm scales (Section 4.2), with consideration of the agri-food scale. Diversification in cropping systems generates complex crop planning (in time and space) and management, as it is supported by biological interactions between several living organisms (plants, trees, soil microorganisms, pest antagonists, etc.). But it also affects the whole agri-food system and the relationships between farmers and other stakeholders. This paradigm shift requires a move towards new research approaches, including participatory research. In this section, we illustrate this new trend in research, using examples drawn from fruit and vegetable pest management systems.

4.3.1 A necessity for working with stakeholders from the whole agri-food system

Although orchards are perennial systems and vegetables are annual ones, both have a similar socioeconomic organisation. The potential to sell fruit and vegetables highly depends on their visual appearance because they are mostly sold as fresh produce to exporters and national distributors that apply strict quality standards. In France, apples with disease spots, lettuces with aphids or curvy cucumbers are nearly impossible to sell in long supply chains, even if they present no risk for human health. Sociologists and agricultural scientists carried out empirical analyses of the agri-food system, which revealed that, for several decades, the agricultural sector has been locked into very specialised systems, the health of which can be maintained only with high level and frequent pesticide use. The strategies of major stakeholders from the agri-food system favour intensive specialised cropping systems, based on high yields, intensive use of synthetic inputs instead of biodiversity, which precludes any increase in crop diversification. Several limits to the development of cropped biodiversity have been identified. Meynard et al. (2018) point out the insufficient breeding activity and the lack of processing equipment to develop leguminous crops in arable systems. Pissonnier et al. (2016) and Boulestreau et al. (2021), respectively for apple and vegetable production, point out the lack of commercial outlets to sell diversification crops (e.g. spring onions). Della Rossa et al. (2020), comparing the sugarcane, banana and vegetable sectors in Martinique, point out that farmers' ability to innovate depends heavily on the quality of coordination in the supply chain. Since alternative cropping practices generate extra production costs for farmers, their implementation is blocked as long as the added value produced along the supply chain is unevenly distributed, particularly when farmers have no financial returns for their efforts. Therefore, scientists need to work not only with farmers but also with a wide range of stakeholders from the agri-food system who influence food production, processing, distribution, preparation and consumption within a given territory (Boulestreau et al., 2021). For the fruit and vegetable sectors, it is of particular interest to include stakeholders from the retail end of the supply chain.

4.3.2 Two examples of participatory design of alternative farming systems

In France, long supply chains through wholesalers and distributors are responsible for high visual quality requirements, to provide fruits and vegetables to consumers with no visual defaults, which promotes pesticide use despite their negative effects on the environment and human health. Similar trends occur at the European level, despite some specificities due to national contexts. Therefore, we now briefly summarise two attempts to take into account stakeholders from the French agri-food system level, in designing and assessing complex agronomic innovations.

Diversified vegetable cropping systems under polythene tunnels were designed by scientists and stakeholders (Lefèvre et al., 2020), with the target of complying with both food value chain specifications and agroecological crop protection principles. In this study, the four crop protection strategies designed and assessed were based on a high level of crop biodiversity (range of cultivars and species in the crop rotation, association of several crops and companion plants in the field, use of biocontrol methods, etc.). The originality of the study was that each strategy targeted a specific food value chain, characterised by two combined factors: organic versus conventional farming, crossed with local direct sale versus super- and hypermarket value chain. This led to technical specifications for the design, in order to produce vegetables that fulfilled specific characteristics as regards visual quality (calibre and visual defaults), temporality (date of harvest) and volumes. The farmers and technical advisers involved in the research project provided information on the functioning of the agri-food system that had to be taken into account, pertaining to work constraints and operational/practical skills to crop certain species that were under study by scientists. It was shown that the nature of trade-offs in crop management, and in particular in agroecological crop protection, depended heavily on the food value chain targeted. This supported the idea that considering the agri-food system was useful, even for agronomic innovation.

In the diversified orchard cropping systems that maximise ecological pest elimination, presented in Section 4.2, the agri-food system was not initially considered in the co-design process. Only the number of trees per cultivar of the major species was selected to produce sufficient fruit volumes that would supply fruit-vegetable boxes or direct sales on the farm. As shown above, fruit species, cultivar diversity and their spatial arrangement were chosen for their pest-suppressive effects. For instance, apple cultivars were mixed on the tree-row, and apple-tree-rows were separated from one another with other species, that were chosen to provide multifunctional services (e.g. hazelnut not only as a companion plant through hosting natural enemies but also as a production plant). These ecological choices raise penetrating questions about the ways to get commercial income from such products which are either new on the farm and/or in the region (e.g. nut fruits, fig or pomegranate), or produced in small quantities, and/or do not comply with current sale standards. Fruit production resulting from such a mixed and zero-pesticide orchard is expected to be (i) heterogeneous in fruit visual aspect; (ii) with a high probability of visual defaults (disease spots) and (iii) in limited quantities for each product. All these characteristics are known to be difficult to sell in long supply chains, which focus on large and homogeneous batches of fruits. Similar problems arise in agroforestry systems mixing fruit trees and vegetables (Paut et al., 2020). A single supply chain is unlikely to support the selling of fruits and vegetables from highly diversified production systems.

For both previous examples, commercial outlets issues need to be addressed in order to take into account additional production costs of agroecological practices and enable farmers to make these diversified cropping systems profitable. Several fresh fruit or vegetable supply chains for a single farm, as well as processing and perhaps non-food use outlets, must therefore be considered. This supports the need to redesign simultaneously the agricultural and agri-food systems (Meynard et al., 2017), and to consider both diversification of production and marketing channels in the design process. At the same time, it raises profound questions on the coordination within the agri-food system among farmers, marketers, distributors, processors and finally consumers.

4.3.3 Conclusion

Designing complex and diversified cropping systems needs several kinds of stakeholders (Fig. 9) from the agri-food system to be involved: (i) the farmers (to identify the range of farm goals and constraints); (ii) the actors upstream in the supply chain (to supply specific cultivars and species during the required period as well as companion plants and some ecological infrastructures); (iii) the actors downstream in the supply chain (to provide enough outlets with enough economic returns to compensate for farmers' extra costs) and even (iv) research and development stakeholders (to provide technical information on crop association or on the possible trade-offs between varied crop management decisions, and; to accompany and train farmers in their cropping decisions). The stakeholders involved should be chosen after an in-depth territory analysis, to identify the main sociotechnical lock-in effects, that is the characteristics or processes in the current agri-food system that impede change towards no-pesticide cropping systems. This sociotechnical diagnosis at the territorial

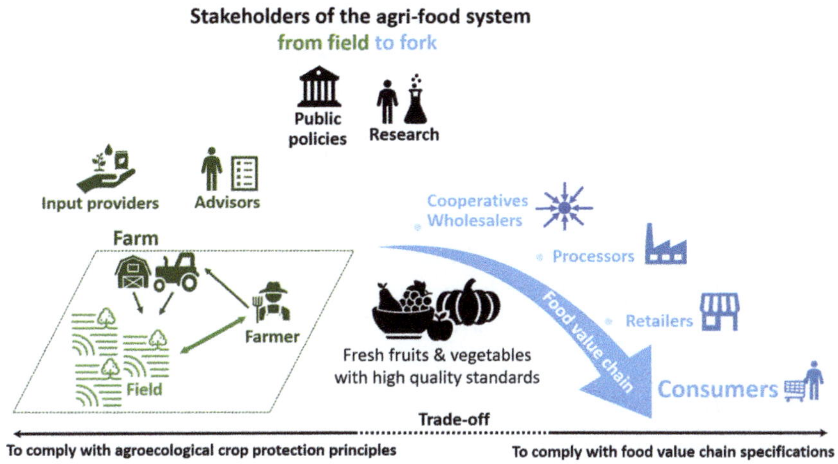

Figure 9 In the co-design process, all stakeholders from the agri-food systems have to be considered in order to comply with both the agroecological crop protection principles and the food value chain specifications.

level also aims at identifying potential unlocking levers that could favour change and the stakeholders able to promote them (Meynard et al., 2018). Tools are presently being developed to support such co-design (Boulestreau et al., 2021; Della Rossa et al., 2020; IDEAS platform https://www6.inrae.fr/ideas-agri-food/Plateforme-d-appui/Nos-outils) at larger scales than the field. Specific methods will be required to support the collective and participatory design of new farming systems, and in particular to help stakeholders and scientists to be creative, to visualise the system under construction (e.g. serious games, 2D or 3D models to represent spatial layouts) and assess its performance, especially the multiple trade-offs and the criteria to take into account (Le Bellec et al., 2012). It will then also require the design of new experimentation processes that could combine agronomic trials at experimental stations with satellite trials on farms, able to take into account a diversity of cropping conditions and to favour farmers' transition towards pesticide-free systems (Navarrete et al., 2018).

5 Where to look for further information

5.1 Further reading

- Deguine, J.-P., Jacquot, M., Allibert, A., Chiroleu, F., Graindorge, R., Laurent, P., Lambert, G., Albon, B., Marquier, M., Gloanec, C., Vanhuffel, L., Vincenot, D. and Aubertot, J.-N. 2018. *Agroecological Protection of Mango Orchards* in La Réunion, in: Gaba, S., Smith, B. and Lichtfouse, E. (Eds.), *Sustainable Agriculture Reviews* 28. Springer International Publishing, Cham, pp. 249–307. https://doi.org/10.1007/978-3-319-90309-5_8.

- Deguine, J.-P., Gloanec, C., Laurent, P., Ratnadass, A. and Aubertot, J.-N. (Eds.). 2017. *Agroecological Crop Protection*. Springer, Quae. 249 pp. ISBN 978-94-024-1184-3.
- Messelink, G. J., Bennison, J., Alomar, O., Ingegno, B. L., Tavella, L., Shipp, L., Palevsky, E. and Wäckers, F. L. 2014. Approaches to conserving natural enemy populations in greenhouse crops: current methods and future prospects. *BioControl* 59, 377–393. https://doi.org/10.1007/s10526-014-9579-6.

5.2 Websites

- http://www.fao.org/agriculture/crops/thematic-sitemap/theme/pests/ipm/more-ipm/en/.
- https://ec.europa.eu/food/plant/pesticides/sustainable_use_pesticides/ipm_en.
- IOBC: international Organization for Biological and Integrated control: https://www.iobc-wprs.org/pub/#bulletin.
- https://www6.inrae.fr/tatabox/content/download/3588/34893/version/1/file/Guide_methodo_tatabox-avril2018-web.pdf.
- IDEAS platform: https://www6.inrae.fr/ideas-agri-food/Plateforme-d-appui/Nos-outils.
- French National Action Plan Ecophyto: https://ecophytopic.fr/pic/pour-aller-plus-loin/ecophyto.

6 References

Amarawardana, L., Bandara, P., Kumar, V., Pettersson, J., Ninkovic, V. and Glinwood, R. (2007). Olfactory response of *Myzus persicae* (Homoptera: Aphididae) to volatiles from leek and chive: Potential for intercropping with sweet pepper. *Acta Agric. Scand. B* 57(1), 87–91.

Amiot, M. J., Fleuriet, A., Cheynier, V. and Nicolas, J. (1997). Phenolic compounds and oxidative mechanisms in fruit and vegetables. In: Tomas-Barberan, F. A. and Robins, R. J. (Eds), *Phytochemistry of Fruit and Vegetables*. Clarendon Press, Oxford, pp. 51–85.

Ando, K., Grumet, R., Terpstra, K. and Kelly, J. D. (2007). Manipulation of plant architecture to enhance crop disease control. *CAB Rev.: Perspect. Agric. Vet. Sci. Nutr. Nat. Resour.* 2(26).

Andres, C., Comoé, H., Beerli, A., Schneider, M., Rist, S. and Jacobi, J. (2016), Chapter 3: Cocoa in monoculture and dynamic agroforestry. In: Lichtfouse, E. (Ed.), *Sustainable Agriculture Reviews*. Springer International Publishing, Switzerland 19, 121–153.

Appel, O. N. and Stipes, R. J. (1984). Canker expansion on water-stressed pin oak colonised by *Endothia gyrosa*. *Plant Dis.* 68, 851–853.

Barkai-Golan, R. (Ed.). (2001). *Postharvest Diseases of Fruits and Vegetables: Development and Control*. Elsevier Science (1st Edition), 442 pp. ISBN: 9780444505842. https://doi.org/10.1016/B978-0-444-50584-2.X5000-X.

Ben Issa, R., Gomez., L. and Gautier, H. (2017). Companion plants for aphid pest management. *Insects* 8(4), 112–131.

Boulestreau, Y., Casagrande, M. and Navarrete, M. (2021). Analyzing barriers and levers for practice change: A new framework applied to vegetables' soil pest management. *Agron. Sustain. Dev.* 41(3), 44.

Brown, M. W. (2001). Flowering ground cover plants for pest management in peach and apple orchards. *Bull. IOBC/WPRS* 24, 379–382.

Bussi, C. and Amiot, M. J. (1998). Effects of nitrogen and potassium fertilization on the growth, yield and pitburn of apricot (cv. Bergeron). *J. Hortic. Sci. Biotechnol.*, 73(3), 387–392.

Bussi, C., Besset, J. and Girard, T. (2003). Effects of fertilizer rates and dates of application on apricot (cv Bergeron) cropping and pitburn. *Sci. Hort.* 98(2), 139–147.

Bussi, C., Parveaud, C.-E., Mercier, V. and Lescourret, F. (2016). Effects of irrigation deprivation and ground cover (*Trifolium repens*) in the tree row on brown rot incidence in peach. *Crop Prot.* 88, 37–44.

Cahenzli, F., Sigsgaard, L., Daniel, C., Herz, A., Jamar., L., Kelderer, M., Kramer Jacobsen, S. K., Kruczyńska, D., Matray., S., Porcel, M., Sekrecka, M., Świergiel, W., Tasin, M., Telfser, J. and Pfiffner, L. (2019). Perennial flower strips for pest control in organic apple orchards–A pan-European study. *Agric. Ecosyst. Environ.* 278, 43–53.

Casagrande, M., Alletto, L., Naudin, C., Lenoir, A., Siah, A. and Celette, F. (2017). Enhancing planned and associated biodiversity in French farming systems. *Agron. Sustain. Dev.* 37(6), 57.

Dancewicz, K. and Gabryś, B. (2008). Effect of extracts of garlic (*Allium sativum* L.), wormwood (*Artemisia absinthium* L.) and tansy (*Tanaceum vulgare* L.) on the behaviour of the peach potato aphid *Myzus persicae* (Sulz.) during the settling on plants. *Pestycydy/Pestic.* 3–4, 93–99.

Dardouri, T., Gautier, H., Ameline, A., Costagliola, G., Schoeny, A. and Gomez, L. (2021). Non-host volatiles disturb the feeding behavior and reduce the fecundity of the green peach aphid, *Myzus persicae*. *Pest Manage. Sci.*, 77(4), 1705–1713.

Dardouri, T., Gautier, H., Ben Issa, R., Costagliola, G. and Gomez, L. (2019a). Repellence of *Myzus persicae* (Sulzer): Evidence of two modes of action of volatiles from selected living aromatic plants. *Pest Manage. Sci.* 75(6), 1571–1584.

Dardouri, T., Gomez, L., Schoeny, A., Costagliola, G. and Gautier, H. (2019b). Behavioural response of green peach aphid *Myzus persicae* (Sulzer) to volatiles from different rosemary (*Rosmarinus officinalis* L.) clones. *Agric. For. Entomol.* 21(3), 336–345.

Deletre, E., Schatz, B., Bourguet, D., Chandre, F., Williams, L., Ratnadass, A. and Martin, T. (2016). Prospects for repellent in pest control: Current developments and future challenges. *Chemoecology* 26(4), 127–142.

Della Rossa, P., Le Bail, M., Mottes, C., Jannoyer, M. and Cattan, P. (2020). Innovations developed within supply chains hinder territorial ecological transition: The case of a watershed in Martinique. *Agron. Sustain. Dev.* 40(1), 10.

Demestihas, C., Plénet, D., Génard, M., Raynal, C. and Lescourret, F. (2017). Ecosystem services in orchards. A review. *Agron. Sustain. Dev.* 37(2), 12.

Durán Zuazo, V. H., Pleguezuelo, C. R. R., Francia Martínez, J. R., Martínez Raya, A., Arroyo Panadero, L., Càrceles Rodríguez, B. and Navarro Moll, M. C. (2008). Benefits of plants strips for sustainable mountain agriculture. *Agron. Sustain. Dev.* 28(4), 497–505.

Ebel, R. C., Proebsting, E. L. and Evans, R. G. (1995). Deficit irrigation to control vegetative growth in apple and monitoring fruit growth to schedule irrigation. *HortSci.* 30(6), 1229–1232.

Finch, S. and Collier, R. H. (2012). The influence of host and non-host companion plants on the behaviour of pest insects in field crops. *Entomol. Exp. Appl.* 142(2), 87–96.

Fischer, I. H., Soares-Colletti, A. R., Palharini, M. CdA., Parisi, M. C. M. and Amorim, L. (2017). Temporal progress and spatial patterns of quiescent diseases in guava influenced by sanitation practices. *Sci. Agric.* 74(1), 68–76.

Galindo, A., Rodriguez, P., Collado-Gonzalez, J., Cruz, Z. N., Torrecillas, E., Ondono, S., Corell, M., Moriana, A. and Torrecillas, A. (2014). Rainfall intensifies fruit peel cracking in water stressed pomegranate trees. *Agric. For. Meteorol.* 194, 29–35.

Garcia-Salazar, C., Gut, L. J. and Whalon, M. E. (2007). Hedgerow barriers and other reduced-risk controls for managing Oriental fruit moth, *Grapholita molesta* (Busck) (Lepidoptera: Tortricidae) in apple. *Renew. Agric. Food Syst.* 22(3), 181–188.

Gibert, C., Chadoeuf, J., Vercambre, G., Genard, M. and Lescourret, F. (2007). Cuticular cracking on nectarine fruit surface. Spatial distribution and development in relation to irrigation and thinning. *J. Am. Soc. Hort. Sci.* 132(5), 583-591.

Gibert, C., Lescourret, F., Genard, M., Vercambre, G. and Perez-Pastor, A. (2005). Modelling the effect of fruit growth on surface conductance to water vapour diffusion. *Ann. Bot.* 95(4), 673–683.

Girona, J., Mata, M. and Marsal, J. (2005). Regulated deficit irrigation during the kernel-filling period and optimal irrigation rates in almond. *Agric. Water Manage.* 75(2), 152–167.

Gomez, C., Brun, L., Chauffour, D. and Le Vallee, D. D. L. (2007). Effect of leaf litter management on scab development in an organic apple orchard. *Agric. Ecosyst. Environ.* 118(1–4), 249–255.

Gomez, C. and Mercier, V. (2008). Effet de l'enherbement total sur le développement des monilioses. *Arboric. Fruitière* 628, 20–23.

Gomez, J. A., Guzman, M. G., Giraldez, J. V. and Fereres, E. (2009). The influence of cover crops and tillage on water and sediment yield, and on nutrient, and organic matter losses in an olive orchard on a sandy loam soil. *Soil Till. Res.* 106(1), 137–144.

Gomez, L., Vercambre, G. and Jordan, M. O. (2020). Spatial-temporal management of nitrogen and carbon on the peach tree (*Prunus persicae* L. Batsch.). *Sci. Hort.* 273.

Grechi, I., Sauge, M. H., Sauphanor, B., Hilgert, N., Senoussi, R. and Lescourret, F. (2008). How does winter pruning affect peach tree - *Myzus persicae* interactions? *Entomol. Exp. Appl.* 128(3), 369-379.

Heil, M. and Silva Bueno, J. C. (2007). Within-plant signaling by volatiles leads to induction and priming of an indirect plant defense in nature. *Proc. Natl. Acad. Sci. U. S. A.* 104(13), 5467-5472.

Hipps, N. A., Hazelden, J. and Faiall, G. B. N. (1990). Control of erosion in mature orchard. *Soil Use Manage.* 6(1), 32–35.

Holb, I. J. (2006). Effect of six sanitation treatments on leaf litter density, ascospore production of *Venturia inaequalis* and scab incidence in integrated and organic apple orchards. *Eur. J. Plant Pathol.* 115(3), 293–307.

Holb, I. J. and Kunz, S. (2016). Integrated control of apple scab and powdery mildew in an organic apple orchard by combining potassium carbonates with wettable sulfur, pruning, and cultivar susceptibility. *Plant Dis.* 100(9), 1894-1905.

Ikeura, H., Kobayashi, F. and Hayata, Y. (2012). Repellent effect of herb extracts on the population of wingless green peach aphid. *Myzus persicae* Sulzer (Hemiptera: Aphididae). *J. Agric. Sci.* 4(5), 139-144.

Irvin, N. A., Scarratt, S. L., Wratten, S. D., Frampton, C. M., Chapman, R. B. and Tylianakis, J. M. (2006). The effects of floral understoreys on parasitism of leafrollers (Lepidoptera: Tortricidae) on apples in New Zealand. *Agr. Forest Entomol.* 8(1), 25-34.

Irving, D. E. and Drost, J. H. (1987). Effects of water deficit on vegetative growth, fruit growth and fruit quality in Cox's Orange Pippin Apple. *J. Hortic. Sci.* 62(4), 427-432.

Jordan, M. O., Sauge, M. H. and Vercambre, G. (2020). Chemical and growth traits of the peach tree may induce higher infestation rates of the green peach aphid, *Myzus persicae* (Sulzer). *Pest Manage. Sci.* 76(2), 797-806.

Kahn, B. A., Rebek, E. J., Brandenberger, L. P., Reed, K. and Payton, M. E. (2017). Companion planting with white yarrow or with feverfew for squash bug. *Anasa tristis* (Hemiptera: Coreidae). Management on summer squash. *Pest Manage. Sci.* 73(6), 1127-1133.

Kotze, W. A. G. and Bothma, K. (1989). Factors that have an effect on the incidence of pitburn in Kakamas peach. *Deciduous Fruit Grower* 39, 159-161.

Kuehrt, U., Samietz, J., Hoehn, H. and Dorn, S. (2006). Modelling the phenology of codling moth: Influence of habitat and thermoregulation. *Agric. Ecosyst. Environ.* 117(1), 29-38.

Labeyrie, B., Sagnes, J.-L., Castel, L., Dubreuil, N., Kreiter, P., Millan, M., Plénet, P., Simon, S., Zavagli, F., Rougier, M., Emonet, E., Longis, S. and Brun, V. (2018). Réseau DEPHY EXPE : Synthèse des résultats à l'échelle nationale - filière Arboriculture. *Cellule d'Animation Nationale DEPHY Ecophyto*, 64 p. https://ecophytopic.fr/sites/default/files/Synthèse_EXPE_Arbo.pdf.

Lampinen, B. D., Shackel, K. A., Southwick, S. M., Olson, B., Yeager, J. T. and Goldhamer, D. (1995). Sensitivity of yield and fruit quality of french prune to water deprivation at different fruit growth stages. *J. Am. Soc. Hort. Sci.* 120(2), 139-147.

Landis, D. A., Wratten, S. D. and Gurr, G. M. (2000). Habitat management to conserve natural enemies of arthropod pests in agriculture. *Annu. Rev. Entomol.* 45, 175-201.

Larson, K. C. and Whitham, T. G. (1997). Competition between gall aphids and natural plant sinks: Plant architecture affects resistance to galling. *Oecologia* 109(4), 575-582.

Laurens, F., Aranzana, M. J., Arús, P., Bassi, D., Bink, M., Bonany, J., Caprera, A., Corelli-Grappadelli, L., Costes, E., Durel, C. E., Mauroux, J. B., Muranty, H., Nazzicari, N., Pascal, T., Patocchi, A., Peil, A., Quilot-Turion, B., Rossini, L., Stella, A., Troggio, M., Velasco, R. and van de Weg, E. (2018). An integrated approach for increasing breeding efficiency in apple and peach in Europe. *Hortic. Res.* 5, 11.

Lauri, P. É. (2008). Trends in apple training in France–An architectural and ecophysiological perspective. *Acta Hortic.* 772(772), 483-490.

Lauri, P. É. (2021). Tree architecture and functioning facing multispecies environments - we have gone only halfway in fruit-trees. *Am. J. Bot.* 108(1), 3-7.

Lauri, P. É., Barkaoui, K., Ater, M. and Rosati, A. (2019). Agroforestry for fruit trees in the temperate Europe and dry Mediterranean. In: Mosquera-Losada, M. R. and Prabhu, R. (Eds), *Agroforestry for Sustainable Agriculture*. Burleigh Dodds Science Publishing, Cambridge, UK, pp. 385-418.

Lauri, P. É. and Corelli-Grappadelli, L. (2014). Tree architecture, flowering and fruiting–Thoughts on training, pruning and ecophysiology. *Acta Hortic.* 1058, 291-298.

Lauri, P. É, Costes, E., Regnard, J. L., Brun, L., Simon, S., Monney, P. and Sinoquet, H. (2009). Does knowledge on fruit tree architecture and its implications for orchard management improve horticultural sustainability? An overview of recent advances in the apple. *Acta Hortic.* 817(817), 243–250.

Lauri, P. É., Hucbourg, B., Ramonguilhem, M. and Joannin, R. (2016a). A step further–Optimizing the natural apple tree habit with the Salsa concept. *Acta Hortic.* 1130, 151–156.

Lauri, P. É., Mézière, D., Dufour, L., Gosme, M., Simon, S., Gary, C., Jagoret, P., Wery, J. and Dupraz, C. (2016b). Fruit-trees in agroforestry systems - Review and prospects for the temperate and Mediterranean zones. *Proceedings of the 3rd European Agroforestry Conference*, pp. 106–109.

Lauri, P. É. and Laurens, F. (2005). Architectural types in apple (*Malus Xdomestica* Borkh.). In: Dris, R. (Ed.), *Crops: Growth, Quality and Biotechnology*. World Food Limited, Helsinki, Finland, pp 1300–1314.

Lauri, P. É., Pitchers, B., Dufour, L. and Simon, S. (2020). Apple farming systems - current initiatives and some prospective views on how to improve sustainability. *Acta Hortic.* 1281(1281), 307–322.

Lauri, P. É. and Simon, S. (2019). Advances and challenges in sustainable apple cultivation. In: Lang, G. A. (Ed.), *Achieving Sustainable Cultivation of Temperate Zone Tree Fruits and Berries Volume 2: Case Studies*. Burleigh Dodds Science Publishing, Cambridge, UK. ISBN: 978 1 78676 212 2; www.bdspublishing.com, pp. 261–288.

Leach, H., Moses, J., Hanson, E., Fanning, P. and Isaacs, R. (2018). Rapid harvest schedules and fruit removal as non-chemical approaches for managing spotted wing Drosophila. *J. Pest Sci.* 91(1), 219–226.

Le Bellec, F., Rajaud, A., Ozier-Lafontaine, H., Bockstaller, C. and Malezieux, E. (2012). Evidence for farmers' active involvement in co-designing citrus cropping systems using an improved participatory method. *Agron. Sustain. Dev.* 32(3), 703–714.

Lechenet, M., Deytieux, V., Antichi, D., Aubertot, J. N., Barberi, P., Bertrand, M., Cellier, V., Charles, R., Colnenne-David, C., Dachbrodt-Saaydeh, S., Debaeke, P., Doré, T., Farcy, P., Fernandez-Quintanilla, C., Grandeau, G., Hawes, C., Jouy, L., Justes, E., Kierzek, R., Kudsk, P., Lamichhane, J. R., Lescourret, F., Mazzoncini, M., Melander, B., Messéan, A., Moonen, A., Newton, A. C., Nolot, J., Panozzo, S., Retaureau, P., Sattin, M., Schwarz, J., Toqué, C., Vasileiadis, V. P. and Munier-Jolain, N. (2017). Diversity of methodologies to experiment Integrated Pest Management in arable cropping system: Analysis and reflections based on a European network. *Eur. J. Agron.* 83, 86–99.

Lefèvre, A., Perrin, B., Lesur-Dumoulin, C., Salembier, C. and Navarrete, M. (2020). Challenges of complying with both food value chain specifications and agroecology principles in vegetable crop protection. *Agric. Syst.* 185, 102953.

Lewis, T. L., Martin, D., Cerny, J. and Ratkowsky, D. A. (1977). The effect of increasing the supply of nitrogen, phosphorus, calcium and potassium to the roots of Merton Worsester apple on leaf and fruit composition and on the incidence of bitter-pit at harvest. *J. Hortic. Sci.* 52(3), 409–419.

Li, S. H., Huguet, J. G. and Bussi, C. (1989a). Irrigation scheduling in mature peach orchard using tensiometers and dendrometers. *Irrig. Drainage Syst.* 3(1), 1–12.

Li, S. H., Huguet, J. G., Schoch, P. G. and Orlando, P. (1989b). Response of peach tree growth and cropping to soil water deficit at various phenological stages of fruit development. *J. Hortic. Sci.* 64(5), 541–552.

Lobell, D. B., Schlenker, W. and Costa-Roberts, J. (2011). Climate trends and global crop production since 1980. *Science* 333(6042), 616–620.

Lovell, S. T., Dupraz, C., Gold, M., Jose, S., Revord, R., Stanek, E. and Wolz, K. J. (2018). Temperate agroforestry research: Considering multifunctional woody polycultures and the design of long-term field trials. *Agrofor. Syst.* 92, 1397–1415.

Mace, K. C. and Mills, N. J. (2015). Response of walnut aphid populations to increasing foliar nitrogen content. *Agric. For. Entomol.* 17(3), 277–284.

Macheix, J. J., Fleuriet, A. and Billot, J. (1990). *Fruit Phenolics*. CRC Press Press, Boca Raton, FL.

Malézieux, E. (2012). Designing cropping systems from nature. *Agron. Sustain. Dev.* 32(1), 15–29.

Malézieux, E., Crozat, Y., Dupraz, C., Laurans, M., Makowski, D., Ozier-Lafontaine, H., Rapidel, B., Tourdonnet, S. and Valantin-Morison, M. (2009). Mixing plant species in cropping systems: Concepts, tools and models. A review. *Agron. Sustain. Dev.* 29(1), 43–62.

Mari, M., Gregori, R. and Donati, I. (2004). Postharvest control of *Monilinia laxa* and *Rhizopus stolonifer* in stone fruit by paracetic acid. *Postharvest Biol. Technol.* 33(3), 319-325.

Matis, T. I., Parajulee, M. N., Matis, J. H. and Shrestha, R. B. (2008). A mechanistic model based analysis of cotton aphid population dynamics data. *Agric. For. Entomol.* 10(4), 355–362.

McGuire, R. G. and Kelman, A. (1984). Reduced severity of *Erwinia* soft rot in potato tubers with increased calcium content. *Phytopathology* 74(10), 1250–1256.

McKerchar, M., Potts, S. G., Fountain, M. T., Garratt, M. P. D. and Westbury, D. B. (2020). The potential for wildflower interventions to enhance natural enemies and pollinators in commercial apple orchards is limited by other management practices. *Agric. Ecosys Environ.* 301, 107034.

Meagher, R. L. Jr. and Meyer, J. R. (1990). Effects of ground cover management on certain abiotic and biotic interactions in peach orchard ecosystems. *Crop Prot.* 9(1), 65–72.

Mercier, V., Frachon, S. and Demoulin, G. (2003). Monilioses: Périodes de sensibilité des pêches au verger. *Phytoma Défense Végétaux* 558, 38–40.

Mercier, V., Bussi, C., Lescourret, F. and Genard, M. (2009). Effects of different irrigation regimes applied during the final stage of rapid growth on an early maturing peach cultivar. *Irrig. Sci.* 27(4), 297–306.

Mercier, V., Bussi, C., Plenet, D. and Lescourret, F. (2008). Effects of limiting irrigation and of manual pruning on brown rot incidence in peach. *Crop Prot.* 27(3–5), 678–688.

Meynard, J., Charrier, F., Fares, M., Le Bail, M., Magrini, M., Charlier, A. and Messéan, A. (2018). Socio-technical lock-in hinders crop diversification in France. *Agron. Sustain. Dev.* 38(5), 54.

Meynard, J. M., Jeuffroy, M. H., Le Bail, M., Lefèvre, A., Magrini, M. B. and Michon, C. (2017). Designing coupled innovations for the sustainability transition of agri-food systems. *Agric. Syst.* 157, 330–339.

Michon, G., Bompard, J., Hecketsweiler, P. and Ducatillion, C. (1983). Tropical forest architectural analysis as applied to agroforests in the humid tropics: The example of traditional village-agroforests in west Java. *Agrofor. Syst.* 1(2), 117–129.

Miller, S. A., Smith, G. S., Boldingh, H. L. and Johansson, A. (1997). Effects of water stress on fruit quality attributes of kiwifruit. *Ann. Bot.* 81, 73-81.

Mody, K., Eichenberger, D. and Dorn, S. (2009). Stress magnitude matters: Different intensities of pulsed water stress produce non-monotonic resistance responses of host plants to insect herbivores. *Ecol. Entomol.* 34(1), 133–143.

Mutiga, S. K., Gohole, L. S. and Auma, E. O. (2010). Effects of integrating companion cropping and nitrogen application on the performance and infestation of collards by *Brevicoryne brassicae*. *Entomol. Exp. Appl.* 134(3), 234–244.

Naor, A., Hupert, H., Greenblat, Y., Peres, M., Kaufman, A. and Klein, I. (2001). The response of nectarine fruit size and midday stem water potential to irrigation level in stage III and crop load. *J. Am. Soc. Hort. Sci.* 126(1), 140–143.

Navarrete, M., Brives, H., Catalogna, M., Gouttenoire, L., Heinisch, C., Lamine, C., Ollion, E. and Simon, S. (2018). Farmers' involvement in collective experimental designs in a French region, Rhône-Alpes. How do they contribute to farmers' learning and facilitate the agroecological transition? *13th European IFSA Symposium*, Chania.

Nijmeijer, A., Lauri, P. É., Harmand, J. M., Freschet, G., Essobo Nieboukaho, J. D., Kenfack Fogang, P., Enock, S. and Saj, S. (2019). Long-term dynamics of ecosystem functions and services in cocoa agroforestry systems established on savannah and forest. *Agric. Ecosyst. Environ.* 275, 100–111.

Palou, L., Crisosto, C. H., Garner, D. and Basinal, L. M. (2003). Effects of continuous exposure to exogenous ethylene during cold storage on postharvest decay development and quality attributes of stone fruits and table grapes. *Postharvest Biol. Technol.* 27(3), 243–254.

Pålsson, J., Porcel, M., Frimodt Hansen, M., Ofenberg, J., Nardin, T., Larcher, R. and Tasin, M. (2020). Aphid-infested beans divert ant attendance from the rosy apple aphid in apple-bean intercropping. *Sci. Rep.* 10(1), 8209.

Parisi, L., Gros, C., Combe, F., Parveaud, C., Gomez, C. and Brun, L. (2013). Impact of a cultivar mixture on scab, powdery mildew and rosy aphid in an organic apple orchard. *Crop Prot.* 43, 207–212.

Parker, M. L. and Meyer, J. R. (1996). Peach tree vegetative and root growth respond to Orchard floor management. American Society for Horticultural Science Western Region Annual Meeting, Vancouver, Canada, Vancouver, Canada. *Hort. Sci.* 3, 322–325.

Parolin, P., Bresch, C., Desneux, N., Brun, R., Bout, A., Boll, R. and Poncet, C. (2012). Secondary plants used in biological control: A review. *Int. J. Pest Manage.* 58(2), 91–100.

Parveaud, C. E., Gomez, C., Bussi, C. and Capowiez, Y. (2012). Effect of white clover (*Trifolium repens*, cv. Huia) cover crop on agronomic properties and soil biology in an organic peach orchard. *Acta Hort.* 933, 373–380.

Pascual-Villalobos, M. J., Cantó-Tejero, M., Vallejo, R., Guirao, P., Rodríguez-Rojo, S. and Cocero, M. J. (2017). Use of nanoemulsions of plant essential oils as aphid repellents. *Ind. Crops Prod.* 110, 45–57.

Paut, R., Sabatier, R. and Tchamitchian, M. (2020). Modelling crop diversification and association effects in agricultural systems. *Agric. Ecosys Environ.* 288, 106711.

Peet, M. M. (1992). Fruit cracking in tomato. *Hort. Technol.* 2(2), 216–223.

Penvern, S., Chieze, B. and Simon, S. (2018). Trade-offs between dreams and reality: Agroecological orchard co-design. 13th European IFSA Symposium, Chania, p. 10.

Perring, M. A. and Preston, A. P. (1974). The effects of orchard factors on the chemical composition of apples. 3. Some effects of pruning and nitrogen application on Cox's orange Pippin fruit. *J. Hortic. Sci.* 49(1), 85–93.

Pickett, J. A., Woodcock, C. M., Midega, C. A. O. and Khan, Z. R. (2014). Push-pull farming systems. *Curr. Opin. Biotechnol.* 26, 125–132.

Pissonnier, S., Lavigne, C., Toubon, J. F. and Le Gal, P. Y. (2016). Factors driving growers' selection and implementation of an apple crop protection strategy at the farm level. *Crop Prot.* 88, 109–117.

Pitchers, B., Do, F. C., Pradal, C., Dufour, L. and Lauri, P. É. (2021). Apple tree adaptation to shade in agroforestry - an architectural approach. *Am. J. Bot.* 108(5), 732–743.

Plenet, D., Hilaire, C., Blanc, P., Borg, J., Borne, S., Bussi, C., Gallia, V., Greil, M. L., Hostalnou, E., Labeyrie, B., Millan, M., Mercier, V., Montrognon, Y., Monty, D., Mouiren, C., Pinet, C. and Ruesch, J. (2019). EcoPêche: Conception et évaluation multisite de vergers de pêche - nectarine économes en produits phytopharmaceutiques et en intrants. *Innov. Agro* 76, 291–310.

Price, P. W. (1991). The plant vigor hypothesis and herbivore attack. *Oikos* 62(2), 244–251.

Rousselin, A., Bevacqua, D., Vercambre, G., Sauge, M. H., Lescourret, F. and Jordan, M. O. (2018). Rosy apple aphid abundance on apple is shaped by vegetative growth and water status. *Crop Prot.* 105, 1–9.

Sauge, M. H., Grechi, I. and Poëssel, J. L. (2010). Nitrogen fertilization effects on *Myzus persicae* aphid dynamics on peach: Vegetative growth allocation or defensive chemistry effects? *Entomol. Exp. Appl.* 136(2), 123–133.

Sekse, L. (1995). Fruit cracking in sweet cherries (Prunus avium L.). Some physiological aspects. A review. *Sci. Hort.* 63, 135–141.

Shrivastava, G., Rogers, M., Wszelaki, A., Panthee, D. R. and Chen, F. (2010). Plant volatiles-based insect pest management in organic farming. *Crit. Rev. Plant Sci.* 29(2), 123–133.

Sigsgaard, L. and Kramer Jacobsen, S. (2017). Functional agrobiodiversity – a novel approach to optimize pest control in fruit production. *IOBC WPRS Bull.* 122, 26–28.

Simon, S., Alaphilippe, A., Borne, S., Penvern, S., Dufils, A., Ricard, J. M. and Lauri, P. É. (2019). Methodology to co-design temperate fruit tree-based agroforestry systems: Three case studies in Southern France. 4th World Congress on Agroforestry, Montpellier, p. 601.

Simon, S., Bouvier, J. C., Debras, J. F. and Sauphanor, B. (2010). Biodiversity and pest management in orchard systems. A review. *Agron. Sustain. Dev.* 30(1), 139–152.

Simon, S., Lesueur-Jannoyer, M., Plénet, D., Lauri, P. É. and Le Bellec, F. (2017). Methodology to design agroecological orchards: Learnings from on-station and on-farm experiences. *Eur. J. Agron.* 82, 320–330.

Simon, S., Morel, K., Durand, E., Brevalle, G., Girard, T. and Lauri, P. É. (2012). Aphids at crossroads: When branch architecture alters aphid infestation patterns in the apple tree. *Trees* 26(1), 273-282.

Song, B., Jiao, H., Tang, G. and Yao, Y. (2014). Combining repellent and attractive aromatic plants to enhance biological control of three tortricids species (Lepidoptera:Tortricidae) in an apple orchard. *Fla. Entomol.* 97(4), 1679–1689.

Song, B., Tang, G., Sang, X., Zhang, J., Yao, Y. and Wiggins, N. (2013). Intercropping with aromatic plants hindered the occurrence of *Aphis citricolain* an apple orchard system by shifting predator-prey abundances. *Biocontrol Sci. Technol.* 23(4), 381–395.

Stagnari, F., Ramazzotti, S. and Pisante, M. (2009). Conservation agriculture: A different approach for crop production through sustainable soil and water management: A review. *Sustain. Agric. Rev.* 1, 55–83.

Stoeckli, S., Mody, K. and Dorn, S. (2008a). Aphis pomi (Hemiptera: Aphididae) population development, shoot characteristics, and antibiosis resistance in different apple genotypes. *J. Econ. Entomol.* 101(4), 1341–1348.

Stoeckli, S., Mody, K., Gessler, C., Patocchi, A., Jermini, M. and Dorn, S. (2008b). QTL analysis for aphid resistance and growth traits in apple. *Tree Genet. Genomes* 4(4), 833–847.

Sun, H., Zhang, F., Chen, S., Guan, Z., Jiang, J., Fang, W. and Chen, F. (2015). Effects of aphid herbivory on volatile organic compounds of *Artemisia annua* and *Chrysanthemum morifolium*. *Biochem. Syst. Ecol.* 60, 225–233.

Tang, G. B., Song, B. Z., Zhao, L. L., Sang, X. S., Wan, H. H., Zhang, J. and Yao, Y. C. (2012). Repellent and attractive effects of herbs on insects in pear orchards intercropped with aromatic plants. *Agrofor. Syst.* 87(2), 273–285.

Thibout, E. and Auger, J. (1997). Composés soufrés des Allium et lutte contre les insectes. *Acta Bot. Gallica* 144(4), 419–426.

Tscharntke, T., Bommarco, R., Clough, Y., Crist, T. O., Kleijn, D., Rand, T. A., Tylianakis, J. M., Van Nouhuys, Sv and Vidal, S. (2007). Conservation biological control and enemy diversity on a landscape scale. *Biol. Control* 43(3), 294–309.

Tscharntke, T., Klein, A. M., Kruess, A., Steffan-Dewenter, I. and Thies, C. (2005). Landscape perspectives on agricultural intensification and biodiversity–ecosystem service management. *Ecol. Lett.* 8(8), 857–874.

Tustin, D. S., Dayatilake, G. A., Henriod, R. E., Breen, K. C. and Oliver, M. (2011). Changes in fruiting behaviour and vegetative development of 'Scifresh' apple in response to artificial spur extinction using centrifugal training. *Acta Hortic.* 903(903), 603–610.

Van Egmond, F. (1979). Fate of calcium in sugar-beet plant. *Comm. Soil Sci. Plant Ana* 10, 311–323.

Van Noordwijk, M., Coe, R. and Sinclair, F. L. (2019). Agroforestry paradigms. In: van Noordwijk, M. (Ed.), *Sustainable Development through Trees on Farms: Agroforestry in Its Fifth Decade*. World Agroforestry (ICRAF) Southeast Asia Regional Program, Bogor, pp. 1–14.

Vukicevich, E., Lowery, T., Bowen, P., Úrbez-Torres, J. R. and Hart, M. (2016). Cover crops to increase soil microbial diversity and mitigate decline in perennial agriculture. A review. *Agron. Sustain. Dev.* 36(3), 36–48.

Wäckers, F. L., Romeis, J. and Van Rijn, P. (2007). Nectar and pollen feeding by insect herbivores and implications for multitrophic interactions. *Annu. Rev. Entomol.* 52, 301–323.

Wezel, A., Casagrande, M., Celette, F., Vian, J.-F., Ferrer, A. and Peigné, J. (2014). Agroecological practices for sustainable agriculture. A review. *Agron. Sustain. Dev.* 34(1), 1–20.

Wijnands, F., Malavolta, C., Alaphilippe, A., Gerowitt, B. and Baur, R. (2018a). *General Technical Guidelines for Integrated Production of Annual and Perennial Crops* (4th edn.), IOBC/WPRS. https://www.iobc-wprs.org/ip_ipm/download_documents.html.

Wijnands, F., Malavolta, C., Alaphilippe, A., Gerowitt, B. and Baur, R. (2018b). *Integrated Production IOBC-WPRS Objectives and Principles* (4th edn.), IOBC/WPRS. https://www.iobc-wprs.org/ip_ipm/download_documents.html.

Willaume, M., Lauri, P. É. and Sinoquet, H. (2004). Light interception in apple trees influenced by canopy architecture manipulation. *Trees* 18(6), 705–713.

Wubie, M., Negash, A., Guadie, F., Molla, G., Kassaye, K. and Nagappan, R. (2014). Repellent and insecticidal activity of *Mentha piperita* (L.) plant extracts against cabbage aphid [*Brevicoryne brassicae* Linn. (Homoptera: Aphididae)]. *Am-Eur. J. Sci. Res.* 9, 150–156.

Wyss, E., Niggli, U. and Nentwig, W. (1995). The impact of spiders on aphid populations in a strip-managed apple orchard. *J. Appl. Entomol.* 119(1-5), 473–478.

Chapter 9

Advancing conservation biological control as a component of integrated pest management of horticultural crops

Robbie D. Girling, Tom D. Breeze and Michael P. Garratt, University of Reading, UK

1 Introduction

1.1 What is conservation biological control?

Conservation biological control (CBC) of invertebrate pests can be considered as the application or establishment of interventions in an agroecosystem that promotes the regulation of pests by enhancing the fitness of their natural enemies (Ehler, 1998). Pest outbreaks occur with much greater frequency in agricultural systems than in nature, and within agroecosystems, the frequency of outbreaks has increased with increasing intensification (Singh and Satyanarayana, 2009; Woltz et al., 2012). In natural systems, the inherently greater biodiversity results in more interspecific competition for resources and these diverse habitats ensure that many different niches exist, facilitating the survival of natural enemies, ensuring no one species dominates and that a dynamic balance is achieved (Gutierrez-Arellano and Mulligan, 2018). The development of intensive commercial agricultural systems and landscapes has created the perfect environment for both crop growth and pest development, removing competition and limitations on the resources pest species require;

http://dx.doi.org/10.19103/AS.2021.0095.10

furthermore, the decreased diversity and complexity in these systems have removed many habitats which support wider biodiversity, thereby suppressing natural pest control (Altieri, 1999; Woltz et al., 2012). Therefore, the aim of CBC is to use interventions, particularly habitat modification, to support the establishment and maintenance of natural enemy populations, which use pest species as a trophic resource, that is, facilitating the natural provision of pest regulation services (Power, 2010; Zhang et al., 2007). CBC is differentiated from other forms of biological control in that it aims to promote natural enemies that already occur in the local environment rather than introducing novel control agents (Helyer et al., 2014).

As discussed in the earlier chapters of this book, there is a clear environmental need, increased political will, increasing need for alternatives to synthetic pesticides and a strong social argument for a shift in the management of cropping systems towards a more ecologically sustainable approach (Pretty et al., 2018; Sánchez-Bayo and Wyckhuys, 2019; Tilman et al., 2011). It is considered that CBC, as a component of an IPM approach, is a key method with which to address the challenges that face current agricultural production (Birch et al., 2011). One of the key principles of integrated pest management (IPM) is the reduction in and more targeted use of insecticides (Barzman et al., 2015); therefore, CBC is especially compatible with this approach, in particular, because many natural enemies are more susceptible to insecticides and likely to suffer non-target effects (Devine and Furlong, 2007), which would reduce the effectiveness of any CBC measures.

1.2 How is conservation biological control employed in horticulture?

Measures employed in CBC of arthropod pests focus on the provision of food and shelter for natural enemies. Holland and Ellis (2008) proposed the acronym SAFE (Shelter, Alternative Prey, Floral resources, Environment) as a method of communicating the key resources that are required to promote natural enemies within cropping environments. Shelter or refugia are required because the life cycle of most natural enemy species require that they have access to other habitats, besides the cropping environment, to overwinter, forage or reproduce. Alternative prey is particularly important in systems where the pests are an ephemeral resource, and, therefore, natural enemies require alternative prey or hosts on which to survive during the intervening periods when pests are not present. Floral resources are important for those natural enemies that require either pollen or nectar during at least one part of their life cycles to survive, for example the adult stage of both parasitoid wasps and hoverflies. The term 'Environment' in the SAFE acronym refers to the provision of diverse vegetation untreated by insecticides, which is required by many natural enemies to support different life stages (Holland and Ellis, 2008).

Within this framework, a range of different interventions have been trialled and tested for horticultural crop production, which can be broadly classified as:

1 **In-field approaches**, that is, changes to the management of the cropping system itself to provide greater diversity and the resources that natural enemies need to persist and survive in the system (e.g. intercropping, companion planting, polycultures and within field flower strips; see Fig. 1a and b). Examples of where such approaches have been investigated in horticultural systems are cabbage (Adati et al., 2011; Balmer et al., 2014, 2013), tomato (Abad et al., 2020), bell pepper (Bickerton and Hamilton, 2012), fennel (Ramalho et al., 2012), citrus (Aguilar-Fenollosa et al., 2011a,b,c; Aguilar-Fenollosa and Jacas, 2013; Kong et al., 2005), apple (Albert et al., 2017; Brown and Mathews, 2007, 2008; Brown et al., 2010) and pear (Song et al., 2010, 2011).
2 **Field margins**, that is, changes to the management of the spaces around the edges of fields or cropping areas (e.g. field edge flower or grass strips, uncultivated areas, beetle banks and hedgerows; see Fig. 1c).

Figure 1 (a) A flower strip between the rows of trees in a cherry orchard, grown under protected cropping in the United Kingdom (Image: Zeus Mateos-Fierro); (b) a crab spider on an oxeye daisy, one of the natural enemy species promoted by, and found in, the flower strip shown in (a) (Image: Zeus Mateos-Fierro); (c) a floristic margin next to an apple orchard in the United Kingdom (Image: Michael Garratt).

Field margin manipulations have been studied in a range of horticultural crops including, but not limited to, tomato (Balzan and Moonen, 2014), brassica (Geiger et al., 2009), lettuce (Pascual-Villalobos et al., 2006) and apples (Santos et al., 2018).

3 **Landscape-scale effects**, that is, the management of non-crop vegetation in the wider farming landscape to ensure refugia and food resources and connectivity of resources are sufficient to promote biodiversity and CBC. Research in this area focuses predominantly on the wider landscape around all agricultural systems (Begg et al., 2017; Bianchi et al., 2006; Chaplin-Kramer et al., 2011; Tscharntke et al., 2016; Tscharntke et al., 2005), although specific examples from horticulture include apple (Happe et al., 2019; Happe et al., 2018), grape (Rusch et al., 2016a, 2017; Thomson and Hoffmann, 2013; Thomson et al., 2010; Wilson et al., 2015, 2017) and olive (Villa et al., 2020).

1.3 Why is uptake of conservation biological control in horticultural production systems still limited, and what are some of the broader knowledge gaps?

As highlighted earlier, there has been significant research to date on many varied CBC approaches in different horticultural cropping systems, and there is evidence of a clear need for such sustainable approaches to be adopted in these cropping systems. However, despite the potential benefits of CBC, its uptake on a commercial scale remains limited (Johnson et al., 2021), and more widely, there is strong evidence to suggest that the uptake of CBC and other forms of ecological intensification is not as widespread as it could be (Kleijn et al., 2019). Therefore, given the potential benefits of CBC and the clear need and political will to move towards more sustainable approaches to farming, what are the barriers that have stopped the more widespread adoption of CBC measures as part of horticultural IPM?

A recent comprehensive study by Johnson et al. (2021) identified 150 primary research papers (comprising 247 separate experiments) from all types of agricultural crop, which have investigated the application of CBC approaches using replicated field experiments, with the aim of understanding why the uptake of CBC in all farming systems globally is currently low. Their key conclusions were that the scope of CBC research to date is too limited, lacking detailed consideration of economic benefits (only 10 of the 247 experiments investigated profitability), and, overall, it is focused too heavily upon metrics of pest and natural enemy abundance.

To better understand whether the findings of the Johnson et al. (2021) study were also representative for horticultural production, we used their published database but removed all non-horticultural crops and re-ran their analysis following the same methods, which investigated the metrics measured and success of CBC studies. This resulted in a subset of 100 papers and 167 experiments, which were dominated by the study of Brassica (22%), apple (22%) and grape (11%). No other single crop was featured in more than 8% of studies. Of these studies, 37% were from Europe, although 84% were from temperate locations.

The most common metric measured in papers was the abundances of natural enemies and pests, both of which were measured in approximately 80% of all experiments conducted (Fig. 2). For all metrics used, a high proportion of the studies demonstrated 'uncertain' results. Surprisingly, very few studies investigated the impacts of the CBC interventions on the crops themselves and only a handful displayed a clear positive outcome for crop damage or crop yield, and of the two studies that investigated the economic effect, only one showed a positive economic impact from the intervention (the crop was clementine mandarin).

From their broader data set, Johnson et al. (2021) demonstrated similar outcomes and concluded that the potential barriers to wider adoption of CBC included the fact that levels of 'inconsistent' or 'no' effect were high for many response variables. As a result, they proposed that more research should be conducted to provide a broader body of data from which to understand such

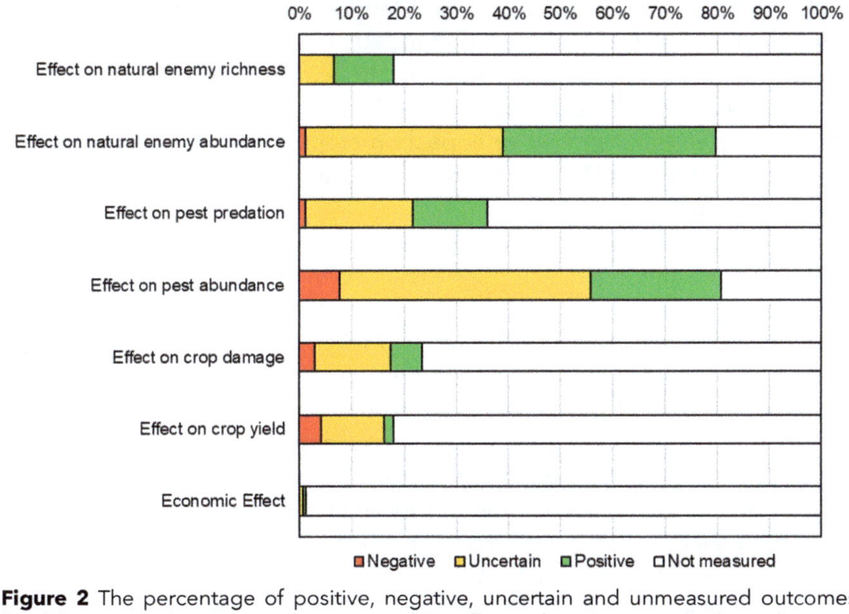

Figure 2 The percentage of positive, negative, uncertain and unmeasured outcomes for different metrics measured to monitor the effects of conservation biological control amendments in 100 field trial studies (including a total of 167 separate experiments) of horticultural crops. Based on a data set by Johnson et al. (2021).

inconsistent or negative effects. However, there are already many studies on some crops, for example those on the brassicaceous crops and apple in this data set (a total of 37 experiments on each), yet inconsistent effects were still common (61% of all recorded effects for brassica were 'uncertain' and 54% for apple). We propose that the inconsistency of effects may not simply be a result of a limited number of studies but that there may be other factors that contribute.

Thus, these results pose two clear questions that researchers in the field must investigate to enhance both the efficacy and uptake of CBC in commercial horticulture:

1 The benefits of CBC on response metrics appear to be varied and uncertain; which factors could be the primary causes of this variation?
2 Given the low number of studies looking at the impacts of CBC measures on the crops themselves, are researchers measuring the appropriate metrics to promote the uptake of CBC amongst practitioners?

2 Effects of scale on the efficacy of conservation biological control

Evidence of the potential benefits of CBC to reduce crop pest damage in horticultural systems has been reported by many studies (Aguilar-Fenollosa et al., 2011c; Balzan, 2017; Balzan and Moonen, 2014; Brown et al., 2010; Cahenzli et al., 2017; Gómez-Marco et al., 2016; Irvin et al., 2006; Salamanca et al., 2018), but as explained previously, outcomes can often be mixed and there are many examples where the results of CBC through habitat management are unclear or even have a negative impact on crop damage (Aguilar-Fenollosa et al., 2011c; Brown and Mathews, 2007; Irvin et al., 2016; Schellhorn and Sork, 1997; Tscharntke et al., 2016).

Such examples could be due to specific factors that prevent CBC action from delivering the benefits intended, such as the impacts of pesticides not allowing natural enemy populations to increase (McKerchar et al., 2020), intra guild predation (Finke and Denno, 2003; Müller and Brodeur, 2002) or that the target natural enemy population is not increased (Larentzaki et al., 2008). Fundamental to CBC is that viable populations of the natural enemies that deliver pest control must exist in the agroecosystem, which, as we stated earlier, requires the provision of the necessary habitat elements that provide supplemental food resources, nesting locations and/or overwintering sites, supporting every life stage of natural enemies (Gurr et al., 2017). Therefore, habitat interventions to promote CBC as a pest control strategy are likely to operate at different spatial and temporal scales to more conventional pest control approaches. This concerns both the application of a CBC intervention and how the impacts of the intervention are assessed.

2.1 Spatial considerations

With regard to active management to improve CBC, such as through the provision of habitats in the form of hedgerows, field margins or beetle banks, it is critical that habitats are both large enough and close enough to farm fields to provide a sufficient increase in natural enemy abundance to deliver pest control services. The abundance of natural enemies is known to decline away from source habitats and depends on habitat quality (Albrecht et al., 2020; Garratt et al., 2017; Woodcock et al., 2016). Furthermore, the density, diversity and function of natural enemies are sensitive to the size of suitable habitats, such as wildflower plantings (Blaauw and Isaacs, 2012), and interventions employed without sufficient coverage may not deliver pest control services effectively (Tscharntke et al., 2016). Introducing in-field habitat interventions (i.e. within the crop itself), such as in the alleyways of fruit orchards, is one way of mitigating these spatial limitations and ensuring uplifts in populations of beneficials, even with only limited spillover of natural enemies. This was shown to improve pest control in apple orchards (Campbell et al., 2017). Similarly, companion cropping alongside cucurbit crops increased the populations of beneficial insects and spiders with the potential to control key pest insects (Qureshi et al., 2009), although impacts on yield were not measured in this case.

Landscape context surrounding crops in terms of the cover of non-cropped land (Chaplin-Kramer et al., 2011; Dainese et al., 2019; Rusch et al., 2016b) and the arrangement and distribution of these elements (Martin et al., 2019) are an important determinant of invertebrate communities and resulting pest control. Therefore, the protection or restoration of these landscape elements is recommended as a solution to improve CBC in agricultural systems (Garibaldi et al., 2019). However, despite the generally positive effects of less intensively managed landscapes being detected across systems as a whole (Dainese et al., 2019), positive responses of natural enemies are not universal and are highly context dependent (Karp et al., 2018). More research is needed to understand to what extent non-crop areas affect CBC in horticultural crops and at what spatial scale this operates, particularly in protected or semi-protected contexts.

Furthermore, landscape context and local CBC interventions, such as floral habitat creation, can interact to determine the relative success of these interventions. Interventions often prove most effective in intermediate landscapes compared to either very simple landscapes, where there are no source populations of natural enemies, or in highly complex landscapes where the response potential is already saturated (Tscharntke et al., 2005). For example, when floral strips of buckwheat were established next to Kale fields in New Zealand, parasitism rates were enhanced, the abundance of pests reduced and crop yield increased in fields in moderately simple landscapes but not in those in highly complex landscapes (Jonsson et al., 2015).

Ultimately, the spatial scale at which CBC management is undertaken (e.g. habitat protection) or interventions implemented (e.g. flowery field margins) depends on the crop system, and importantly, on those species that have a role to play in CBC, either as protagonists or antagonists. The traits of these different actors can be used to anticipate which approaches are likely to deliver better pest control. For example, in a large-scale meta-analysis, which included data on a number of horticultural crops, it was discovered that those species of natural enemies which could fly benefitted more greatly than those species that were unable to fly from a high density of field edges, which promoted spillover into crop fields, in contrast to less mobile ground-dispersing natural enemies which are better able to persist in crop fields and were most abundant in landscapes with few edges (Martin et al., 2019).

2.2 Temporal considerations

Spatial components are not the only considerations when implementing management to exploit CBC in horticultural crop systems. Considering the temporal availability of key resources is also critically important. Very often, the natural enemies that are relied upon to deliver CBC services are longer lived than many pest species, living over several seasons and requiring different resources at different times depending on their life stage. Therefore, it is not only just about how much and where resources are available but also when they are available. Targeted measures that secure the continuity of resources throughout the life cycle of service-providing organisms are therefore needed (Schellhorn et al., 2015). For example, non-crop areas provided critical overwintering habitats for natural enemies in Brussels sprout production systems with herbaceous non-crop habitats, in particular, providing important refugia for predators important for CBC (Geiger et al., 2009).

Another temporal consideration when establishing habitat management approaches to increase populations of locally abundant natural enemies is that interventions such as wildflower strips can take time to establish and deliver benefits. It can take several generations for local populations of beneficial organisms to respond with an increase in local abundance. A classic example concerns not insect natural enemies but insect pollinators, where it has been shown that it takes several seasons for wild bee populations to increase in abundance following the establishment of flower plots adjacent to blueberry crops (Blaauw and Isaacs, 2014). The same lag effect would be expected for habitat effects on natural enemies, although a recent meta-analysis, incorporating data from many different crop systems, demonstrated that while the age since the establishment of flower strips affected pollinators and pollination, no such effect was seen for pest regulation services (Albrecht et al., 2020). This suggests that perhaps natural enemies are better able to respond to such interventions in the short term, although more targeted research is required to establish this effect and to understand how broadly it applies to different groups of natural enemies. For example, the increased abundance of natural enemies in apple trees in response to alleyway plantings of flowers was seen after just 1 year post establishment in cider apple orchards, although benefits to yield and crop quality were not observed (Campbell et al., 2017). But it is unclear whether these benefits would be observed over a greater time period. Furthermore, CBC interventions, including hedgerows and floral plantings, were employed in another study on orchard systems where they increased the abundance of spiders, an important natural enemy in this system. However, the benefits of this were not observed until the subsequent season after abundant spider populations in the previous autumn had reduced the number of aphid fundatrices the following spring, a clear example of a lag effect of an intervention (Cahenzli et al., 2017).

Successfully implementing CBC using habitat management, either through the protection of non-cropped areas or through the establishment of new resource-rich habitats, is knowledge intensive in terms of knowing which approaches will work and where. Importantly, it requires a specific recognition of the spatial and temporal factors which will ultimately determine whether an approach is successful or not. This includes the extent to which effects spill over from different habitats and whether these habitats, in combination with other landscape elements, provide a continuous supply of the necessary resources for beneficials. Both factors are likely to be determined by the ecological and physical traits of natural enemies within the crop system (Martin et al., 2019) and can be used to help target management approaches and floral species selection for horticultural crops (van Rijn and Wäckers, 2016; Wäckers and van Rijn). The time it may take to realise benefits from natural enemy abundance or crop production in response to CBC interventions must also be considered, particularly when being compared with direct approaches with more immediate effects such as pesticide use (Wilson and Tisdell, 2001).

3 Evaluating the effectiveness of conservation biological control management in multiple dimensions

Research to date has therefore developed our understanding of the biological and ecological factors that must be considered when designing CBC management options, but which other factors may contribute to low uptake of CBC in many parts of the world? Several authors have suggested that low uptake is largely due to the focus of much research on the strictly ecological aspects of CBC interventions: changes in pest and predator density, alterations of species community compositions or binary evaluations of changes in pest damage (Chaplin-Kramer et al., 2019; Johnson et al., 2021). Farmers are often highly risk averse and ultimately require a strong economic incentive to undertake such major management changes, particularly if they do not perceive them to be effective (Zhang et al., 2018b). As such, there are growing calls for greater study into the full economic impacts of CBC methods, including formal cost-benefit analyses, to demonstrate the full benefits of CBC measures (Chaplin-Kramer et al., 2019; Johnson et al., 2021; Shields et al., 2019).

A number of studies have undertaken some economic appraisal of the yield impacts observed following CBC interventions in horticultural cropping systems (e.g. Colloff et al. 2013), but many lack an assessment of the intervention costs, and almost all are based on simple 'before vs after' comparisons of interventions, without an assessment of the baseline levels of pest control (but see Rodríguez-San Pedro et al. 2020). Furthermore, unlike some other ecosystem services that arise due to natural ecological processes (e.g. carbon storage) or simple trophic interactions (e.g. pollination), pest regulation by natural enemies is often dependent on a more complex trophic system, involving a wide number of pests and their associated predators. Each of these is likely to have different responses to the environment and to interact with one another at different times of the year. As such, even if other factors can be controlled for, it can be very difficult to link interventions to economic metrics of yield because links between interventions and predator populations, predators and pests, and pests and damage all need to be accounted for separately.

Here, we present a simple best practice guideline for researchers to assess the economic costs and benefits of CBC interventions (Fig. 3). Steps are marked as either *Critical* or *Recommended*. Critical steps will give the researcher the data they need to assess costs and benefits without any intensive economic

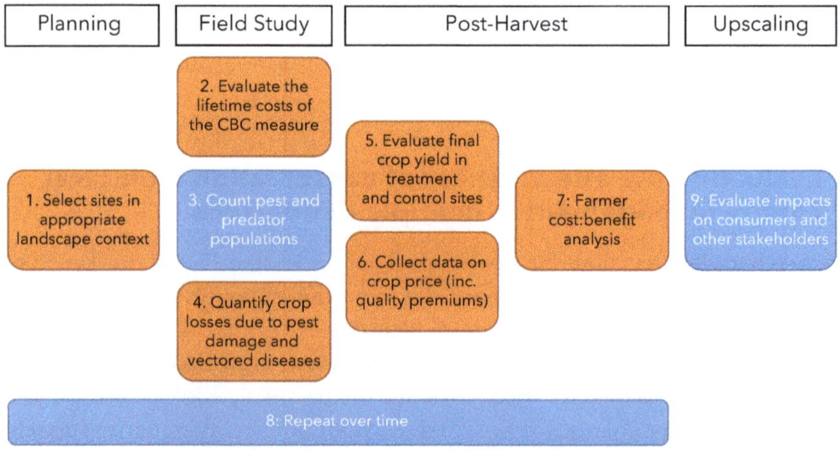

Figure 3 Process diagram for assessing economic benefits of CBC. Cells in orange are critical, while cells in blue are recommended but optional.

background. Recommended steps allow for a deeper appraisal of these values but may be resource limited.

1 *Determine appropriate spatial scales*: As highlighted in Section 2, the CBC treatment site should account for (i) proximity to the CBC intervention and (ii) the overall landscape context, although the specific details of this are likely to vary between crops, systems and regions. Within sites, sampling should be undertaken using a stratified random design to ensure that surveys of pest damage are representative of the whole field/orchard. The sites must be commercial or have the aim of becoming commercial, to ensure relatively representative management. (*Critical*).
2 *Calculate costs of CBC management*: Ideally, these costs should be based on actual management costs experienced in situ over the lifespan of the intervention. However, if this is not possible, then it is important to estimate these costs over time - specifically, the initial establishment costs (e.g. planting) and maintenance costs (e.g. cutting, re-sowing) of the measures, including all relevant materials and labour. Cost assessment should also include any cost reductions from the CBC method, such as reduced pesticide usage. Farming handbooks (e.g. Redman 2020) can provide this information, but direct discussion with land owners is preferable. (*Critical*).
3 *Record counts of pests and predators*: If it is possible, ecological surveys of the abundance and diversity of known crop pests and their predators should be undertaken throughout the study and compared with suitable control sites. Unlike simple observations of pest presence/absence, direct counts of pest and predator populations allow quantifiable links to be drawn between the CBC measures and changes in pest damage (see Point 4) arising from altered natural pest regulation services. Although species level assessments are ideal, responses may be apparent only when considering functional guilds (Gardarin et al., 2018; Staton et al., 2021), which may be more practical for some researchers. (*Recommended*).
4 *Evaluate pest damage*: Levels of pest damage to the crop (e.g. fruit loss/damage or occurrence of disease vectored by pests) should be monitored throughout the growing period of the crop and either compared with damage in untreated fields or from past years of that field as appropriate. Distinction should be drawn between this observed loss and natural losses due to, for example, fruit abortion. Where possible, researchers should look for opportunities to separate out the impacts of specific pests, in case their responses differ (e.g. are either suppressed less or increase) as a result of the CBC measures. (*Critical*).

5 *Evaluate final crop yield*: Evaluating final harvested crop yield should consider both the total weight of marketable crop and any quality parameters (e.g. shape) that may affect the final sale price. (*Critical*).
6 *Determine crop price*: The market sale price of the crop should be as current as possible and account for any difference in prices due to crop quality. Ideally, this should draw from industry data; however, national statistical agencies and FAO data (FAOSTAT, 2021) can be used if necessary. If the research is relatively short term, an average of the past 3–5 year prices can be used to account for price fluctuations. (*Critical*).
7 *Conduct a farmer cost-benefit analysis:* Benefits are simply calculated as the total output (yield × quality × price) of the CBC system compared with an untreated control. From then, it is possible to subtract the costs of the CBC measures from the difference to give an initial estimate of net benefits. While projecting costs and benefits into the future, it is important to include a measure of variance in these values over time and apply a discounting rate (a projection of inflation in the future, representing the decreasing value of currency over time) to future years. (*Critical*).
8 *Determine the appropriate temporal scale*: Many CBC measures are unlikely to have an immediate effect, as populations of predators may take some time to grow and habitat modification measures often take time to establish (see Section 2). It is preferable for a study to be undertaken over multiple years, ideally over the lifespan of the CBC measure; monitoring the change in cost-benefit each year can identify: (i) when, if at all, the measure will become profitable and (ii) how the CBC measure affects the stability of economic output over time. (*Recommended*).
9 *Evaluate consumer impacts*: If the researcher is interested in upscaling the results to a national or regional scale, then they may wish to consider evaluating the impacts on consumers as well as producers. This can be achieved using partial equilibrium modelling (see e.g. Zhang et al. 2018a) wherein the change in the total crop output resulting from the mass implementation of the CBC measure is translated to a change in consumer price. This requires more advanced economic modelling but can give a measure of the net societal benefits of the CBC measures, which may incentivise policy support. (*Recommended*).

Although income is a major concern for many farmers, economic arguments alone may not be enough to fully promote CBC methods. A number of sociological factors have been identified as potential barriers to alternative pest management in general. Farmers may be reluctant to deviate too strongly from their neighbours for fear of becoming a pest reservoir (Wilson and

Tisdell, 2001). At the same time, however, landscape-scale management may not be beneficial to all participants, particularly in mixed landscapes where some landowners, such as pasture or arable farmers, may face little or very different pest pressures compared to horticultural growers. Landowners may also find certain CBC measures, such as flower-rich field margins, unsightly and poorly reflective upon their abilities as farmers (Burton et al., 2008) or lack the knowledge to properly implement them (Mankad et al., 2017).

Addressing these barriers requires additional data gathering outside of a standard ecological study. Ideally, before any research is conducted, CBC measures should be co-developed with groups of local stakeholders in order to maximise their collective willingness to take them up and identify barriers (e.g. Giles et al. 2017; Husson et al. 2016). On a wider scale, farmer surveys, using psychological frameworks such as the theory of planned behaviour (Ajzen, 1991), can also help identify wider perceived and observable barriers to wider uptake (Mankad et al., 2017). Finally, additional economic analysis, drawing from the cost–benefit work outlined earlier, can also identify the points at which subsidies will make otherwise unprofitable CBC systems more economically viable, particularly in the earlier years when they are initially establishing (Yang et al., 2020).

4 Benefits of conservation biological control beyond integrated pest management

By their very nature CBC management approaches can change the nature of the cropping environment through adaptive management at the field scale (Larentzaki et al., 2008), habitat creation in and around crop fields (Albrecht et al., 2020; Campbell et al., 2017) or larger-scale alteration to landscape context (Jonsson et al., 2015; Martin et al., 2019). These are put in place to manipulate populations of natural enemies and pests to support sustainable crop production through IPM but can inevitably have impacts beyond this.

One area where there are clear synergies concerns management approaches to increase populations of wild pollinators and the pollination services they provide to crops. Approaches to boost pollinators are often equivalent to those for CBC, including reducing harmful inputs, the establishment of floral resources through flower margins and protection and management of non-cropped habitats such as hedgerows (Kovács-Hostyánszki et al., 2017). Numerous studies have highlighted the co-benefits of such approaches to both pollinators and natural enemies of pests (Albrecht et al., 2020; Garratt et al., 2017; Wratten et al., 2012). In fact, by considering these co-benefits and their positive impact on the yield of tomato crops delivered by hedgerows, the economic return on investment and breakeven point for hedgerow establishment was reduced from 16 to 7 years (Morandin et al.,

2016). Despite the similarities in the habitat requirements of wild pollinators and many natural enemies, inevitably they are not exactly the same, and features, such as wildflower plots, can be tailored to support different functional groups of beneficial organisms. As a consequence, some compromises or trade-offs are likely when managing to promote both pest regulation and pollination (Campbell et al., 2012; Campbell et al., 2017). However, given the obvious similarities and potential synergies that the management of pollination and wider IPM practices offer, they can, and should, be better integrated (Egan et al., 2020).

There are many other potential benefits arising from implementing CBC practices including direct impacts from in-field crop management approaches such as reduced or more targeted pesticide spraying to protect natural enemies (Ruberson et al., 1998) and the associated environmental benefits (Aktar et al., 2009) to reduced tillage to improve biocontrol (Roger-Estrade et al., 2010) benefiting multiple soil physical, chemical and biological properties (Busari et al., 2015). Particularly, diverse benefits can be realised through the introduction of non-crop habitats such as flower plots and hedgerows including species conservation (Requier and Leonhardt, 2020), improved soil and water quality (Montgomery et al., 2020) and enhancing rural aesthetics (Wratten et al., 2012). By the same measure, management approaches implemented for other reasons can deliver CBC benefits; for example field margins to support conservation of birds benefitted natural enemies of pests as an unintended side effect (Olson and Wäckers, 2007). When considering implementing CBC actions, these multiple benefits should be taken into consideration, particularly when comparing the benefits against more conventional approaches such as pesticide application. Furthermore, these co-benefits should be factored into calculations of any economic analysis of CBC measures.

5 Case studies of conservation biological control in practice

The application of CBC as a component of an IPM approach is met with a range of challenges that are often specific to the growing system and crop. Therefore, detailed research is often required to provide a clearer understanding of the challenges each system/crop faces and to optimise the different CBC methods, with respect to both their efficacy and cost-effectiveness, that can be applied. Here we provide two case study examples of research that has contributed towards the inclusion of CBC in commercial practice. These case studies highlight the application of CBC in an organic field vegetable horticultural production system and in a conventional protected-cropping stone fruit orchard production system.

5.1 Case study 1: Organic lettuce production in California

Globally, aphids are a key pest of field vegetable production. They can damage their host plants, resulting in yield reductions, in a range of ways, including by depriving the plant of nutrients through feeding on the plants phloem, by the transmission of viruses during feeding (Tomlinson, 1987) and by the honeydew they produce while feeding, encouraging the growth of sooty mould (Dedryver et al., 2010). Their presence and the damage they cause can also reduce the marketability of produce for a range of horticultural crops (AHDB, 2021). During the main growing season, aphids reproduce parthenogenically, meaning they reproduce asexually, and are viviparous, meaning that they give birth to live young (Hardie, 2017). They, therefore, have a very fast rate of increase, which means that their control is challenging.

Conventional horticultural production systems have commonly relied-upon insecticides to control aphid populations; however, their fast rate of increase has resulted in the development of insecticide resistance in multiple aphid species (Foster et al., 2017). Furthermore, the effectiveness of using insecticides to control the transmission of viruses, through the control of the vector, is mixed and often ineffective (Perring et al., 1999). Therefore, there is increasing demand for alternative approaches to controlling aphid pests, which can be adopted as part of an IPM approach (Dedryver et al., 2010).

Organic production has promoted the development of a variety of alternative approaches to pest control, including CBC, although, for many of these pest management strategies, there is a paucity of rigorous scientific evidence to substantiate their benefits (Zehnder et al., 2007). However, one case study in which the benefits of an organic pest management strategy have been successfully quantified is the application of CBC to control aphid populations in lettuce production on the central coast of California (Brennan, 2013). The predominant aphid pest in lettuce production in central California is the currant-lettuce aphid, *Nasonovia ribisnigri* Mosley (Smith et al., 2008). Controlling this aphid species in lettuce production can be challenging because it aggregates and feeds upon the interior leaves of the lettuce plants (Liu, 2004).

In central California, it is well established in the organic farming community that by intercropping lettuce plants with sweet alyssum, *Lobularia maritima* (L.) Desv. (Fig. 4), it is possible to control *N. ribisnigri* (Gillespie et al., 2011). The mechanism for this control is that the intercropped alyssum acts as an insectary plant; that is, by having abundant flowers that produce pollen and nectar, it attracts beneficial insects, in particular, hoverflies, which feed on these floral resources (Colley and Luna, 2000). The adult hoverflies mate, and, subsequently, the females forage in the local vicinity of the insectary plants, searching for patches of aphids in which to lay their eggs. The larvae that hatch

Figure 4 Additive intercropping of the white flowering plant sweet alyssum (*Lobularia maritima* (L.) Desv.) into organic romaine lettuce fields in (a) the USDA-ARS organic research farm in Salinas CA; (b) a commercial organic farm in the region (Image: Eric Brennan, USDA-ARS).

from these eggs are voracious predators and specialise in feeding on aphids, eating up to 168 currant-lettuce aphids per day (Hopper et al., 2011). Therefore, sweet alyssum has been used as an insectary plant in organic lettuce fields on the Californian Central Coast for many years (Bugg et al., 2008).

A range of different approaches have been used by growers in the region to incorporate these insectary plants, from strip cropping by interplanting full beds of sweet alyssum across the field of lettuce to interspersing plants within lettuce beds. As a result, the land used to plant these insectary plants displaces lettuce plants, which reduces the cropping area by between 5% and 10%. This means that there is an economic cost through crop displacement, in addition to the costs of establishment and maintenance of this CBC measure (Bugg et al., 2008; Colfer, 2004). Therefore, a study was conducted by Brennan (2013) to investigate different methods that could be used to optimise the quantity and arrangement of sweet alyssum and minimise the displacement of organic romaine lettuce, by investigating the effect of different methods of incorporating alyssum on plant biomass/yield. Brennan trialled a series of different replacement intercropping approaches - that is, various numbers of lettuce transplants were replaced by alyssum transplants (replacing between 2% and 8% of lettuce transplants) - and novel additive intercropping approaches - that is, alyssum transplants were added to lettuce transplants without displacing them, in comparison to a lettuce monoculture (control). Over two growing seasons, the biomass of both lettuce and alyssum was recorded, as was the flower production in the alyssum. All but one of the treatments (replacement with 4% of lettuce transplants displaced in a symmetrical pattern) resulted in decreases in romaine lettuce head dry weight biomass relative to those plants in the lettuce monoculture treatment, indicating that the inclusion of alyssum came at a yield cost. The additive treatments, where plants were at a greater density per unit area and therefore under greater competition, resulted in a decrease in the dry weight biomass of alyssum plants; however, this was countered by an increase in the number of inflorescences per gram (dry weight) of alyssum. The novel additive treatments appeared to offer the best option for growers because there was no effect on the number of heads grown in comparison to the lettuce monoculture control. Whilst there was a reduction in head dry weight of plants grown in the additive compared to the replacement treatments, it was proposed that in commercial production, this may have minimal impact after trimming and packing are considered.

Further research by Brennan (pers. comm.) has refined the additive methods (see Fig. 4a) so that alyssum plants are only included in 'insectary beds', which are separated by ten lettuce-only beds. Alyssum plants are planted in only one line of each insectary bed at a density of one alyssum plant between every five lettuce plants. This method has been shown to recruit sufficient hoverflies to maintain good control of aphid pests, and it has been demonstrated that the insectary beds produce yields of equal marketable weight to lettuce-only beds. This suggests that it is possible to gain all the benefits of pest control by using alyssum in a CBC approach, without having any negative effect on yield, and therefore the cost incurred is only on purchasing and maintaining the alyssum transplants. This is a significant advance on the traditional replacement methods where up to 10% of the lettuce plants were being substituted for alyssum and has resulted in adoption by a number of growers in the region (Fig. 4b).

5.2 Case study 2: Conventional sweet cherry production in the United Kingdom

Sweet cherry (*Prunus avium* L.) presents several challenges for incorporating CBC as part of an effective IPM programme. Firstly, it is a high-value crop where little pest damage can be tolerated; secondly, it is often intensively produced under protected or semi-protected conditions (Lang, 2014) where the influx of naturally occurring pest enemies may be constrained, and, finally, chemical pest control options are widely used in cherry production (Daniel and Grunder, 2012), which may not always be compatible with CBC.

As demonstrated in Section 1, the introduction of wildflower habitats can increase both the abundance of natural enemies and the pest regulation services they provide, including in apples (Campbell et al., 2017; McKerchar et al., 2020) and blueberry (Blaauw and Isaacs, 2014). However, such habitat interventions have rarely, if ever, been tested in semi-protected cultivation because of additional barriers to their adoption including (i) increased watering and maintenance costs for flower habitats under plastic, (ii) changes in microclimate increasing the risk of pathogens and (iii) inconvenience to grower operations, including spraying and picking, for which alleyways are usually kept mown short.

Effective CBC tools are needed in such intensive systems if they are to become less reliant on high inputs of pesticide, particularly in the face of increasing limitations due to changing legislation (e.g. thiacloprid for aphid control) (Daniel and Grunder, 2012). These CBC tools need to be compatible with grower operations, suitable in protected-cropping systems, and deliver benefits that are important to growers; thus, co-development with growers is essential (Cullen et al., 2008; Kleijn et al., 2019; Simon et al., 2017).

Working closely with growers, a recent study looked to address this challenge directly in sweet cherry (Mateos-Fierro et al., 2021). The study investigated the feasibility of establishing wildflower habitats in the alleyways of sweet cherry orchards, measured how the cutting of these habitats affected their quality as a CBC resource and quantified to what extent wildflower habitats delivered not only increases in natural enemy numbers but also improved pest regulation. Using relatively standard management approaches, wildflower strips were successfully established in cherry orchards (Mateos-Fierro et al., 2018), achieving good coverage of a variety of sown flower species (Fig. 1a). Growers were engaged in the development of the CBC amendments and suggested modifications for trial, for example, in season cutting of the flower strips.

Wildflower treatments almost doubled the abundance of natural enemies in alleyways and increased abundance in cherry trees by ~15% compared to the standard alleyway management, although importantly benefits were only seen from year 2 of the study (Fig. 5a–d) (Mateos-Fierro et al., 2021). Wildflower strips increased predation of aphids (measured using bait cards) in cherry trees by 25% early in the season (Fig. 5e–f). No difference in natural enemy abundance, richness or pest control was recorded between wildflower strips that were left uncut and those that were actively managed in a way preferred by growers, specifically involving regular cutting to maintain a sward height of 20 cm. Furthermore, these differences in natural enemy abundance and predation rates between wildflower and control treatments were detected despite the continued use of pesticides by growers.

This study demonstrated that even in intensive production systems, habitat creation for CBC can be effective. Engaging with growers during the development of the CBC amendments aimed to encourage greater uptake of successful measures into commercial production. As with supporting CBC in any production system, identifying the link between ecosystem services and the factors that farmers view as most important may positively influence communication and the potential of adoption (Bardenhagen et al., 2020).

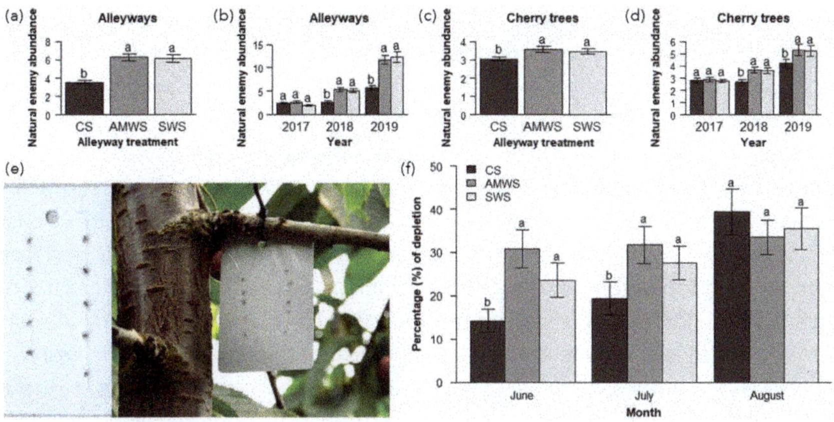

Figure 5 Mean of natural enemy abundance (±SE) recorded throughout the 3-year study according to (a) and (c) alleyway treatment, and (b) and (d) year, and the effect of alleyway treatment in either the alleyways or the cherry trees. (e) Bait cards with ten dead aphids glued to the surface to assess predation rate. (f) Mean percentage (±SE) of *Acyrthosiphon pisum* aphids (dead) depleted from bait cards. The same superscript letters indicate no significant difference (Tukey test, $P > 0.05$). CS, control strips; AMWS, actively managed wildflower strips; SWS, standard wildflower strips (adapted from Mateos-Fierro et al., 2021).

6 Conclusion and future trends in research

Whilst CBC has been shown, by a number of studies, to be an effective method of pest control for a range of horticultural crops, when considered as a whole, the effects measured to date have been inconsistent and, furthermore, uptake has been low. Understanding the cause of such inconsistency in results is therefore critical. It is clear that spatial and temporal scales matter for CBC because it relies on existing biodiversity as opposed to conventional pest control approaches, which often have rapid effects at a local scale (e.g. pesticide application). Habitat interventions for CBC need to be employed with sufficient coverage and sufficiently close to the crop to ensure benefits are delivered, because declining benefits at increasing distances from interventions are common. The local landscape is critical because non-crop areas can provide important habitats for natural enemies in agroecosystems and landscape context can modify the effects of habitat interventions, which often prove most effective in landscapes with an intermediate amount of non-crop area. The effects of local landscape and management interventions on CBC are very context dependent, depending on the cropping system, and are influenced by the traits of the species involved (both pests and natural enemies). Management interventions to improve CBC need to provide resources for all life stages of natural enemies, which are often longer lived and have a more complex life history than pests. The time it takes to realise benefits to natural enemy abundance or crop production from CBC interventions must be taken into account, particularly when being compared to direct approaches with more immediate effects such as pesticides.

We propose that in order to improve uptake of CBC measures, future studies could make use of the framework we set out in our simple best practice guideline (Fig. 3), which will assist researchers in assessing the economic costs and benefits of CBC interventions. It is critical that, moving forward, a clearer economic understanding is developed for all proposed CBC interventions. This is particularly important given the understandably risk-averse nature of many farmers, especially with respect to making major system changes. In order to achieve wider uptake, particularly in the light of the many inconsistent studies to date, which in itself is likely to result in farmers questioning the effectiveness of such measures, it is important that evidence of strong and clear economic incentives of system change is provided.

CBC management approaches often involve changing the nature of the cropping environment and so inevitably have impacts beyond CBC alone. One area where there are clear synergies concerns managing for CBC and managing for wild pollinators, and in particular, those approaches used to boost pollinator populations (e.g. reducing harmful inputs, establishment of floral resources, protection of non-cropped habitats) often deliver benefits for CBC and vice versa. There are many other benefits that CBC practices can deliver, including reduced negative environmental impacts, improved soil health and positive effects on the rural aesthetic. These multiple benefits should be taken into consideration when deciding whether to implement CBC actions, particularly when compared with more conventional approaches such as pesticide application, and should form a key part of any economic analyses that are conducted to evaluate the cost-effectiveness of CBC approaches.

7 Where to look for further information

Most text books on this subject tend to be dominated by discussion of classical and inoculative biological control rather than providing a focus on conservation biological control *per se*. For a wider introduction to biological control as a whole Helyer et al 2014 provides a good starting point.

However, to understand conservation biological control in more depth there are a number of journal articles that introduce, summarise and discuss different aspects of the subject. These include:

- Begg, G. S., Cook, S. M., Dye, R., Ferrante, M., Franck, P., Lavigne, C., Lövei, G. L., Mansion-Vaquie, A., Pell, J. K., Petit, S., Quesada, N., Ricci, B., Wratten,

S. D. and Birch, A. N. E. (2017). 'A functional overview of conservation biological control', *Crop Protection* 97: 145–158.

- Chaplin-Kramer, R., O'Rourke, M., Schellhorn, N., Zhang, W., Robinson, B. E., Gratton, C., Rosenheim, J. A., Tscharntke, T. and Karp, D. S. (2019). 'Measuring what matters: Actionable information for conservation biocontrol in multifunctional landscapes', *Frontiers in Sustainable Food Systems* 3: 60.
- Dainese, M. et al. (2019), 'A global synthesis reveals biodiversity-mediated benefits for crop production', *Science advances* 5(10): eaax0121. doi: https://doi.org/10.1126/sciadv.aax0121.
- Holland, J. M., Bianchi, F. J., Entling, M. H., Moonen, A.-C., Smith, B. M. and Jeanneret, P. (2016). 'Structure, function and management of semi-natural habitats for conservation biological control: a review of European studies', *Pest Management Science* 72: 1638–1651.
- Martin, E. A. et al. (2019). 'The interplay of landscape composition and configuration: New pathways to manage functional biodiversity and agroecosystem services across Europe', *Ecology Letters* 22(7): 1083–1094. doi: https://doi.org/10.1111/ele.13265.
- Shields, M. W., Johnson, A. C., Pandey, S., Cullen, R., González-Chang, M., Wratten, S. D. and Gurr, G. M. (2019). 'History, current situation and challenges for conservation biological control', *Biological Control* 131: 25–35.

Two websites which act as broader hubs of knowledge on this and related topics are:

- Agricology - a platform to share practical information on sustainable approaches with farmers and growers (https://www.agricology.co.uk/).
- FAO Agroecology - a knowledge hub providing information about agroecological approaches (http://www.fao.org/agroecology).

There are a number of centres of expertise, where research into conservation biological control is ongoing, these include:

- The Centre for Agri-Environment Research at the University of Reading in the UK (http://www.reading.ac.uk/caer/).
- The Centre for Biological Control at the Swedish University of Agricultural Sciences (https://www.slu.se/en/Collaborative-Centres-and-Projects/centre-for-biological-control-cbc).
- The Research Institute of Organic Agriculture (FiBL), which has locations in Switzerland, Germany, Austria, France and a representation in Brussels, Belgium (https://www.fibl.org/en/).

8 References

Abad, M. K. R., Fathi, S. A. A., Nouri-Ganbalani, G. and Amiri-Besheli, B. 2020. Influence of tomato/clover intercropping on the control of Helicoverpa armigera (Hubner). *International Journal of Tropical Insect Science* 40(1), 39–48.

Adati, T., Susila, W., Sumiartha, K., Sudiarta, P., Toriumi, W., Kawazu, K. and Koji, S. 2011. Effects of mixed cropping on population densities and parasitism rates of the diamondback moth, plutella xylostella (Lepidoptera: Plutellidae). *Applied Entomology and Zoology* 46(2), 247–253.

Aguilar-Fenollosa, E., Ibanez-Gual, M. V., Pascual-Ruiz, S., Hurtado, M. and Jacas, J. A. 2011a. Effect of ground-cover management on spider mites and their phytoseiid natural enemies in clementine mandarin orchards (I): Bottom-up regulation mechanisms. *Biological Control* 59(2), 158–170.

Aguilar-Fenollosa, E., Ibanez-Gual, M. V., Pascual-Ruiz, S., Hurtado, M. and Jacas, J. A. 2011b. Effect of ground-cover management on spider mites and their phytoseiid natural enemies in clementine mandarin orchards (II): Top-down regulation mechanisms. *Biological Control* 59(2), 171–179.

Aguilar-Fenollosa, E., Pascual-Ruiz, S., Hurtado, M. A. and Jacas, J. A. 2011c. Efficacy and economics of ground cover management as a conservation biological control strategy against Tetranychus urticae in clementine mandarin orchards. *Crop Protection* 30(10), 1328–1333.

Aguilar-Fenollosa, E. and Jacas, J. A. 2013. Effect of ground cover management on Thysanoptera (thrips) in clementine mandarin orchards. *Journal of Pest Science* 86(3), 469–481.

AHDB 2021. *Encyclopaedia of Pests and Natural Enemies in Field Crops*. Warwickshire, UK: Agriculture and Horticulture Development Board.

Ajzen, I. 1991. The theory of planned behavior. *Organizational Behavior and Human Decision Processes* 50(2), 179–211.

Aktar, M. W., Sengupta, D. and Chowdhury, A. 2009. Impact of pesticides use in agriculture: Their benefits and hazards. *Interdisciplinary Toxicology* 2(1), 1–12.

Albert, L., Franck, P., Gilles, Y. and Plantegenest, M. 2017. Impact of agroecological infrastructures on the dynamics of Dysaphis plantaginea (Hemiptera: Aphididae) and its natural enemies in apple orchards in Northwestern France. *Environmental Entomology* 46(3), 528–537.

Albrecht, M., Kleijn, D., Williams, N. M., Tschumi, M., Blaauw, B. R., Bommarco, R., Campbell, A. J., Dainese, M., Drummond, F. A., Entling, M. H., Ganser, D., Arjen de Groot, G., Goulson, D., Grab, H., Hamilton, H., Herzog, F., Isaacs, R., Jacot, K., Jeanneret, P., Jonsson, M., Knop, E., Kremen, C., Landis, D. A., Loeb, G. M., Marini, L., McKerchar, M., Morandin, L., Pfister, S. C., Potts, S. G., Rundlöf, M., Sardiñas, H., Sciligo, A., Thies, C., Tscharntke, T., Venturini, E., Veromann, E., Vollhardt, I. M. G., Wäckers, F., Ward, K., Westbury, D. B., Wilby, A., Woltz, M., Wratten, S. and Sutter, L. 2020. The effectiveness of flower strips and hedgerows on pest control, pollination services and crop yield: A quantitative synthesis. *Ecology Letters* 23(10), 1488–1498.

Altieri, M. A. 1999. The ecological role of biodiversity in agroecosystems. *Agriculture, Ecosystems and Environment* 74(1–3), 19–31.

Balmer, O., Géneau, C. E., Belz, E., Weishaupt, B., Förderer, G., Moos, S., Ditner, N., Juric, I. and Luka, H. 2014. Wildflower companion plants increase pest parasitation and yield in cabbage fields: Experimental demonstration and call for caution. *Biological Control* 76, 19–27.

Balmer, O., Pfiffner, L., Schied, J., Willareth, M., Leimgruber, A., Luka, H. and Traugott, M. 2013. Noncrop flowering plants restore top-down herbivore control in agricultural fields. *Ecology and Evolution* 3(8), 2634–2646.

Balzan, M. V. 2017. Flowering banker plants for the delivery of multiple agroecosystem services. *Arthropod-Plant Interactions* 11(6), 743–754.

Balzan, M. V. and Moonen, A.-C. 2014. Field margin vegetation enhances biological control and crop damage suppression from multiple pests in organic tomato fields. *Entomologia Experimentalis et Applicata* 150(1), 45–65.

Bardenhagen, C. J., Howard, P. H. and Gray, S. A. 2020. Farmer mental models of biological pest control: Associations With adoption of conservation practices in blueberry and cherry orchards. *Frontiers in Sustainable Food Systems* 4, 54. https://doi.org/10.3389/fsufs.2020.00054.

Barzman, M., Bàrberi, P., Birch, A. N. E., Boonekamp, P., Dachbrodt-Saaydeh, S., Graf, B., Hommel, B., Jensen, J. E., Kiss, J., Kudsk, P., Lamichhane, J. R., Messéan, A., Moonen, A.-C., Ratnadass, A., Ricci, P., Sarah, J.-L. and Sattin, M. 2015. Eight principles of integrated pest management. *Agronomy for Sustainable Development* 35(4), 1199–1215.

Begg, G. S., Cook, S. M., Dye, R., Ferrante, M., Franck, P., Lavigne, C., Lövei, G. L., Mansion-Vaquie, A., Pell, J. K., Petit, S., Quesada, N., Ricci, B., Wratten, S. D. and Birch, A. N. E. 2017. A functional overview of conservation biological control. *Crop Protection* 97, 145–158.

Bianchi, F. J. J. A., Booij, C. J. H. and Tscharntke, T. 2006. Sustainable pest regulation in agricultural landscapes: A review on landscape composition, biodiversity and natural pest control. *Proceedings. Biological Sciences* 273(1595), 1715–1727.

Bickerton, M. W. and Hamilton, G. C. 2012. Effects of intercropping With flowering plants on predation of Ostrinia nubilalis (Lepidoptera: Crambidae) eggs by generalist predators in bell peppers. *Environmental Entomology* 41(3), 612–620.

Birch, A. N. E., Begg, G. S. and Squire, G. R. 2011. How agro-ecological research helps to address food security issues under new IPM and pesticide reduction policies for global crop production systems. *Journal of Experimental Botany* 62(10), 3251–3261.

Blaauw, B. R. and Isaacs, R. 2012. Larger wildflower plantings increase natural enemy density, diversity, and biological control of sentinel prey, without increasing herbivore density. *Ecological Entomology* 37(5), 386–394.

Blaauw, B. R. and Isaacs, R. 2014. Flower plantings increase wild bee abundance and the pollination services provided to a pollination-dependent crop. *Journal of Applied Ecology* 51(4), 890–898.

Brennan, E. B. 2013. Agronomic aspects of strip intercropping lettuce with alyssum for biological control of aphids. *Biological Control* 65(3), 302–311.

Brown, M. W. and Mathews, C. R. 2007. Conservation biological control of rosy apple aphid, Dysaphis plantaginea (Passerini), in Eastern North America. *Environmental Entomology* 36(5), 1131–1139.

Brown, M. W., Mathews, C. R. and Krawczyk, G. 2010. Extrafloral nectar in an apple ecosystem to enhance biological control. *Journal of Economic Entomology* 103(5), 1657–1664.

Brown, M. W. and Mathews, C. R. 2008. Conservation biological control of spirea aphid, Aphis spiraecola (Hemiptera: Aphididae) on apple by providing natural alternative food resources. *European Journal of Entomology* 105(3), 537–540.

Bugg, R. L., Colfer, R. G., Chaney, W. E., Smith, H. A. and Cannon, J. 2008. *Flower Flies (Syrphidae) and Other Biological Control Agents for Aphids in Vegetable Crops*. Oakland CA: Division of Agriculture and Natural Resources, University of California.

Burton, R. J. F., Kuczera, C. and Schwarz, G. 2008. Exploring farmers' cultural resistance to voluntary Agri-environmental schemes. *Sociologia Ruralis* 48(1), 16–37.

Busari, M. A., Kukal, S. S., Kaur, A., Bhatt, R. and Dulazi, A. A. 2015. Conservation tillage impacts on soil, crop and the environment. *International Soil and Water Conservation Research* 3(2), 119–129.

Cahenzli, F., Pfiffner, L. and Daniel, C. 2017. Reduced crop damage by self-regulation of aphids in an ecologically enriched, insecticide-free apple orchard. *Agronomy for Sustainable Development* 37(6), 65.

Campbell, A. J., Biesmeijer, J. C., Varma, V. and Wäckers, F. L. 2012. Realising multiple ecosystem services based on the response of three beneficial insect groups to floral traits and trait diversity. *Basic and Applied Ecology* 13(4), 363–370.

Campbell, A. J., Wilby, A., Sutton, P. and Wäckers, F. 2017. Getting more Power from your flowers: Multi-functional flower strips enhance pollinators and pest control agents in apple orchards. *Insects* 8(3), 101. https://doi.org/10.3390/insects8030101.

Chaplin-Kramer, R., O'rourke, M., Schellhorn, N., Zhang, W., Robinson, B. E., Gratton, C., Rosenheim, J. A., Tscharntke, T. and Karp, D. S. 2019. Measuring what matters: Actionable information for conservation biocontrol in multifunctional landscapes. *Frontiers in Sustainable Food Systems* 3, 60. https://doi.org/10.3389/fsufs.2019.00060.

Chaplin-Kramer, R., O'rourke, M. E., Blitzer, E. J. and Kremen, C. 2011. A meta-analysis of crop pest and natural enemy response to landscape complexity. *Ecology Letters* 14(9), 922–932.

Colfer, R. 2004. Using habitat management to improve biological control on commercial organic farms in California. Proceedings of the California Conference on Biological Control IV, University of California, Berkeley, CA, pp. 55–62.

Colley, M. R. and Luna, J. M. 2000. Relative attractiveness of potential beneficial insectary plants to aphidophagous hoverflies (Diptera: Syrphidae). *Environmental Entomology* 29(5), 1054–1059.

Colloff, M. J., Lindsay, E. A. and Cook, D. C. 2013. Natural pest control in citrus as an ecosystem service: Integrating ecology, economics and management at the farm scale. *Biological Control* 67(2), 170–177.

Cullen, R., Warner, K. D., Jonsson, M. and Wratten, S. D. 2008. Economics and adoption of conservation biological control. *Biological Control* 45(2), 272–280.

Dainese, M., Martin, E. A., Aizen, M. A., Albrecht, M., Bartomeus, I., Bommarco, R., Carvalheiro, L. G., Chaplin-Kramer, R., Gagic, V., Garibaldi, L. A., Ghazoul, J., Grab, H., Jonsson, M., Karp, D. S., Kennedy, C. M., Kleijn, D., Kremen, C., Landis, D. A., Letourneau, D. K., Marini, L., Poveda, K., Rader, R., Smith, H. G., Tscharntke, T., Andersson, G. K. S., Badenhausser, I., Baensch, S., Bezerra, A. D. M., Bianchi, F. J. J. A., Boreux, V., Bretagnolle, V., Caballero-Lopez, B., Cavigliasso, P., Ćetković, A., Chacoff, N. P., Classen, A., Cusser, S., Da Silva E Silva, F. D., De Groot, G. A., Dudenhöffer, J. H., Ekroos, J., Fijen, T., Franck, P., Freitas, B. M., Garratt, M. P. D., Gratton, C., Hipólito, J., Holzschuh, A., Hunt, L., Iverson, A. L., Jha, S., Keasar, T., Kim, T. N., Kishinevsky, M.,

Klatt, B. K., Klein, A. M., Krewenka, K. M., Krishnan, S., Larsen, A. E., Lavigne, C., Liere, H., Maas, B., Mallinger, R. E., Martinez Pachon, E., Martínez-Salinas, A., Meehan, T. D., Mitchell, M. G. E., Molina, G. A. R., Nesper, M., Nilsson, L., O'rourke, M. E., Peters, M. K., Plećaš, M., Potts, S. G., Ramos, D. L., Rosenheim, J. A., Rundlöf, M., Rusch, A., Sáez, A., Scheper, J., Schleuning, M., Schmack, J. M., Sciligo, A. R., Seymour, C., Stanley, D. A., Stewart, R., Stout, J. C., Sutter, L., Takada, M. B., Taki, H., Tamburini, G., Tschumi, M., Viana, B. F., Westphal, C., Willcox, B. K., Wratten, S. D., Yoshioka, A., Zaragoza-Trello, C., Zhang, W., Zou, Y. and Steffan-Dewenter, I. 2019. A global synthesis reveals biodiversity-mediated benefits for crop production. *Science Advances* 5(10), eaax0121.

Daniel, C. and Grunder, J. 2012. Integrated management of European cherry fruit fly Rhagoletis cerasi (L.): Situation in Switzerland and Europe. *Insects* 3(4), 956–988.

Dedryver, C. A., Le Ralec, A. and Fabre, F. 2010. The conflicting relationships between aphids and men: A review of aphid damage and control strategies. *Comptes Rendus Biologies* 333(6–7), 539–553.

Devine, G. J. and Furlong, M. J. 2007. Insecticide use: Contexts and ecological consequences. *Agriculture and Human Values* 24(3), 281–306.

Egan, P. A., Dicks, L. V., Hokkanen, H. M. T. and Stenberg, J. A. 2020. Delivering integrated pest and pollinator management (IPPM). *Trends in Plant Science* 25(6), 577–589.

Ehler, L. 1998. Chapter 1.Conservation biological control: Past, present, and future. In: Barbosa, P. (Ed.), *Conservation Biological Control*. San Diego, CA: Academic Press.

FAOSTAT. 2021. Food and Agriculture Organization of the United Nations. (2021). FAOSTAT. Rome, Italy: FAO. http://fenix.fao.org/faostat/internal/en/#home [Online].

Finke, D. L. and Denno, R. F. 2003. Intra-guild predation relaxes natural enemy impacts on herbivore populations. *Ecological Entomology* 28(1), 67–73.

Foster, S. P., Devine, G. and Devonshire, A. L. 2017. Insecticide resistance. In: Van Emden, H. F. and Harrington, R. (Eds), *Aphids as Crop Pests* (2nd edn.). Wallingford, UK: CABI Publishing.

Gardarin, A., Plantegenest, M., Bischoff, A. and Valantin-Morison, M. 2018. Understanding plant-arthropod interactions in multitrophic communities to improve conservation biological control: Useful traits and metrics. *Journal of Pest Science* 91(3), 943–955.

Garibaldi, L. A., Pérez-Méndez, N., Garratt, M. P. D., Gemmill-Herren, B., Miguez, F. E. and Dicks, L. V. 2019. Policies for ecological intensification of crop production. *Trends in Ecology and Evolution* 34(4), 282–286.

Garratt, M. P. D., Senapathi, D., Coston, D. J., Mortimer, S. R. and Potts, S. G. 2017. The benefits of hedgerows for pollinators and natural enemies depends on hedge quality and landscape context. *Agriculture, Ecosystems and Environment* 247, 363–370.

Geiger, F., Wäckers, F. L. and Bianchi, F. J. J. A. 2009. Hibernation of predatory arthropods in semi-natural habitats. *BioControl* 54(4), 529–535.

Giles, K. L., Mccornack, B. P., Royer, T. A. and Elliott, N. C. 2017. Incorporating biological control into IPM decision making. *Current Opinion in Insect Science* 20, 84–89.

Gillespie, M., Wratten, S., Sedcole, R. and Colfer, R. 2011. Manipulating floral resources dispersion for hoverflies (Diptera: Syrphidae) in a California lettuce agro-ecosystem. *Biological Control* 59(2), 215–220.

Gómez-Marco, F., Urbaneja, A. and Tena, A. 2016. A sown grass cover enriched with wild forb plants improves the biological control of aphids in citrus. *Basic and Applied Ecology* 17(3), 210–219.

Gurr, G. M., Wratten, S. D., Landis, D. A. and You, M. 2017. Habitat management to suppress pest populations: Progress and prospects. *Annual Review of Entomology* 62, 91–109.

Gutierrez-Arellano, C. and Mulligan, M. 2018. A review of regulation ecosystem services and disservices from faunal populations and potential impacts of agriculturalisation on their provision, globally. *Nature Conservation* 30, 1–39.

Happe, A. K., Alins, G., Bluthgen, N., Boreux, V., Bosch, J., Garcia, D., Hamback, P. A., Klein, A. M., Martinez-Sastre, R., Minarro, M., Muller, A. K., Porcel, M., Rodrigo, A., Roquer-Beni, L., Samnegard, U., Tasin, M. and Mody, K. 2019. Predatory arthropods in apple orchards across Europe: Responses to agricultural management, adjacent habitat, landscape composition and country. *Agriculture, Ecosystems and Environment* 273, 141–150.

Happe, A. K., Roquer-Beni, L., Bosch, J., Alins, G. and Mody, K. 2018. Earwigs and woolly apple aphids in integrated and organic apple orchards: Responses of a generalist predator and a pest prey to local and landscape factors. *Agriculture, Ecosystems and Environment* 268, 44–51.

Hardie, J. 2017. Life cycles and polyphenism. In: Van Emden, H. F. and Harrington, R. (Eds), *Aphids as Crop Pests* (2nd edn.). Wallingford, UK: CABI Publishing.

Helyer, N., Cattlin, N. D. and Brown, K. C. 2014. *Biological Control in Plant Protection: A Colour Handbook* (2nd edn.). London: CRC Press Press.

Holland, J. M. and Ellis, S. 2008. *Beneficials on Farmland - Identification and Management Guidelines*. London: Home-Grown Cereals Authority.

Hopper, J. V., Nelson, E. H., Daane, K. M. and Mills, N. J. 2011. Growth, development and consumption by four syrphid species associated with the lettuce aphid, Nasonovia ribisnigri, in California. *Biological Control* 58(3), 271–276.

Husson, O., Tran Quoc, H., Boulakia, S., Chabanne, A., Tivet, F., Bouzinac, S., Lienhard, P., Michellon, R., Chabierski, S., Boyer, J., Enjalric, F., Rakotondramanana, U., Moussa, N., Jullien, F., Balarabe, O., Rattanatray, B., Castella, J.-C., Charpentier, H. and Séguy, L. 2016. Co-designing innovative cropping systems that match biophysical and socio-economic diversity: The DATE approach to Conservation Agriculture in Madagascar, Lao PDR and Cambodia. *Renewable Agriculture and Food Systems* 31(5), 452–470.

Irvin, N. A., Bistline-East, A. and Hoddle, M. S. 2016. The effect of an irrigated buckwheat cover crop on grape vine productivity, and beneficial insect and grape pest abundance in southern California. *Biological Control* 93, 72–83.

Irvin, N. A., Scarratt, S. L., Wratten, S. D., Frampton, C. M., Chapman, R. B. and Tylianakis, J. M. 2006. The effects of floral understoreys on parasitism of leafrollers (Lepidoptera: Tortricidae) on apples in New Zealand. *Agricultural and Forest Entomology* 8(1), 25–34.

Johnson, A. C., Liu, J., Reynolds, O., Furlong, M. J., Mo, J., Rizvi, S. and Gurr, G. M. 2021. Conservation biological control research is strongly uneven across trophic levels and economic measures. *Pest Management Science* 77(5), 2165–2169.

Jonsson, M., Straub, C. S., Didham, R. K., Buckley, H. L., Case, B. S., Hale, R. J., Gratton, C. and Wratten, S. D. 2015. Experimental evidence that the effectiveness of conservation biological control depends on landscape complexity. *Journal of Applied Ecology* 52(5), 1274–1282.

Karp, D. S., Chaplin-Kramer, R., Meehan, T. D., Martin, E. A., Declerck, F., Grab, H., Gratton, C., Hunt, L., Larsen, A. E., Martínez-Salinas, A., O'rourke, M. E., Rusch, A., Poveda, K., Jonsson, M., Rosenheim, J. A., Schellhorn, N. A., Tscharntke, T., Wratten, S. D.,

Zhang, W., Iverson, A. L., Adler, L. S., Albrecht, M., Alignier, A., Angelella, G. M., Zubair Anjum, M., Avelino, J., Batáry, P., Baveco, J. M., Bianchi, F. J. J. A., Birkhofer, K., Bohnenblust, E. W., Bommarco, R., Brewer, M. J., Caballero-López, B., Carrière, Y., Carvalheiro, L. G., Cayuela, L., Centrella, M., Ćetković, A., Henri, D. C., Chabert, A., Costamagna, A. C., De La Mora, A., De Kraker, J., Desneux, N., Diehl, E., Diekötter, T., Dormann, C. F., Eckberg, J. O., Entling, M. H., Fiedler, D., Franck, P., Frank Van Veen, F. J., Frank, T., Gagic, V., Garratt, M. P. D., Getachew, A., Gonthier, D. J., Goodell, P. B., Graziosi, I., Groves, R. L., Gurr, G. M., Hajian-Forooshani, Z., Heimpel, G. E., Herrmann, J. D., Huseth, A. S., Inclán, D. J., Ingrao, A. J., Iv, P., Jacot, K., Johnson, G. A., Jones, L., Kaiser, M., Kaser, J. M., Keasar, T., Kim, T. N., Kishinevsky, M., Landis, D. A., Lavandero, B., Lavigne, C., Le Ralec, A., Lemessa, D., Letourneau, D. K., Liere, H., Lu, Y., Lubin, Y., Luttermoser, T., Maas, B., Mace, K., Madeira, F., Mader, V., Cortesero, A. M., Marini, L., Martinez, E., Martinson, H. M., Menozzi, P., Mitchell, M. G. E., Miyashita, T., Molina, G. A. R., Molina-Montenegro, M. A., O'Neal, M. E., Opatovsky, I., Ortiz-Martinez, S., Nash, M., Östman, Ö., Ouin, A., Pak, D., Paredes, D., Parsa, S., Parry, H., Perez-Alvarez, R., Perović, D. J., Peterson, J. A., Petit, S., Philpott, S. M., Plantegenest, M., Plećaš, M., Pluess, T., Pons, X., Potts, S. G., Pywell, R. F., Ragsdale, D. W., Rand, T. A., Raymond, L., Ricci, B., Sargent, C., Sarthou, J.-P., Saulais, J., Schäckermann, J., Schmidt, N. P., Schneider, G., Schüepp, C., Sivakoff, F. S., Smith, H. G., Whitney, K. S., Stutz, S., Szendrei, Z., Takada, M. B., Taki, H., Tamburini, G., Thomson, L. J., Tricault, Y., Tsafack, N., Tschumi, M., Valantin-Morison, M., Trinh, M. V., van der Werf, W., Vierling, K. T., Werling, B. P., Wickens, J. B., Wickens, V. J., Woodcock, B. A., Wyckhuys, K., Xiao, H., Yasuda, M., Yoshioka, A. and Zou, Y. 2018. Crop pests and predators exhibit inconsistent responses to surrounding landscape composition. Proceedings of the National Academy of Sciences.

Kleijn, D., Bommarco, R., Fijen, T. P. M., Garibaldi, L. A., Potts, S. G. and Van Der Putten, W. H. 2019. Ecological intensification: Bridging the gap between science and practice. *Trends in Ecology and Evolution* 34(2), 154–166.

Kong, C. H., Hu, F., Xu, X. H., Zhang, M. X. and Liang, W. J. 2005. Volatile allelochemicals in the Ageratum conyzoides intercropped citrus orchard and their effects on mites Amblyseius newsami and Panonychus citri. *Journal of Chemical Ecology* 31(9), 2193–2203.

Kovács-Hostyánszki, A., Espíndola, A., Vanbergen, A. J., Settele, J., Kremen, C. and Dicks, L. V. 2017. Ecological intensification to mitigate impacts of conventional intensive land use on pollinators and pollination. *Ecology Letters* 20(5), 673–689.

Lang, G. A. 2014. Growing sweet cherries UNDER plastic covers and tunnels: Physiological aspects and practical considerations. International Society for Horticultural Science (ISHS), Leuven, Belgium, pp. 303–312.

Larentzaki, E., Plate, J., Nault, B. A. and Shelton, A. M. 2008. Impact of straw mulch on populations of onion thrips (Thysanoptera: Thripidae) in onion. *Journal of Economic Entomology* 101(4), 1317–1324.

Liu, Y. B. 2004. Distribution and population development of Nasonovia ribisnigri (Homoptera: Aphididae) in iceberg lettuce. *Journal of Economic Entomology* 97(3), 883–890.

Mankad, A., Loechel, B. and Measham, P. F. 2017. Psychosocial barriers and facilitators for area-wide management of fruit fly in southeastern Australia. *Agronomy for Sustainable Development* 37(6), 67.

Martin, E. A., Dainese, M., Clough, Y., Báldi, A., Bommarco, R., Gagic, V., Garratt, M. P. D., Holzschuh, A., Kleijn, D., Kovács-Hostyánszki, A., Marini, L., Potts, S. G., Smith, H. G., Al Hassan, D., Albrecht, M., Andersson, G. K. S., Asís, J. D., Aviron, S., Balzan, M. V., Baños-Picón, L., Bartomeus, I., Batáry, P., Burel, F., Caballero-López, B., Concepción, E. D., Coudrain, V., Dänhardt, J., Diaz, M., Diekötter, T., Dormann, C. F., Duflot, R., Entling, M. H., Farwig, N., Fischer, C., Frank, T., Garibaldi, L. A., Hermann, J., Herzog, F., Inclán, D., Jacot, K., Jauker, F., Jeanneret, P., Kaiser, M., Krauss, J., Le Féon, V., Marshall, J., Moonen, A. C., Moreno, G., Riedinger, V., Rundlöf, M., Rusch, A., Scheper, J., Schneider, G., Schüepp, C., Stutz, S., Sutter, L., Tamburini, G., Thies, C., Tormos, J., Tscharntke, T., Tschumi, M., Uzman, D., Wagner, C., Zubair-Anjum, M. and Steffan-Dewenter, I. 2019. The interplay of landscape composition and configuration: New pathways to manage functional biodiversity and agroecosystem services across Europe. *Ecology Letters* 22(7), 1083–1094.

Mateos-Fierro, Z., Fountain, M. T., Garratt, M. P. D., Ashbrook, K. and Westbury, D. B. 2021. Active management of wildflower strips in commercial sweet cherry orchards enhances natural enemies and pest regulation services. *Agriculture, Ecosystems and Environment* 317, 107485.

Mateos-Fierro, Z., Garratt, M. P. D., Fountain, M. T., Ashbrook, K. and Westbury, D. B. 2018. Wildflower strip establishment for the delivery of ecosystem services in sweet cherry orchards. *Aspects of Applied Biology: Ecosystem and Habitat Management: Research, Policy, Practice* 139, 179–186.

McKerchar, M., Potts, S. G., Fountain, M. T., Garratt, M. P. D. and Westbury, D. B. 2020. The potential for wildflower interventions to enhance natural enemies and pollinators in commercial apple orchards is limited by other management practices. *Agriculture, Ecosystems and Environment* 301, 107034.

Montgomery, I., Caruso, T. and Reid, N. 2020. Hedgerows as ecosystems: Service delivery, management, and restoration. *Annual Review of Ecology, Evolution, and Systematics* 51(1), 81–102.

Morandin, L. A., Long, R. F. and Kremen, C. 2016. Pest control and pollination cost–benefit analysis of hedgerow restoration in a simplified agricultural landscape. *Journal of Economic Entomology* 109(3), 1020–1027.

Müller, C. B. and Brodeur, J. 2002. Intraguild predation in biological control and conservation biology. *Biological Control* 25(3), 216–223.

Olson, D. M. and Wäckers, F. L. 2007. Management of field margins to maximize multiple ecological services. *Journal of Applied Ecology* 44(1), 13–21.

Pascual-Villalobos, M. J., Lacasa, A., González, A., Varó, P. and García, M. J. 2006. Effect of flowering plant strips on aphid and syrphid populations in lettuce. *European Journal of Agronomy* 24(2), 182–185.

Perring, T. M., Gruenhagen, N. M. and Farrar, C. A. 1999. Management of plant viral diseases THROUGH chemical control of insect vectors. *Annual Review of Entomology* 44, 457–481.

Power, A. G. 2010. Ecosystem services and agriculture: Tradeoffs and synergies. *Philosophical Transactions of the Royal Society of London. Series B, Biological Sciences* 365(1554), 2959–2971.

Pretty, J., Benton, T. G., Bharucha, Z. P., Dicks, L. V., Flora, C. B., Godfray, H. C. J., Goulson, D., Hartley, S., Lampkin, N., Morris, C., Pierzynski, G., Prasad, P. V. V., Reganold, J., Rockström, J., Smith, P., Thorne, P. and Wratten, S. 2018. Global assessment of agricultural system redesign for sustainable intensification. *Nature Sustainability* 1(8), 441–446.

Qureshi, S. A., Midmore, D. J., Syeda, S. S. and Reid, D. J. 2009. A comparison of alternative plant mixes for conservation bio-control by native beneficial arthropods in vegetable cropping systems in Queensland Australia. *Bulletin of Entomological Research* 100(1), 67–73.

Ramalho, F. S., Fernandes, F. S., Nascimento, A. R. B., Júnior, J. L. N., Malaquias, J. B. and Silva, C. A. 2012. Assessment of fennel aphids (Hemiptera: Aphididae) and their predators in fennel intercropped with cotton with colored fibers. *Journal of Economic Entomology* 105(1), 113–119, 7.

Redman, G. 2020. *John Nix Farm Management Pocketbook* (51st edn.). Leicestershire, UK: Melton Mowbray.

Requier, F. and Leonhardt, S. D. 2020. Beyond flowers: Including non-floral resources in bee conservation schemes. *Journal of Insect Conservation* 24(1), 5–16.

Rodríguez-San Pedro, A., Allendes, J. L., Beltrán, C. A., Chaperon, P. N., Saldarriaga-Córdoba, M. M., Silva, A. X. and Grez, A. A. 2020. Quantifying ecological and economic value of pest control services provided by bats in a vineyard landscape of central Chile. *Agriculture, Ecosystems and Environment* 302, 107063.

Roger-Estrade, J., Anger, C., Bertrand, M. and Richard, G. 2010. Tillage and soil ecology: Partners for sustainable agriculture. *Soil and Tillage Research* 111(1), 33–40.

Ruberson, J., Nemoto, H. and Hirose, Y. 1998. Chapter 11. Pesticides and conservation of natural enemies in pest management. In: Barbosa, P. (Ed.), *Conservation Biological Control*. San Diego, CA: Academic Press.

Rusch, A., Binet, D., Delbac, L. and Thiery, D. 2016a. Local and landscape effects of agricultural intensification on Carabid community structure and weed seed predation in a perennial cropping system. *Landscape Ecology* 31(9), 2163–2174.

Rusch, A., Chaplin-Kramer, R., Gardiner, M. M., Hawro, V., Holland, J., Landis, D., Thies, C., Tscharntke, T., Weisser, W. W., Winqvist, C., Woltz, M. and Bommarco, R. 2016b. Agricultural landscape simplification reduces natural pest control: A quantitative synthesis. *Agriculture, Ecosystems and Environment* 221, 198–204.

Rusch, A., Delbac, L. and Thiery, D. 2017. Grape moth density in Bordeaux vineyards depends on local habitat management despite effects of landscape heterogeneity on their biological control. *Journal of Applied Ecology* 54(6), 1794–1803.

Salamanca, J., Souza, B. and Rodriguez-Saona, C. 2018. Cascading effects of combining synthetic herbivore-induced plant volatiles with companion plants to manipulate natural enemies in an agro-ecosystem. *Pest Management Science* 74, 2133 -2145.

Sánchez-Bayo, F. and Wyckhuys, K. A. G. 2019. Worldwide decline of the entomofauna: A review of its drivers. *Biological Conservation* 232, 8-27.

Santos, L. A. O., Botelho Costa, M., Lavigne, C., Fernandes, O. A., Bischoff, A. and Franck, P. 2018. Influence of the margin vegetation on the conservation of aphid biological control in apple orchards. *Journal of Insect Conservation* 22(3–4), 465–474.

Schellhorn, N. A., Gagic, V. and Bommarco, R. 2015. Time will tell: Resource continuity bolsters ecosystem services. *Trends in Ecology and Evolution* 30(9), 524–530.

Schellhorn, N. A. and Sork, V. L. 1997. The impact of weed diversity on insect population dynamics and crop yield in collards, Brassica oleraceae (Brassicaceae). *Oecologia* 111(2), 233–240.

Shields, M. W., Johnson, A. C., Pandey, S., Cullen, R., González- Chang, M., Wratten, S. D. and Gurr, G. M. 2019. History, current situation and challenges for conservation biological control. *Biological Control* 131, 25–35.

Simon, S., Lesueur-Jannoyer, M., Plénet, D., Lauri, P. É. and Le Bellec, F. 2017. Methodology to design agroecological orchards: Learnings from on-station and on-farm experiences. *European Journal of Agronomy* 82, 320–330.

Singh, T. V. K. and Satyanarayana, J. 2009. Insect outbreaks and their management. In: Peshin, R. and Dhawan, A. K. (Eds), *Integrated Pest Management: Innovation-Development Process* (vol. 1). Dordrecht: Springer Netherlands.

Smith, H. A., Chaney, W. E. and Bensen, T. A. 2008. Role of syrphid larvae and other predators in suppressing aphid infestations in organic lettuce on California's central coast. *Journal of Economic Entomology* 101(5), 1526–1532, 7.

Song, B. Z., Wu, H. Y., Kong, Y., Zhang, J., Du, Y. L., Hu, J. H. and Yao, Y. C. 2010. Effects of intercropping with aromatic plants on the diversity and structure of an arthropod community in a pear orchard. *BioControl* 55(6), 741–751.

Song, B. Z., Zhang, J., Hu, J. H., Wu, H. Y., Kong, Y. and Yao, Y. C. 2011. Temporal dynamics of the arthropod community in pear orchards intercropped with aromatic plants. *Pest Management Science* 67(9), 1107–1114.

Staton, T., Walters, R. J., Smith, J., Breeze, T. D. and Girling, R. D. 2021. Evaluating a trait-based approach to compare natural enemy and pest communities in agroforestry vs. arable systems. *Ecological Applications: A Publication of the Ecological Society of America* 31(4), e02294.

Thomson, L. J. and Hoffmann, A. A. 2013. Spatial scale of benefits from adjacent woody vegetation on natural enemies within vineyards. *Biological Control* 64(1), 57–65.

Thomson, L. J., McKenzie, J., Sharley, D. J., Nash, M. A., Tsitsilas, A. and Hoffmann, A. A. 2010. Effect of woody vegetation at the landscape scale on the abundance of natural enemies in Australian vineyards. *Biological Control* 54(3), 248–254.

Tilman, D., Balzer, C., Hill, J. and Befort, B. L. 2011. Global food demand and the sustainable intensification of agriculture. *Proceedings of the National Academy of Sciences of the United States of America* 108(50), 20260–20264.

Tomlinson, J. A. 1987. Epidemiology and control of virus diseases of vegetables. *Annals of Applied Biology* 110(3), 661–681.

Tscharntke, T., Karp, D. S., Chaplin-Kramer, R., Batáry, P., Declerck, F., Gratton, C., Hunt, L., Ives, A., Jonsson, M., Larsen, A., Martin, E. A., Martínez-Salinas, A., Meehan, T. D., O'rourke, M., Poveda, K., Rosenheim, J. A., Rusch, A., Schellhorn, N., Wanger, T. C., Wratten, S. and Zhang, W. 2016. When natural habitat fails to enhance biological pest control – five hypotheses. *Biological Conservation* 204, 449–458.

Tscharntke, T., Klein, A. M., Kruess, A., Steffan-Dewenter, I. and Thies, C. 2005. Landscape perspectives on agricultural intensification and biodiversity – ecosystem service management. *Ecology Letters* 8(8), 857–874.

van Rijn, P. C. J. and Wäckers, F. L. 2016. Nectar accessibility determines fitness, flower choice and abundance of hoverflies that provide natural pest control. *Journal of Applied Ecology* 53(3), 925–933.

Villa, M., Santos, S. A. P., Sousa, J. P., Ferreira, A., Da Silva, P. M., Patanita, I., Ortega, M., Pascual, S. and Pereira, J. A. 2020. Landscape composition and configuration affect the abundance of the olive moth (Prays oleae, Bernard) in olive groves. *Agriculture, Ecosystems and Environment* 294, 106854. https://doi.org/10.1016/j.agee.2020.106854.

Wäckers, F. L. and van Rijn, P. C. J. 2012. Pick and mix: Selecting flowering plants to meet the requirements of target biological control insects. In: Gurr, G.M., Wratten, S.D., Snyder, W. E., and Read, D. M. Y. (Eds), *Biodiversity and Insect Pests*. https://doi.org/10.1002/9781118231838.ch9.

Wilson, C. and Tisdell, C. 2001. Why farmers continue to use pesticides despite environmental, health and sustainability costs. *Ecological Economics* 39(3), 449–462.

Wilson, H., Miles, A. F., Daane, K. M. and Altieri, M. A. 2015. Landscape diversity and crop vigor influence biological control of the western grape leafhopper (E. elegantula Osborn) in vineyards. *PLoS ONE* 10(11), e0141752. doi:10.1371/journal.pone.0141752.

Wilson, H., Miles, A. F., Daane, K. M. and Altieri, M. A. 2017. Landscape diversity and crop vigor outweigh influence of local diversification on biological control of a vineyard pest. *Ecosphere* 8(4), e01736.

Woltz, J. M., Werling, B. P. and Landis, D. A. 2012. Natural enemies and insect outbreaks in agriculture: A landscape perspective. In: Barbosa, P., Letourneau, D. K. and Agrawal, A. A. (Eds), *Insect Outbreaks Revisited*. https://doi.org/10.1002/9781118295205.ch17.

Woodcock, B. A., Bullock, J. M., McCracken, M., Chapman, R. E., Ball, S. L., Edwards, M. E., Nowakowski, M. and Pywell, R. F. 2016. Spill-over of pest control and pollination services into arable crops. *Agriculture, Ecosystems and Environment* 231, 15–23.

Wratten, S. D., Gillespie, M., Decourtye, A., Mader, E. and Desneux, N. 2012. Pollinator habitat enhancement: Benefits to other ecosystem services. *Agriculture, Ecosystems and Environment* 159, 112–122.

Yang, Y., Chen, Y., Yu, Z., Li, P. and Li, X. 2020. How does improve farmers' attitudes toward ecosystem services to support sustainable development of agriculture? Based on environmental Kuznets curve theory. *Sustainability* 12(20), 8655.

Zehnder, G., Gurr, G. M., Kühne, S., Wade, M. R., Wratten, S. D. and Wyss, E. 2007. Arthropod pest management in organic crops. *Annual Review of Entomology* 52, 57–80.

Zhang, H., Garratt, M. P. D., Bailey, A., Potts, S. G. and Breeze, T. 2018a. Economic valuation of natural pest control of the summer grain aphid in wheat in South East England. *Ecosystem Services* 30, 149–157.

Zhang, H., Potts, S. G., Breeze, T. and Bailey, A. 2018b. European farmers' incentives to promote natural pest control service in arable fields. *Land Use Policy* 78, 682–690.

Zhang, W., Ricketts, T. H., Kremen, C., Carney, K. and Swinton, S. M. 2007. Ecosystem services and dis-services to agriculture. *Ecological Economics* 64(2), 253–260.

Chapter 10

Assessing the economics of integrated pest management for horticultural crops

Philip R. Crain and David W. Onstad, Corteva Agriscience, USA

1 Introduction

Assessing the value and strategic implications of integrated pest management (IPM) is a challenging but necessary exercise. Economic value, which can be defined as an operating loss or gain on a crop in a given period of time, is a very real observation (i.e. one point of data) that carries significant consequences for farmers. Economic value, however, is a multifaceted, variable, and dynamic concept. For example, the loss or profit in a given year at a given location is subject to many forces, some sustained and some ephemeral. The challenge to IPM is to help optimize economic value while designing robust pest management strategies for farmers and other stakeholders, including those concerned about environmental impacts. These strategies should address known risks and be flexible enough to quickly adapt to emerging risks.

Optimized economic value is obtained when farmers can meet their short-term growth targets that sustain and improve their livelihood while also addressing the need to produce more in environmentally and financially sustainable ways. Concepts in IPM, like economic injury level and economic threshold, are conceptual models that seek to optimize economic value in the face of pests. By only using control tactics when the economic threshold had been reached, in theory, growers and control decision makers were being economical in their decisions. However, what if the control tactic was

http://dx.doi.org/10.19103/AS.2021.0095.12

more expensive than a change in the design of the IPM system? What if a new technology, for example exclusion nets, provided a more reliable, ecologically sustainable, and more valuable control compared to the overuse of a chemical insecticide? This highlights a significant dilemma facing growers and pest control advisors, which is: when should we 'think inside the box' of current pest management paradigms, and when should we explore 'outside the box' to unlock better value or sustainability?

'Thinking inside the box' tends to have a negative connotation. It is somehow viewed as less inspired than those solutions that 'think outside the box.' There have been many great successes that have resulted from thinking inside the box. 'The box' in this metaphor represents a system or framework that is useful in describing the attributes of a system and the most common decisions that are confronted. New control tactics, which may arise from inside the box thinking, address a known need to manage risk and create value for stakeholders. Both current and new control tactics are less likely to create value when new risks arise. The emergence of unknown risks may have been driven by the selection pressure exerted by control tactics or it could represent new pest invasions. To manage new risks, stakeholders need to think differently, consider disruptive technologies, and explore system design changes to find new solutions that maintain economic value. Economic thinking is critical to making good decisions that not only balance potential benefits but also realize the potential costs of decisions (i.e. optimizing economic value).

Economic thinking that focuses on the return on investment of control strategies in agricultural systems and specifically IPM is difficult, however. The complex and chaotic nature of ecological interactions lends itself to rule of thumb concepts that simplify the complexity. Simplifying complex systems can be too easily interpreted as a disruptive innovation. It distills difficult *n*-dimensional systems into a set of manageable, useful choices. This provides benefit to practitioners because it allows them to measure specific attributes of the system that can guide decisions that are likely better. The risk is that these systems become static. Because agricultural systems are dynamic, static frameworks are useful only for a specific window of time. IPM practitioners must consciously think about the complexity and dynamic nature of systems to build conceptual models that distill the most important system attributes into a series of decisions that can positively impact the outcome of farmers and society.

Despite the dynamic and complex nature of agricultural problems, the basic components of economic thinking are extremely helpful to promote careful consideration of pest management strategies. In its most basic sense, economic thinking is confronting the assumptions that stakeholders hold about the benefit (i.e. revenue) and costs of decisions they take within their system. The number of unknowns at the start of a season is very high, and each decision a farmer takes should address an unknown. Each decision has a potential positive and negative impact. It is critical to weigh these concepts to make good pest management decisions. Of course, the set of unknowns that exist before a single season, and which focus on a single season, is different from the set of unknowns that dictate more long-term questions like resistance management (Onstad 2013). Economic thinking can also be used to understand how to focus on short- and long-term goals to maintain viability and sustainability of operations.

In this chapter, we introduce the concepts of economic thinking, describing the costs and potential benefits of IPM control solutions as well as discussing time horizons, stakeholders, and economic injury level and economic thresholds. Next, we will look at some case studies, with a focus on horticulture, which exemplify how to measure ecological complexity through economic thinking. Throughout the discussion of economic principles and case studies, we will highlight opportunities to consider control tactics versus IPM system design changes and the implications for economic decision-making. Finally, we will address the future of economic thinking in IPM with a challenge that creates more value for society and growers in the future through innovation and new technology that is becoming available.

2 Concepts of economic thinking

Economics is broadly the study of human choice under the limits of scarce resources (Sowell 2014). More specifically, economics is often defined as the branch of knowledge that focuses on the production, consumption, and transfer of wealth. While economics does deal with value, wealth, and currency, it is not limited to viewing components of a system as either costs or benefits financially. Economics also must consider human values and how they are reflected in decision-making. This generally describes normative economics, or the analysis of what 'ought to be' (one can also focus on positive economics or 'what is'). The goods and services that people use carry with them assumptions for how they are valued and how they are perceived. For example, entomologists may recognize that an arthropod pest is a subjective, human-centric definition that only applies to specific places at specific times. Insects can be beautiful and indicative of a healthy ecosystem in some contexts but hurtful in another. For example, European honeybees (*Apis mellifera*) are generally regarded as beneficial insects due to the pollination services they provide, but they may be considered pests if a hive is too close to humans or if they are disrupting the natural bees present in an ecosystem.

To perform an economic analysis, it is important at the outset to rigorously define the problem. Components of a problem are (1) identifying the stakeholders involved and what is the primary perspective being taken, (2) aligning the goals of the analysis with the perspective of the primary stakeholder(s), and (3) defining the time period and spatial scale for the analysis as much as possible to expose any risks or dependencies. As mentioned previously, the decisions, benefits, and costs that characterize a single year or crop-growing season may be very different from those associated with a multiyear or long-term outlook. Economic thinking helps to align stakeholders' expectations, costs, benefits, and time horizons so that decision makers can use this information to allocate resources and manage pests to achieve multiple objectives in the lives and well-being of people and economies (National Research Council 1999, 2005)

To illustrate the scope of this challenge, consider two different examples. First, vineyard owners have an established vineyard and are looking to maximize their return for the upcoming year. Because vineyards are perennial crops, there is no decision to take on what crop should be planted, but instead, their focus will look to predicting pest incidence and what tools they can use to control those problems before they cause economic damage. Along with those decisions, they must consider how they will market their product and if that constrains their possible decisions. For example, organic production will have a very different toolkit for controlling pests relative to conventional vineyards. Grapes for wine production may have different considerations versus table grapes. Many decisions may be taken before the season begins, but many decisions will also occur within the season based on the weather and conditions experienced by their crop that year. To create profit, the owners must generate enough revenue to cover the costs of any control treatments, labor, equipment or land costs, and many other factors. A second stakeholder could be a consumer of the products from this vineyard. Being focused on environmental impacts and long-term sustainability, consumers could create secondary considerations for the vineyard grower in the long-term. Maybe they need to shift the acres in their production to more organic to accommodate consumer demand? Maybe grape demand is falling, and they will need to confront the costs associated with picking a new crop? These are different questions arising from various stakeholders with different time horizons. Defining problems within these spaces is critical to solving the right problem with economic thinking.

In IPM, stakeholders range from growers to consultants or advisors to input and processing companies to product consumers. All these individuals are influenced by IPM decisions. Growers typically bear the costs of any control tactic used, while processing companies and consumers may define market opportunities and their value (e.g. premium prices for organic production vs. conventional). As new market opportunities become available, growers must carefully consider the return they may generate for the cost they will incur and the time horizon for both. For example, while organic production is more desirable for many consumers and may lead to higher prices for produce, the cost of organic production is generally higher with the need to use more control tactics in combination coupled with less efficacy for each individual tactic. Akin and Kara (2021) found that willingness to continue organic almond production in Turkey decreased when orchard size increased because the value of additional production in an additional hectare of land was more valuable to growers than the payments they received to maintain organic production (organic production has lower yields relative to conventional production in this system). Walsh et al. (2011) performed an economic assessment of an experimental orchard with apple and pear trees that included both organic and conventional growing practices. Their analysis found that organic production required more labor and inputs relative to conventional production and estimated yield was also lower in organic blocks. Understanding the relationships among stakeholders and the focal perspective they hold allow us to solve problems, but stakeholders and their perspectives may change (e.g. a grower may consider different cropping systems if they add more hectares to their farm) as markets evolve.

Time horizon is the length of time over which a problem will be assessed. The decision to focus on a single year, multiple years, or a generation can have important ramifications on the decisions made, relative to their costs and benefits. Longer time horizons are not only associated with increased uncertainty but also provide more time to potentially reap benefits from control tactics or system design changes that are more costly upfront but provide ample returns after initiation. However, short-term profit generation is key to providing additional resources that growers can use to increase opportunities for future investments that may have more far-reaching outcomes. For example, purchasing additional land can help increase revenue but requires some free cash for deposits as well as financing loan payments. It is important to recognize that many growers are under significant short-term pressure to limit costs while maintaining or increasing their profitability but that social and environmentally sustainable solutions have long-term cycles that need to also be addressed in viable ways.

Within the context of horticultural cropping systems, and agriculture more generally, readers may be aware of the concepts of economic injury levels (EIL) and economic thresholds (ET). These concepts have been implemented in many systems and represent a simple interpretation of agriculture through an economic perspective. EIL is the lowest population density that will cause economic damage, while the ET is the population density where control measures should be applied to stop a pest from reaching or exceeding the EIL (Stern et al. 1959). These definitions allow for empirical studies to quantify both EIL and ET which, in theory, can lead to effective, economic control. This system is often focused on single-season dynamics and control versus system design opportunities and multiyear cropping systems. It is important to maintain the long-term view as well, especially if the crop of interest is a perennial crop where design changes may be more tractable, given the upfront costs and long-term characteristics of a particular cropping system (Hoyt and Gilpatrick 1976, Westigard 1979).

Complementary to the ET and EIL concepts, economic thinking should also allow for the consideration of system design changes. System design considers the whole agroecosystem and challenges whether the structures and/or components of a system (e.g. land, crops, cropping systems, etc.) could be changed to achieve desired outcomes more economically (Onstad and Crain 2019). Barzman et al. (2015) introduced the eight principles of pest management and stressed the importance to design cropping systems such that they only use harsh crop protection products when they best fit the needs of a grower. The first principle focuses on prevention and suppression and discusses how crop choice, management practices, and structural layout can suppress many pest problems before they become significant issues. Sciarretta et al. (2011) evaluated the effect of site-specific IPM to manage *Lobesia botrana* in grapes in central Italy and concluded that identifying and managing spatial hotspots can be more cost-effective than uniform management practices and that monitoring border olive groves helped inform grape management. Understanding the agroecosystem and pest dynamics in both time and space should allow stakeholders to identify when system design changes are economical versus their current practices.

3 Economic impact of horticultural crops and case studies on the complexity of integrated pest management

3.1 Overview of horticultural economics

Armed with the concepts of economic thinking, we will review some general economics for horticulture as well as use several case studies to dive more deeply into the themes of design and control, stakeholder engagement, and time horizons. The following sections will highlight the scope and value of horticulture in a few geographies and then transition to several case studies that can serve as examples for the themes above.

Fruit and vegetable production in Europe represented nearly 14% of the total value of agricultural production in 2018 or over €57B (Rossi 2019). Fruits accounted for 40% of the total value and vegetables the remaining 60%. Of total production tonnes, tomatoes and apples are the most significant contributors, and each represents approximately 20% of the total value of vegetables and fruits. The costs associated with pest management to produce those quantities are difficult to estimate because many control options are not easily quantifiable or tracked by government agencies or industry partners (e.g. agricultural practices that contribute to cultural control). However, Wesseler et al. (2015) measured the input costs associated with plant protection agents, which largely reflected the costs of chemical crop protection products, and found that plant protection agents accounted for more than €7B in 2010 in Europe.

In the United States, vegetable production totaled more than $19B in 2019 (Lucier and Davis 2020). Fruit, tree nut, and berry production totaled more than $28B in 2017. Again, it was difficult to quantify the costs of inputs, but the National Road Map for Integrated Pest Management (FIPMCC 2018) suggested that cost-benefit analyses be holistic and not limited to strictly financial metrics.

A review of agriculture in Brazil by Oliveira et al. (2014) predicted total crop losses as well as spending on crop protection products to protect yields against insects across broadacre row crops and specialty crops. Sugarcane had the highest economic loss (combining use as sugar and ethanol, ~$6B), followed by other crops like corn (~$2B) and soybean (~$2B). The total loss for these crops is proportional to the large acreage they cover. If costs are evaluated per hectare, then apples ($4281/ha), tomatoes ($3807/ha), tobacco ($2730/ha), and garlic ($2655/ha) are most susceptible to yield loss from insects. In total, insecticide sales alone totals nearly $3B with a significant portion of that spent focusing on soybeans (~$1.3B).

3.2 Case studies

The invasion of *Drosophila suzukii*, the spotted wing Drosophila (SWD), has led to many efforts to build economic IPM plans to control the pest and limit or slow its geographical spread (Asplen et al. 2015). SWD prefer ripening fruit and laying eggs under the fruit's skin, resulting in unmarketable fruit (Atallah et al. 2014). A study by Mazzi et al. (2017) discussed different IPM tactics for controlling SWD and documented their potential costs in sweet cherries in Switzerland. They surveyed growers in 2015 and 2016 and learned that the most widely used tactics to manage damage from SWD were (1) insecticide applications and (2) harvest practices aimed at collecting the majority of fruits, reducing any unharvested fruits that may harbor SWD. They also mentioned exclusion nets were prevalent in a significant proportion of survey respondents (approximately 1/3). The costs of the different methods were significant, and they called for increased stakeholder engagement to focus on both the value created by control measures and the science needed to technically vet the solutions.

A second study by Santoiemma et al. (2020) also examined which control tactics best reduce *D. suzukii* infestations in sweet cherry orchards in Italy. Similar to Mazzi et al. (2017), Santoiemma et al. (2020) found that several orchard management practices provide good protection against SWD damage; specifically, the high frequency of mowing interrow vegetation and good harvest practices, covering both the timing of harvest and completeness of removing all fruit from the field, correlated with less damaged fruit. They did not see a reduction in fruit damage based on the use of insecticides.

Both the previous studies outline the challenge of managing SWD, the tactics that seem to work best, and even the costs to implement those tactics. However, there also needs to be careful consideration of the benefits to growers (and other stakeholders) for implementing these tactics. An economic analysis by Del Fava et al. (2017) used partial budget analysis in a cost-benefit analysis to investigate the expected net present value of different IPM systems following the invasion of SWD into soft fruit crops in Northern Italy in 2009. Initially, a conventional IPM strategy was employed while fly population densities were low. However, as populations continued to increase, the value of current control tactics was insufficient. Growers opted to introduce exclusion nets, despite significant implementation cost, and Del Fava et al. (2017) showed that exclusion nets resulted in increases of profit up to €2.5 million.

A similar series of studies on the area-wide pest management effort to control Queensland fruit fly, *Bactrocera tryoni,* in Australia, provides an opportunity to think more about the benefits of different control measures and how those benefits influence resource allocation and opportunities for success. The Queensland fruit fly is a significant economic pest of horticultural goods (Dominiak 2019). *B. tryoni* is endemic to Australia but is a key target for exports due to the risk of its invasion and establishment in other areas, leading to the development of a code of practice for the management of Queensland fruit fly (Dominiak et al. 2015, Standing Committee on Agriculture and Resource Management 1996). Areas of Australia (Sunraysia, or northwestern Victoria and southwestern New South Wales) have been designated as pest-free areas and exclusion zones, and these areas are granted more access to export and domestic markets relative to other areas where the pest is endemic. An economic analysis by Florec et al. (2013) evaluated the economics of area-wide pest management against this pest focusing on the eradication of Queensland fruit fly outbreaks as well as post-harvest treatments. The analysis found that reducing the probability of outbreaks by enforcing stricter control at roadblock sites, thus reducing post-harvest disinfection insecticide treatments, was more economical compared to better surveillance or faster eradication. The authors concluded that improving the ability to identify, quantify, and separate produce needing these treatments by geographical area is important to effectively managing the budget for management. Although economic benefits are difficult to quantify and require aligned assumptions among stakeholders, decisions supported by science and expected benefits will allow for allocation of resources in areas that are most likely to impact growers and society.

In cases where pest management is particularly challenging, it is worthwhile to consider innovative control techniques or system design changes that may provide unique benefits. Nikolouli et al. (2020) investigated the potential for sterile and incompatible insect techniques to aid in the suppression of SWD. Using irradiation and *Wolbachia* sp., the authors documented high levels of sterilization or incompatibility without significantly affecting the life history or behavior of sterile males. The results suggest these techniques may contribute to an IPM program, but the economic costs and benefits need to be further quantified based on limited field trials. Although not an example from horticulture, Alphey et al. (2011) found that the cost-effectiveness for using sterile insect technique coupled with the genetic control technology RIDL (release of insects carrying a dominant lethal) was much higher than conventional vector-borne disease-control strategies. Novel changes to the design of a system can dramatically improve economic pest management. However, during the early technical validation of new or innovative IPM practices, there should be funding to support the testing and scaling of technical solutions at grower operations to bring better alignment with technical solutions and practical operations.

Agriculture represents an ecological system that can be manipulated by humans, but it is also subject to the uncertainty of weather, evolution, and species interactions. Additionally, agriculture is governed by consumer demands, market trends, and the need to increase sustainability. A study by Rodriguez Sousa et al. (2020) examined the economics of ecological farming in olive production in southwestern Spain and compared it to a more conventional farming practice. They used an analytic hierarchy process to assess both the financial returns and the environmental and societal benefits of both cropping systems. The study reported that integrated (or more conventional) olive farms saw higher financial returns but that ecological-farmed olives brought higher environmental and societal benefits. In the aggregate analysis, the externalities were considered more valuable than the difference in on-farm income (~\$100/ha advantage to integrated farms). Importantly, environmental and societal benefits are enjoyed by many people, but the costs are transferred to growers. We should not only continue to support transdisciplinary analyses but also find ways to drive economic well-being for the growers. Can ecological farms also serve as agroecotourism opportunities? Are farmers willing to incur the costs to make such an opportunity a reality? We advocate for economic analyses and broad stakeholder engagement to drive collaboration and better decision-making for the near and long-term.

Another consideration for economic analysis is cost of sampling to make a decision. It is not enough to consider only the costs and benefits of a treatment, but it must holistically include the data and labor needed to support a decision. A study by Figueiredo et al. (2021) took an interesting approach to improving the sampling of *Helicoverpa armigera* in tomatoes to determine if an insect control treatment is needed. They developed a sequential sampling procedure that had several criteria that either stopped sampling sooner or continued sampling to obtain acceptable precision for when to treat infestations in processing tomatoes. For a given economic threshold, the standard approach required sampling 60 plants to inform a treatment decision. By sequentially sampling the field, using the knowledge of the number of plants already sampled as well as the number of eggs, and mathematical calculations (Iwao's method based on Taylor's power law regression), scouts could stop sampling a field much sooner, sampling as few as 20 plants. By providing criteria to stop sampling and make decisions sooner, Figueiredo et al. (2021) allow scouts to be more efficient with their time. However, the cost of sampling is often ignored (Onstad et al. 2019). Building a more flexible framework that allows for more efficient decisions is an important step. Another advancement would be the coupling of sensing technology or prediction methods to further create efficiency for the grower or pest-management decision maker.

4 Conclusion

Optimizing economic value is elusive. It requires intentional focus on behalf of all stakeholders, clear and frequent communication, data collection and analysis, and an understanding of who bears costs and reaps benefits. Just as natural selection creates genotypes with desirable phenotypes, the pressures of economic viability will select for approaches that optimize value. We can expedite that optimization by employing economic thinking.

5 Future trends in research

The rise of internet of things systems, cellular connectivity, and machine learning has changed the way many people think about sensing, monitoring, and analyzing data. In earnest, IPM is the simplification of a complex system to a set of decisions that take data as input and generate sustainable revenue as an output. It is exciting to consider the impact of innovation on horticulture and agriculture more broadly, but as we have mentioned earlier, innovation comes with a cost and requires a solution that solves a problem with potentially different time horizons and stakeholders. We must help innovation in IPM succeed by making it applicable to all stakeholders in both the short and the long term.

Innovation is challenging for stakeholders to consider because disruptive innovation challenges the standard cost–benefit paradigm of cropping systems by often creating new ways of capturing value. A study by Djian-Caporalino et al. (2019) focused on evaluating several different cropping systems to understand the impact to sustainable management of root-knot nematodes in vegetable systems. By combining cultural practices, including planting nematicidal cover crops, with resistant varieties of vegetables, they found several solutions that were amenable to growers. Djian-Caporalino et al. (2019) established that more work was needed to gain efficiency for farmers, suggesting that future work should focus on innovative application of science. Using science to understand agroecological systems is critical, but understanding the costs that translate science in practical tools or methods is challenging. Economic analyses as we discussed in the previous section can help identify the areas where we can allocate resources that not only provide the best opportunity to create significant innovation but also generate profitability. Understanding the technical characteristics and attributes of new technologies as well as market opportunity (i.e. economic viability) requires intentional and dedicated thought.

A review by Fernández et al. (2018) highlights the shifts in greenhouse-based horticulture and how many of those factors are addressing emerging trends in technological and agricultural innovation. Greenhouse-based cropping systems provide a unique case to understand the benefits and costs of IPM in a more controlled environment. While the article focuses on physical structures, crops, water, and fertilization, it also discusses the need to continue innovating in IPM. The authors highlight innovations ranging from exclusion netting to biological control to a modern take on the traditional push-pull systems of IPM (see Harrison et al. (2019) for a background on push-pull systems and discussion of some agroecological management options more broadly than controlled environments). These innovations help agriculture accomplish the goal of becoming more sustainable, which is a critical task, given the importance of climate change, but they must also be profitable to the growers who must bear the cost of new operating models. Economic analyses to help understand the risks surrounding the establishment of innovative and sustainable technology, especially in the first few years, are critical. Early investment is needed to validate the science and applicability of innovative solutions and address the path from science to practical tools and methods. This early period of innovation and discovery may not result in near-term, economically viable changes for farmers and should be supported by stakeholders who would benefit from long-term adoption.

As technology and innovation is used more widely in agriculture, it will represent the opportunity to re-evaluate economic principles in IPM. As Rodriguez Sousa et al. (2020) demonstrated, value is strictly defined as a financial metric but is more encompassing to cover sustainability and quality of life. Similarly, Dara (2019) has suggested a new IPM paradigm that considers not only economic viability but also environmental safety and social acceptability. As we strive to be more sustainable, both in terms of social awareness and environment, technology and innovation offer the opportunity to collect data in new ways and apply analytics to solve complex problems. Financial viability should underlie solutions because farmers and growers must take on unique risks, so technology and innovation need to create financial benefit to support the move to broader economic criteria, including externalities.

There are many opportunities to change the design of systems, engage with stakeholders early, and define agricultural problems well to facilitate innovative solutions. The founders of IPM were aware of the challenges posed by simplifying complex systems and created important frameworks that helped turn data into action that was economical. Their initial definitions allowed for significant advancement with respect to any single control tactic within the IPM design. With the introduction of new data sources, the framework (design) of IPM needs to evolve to utilize the value of new data streams. However, the new data streams are also complex. Egea et al. (2020) discussed bioeconomy and circular economies that are driving change to greenhouse horticulture in Spain. They not only highlighted that much work has been done to create a space for sustainable and economic transformation but also signaled that there are significant needs to understand the science behind implementing bioeconomies. This work must be transdisciplinary incorporating empirical horticulture scientists, economists, data scientists and engineers, and many others. The next challenge for the stakeholders in horticulture is to take the new data streams and machine learning analytics to simplify farming operations or improve efficiency in a tractable way. For growers, the most tractable solutions reduce management complexity and increase the profit that allows them to better meet their future needs as well as society's.

6 Where to look for further information

There is a significant opportunity for individuals who wish to pursue and apply an economic perspective to IPM. There is a wide range of literature that can be used to build a good base from which many research directions can flourish.

A work that supports many of the ideas introduced here was written by the authors and is titled *The Economics of Integrated Pest Management of Insects*. Although the book focuses on insects, it broadly considers the elements of economic decision-making such as time horizon, stakeholder perspectives, and design versus control.

Agricultural economics is a discipline that will provide a rich area of exploration for those wishing to learn more. Of note, however, not all agricultural economists will consider the nuances of the economics of pest control. Notably, in the book edited by David Onstad titled *Insect Resistance Management*, Onstad and Paul Mitchell (University of Wisconsin-Madison, USA) wrote on the value of susceptibility of populations. Additional work by George Frisvold (University of Arizona, USA) and Terry Hurley (University of Minnesota) has looked at the applied economics of resistance management and integrated pest management, specifically area-wide pest management. Zack Brown (North Carolina State University, USA) is another researcher who has examined the economics of IPM and IRM.

7 References

Akin, S. and Kara, A. (2021). Factors effective on reaching goals and target groups of agricultural supports: a case study in Diyarbakir Province, Turkey. *Ciência Rural* 51(1). https://doi.org/10.1590/0103-8478cr20200260.

Alphey, N., Alphey, L. and Bonsall, M. B. (2011). A model framework to estimate impact and cost of genetics-based sterile insect methods for dengue vector control. *PLoS ONE* 6(10), e25384.

Anon. (2018). A national roadmap for integrated pest management. United States Department of Agriculture. Available at: https://www.usda.gov/media/press-releases/2018/10/24/usda-announces-update-national-road-map-integrated-pest-management (accessed 16 April 2021).

Asplen, M. K., Anfora, G., Biondi, A., Choi, D. S., Chu, D., Daane, K. M., Gibert, P., Gutierrez, A. P., Hoelmer, K. A., Hutchison, W. D., Isaacs, R., Jiang, Z., Kárpáti, Z., Kimura, M. T., Pascual, M., Philips, C. R., Plantamp, C., Ponti, L., Vétek, G., Vogt, H., Walton, V. M., Yu, Y., Zappalà, L. and Desneux, N. (2015). Invasion biology of spotted wing Drosophila (Drosophila suzukii): a global perspective and future priorities. *Journal of Pest Science* 88(3), 469–494.

Atallah, J., Teixeira, L., Salazar, R., Zaragoza, G. and Kopp, A. (2014). The making of a pest: the evolution of a fruit-penetrating ovipositor in Drosophila suzukii and related species. *Proceedings. Biological Sciences* 281(1781), 20132840.

Barzman, M., Bàrberi, P., Birch, A. N. E., Boonekamp, P., Dachbrodt-Saaydeh, S., Graf, B., Hommel, B., Jensen, J. E., Kiss, J., Kudsk, P., Lamichhane, J. R., Messéan, A., Moonen, A., Ratnadass, A., Ricci, P., Sarah, J. and Sattin, M. (2015). Eight principles of integrated pest management. *Agronomy for Sustainable Development* 35(4), 1199–1215.

Dara, S. K. (2019). The new integrated pest management paradigm for the modern age. *Journal of Integrated Pest Management* 10(1), 12.

Del Fava, E., Ioriatti, C. and Melegaro, A. (2017). Cost–benefit analysis of controlling the spotted wing drosophila (Drosophila suzukii (Matsumura)) spread and infestation of soft fruits in Trentino, Northern Italy. *Pest Management Science* 73(11), 2318–2327.

Djian-Caporalino, C., Navarrete, M., Fazari, A., Baily-Bechet, M., Marteu, N., Dufils, A., Tchamitchian, M., Lefevre, A., Pares, L., Mateille, T., Tavoillot, J., Palloix, A., Sage-Palloix, A.-M., Vedie, H., Goillon, C., and Castagnone-Sereno, P., and Castagnone-Sereno, P. (2019). Design and assessment of innovative Mediterranean vegetable cropping systems to manage root-knot nematodes description of the subject. *Biotechnologie, Agronomie, Société et Environnement* 23(1), 7–21.

Dominiak, B. C. (2019). Components of a systems approach for the management of Queensland fruit fly Bactrocera tryoni (Froggatt) in a post dimethoate fenthion era. *Crop Protection* 116, 56–67.

Dominiak, B. C., Wiseman, B., Anderson, C., Walsh, B., McMahon, M. and Duthie, R. (2015). Definition of and management strategies for areas of low pest prevalence for Queensland fruit fly Bactrocera tryoni Froggatt. *Crop Protection* 72, 41–46.

Egea, F. J., López-Rodriguez, M. D., Ona-Burgos, P., Castro, A. J., and Glass, R. (2020). Bioeconomy as a transforming driver in intensive greenhouse horticulture in SE Spain. *New biotechnology*, 61, 50–56. https://doi.org/10.1016/j.nbt.2020.11.010.

Federal Integrated Pest Management (IPM) Coordinating Committee (FIPMCC). (2018). A National Roadmap for Integrated Pest Management. Revised September 2018. PDF available at: https://ars.usda.gov/ARSUserFiles/OPMP/IPMRoad-Map-FINAL.pdf.

Fernández, J. A., Orsini, F., Baeza, E., Oztekin, G. B., Muñoz, P., Contreras, J. and Montero, J. I. (2018). Current trends in protected cultivation in mediterranean climates. *European Journal of Horticultural Science* 83(5), 294–305.

Figueiredo, E., Gonçalves, C., Duarte, S., Godinho, M. C., Mexia, A. and Torres, L. (2021). Risk assessment for tomato fruitworm in processing tomato crop-egg location and sequential sampling. *Insects* 12(1), 13.

Florec, V., Sadler, R. J., White, B. and Dominiak, B. C. (2013). Choosing the battles: the economics of area wide pest management for Queensland fruit fly. *Food Policy* 38, 203–213.

Harrison, R. D., Thierfelder, C., Baudron, F., Chinwada, P., Midega, C., Schaffner, U. and van den Berg, J. (2019). Agro-ecological options for fall armyworm (Spodoptera frugiperda JE Smith) management: providing low-cost, smallholder friendly solutions to an invasive pest. *Journal of Environmental Management* 243, 318–330.

Hoyt, S. C. and Gilpatrick, J. D. (1976). Pest management on deciduous fruits: multidisciplinary aspects. In: J. Lawrence Apple and R. F. Smith (Eds.), *Integrated Pest Management* (pp. 133–147). Springer, Boston, MA.

Lucier, G. and Davis, W. (2020). Vegetables and pulses outlook. United States Department of Agriculture. Available at: https://www.ers.usda.gov/topics/crops/vegetables-pulses/market-outlook/#:~:text=Vegetables%20and%20Pulses%20Outlook%3A%20December%202020&text=Total%20domestically%2Dgrown%20vegetable%20and,dry%20pea%2C%20and%20lentil%20crops (accessed 16 April 2021).

Mazzi, D., Bravin, E., Meraner, M., Finger, R. and Kuske, S. (2017). Economic impact of the introduction and establishment of Drosophila suzukii on sweet cherry production in Switzerland. *Insects* 8(1), 18.

National Research Council (1999). *Perspectives on Biodiversity: Valuing Its Role in an Everchanging World*. Washington, DC: The National Academies Pressd.

National Research Council (2005). *Valuing Ecosystem Services: Toward Better Environmental Decision-Making*. National Academies Press.

Nikolouli, K., Sassù, F., Mouton, L., Stauffer, C. and Bourtzis, K. (2020). Combining sterile and incompatible insect techniques for the population suppression of Drosophila suzukii. *Journal of Pest Science* 93(2), 647–661.

Oliveira, C. M., Auad, A. M., Mendes, S. M. and Frizzas, M. R. (2014). Crop losses and the economic impact of insect pests on Brazilian agriculture. *Crop Protection* 56, 50–54.

Onstad, D. W. (Ed.) (2013). *Insect Resistance Management: Biology, Economics, and Prediction*. London: Academic Press.

Onstad, D. W. and Crain, P. (Eds.) (2019). *The Economics of Integrated Pest Management of Insects*. CABI Publishing.

Onstad, D. W., Bueno, A. D. F. and Favetti, B. M. (2019). Economic thresholds and sampling in integrated pest management. *The Economics of Integrated Pest Management of Insects* 122.

Rossi, R. (2019). The EU fruit and vegetable sector: main features, challenges and prospects. European Parliament Think Tank. Available at: https://www.europarl.europa.eu/thinktank/en/document.html?reference=EPRS_BRI(2019)635563 (accessed 16 April 2021).

Santoiemma, G., Tonina, L., Marini, L., Duso, C. and Mori, N. (2020). Integrated management of Drosophila suzukii in sweet cherry orchards. *Entomologia Generalis* 40(3), 297–305.

Sciarretta, A., Zinni, A. and Trematerra, P. (2011). Development of site-specific IPM against European grapevine moth Lobesia botrana (D. & S.) in vineyards. *Crop Protection* 30(11), 1469–1477.

Sousa, A. R., Parra-López, C., Sayadi-Gmada, S., Barandica, J. M. and Rescia, A. J. (2020). A multifunctional assessment of integrated and ecological farming in olive agroecosystems in southwestern Spain using the analytic hierarchy process. *Ecological Economics* 173, 106658.

Sowell, T. (2014). *Basic Economics*. Hachette, UK.

Standing Committee on Agriculture and Resource Management (1996). *Code of Practice for Management of Queensland Fruit Fly*. Department of Primary Industries, Canberra.

Stern, V. M. R. F., Smith, R. F., van den Bosch, R. and Hagen, K. S. (1959). The integration of chemical and biological control of the spotted alfalfa aphid: the integrated control concept. *Hilgardia* 29(2), 81–101.

Walsh, C. S., Ottesen, A. R., Newell, M. J., Hanson, J. C. and Leone, E. H. (2011). The effect of organic and conventional management programs on apple and Asian pear tree growth, productivity, expenses and revenues in a hot, humid climate. *Acta Horticulturae* 903, 665–672.

Wesseler, J. H. H., Bonanno, A., Drabik, D., Materia, V. C., Malaguti, L., Meijer, M. and Venus, T. J. (2015). *Overview of the Agricultural Inputs Sector in the EU*. Brussels: European Union.

Westigard, P. H. (1979). Integrated pest management of insects and mites of pear. In: D. J. Boethal (Ed.) *Pest Management Programs for Deciduous Tree Fruits and Nuts* (pp. 151–202). Springer, Boston, MA.

Chapter 11

Encouraging integrated pest management uptake in horticultural crop production

Norma R. Samuel, Associate District Extension Director and Urban Horticulture Extension Agent, Institute of Food and Agricultural Sciences, University of Florida, USA; and Oscar E. Liburd, University of Florida-Gainesville, USA*

1 Introduction

Integrated pest management (IPM) is one of the most widely used terms in agriculture. The concept involves using various strategies in a management approach for protecting crops, landscape, ornamental plants, non-target organisms, and the environment. IPM utilizes tools such as effective monitoring and identification of pests and beneficial arthropods; economic thresholds; cultural or ecological management, which involves the use of preventative

* Oscar E. Liburd, oeliburd@ufl.edu, Professor of Fruit and Vegetable Entomology, Entomology and Nematology Department, University of Florida, Gainesville, USA.

http://dx.doi.org/10.19103/AS.2021.0095.13

tactics, modification of crop and pest environment, and host plant resistance; and biological control. The use of biorational pesticides is usually introduced as a last resort with the ultimate goal to protect the environment and make important economic decisions.

The concept of IPM has been around since the late 1950s. In the United States, IPM programs have saved over $1.3 billion in pesticide costs in selected orchard (oranges and almonds), field (cotton), and vegetable (processed tomatoes) crops (Mullen et al., 2005). However, it is still considered to be an innovation that must be reinvented and reintroduced to audiences to tackle traditional pests as they develop resistance to pesticides, to protect human and environmental health, and to address newly introduced pest species.

IPM as an innovation can be introduced to growers/farmers by their peers or by change agents, persons charged with influencing the adoption of an innovation. In the horticultural/agricultural world, change agents may be public sector (government) or non-governmental organization extension professionals or researchers, or representatives of input supply companies. There are many terms for extension services. This includes cooperative extension services, agricultural extension services, and rural advisory services; but regardless of the name, the main purpose is to use innovation(s) to improve the target audience's livelihoods and ultimately national productivity.

Rivera et al. (2001) indicated that the World Bank and the Food and Agriculture Organization (FAO) recognize extension as part of a larger system referred to as the Agricultural Knowledge and Information System for Rural Development or simply the Agricultural Knowledge System (AKS), which shows the interconnectivity between extension, research, and education, with the farmers being the common connection. Although this chapter will focus on the role of extension in encouraging uptake of IPM in horticultural crop production, it is important to keep in mind that research helps generate new IPM innovations or improve existing ones and that education creates new talent for the field of IPM.

The diffusion and adoption of IPM approaches have been the subject of research to determine which variables within the social system influence the process. Chitere (1998) studied the diffusion and adoption of IPM technologies of farmers in Kenya, and the rate of adoption between farmers participating in the project and non-participating farmers and found there was no direct correlation between uptake of IPM technologies and formal education. Exposure of farmers with varying levels of literacy to 'functional education' such as field trials, and connection to extension workers on the project, increased the rate and quality of adoption. In contrast, Muriithi et al. (2020) found that the rate of early adoption of IPM strategies for the suppression of mango fruit flies in East Africa increased with education level. This is consistent with the findings of Shiferaw et al. (2014) on the adoption of technologies.

The adoption of IPM information is critical for growers to obtain maximum benefits. In a recent article, Magarey et al. (2019) indicated that despite the many successes that have been achieved over the last 50 years, there are still significant challenges to the delivery and adoption of IPM information. They proposed a social ecological system to address some of the delivery and adoption issues. Rao et al. (2011) found that age, education, participation in community-based organizations, ability to recognize the pest insects, and farm size were important factors in studies to determine the adoption and impact of IPM technologies in India. The adoption of IPM techniques resulted in a reduction in pesticide usage and increased net returns for the growers.

Kaine and Bewsell (2008) reviewed numerous studies and found many inconsistencies in whether demographic factors influenced the rate of adoption of IPM practices. In studying apple growers in Australia, Kaine and Bewsell (2008) examined factors that affected the adoption of IPM practices. Contextual factors related to the situation on the farm – topography, climate, crop mix, orchard size, spatial isolation, pest pressure; available options for controlling pests; and markets – fresh, process, or export, affected IPM adoption. In contrast, age, education, and orchard size did not have any bearing on the pest management practices adopted. Rather, significant relationships existed between location and pest intensity and adoption of IPM practices. Since multiple variables are at play to influence the diffusion of IPM as an innovation, the extension professional who is the change agent should develop a well-designed program in line with the values and needs of the growers and other stakeholders within the social system. The use of a logic model to give a graphical representation of the new or existing program is advised (Suber, 2019).

In this chapter, we will (i) discuss Rogers' (1963) *Diffusion of Innovation Theory* as it relates to adoption of IPM practices, (ii) use the University of Wisconsin's (2009) logic model to demonstrate how logic models can be used to develop an IPM program plan, (iii) identify a few approaches used by extension agents to deliver IPM innovations, (iv) discuss the materials for building an IPM toolkit, (v) describe useful resources to aid in the diffusion of IPM technologies, and (vi) consider barriers to the adoption of IPM in horticultural crop production and how to overcome these barriers.

2 Rogers' *Diffusion of Innovation Theory* in relation to integrated pest management

Extension and advisory services are known for their success in disseminating innovative technologies to initiate human behavior change (Rogers, 1963). Frontline extension professionals typically enter the workplace with subject matter expertise related to their position, such as agronomy, horticulture, pest management, and little to no training in andragogy, adult learning, and behavioral change. An extension professional's role is to influence behavior change that would result in positive outcomes to improve people's livelihoods in the communities served. An understanding of Rogers' (1963) *Diffusion of Innovation Theory* would help the extension professional develop some perspective on strategically designing extension programs to effect intended change.

Rogers (2003) defined diffusion as 'the process in which an innovation is communicated through certain channels over time among the members of a social system'. Communication is a two-way flow of information between individuals to share knowledge. Diffusion involves sharing a new idea or innovation and is achieved not after a single interaction but through a series of interactions that eventually lead to adoption. Innovation has four main elements, as described by Rogers (2003).

- *Innovation:* It is an idea that may be perceived to be new. It could be a technology that has been around for a long time, but if introduced to an area where it was previously unknown, it becomes an innovation. The perception of the members of the social system with regard to the attributes of the innovation will affect the rate of adoption (Rogers, 2003).
    - *Relative advantage* – This determines to what extent is the innovation perceived to be better than the one currently being used. Overhead irrigation is widely used in some areas. The change agent must be able to convey to the grower the benefits of adopting micro-irrigation as an innovation, given the high initial cost of micro-irrigation equipment and installation. The added benefits include reduced incidence of diseases; fewer weeds; a smaller amount of water used, which can be a major selling point in drought-stricken areas; administering fertilizer through the irrigation system; and reduced water runoff.
    - *Compatibility* – Compatibility is the perceived fit of the innovation with the values and norms of the social system. Individuals within a community that values protecting the environment would be more likely to quickly transition from broad-spectrum insecticides to more targeted products that are less toxic, given similar price points and other variables.
    - *Complexity* – Complexity is the perception of ease or difficulty of use or implementation of the innovation. Innovations perceived to be easy to use will diffuse more quickly than those perceived to be difficult. It is better to introduce more difficult innovations in steps. In IPM actions, thresholds are used to determine when control is warranted. Instead of immediately introducing the concept, determine where the target audience is at with scouting and monitoring for different pests. use of sticky traps, how to assess the level of damage per plant, and so on. Then, combine all these measures to introduce the action threshold.

- *Trialability* - It is the ease with which the innovation can be trialed on a limited basis. The introduction of a pest-resistant plant variety can be trialed by a grower at the same time as the traditional variety, to compare results. The use of field trials helps to convey the practicability of a particular innovation.
- *Observability* - This relates to the visibility of the results of the innovation to members of the social system. The more visible the results, the higher the likelihood of the innovation being adopted. Extension uses field days as an observability tool for members of the community to see the results of field trials conducted on research stations and or farm demonstrations. This shows the innovation is not just theory but will work under local conditions.

- *Communication channels:* This refers to how messages get transmitted from one person to another. Communication channels targeting large numbers such as mass media or social media are effective in bringing awareness to the innovation, whereas interpersonal communication is more effective in realizing change in attitudes and eventual adoption of an innovation (Rogers, 2003).
- *Time:* The time variable in the diffusion process starts from the point where the person is first introduced to the innovation, the timeframe within which the individual(s) adopt the innovation (early or late), and the timeframe of adoption within the social system at large. Adopters can be grouped into five categories based on the rate of adoption of innovation (Rogers, 2003) (see Fig. 1).
  - *Innovators* - They are well connected outside of the social system, well informed, and risk-takers with available financial capital; approximately 2.5% of the population.
  - *Early adopters* - They are opinion leaders, well respected within the social system, share thoughts on innovation with near peers; 13.5% of the population.

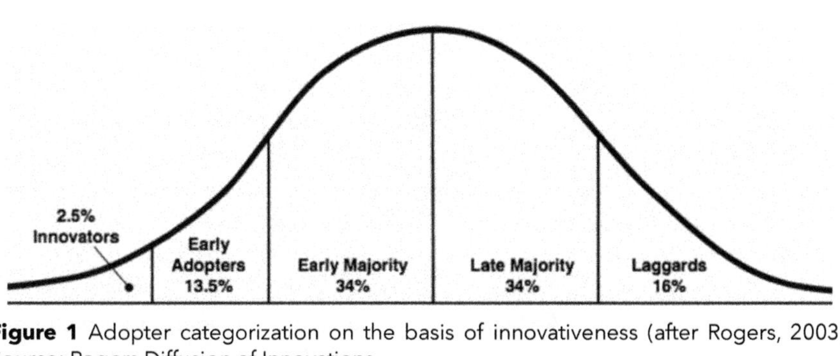

Figure 1 Adopter categorization on the basis of innovativeness (after Rogers, 2003). Source: Rogers Diffusion of Innovations.

  - *Early majority* - They adopt innovation before the majority of persons within the system, are thoughtful in willingness to adopt and there are few opinion leaders in this group; approximately 34% of the population.
  - *Late majority* - These are the skeptics, with limited finances. They adopt innovation because of pressures within the system; approximately 34% of the population.
  - *Laggards* - The last group to adopt an innovation. Use the past as a frame of reference; 16% of the population.
- *Social System:* This refers to a network of persons with common values or norms, such as entomologists within a department at a university or nursery growers within a specific locale. Opinion leaders and change agents drive adoption within a social system. Opinion leaders influence others' attitudes within the system to which they hold membership. Change agents on the other hand influence others' decisions towards adopting innovation and may be from outside of the system (Rogers, 2003).

3 Using a logic model to develop an integrated pest management program plan

The effectiveness or speed of uptake of IPM practices in horticultural crop production requires a well-designed program/series of activities geared to influence the adoption of practices by a particular target audience. The University of Wisconsin's (2009) logic model (Fig. 2) or a similar framework can be used as a model to develop a comprehensive IPM outreach strategy before conducting any activities. Logic model components include a situation statement, inputs, outputs, outcomes/impacts, evaluation, assumptions, and external factors.

Situation statement: It may be necessary to conduct a literature search and/or a needs assessment to gather information on the need(s) to be addressed. A demand-driven extension project is likely to get early buy-in of program participants, as they are involved in the decision-making process. The situation statement should be a brief statement outlining the problem/situation including SMART (Specific, Measurable, Achievable, Realistic, and Time Bound) program objectives.

Inputs: These are the items needed to execute the project. These may include human capital, finance, training, and equipment.

Outputs: There are two general sub-headings under output: (i) Activities - Several targeted activities should be conducted to elicit the expected outcomes/impacts. Activities may include developing materials, workshops, field days, and media outreach. (ii) Participation - Identify the

Program: Horticulture Integrated Pest Management
Situation: Agriculture is the 2nd largest industry in Florida. Over 90% of Florida's farms are small farms. Florida is also a leader in ornamental plant production, and with tourism being the number 1 industry in the state, landscape aesthetics are of great importance to maintaining the Florida image. Significant pest problems are associated with food, ornamental plants, and landscape products. Protection of plant, animal, and human health using Integrated Pest Management is challenging in Florida due to the mild climate and global agricultural markets that cause the state to be susceptible to introduction of invasive pests. More effective, safe, and sustainable IPM options are needed, along with training required for adoption of practices across the state. IPM employs the effective management of pests using a variety of techniques that minimize risks to human health and the environment while also maintaining or increasing the profitability of the grower.

Inputs	Outputs		Outcomes -- Impact		
	Activities	*Participation*	*Short*	*Medium*	*Long*
Funds: - Grants - In kind - Donations Extension professionals Industry partners Research	Needs assessment & collect baseline data Experiential IPM Educational Workshops Develop resource and presentation materials Institutional and on-farm research and field days Marketing – mass media, social media, websites, email lists, etc. Field visits Plant clinics	Nursery workers Landscapers Farmers & Farmer orgs Commodity groups	Increased knowledge on potential economic, health and environmental benefits of implementing an IPM program Increased understanding of IPM practices Increased ability to identify, select and implement/tailor IPM control tactics to their specific situation.	Increased use of reduced risk IPM practices	Reduced pesticide exposure and poisoning for workers and consumers Increase pesticide safety (measure reduced exposure, medical cost, time off work, proper calibration) Reduced cost of biorational pesticides and other IPM products Value added to IPM crop Restore critical areas by reducing nutrient runoff and leachate, reducing pesticide use
		Government agencies	Increased awareness of need to provide policies that support IPM	Implement a national/local IPM plan	
		Residents	Increased awareness of benefits of IPM; and willingness to support IPM-branded goods or services.	Purchase IPM branded goods or services.	

Assumptions:
Target audience will be willing to implement IPM practices. Recommended pesticide products will be available on the market for growers to purchase and use

External Factors:
Weather, policy changes, availability of IPM, materials, cost of IPM materials

Evaluation Questions: What do you want to know?					
Inputs	Activities	Target Audience	Short-Term	Medium Term	Long-Term
Were the inputs sufficient, timely? Did they meet the program goals? Were they sufficient given the situation?	Did all activities occur as intended? Quality of the intervention: appropriate content; useable resources; well-planned workshops, plant clinics, field visits?	Did the target audience participate? Who did not? What were their reactions? Who else was reached?	Based on pre and post-test scores and or follow-up surveys: Did knowledge increase? Dis understanding of IPM tactics increase? What else happened?	Is target audience implementing IPM tactics? Anything else?	Is there an increase in IPM product use? Has reports of pesticide poisoning decreased? Has the # of pesticide applications decreased? Cost savings? Social acceptance to IPM-branded products? Other benefits? Negative consequences?
Success Indicators: How will you know it?					
# faculty/staff invested. Amount of money. Delivery timetable.	# of workshops, plant clinics, field visits, research trials. # of educational materials and marketing produced. Participant feedback on quality	Actual vs. desired attendance. %/# attended each session. # of followers, visits, likes, engagement on social media pages	#/% with increased knowledge of specific IPM practices? Additional outcomes: +/-	#/% of target audience implementing IPM practices? #/% monitoring the effectiveness of IPM practices? Additional outcomes: +/-	#/% increasing use of IPM products? # of reports of pesticide poisonings/exposure? Sales trend of IPM-branded vs. conventional commodities. Dollars saved per grower? Additional outcomes: +/-

This logic model was adapted with permission from unpublished Pesticides, IPM & Environment logic model by Hannah Wooten, University of Florida (2020) and Water Quality Program: Logic model, evaluation questions, indicators from Ellen Taylor-Powell, (2002) University of Wisconsin Extension
*Examples of IPM logic models can also be found at: https://logicmodels.ipmcenters.org/samplelogicmodels.cfm

Figure 2 Logic model used to improve the planning and efficiency of extension programs.

group(s) who will be the target of this program. For example, growers, input suppliers, policy makers, and extension professionals.

Outcomes/impacts: There are three groupings of outcomes and impacts: short, medium, and long term. (i) Short-term outcomes relate to knowledge, attitude, skills, and abilities. For example, an increase in knowledge gain in IPM practices to control pests on a specific crop; a change in attitude, such as using broad-spectrum insecticides as a last line of defense; or increased skills in how to scout for pests. (ii) Medium-term outcomes relate to adoption or practice change. For example, a grower uses the least toxic pesticides first and broad-spectrum pesticides as a last resort or a grower regularly calibrates spray equipment after a series of interventions. (iii) Long-term outcomes refer to as impacts related to the social/health, economic, and environmental impacts of the practices adopted. For example, health – reduced incidence of pesticide poisoning; economic – lower cost of production and increased profits; and environmental – less pollution of waterways.

Evaluation: Monitoring and evaluation (M&E) of the IPM program should not be an afterthought. An M&E plan should be developed during program design. It is important to determine the outcome indicators that would be used to determine if targets are being reached. Evaluation can be formative – takes place while the project is ongoing or summative – takes place at the end of the project. A good evaluation plan utilizes a combination of formative and summative evaluation methods, and tweaks are made to the project along the way based on findings. Evaluation tools may include pre- and post-tests to assess knowledge gain for a workshop conducted, along with questions to assess instructors and facilities. Follow-up surveys are useful to determine which practices were adopted as a result of program participation. It is important to conduct the follow-up survey after program participants have had enough time to implement the practices taught. A follow-up survey conducted at 6 months, when practices are unlikely to have been adopted until 9 months later, would not yield valuable data. Observation of growers is also a valuable evaluation tool. Outcome indicators may be the number of IPM training sessions conducted, persons trained, and factsheets developed and/or distributed.

Assumptions: What assumptions can you make that may influence the program? An example of an assumption is that the grower will be willing to adopt IPM practices, or recommended pesticide products will be available on the market for growers to purchase and use.

External factors: These are factors beyond your control that can potentially influence the program. For example, the weather, policy changes, availability of IPM materials, cost of IPM materials, and so on.

Once the situation statement has been developed, the direction of completion of the logic model is a matter of preference. Some prefer working left to right, starting with inputs and ending with long-term outcomes/impacts. Others find it easier to figure out what the expected long-term impacts of the intervention would be and then work back to inputs. A well-developed logic model positions the extension professional to implement an IPM program that would encourage uptake of introduced and or modified IPM innovations.

4 Extension approaches to delivering integrated pest management technologies

A demand-driven (bottom-up) extension program where the activities are developed based on the needs identified by the target audience will be more impactful than a top-down program (Parkinson, 2009). Alternatively, a top-down approach is more effective to deal with crisis situations, such as a newly introduced invasive species. The extension approaches outlined here, workshops, field visits, field days, farmer field schools (FFS), diagnostic services, and E-extension, are by no means exhaustive and fall under the activities section of the logic model. Inclusion of multiple approaches in the program design positions the change agent/extension professional to have multiple interactions with the grower or potential adopter, to enhance diffusion and uptake of IPM innovation(s).

4.1 Workshops

Workshops are widely used as an educational outreach method in extension programs for groups of varying sizes. LeBude et al. (2017) found workshops to be an effective tool to influence IPM behavior/practice change. They conducted a series of 3-day 4-h workshops in the Southeastern United States for nursery growers and their follow-up surveys revealed that 60% of the growers adopted nursery IPM production practices as a result of attending the workshops (LeBude et al., 2017). The format of the workshop and delivery methods depend on the characteristics of the target audience, meeting location, and available technology. The extension professional should adapt the content and delivery style accordingly.

As a rule of thumb, speakers should keep workshop presentations and other communications to a grade level on par with that of the target audience. If the audience has formal education beyond high school or is familiar with the subject, the presentation could be more technical or advanced. A less-educated audience may need to be introduced to the topic in stages and with more visuals and hands-on than text. In some instances, the audience may be heterogeneous, with persons who are more highly qualified or knowledgeable about the subject than others. The use of cooperative learning where participants are taught in small groups is very effective with heterogeneous groups (Groseta & Myers, 2006). It is important to try to capture this type of information during the registration and prepare accordingly so that each attendee will benefit from the training.

Trainers should examine the IPM technologies that they plan to teach and determine which can be taught in a classroom setting and which require outdoor hands-on practice and experiential activities. Concepts such as sprayer calibration and scouting can be taught indoors with, for example, PowerPoint presentations and followed up in the field or greenhouse with a practical component. When teaching about pesticide labels, an exercise involving reading an actual label and answering questions about the label's information, including calculating application rates, is useful. Weed identification can include an exercise to identify weed specimens brought to the classroom or out in the field.

In the United States, individuals who apply pesticides must undertake specialized training and pass the required tests before a certificate is given permitting them to apply pesticides to designated areas. Rebek et al. (2013) used the conceptual and experiential learning approach to teach IPM to greenhouse growers in Oklahoma and tested students before and after the course to measure knowledge gain and intent to adopt IPM practices. This approach resulted in significant gains in both knowledge gain and intent to adopt IPM practices. Additional tips for successful workshop presentations are shown in Box 1.

Box 1 Additional Tips for Successful Workshop Presentations

- Promote at venues, on mass and or social media platforms likely used by the target audience to get maximum participation.
- Select a classroom with appropriate lighting and internet capability.
- Keep text on presentation slides to a minimum.
- Incorporate high-quality photos or other graphics. Cite the source of any photos you do not own. Get permission to use copyrighted photos. Photos from projects that you have worked on will make a stronger impact than those pulled from elsewhere.
- Keep the audience engaged by adding videos and hands-on activities. Allow time for participants to ask questions and or share experiences.
- Start and end on time. Do a trial run if needed.
- Schedule time for breaks if more than 2 h. Provide refreshments and or lunch based on length of workshop and available budget.
- Introduce technologies that are relevant to the local situation.
- Evaluate the session using evaluation tools such as pre- and post-session tests, post-session reflective surveys, observations by the presenter/facilitator.

Figure 3 University of Florida IPM Workshop participants getting hands-on insect identification experience.

4.2 Field visits

Face-to-face interaction is an integral part of every extension program. Field visits provide an opportunity for the extension professional to develop a personal relationship with the client (Fig. 3), see/understand first-hand conditions or context in the field or greenhouse, share new and or innovative technologies specific to the client's situation, and identify challenges and or opportunities to be addressed. The extension professional, especially those who lack field experience, should also use field visits as an opportunity to learn from the client.

4.3 Field days

Clients and/or professionals are invited to the field to see the latest research or innovative ideas such as new varieties or production practices being conducted by the client or researchers. Seeing that an application works in the field provides an opportunity for the client to gain that functional knowledge mentioned by Chitere (1998) that would encourage adoption and possible scaling up of practices being showcased. The most successful field days incorporate presentation(s) by an expert, for example, a researcher, allow time for farmer-to-farmer interaction, and include an evaluation (Comito et al., 2017).

4.4 Farmer field schools

According to the FAO (2021), FFS is an experiential learning and a non-formal adult teaching method that blends local knowledge and new innovations such as IPM. Trained facilitators regulate discussions throughout the growing season where participants discuss observations of practices being implemented in their fields and suitability to local conditions. Participants develop critical thinking skills and become empowered as a community through this process. FFS is used in 90 countries (FAO, 2021) and was the approach used for the implementation of a successful national IPM campaign in Indonesia (Thorburn, 2015).

Van den Berg et al. (2020) conducted a global survey to assess FFS design, implementation, and M&E. The study population was the regional and national focal points who were registered in the FAO's database who in turn liaise with their team to respond to the survey questions. The following guidance on FFS was provided (van den Berg et al., 2020).

- Success depends on having competent facilitators.
- Seek the input of farmers to develop curriculum content.
- Ensure adequate support is provided to FFS that addresses multiple crops and or livestock systems.
- Ensure the provision of adequate technical training for farmer facilitators.
- Have appropriate M&E that will monitor environmental, social, and economic impacts.
- Use M&E to maintain program integrity by ensuring adherence to original FFS educational methods, rather than lectures and demonstrations (van den et al., 2020).

4.5 Diagnostic services

Accurate identification of the pest is needed for an effective IPM program. Growers who lack skills or the resources to hire experts with identification skills can rely on the extension system to fill the gap and provide a combination of avenues to address the need.

4.5.1 Plant clinics

These are analogous to visiting the doctor and being asked a battery of questions to get to a diagnosis. Typically, when the extension professional(s) at the plant clinic receive a difficult sample to diagnose they have access to specialists with whom they can consult to receive a diagnosis for the client either immediately or as a follow-up contact. In the United States, many extension offices run plant clinics where growers can call, email, or bring in plant material or pest problems for identification and control recommendations. See Box 2 for information on the condition of plant samples for disease diagnosis and insect samples for identification.

According to Bently et al. (2018), the use of plant clinics in Africa, Asia, and Latin America is a relatively new extension approach. These clinics are held in extension offices, markets, or other areas frequented by farmers and staffed by extension workers called plant doctors. The growers receive verbal and written diagnoses, often referred to as prescription forms that reinforce the verbal information given by the plant doctors. In some instances, they are also provided with supporting documentation such as factsheets with instructions for control, to which they can later refer (Bently et al., 2018). Bently et al. (2018) emphasized the importance of providing written prescriptions in the local language spoken by the client to increase the likelihood of adoption.

In a study conducted on the use of plant clinics in Malawi, Nepal, and Costa Rica, Bently et al. (2018) found that the rate of adoption of cultural practices as a pest control measure was 54%. They also found that for the farmers who received recommendations for biorational pesticide applications, the adoption rate was 75%, indicating they are more likely inclined to adopt a spray recommendation over cultural control methods. They attributed the lower response to cultural control measures to the fact that in some instances they were already being used or not practical since they may be out of the farmer's control. Tambo et al. (2021) examined the likelihood of plant clinics improving the adoption of multiple IPM approaches to control the fall armyworm (FAW) by comparing the management practices of plant clinic users versus non-users in Rwanda and Zambia. They found that plant clinic users were more likely to use pesticides in combination with cultural and physical practices in the control of FAW, in comparison to non-users (Tambo et al., 2021).

Box 2 Condition of Plant Samples for Disease Diagnosis and Insect Samples for Identification

Plant Samples for Disease Diagnosis: Freshly collected samples are best. If the sample must be transported over long distances or collected a few days early, the grower should keep the sample in a plastic bag in a cooler with an icepack, or in a refrigerator, to keep it fresh. The sample should show the progression of the disease – from beginning to advanced symptoms. Very little can be determined from a sample that is dead. For samples with suspected root problems, such as root rot or nematodes, it is important to uproot the plant or bring in the entire container.

Questions to ask pertaining to the disease sample brought in for diagnosis include the following:

- Location found.
- Date symptoms were first noticed.
- Age of the sample collected.
- Number of plants affected and pattern/distribution of symptoms in the field.
- Plant parts affected. Which parts were affected first?
- Cultural practices utilized – fertilization and pesticide application dates; products applied, method and rate; irrigation method, times and amounts. Some symptoms caused by improper cultural practices can be easily mistaken for those caused by biotic factors/living organisms.

Supporting photos of diseased plants are very helpful for identification. Photos should include panoramic shots showing the entire field or section of the greenhouse where plants are growing to detect the pattern, shot of a single plant, and close-up photos showing the progression of symptoms.

Insect Specimens for Identification: These may or may not include plant material showing damage, frass/excrement. Specimens can be alive or dead. Soft-bodied insects should be placed in a vial with isopropyl alcohol (70% rubbing alcohol). Caterpillars placed directly in alcohol will turn black, making identification difficult. Thus, growers should drop caterpillars in boiling water that has been removed from heat and allowed to stand a few minutes before placing them in alcohol (Buss, 2019). If the specimen is determined to be a new occurrence in the state or country, another submission with a live specimen, including plant material, is usually requested by the laboratory. The live specimen can be reared out for all stages of development to be observed and identified.

Questions to ask pertaining to the insect sample(s) brought in for diagnosis include the following:

- Location found.
- Date first collected.
- Affected plant parts.
- Symptoms.

4.5.2 Laboratory diagnostic services

The frontline extension professional typically will refer a client to a university, government, commercial diagnostic laboratory, or an herbarium when in doubt on the identification of a pest. The client has to pay a fee for these services. Fees tend to be lower at university and government facilities.

Guidelines for collecting samples for submission to a laboratory are similar to those outlined for in-person identification. Each laboratory will have a

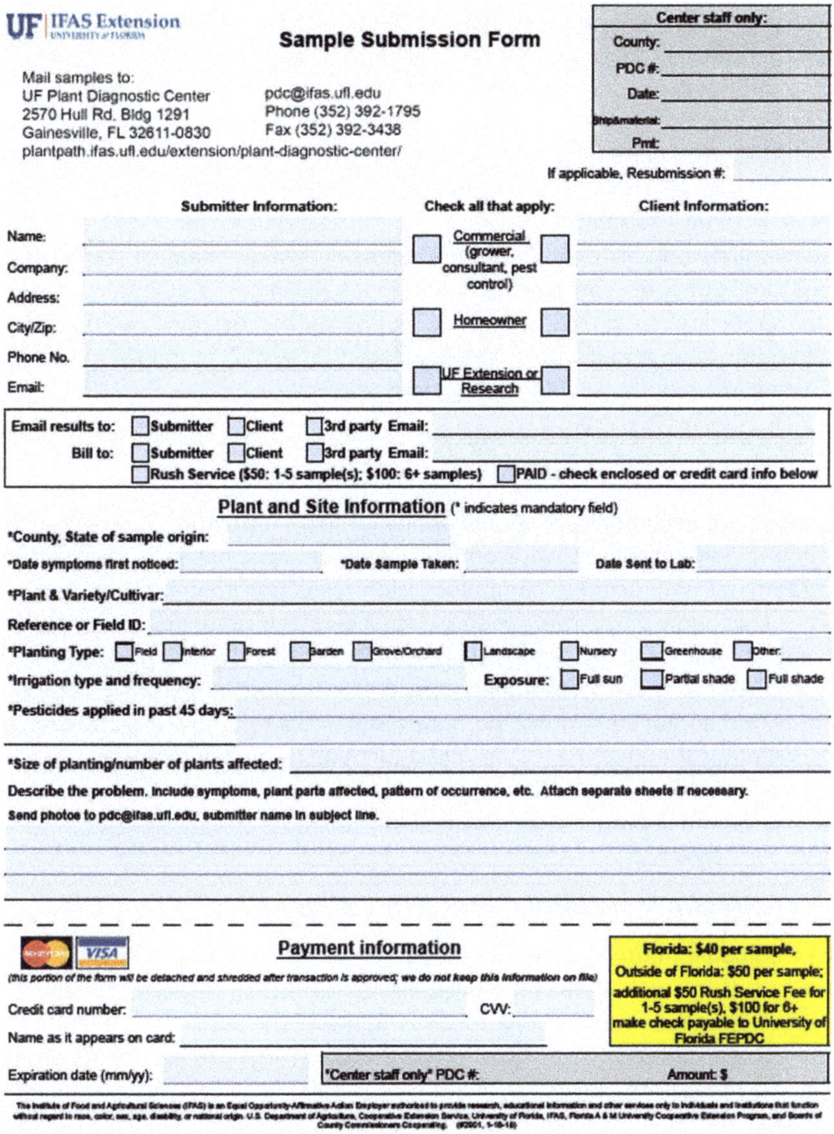

UF | IFAS Extension
UNIVERSITY of FLORIDA

Sample Submission Form

Center staff only:
County:
PDC #:
Date:
Ship&material:
Pmt:

Mail samples to:
UF Plant Diagnostic Center
2570 Hull Rd. Bldg 1291
Gainesville, FL 32611-0830
plantpath.ifas.ufl.edu/extension/plant-diagnostic-center/

pdc@ifas.ufl.edu
Phone (352) 392-1795
Fax (352) 392-3438

If applicable, Resubmission #:

Submitter Information:	Check all that apply:	Client Information:
Name:	Commercial (grower, consultant, pest control)	
Company:		
Address:		
City/Zip:	Homeowner	
Phone No.		
Email:	UF Extension or Research	

Email results to: Submitter Client 3rd party Email:
Bill to: Submitter Client 3rd party Email:
Rush Service ($50: 1-5 sample(s); $100: 6+ samples) PAID - check enclosed or credit card info below

Plant and Site Information (* indicates mandatory field)

*County, State of sample origin:
*Date symptoms first noticed: *Date Sample Taken: Date Sent to Lab:
*Plant & Variety/Cultivar:
Reference or Field ID:
*Planting Type: Field Interior Forest Garden Grove/Orchard Landscape Nursery Greenhouse Other:
*Irrigation type and frequency: Exposure: Full sun Partial shade Full shade
*Pesticides applied in past 45 days:
*Size of planting/number of plants affected:
Describe the problem. Include symptoms, plant parts affected, pattern of occurrence, etc. Attach separate sheets if necessary.
Send photos to pdc@ifas.ufl.edu, submitter name in subject line.

Payment information
(this portion of the form will be detached and shredded after transaction is approved; we do not keep this information on file)

Florida: $40 per sample, Outside of Florida: $50 per sample; additional $50 Rush Service Fee for 1-5 sample(s), $100 for 6+ make check payable to University of Florida FEPDC

Credit card number: CVV:
Name as it appears on card:
Expiration date (mm/yy): *Center staff only* PDC #: Amount: $

The Institute of Food and Agricultural Sciences (IFAS) is an Equal Opportunity-Affirmative Action Employer authorized to provide research, educational information and other services only to individuals and institutions that function without regard to race, color, sex, age, disability, or national origin. U.S. Department of Agriculture, Cooperative Extension Service, University of Florida, IFAS, Florida A & M University Cooperative Extension Program, and Boards of County Commissioners Cooperating. (#2001, 1-18-18)

Figure 4 Plant Disease Diagnostic Form. Source: University of Florida Plant Diagnostic Center.

submission form with detailed information on how to collect and package the sample for submission (Fig. 4). Questions on the form should be answered fully as the information provided will aid in identification and help to determine the conditions under which the pest thrives.

Laboratories generally have special guidelines for collecting and sampling newly introduced invasive species as they may be under regulatory control. Some public laboratories will offer free diagnostic services for the identification of newly introduced invasive species. Early identification and eradication are needed for these pests.

It is advised that plant samples and live specimens be mailed early in the week to ensure samples do not sit at the post office over the weekend, as quality will deteriorate over time. If the extension worker's name is provided on the submission form, some public laboratories will inform them of the findings at the same time the grower is notified. This serves the following two purposes: (1) to notify them about a pest in their jurisdiction and (2) to follow up with the grower in developing an IPM strategy for the pest.

4.5.3 Distance diagnostics through digital imaging

The use of this technology became popular in the United States in the late 1990s. County extension offices were outfitted with stereoscopes and microscopes with cameras connected to computers. High-resolution images taken on these devices at county offices were then sent to a central location for diagnosis and recommendations. Plant pathologists, entomologists, weed scientists, and so on, connected to the system are notified of the submission and a response is provided sometimes within the hour (Hale, n.d.a,b). Another advantage of this system is that the images are used to create a digital library that can be accessed by extension professionals, growers, farmers, and master gardener volunteers who are registered to access the library.

4.6 E-extension and virtual communication

The traditional mode of extension program delivery relies heavily on one on one or small group interactions between the extension professional and the client(s). This limits the clientele reach and, therefore, the potential impact of the work of extension. In situations where there are large groups of farmers and only a few extension personnel, this can be problematic. For example, in India, it is estimated that there are 2879 farmers to 1 extension worker (Saravanan and Bhattacharjee, 2017). E-extension is a virtual digital platform that delivers information to a broad clientele group in a short period of time. In many instances, it is given in the form of a webinar. This allows a speaker from the host institution to share audio and visual (PowerPoint presentations, videos, and webpages) with audiences that can be located anywhere in the world. If the webinar is advertised properly, the audience can be huge. The disadvantage is that it requires an internet connection as a medium to transfer information from the devices of the extension professional to that of the clientele. Thus, it is important for the extension professionals to know the digital tools available to their clientele and the digital penetration. The estimated population of a country or region and the estimated number of persons using the internet is used to calculate the digital penetration of that country or region (Internet World Stats, 2020). See tips for successful virtual extension training shown in Box 3.

The average global digital penetration is 63.2%. Regions above the world average are North America, 90.3%; Europe, 87.2%; Latin America and the Caribbean, 71.5%; the Middle East, 70.8%; and Oceania, 67.7%. Regions below the world average are Asia, 59.5% and Africa 47.1%. However, Asia and Africa had the first and third highest numbers of persons, respectively, using the internet (Internet World Stats, 2020). Internet World Stats and similar websites provide digital penetration rates for each country in the world. The extension professionals should be aware that irrespective of their national digital penetration numbers, the digital penetration rate for the locale they serve can be higher or much lower than the national average. Access to and the availability of a reliable internet connection will have an impact on the virtual communication tools used for the delivery of IPM innovations.

Given that the use of the internet is so widespread globally, access to digital tools would enhance the capacity of the extension worker to reach a larger number of farmers and the frequency of the interventions compared to traditional methods. The use of information and communication technologies such as emails, social media platforms (e.g. Facebook, Twitter), texts, group chats (e.g. GroupMe, WhatsApp), and plant and pest identification apps have opened new avenues for extension to expand its reach to communicate with clients to provide rapid responses to pest identification and other questions. These tools can be utilized for individual and or group communication to push out answers to questions posed, pest alert notifications, and newsletter updates.

Saravanan and Bhattacharjee (2017) provide an extensive listing of social media platforms and examples with descriptions and tips for integrating these tools into extension programming. Saravanan et al. (2015) outlined the advantages and provided a list of opportunities and challenges. Advantages include capitalizing on a technology already being utilized for communication with family and friends; cost-effectiveness, expanded reach, and accessibility; user-generated content; platforms provide means to track clientele contacts. Challenges cited include the lack of skilled professionals to monitor and or maintain social media platforms; quality control as all users may be able to provide input on a post, some of which may not be accurate; meeting the needs of a heterogeneous audience; issues at organizational level regarding the use of social media platforms; and difficulty in measuring program impacts (Saravanan et al., 2015).

Another virtual communication tool is the virtual delivery of educational workshops using platforms such as Zoom and Teams. Its use by extension professionals exploded in 2020 due to the global pandemic. Virtual workshops allow for continued interaction with extension clientele during the pandemic lockdowns. The feedback received on the use of this tool is mixed. Some clients appreciate not having to travel to another location to attend training sessions and being able to access sessions later if they were recorded. On the other hand, clients in rural communities felt they were being left behind because of a lack of strong enough bandwidth or a lack of the necessary equipment to participate. These virtual tools are expected to become commonplace in extension. However, extension professionals must be strategic in their use and still offer in-person workshops on the same topics to ensure that limited-resource clientele have access to the information.

Box 3 Tips for Successful Virtual Extension Trainings

The previous tips provided for the preparation of slide presentations for in-person classes apply here.

- Provide detailed step-by-step instructions to participants on how to join the session. Suggest early log in to troubleshoot problems. Provide a number or email address to contact if having problems logging in.
- Enable the waiting room. This allows time for presenter(s) and facilitators to go over last-minute details.
- The introduction should include a slide on webinar etiquette to outline expectations: keeping videos and cameras on or off, placing questions in the chat, etc.
- Collaborate with at least one other person to act as a facilitator and provide technical support and monitor the chat. If the facilitator is familiar with the subject being taught, send a copy of the presentation a day or two beforehand for them to become familiar with the presentation. This is especially useful if the presenter loses internet connection, then the facilitator can progress with the training instead of bringing the session to an abrupt close.
- Keep sessions short. If possible, break full-day workshops into two to three hour sessions over multiple days.
- Utilize breakout rooms, whiteboard, polls, and other tools within or compatible with the platform as additional methods of engaging the audience. Use these tools regularly within the presentation for continued engagement. The polls are also useful to determine the level of understanding of a topic and to capture data needed for reporting.

5 Building an integrated pest management toolkit

There are a few items every horticulturist or extension professional should try to have on hand to aid in identification of pest problems in the field or to collect samples for identification later. The use of some of the tools on this list as incentives for attending training provides an opportunity for growers to develop their own IPM toolkit and encourage adoption of IPM practices.

- *Hand lens* - Also referred to as a loupe, these are available in different magnifications and lenses in plastic or glass. Glass lenses are scratch resistant and provide a better quality image. 10× magnification is preferred for field use for the identification of small insects (Skvarla, 2017).
- *Digital or smartphone cameras* - These are ideal for taking photos that can be used for identification.
- *Magnifiers* - Relatively inexpensive magnifiers can be purchased online and are attached to cellphone cameras to provide close-up details on small insects.
- *Ruler* - A ruler is useful to provide scale to the sample. The ruler is placed beside the insect or the weed to give the person making the identification an idea of the size. A penny or some other commonly used object is also useful to provide an idea of the size of the specimen.
- *Alcohol* - Isopropyl alcohol or 70% rubbing alcohol is used for preserving soft-bodied insects or to sterilize tools.
- *Vials* - The vials are of assorted sizes with corks or stoppers for soft-bodied insects.
- *Envelopes* - Small white envelopes are handy for securing moths to keep the scales intact, so the patterns are visible for identification.
- *Plastic bags* - Sandwich and gallon bag sizes are useful for plant cuttings.
- *Secateurs or pocketknife* - These are used to cut specimens such as diseased or insect-infested branches. These are cleaned by sterilizing with isopropyl alcohol or 10% bleach solution after each use.
- *Killing jars* - These are used to kill insects, such as beetles, wasps, and flies after collection. Killing jars can be easily made by adding a 1-inch layer of plaster of Paris paste to the bottom of a wide-mouthed glass jar and allowing it to dry/harden. Enough ethyl acetate or nail polish remover is poured into the jar to saturate the plaster of Paris. A layer or tissue or other paper is placed in the jar to prevent the insects from coming into direct contact with the moist plaster of Paris. The jar is tightly closed and a label is affixed to the jar with the word 'POISON' (Purdue University Extension-Entomology, n.d.).
- *Sweep nets* - These are used for collecting insects in the field and are available in muslin or sailcloth.

- *Tweezers* – Fine point tweezers are useful for picking up or manipulating small insects, moths, and butterflies.
- *Permanent marker* – A brand with ink is selected that is indelible in alcohol to prevent the label from fading over time.
- *Hand spades* – These are used for collecting soil samples or digging up small plant material.
- *Brown paper bags* – These are ideal for collecting soil samples.

6 Useful resources to aid in the diffusion of integrated pest management technologies

- *Extension professionals* – Many extension professionals publish their work online in journal articles, as success stories, presentations, and so on. The concept of not reinventing the wheel holds true in this line of work. Extension professionals are thrilled when others acknowledge that they have done excellent work and ask permission to use it as it is or modify it to meet a local situation. Follow your institution's protocol for requesting permission to use someone else's work and provide recognition to the original author as agreed.
- *Factsheets* – Many extension offices produce factsheets or bulletins to provide detailed information on a single topic or theme. Factsheets may be for a specific location or a large geographic area.
- *Pest ID decks or cards* – Laminated individual cards or cards grouped and held together by a ring are handy for use in the field. No internet connection is needed, and they can easily be wiped clean if soiled.
- *Apps* – There are IPM-related apps available to identify pests and plants, how-to videos, newsletters, and so on, for use on cellular devices, computers, and tablets at no cost or for a fee.
- *Books*: Numerous well-written books are available on the pests of various horticultural crops and control strategies, insects, diseases, weeds, and IPM. The Flint (2012) book on *IPM in Practice – Principles and Methods of Integrated Pest Management* is highly recommended for extension professionals and growers. Publication 3418. University of California, Oakland California.
- *Internet* – The internet has a vast amount of resources that can be utilized to develop/integrate into an IPM plan. It is important for one to be able to conduct targeted online searches and discern which information is factual and unbiased. When conducting an internet search, enter topic keywords or the key characteristics of the item to be identified into the search engine. Adding site:edu narrows the search to only sites ending in edu. These are generally reliable sources of information. Some specific tools to search for on the web to aid in developing an IPM plan include

action thresholds, pest identification, pesticide information, IPM products/tools, weather, and IPM logic models.

7 Barriers to adoption of integrated pest management in horticulture crop production and how to overcome them

The prevention of pest insects and plant pathogens becoming a problem that limits production in a cropping system, together with protecting human and environmental health, is the main goal of IPM. According to Rogers (2003):

> [d]iffusion scholars have found relative advantage to be one of the strongest predictors of an innovation's rate of adoption. "Relative advantage" is a ratio of the expected benefits and the costs of adoption of an innovation. Subdimensions of relative advantage include economic profitability, low initial cost, a decrease in discomfort, social prestige, a saving of time and effort, and immediacy of reward. This latter factor explains in part why preventative innovations generally have an especially slow rate of adoption, ... A preventative innovation is a new idea that an individual adopts now in order to lower the probability of some unwanted future event... The relative advantage of preventative innovations is difficult for change agents to demonstrate to their clients, because the advantages occur at some future and unknown time and may not happen at all. Thus, the relative advantage of a preventative innovation is highly uncertain. (p. 233, para. 2)

In essence, individuals may be hesitant to adopt an innovation for benefits they perceive may not be realized. In addition to determining the level of IPM knowledge of the target audience, the change agent should also assess their attitudes towards IPM and possible barriers to adoption. Recommended IPM strategies should be within the limitations of their situation to facilitate a faster rate of adoption. Possible barriers to IPM adoption include the following.

7.1 Lack of trust

This can be a challenge especially for a new extension professional who is not known to the community or entering a community where relationships have been soured in the past. It is imperative that the change agent commit time to relationship building, identify and connect with gatekeepers within the social system, and or identify areas where they share commonalities and build upon that. One of the authors can recall their first field visit. Upon arrival to the farm in a government-marked vehicle, the farmer said, 'once you are from the government, I don't want to talk to you'. Luckily, there were some diseased tomatoes at the edge of the field, so that was used to strike up a conversation. The farmer indicated that he had been having a problem with them for a while, and the author provided him with an IPM plan that included cultural practices, resistant varieties, and pesticide options, for how to control and prevent in the future. The farmer later reported back on the success of the IPM plan and continued to seek assistance.

7.2 Limited financial resources

The financial investment needed for the adoption of IPM practices varies from minimal to significant. Some practices may be a simple modification of existing practices such as cutting back on the frequency of watering or changing watering times. Others may require financial capital such as additional inputs or added labor to, for example, scout crops could be a barrier to adoption. Muriithi et al. (2020) indicated that farming households with limited income or that are credit constrained are usually in the late adopter category. When working with growers with limited resources, the innovation recommended should be within financial reach or they will not be able to afford to adopt it.

7.3 Unavailability of integrated pest management products

The unavailability of products such as biorational pesticides needed to implement recommended IPM activities is a possible barrier. For example, a recommendation for the use of biorational pesticide options requires that there be availability of these products on the market. Thus, if broad-spectrum insecticides are more widely available and promoted by input suppliers, then that can be a hindrance to adoption.

7.4 Lack of integrated pest management knowledge

Individuals with little or no knowledge of IPM are unable to adopt the innovation due to a lack of awareness of its existence. The role of the change agent is to assess the level of IPM knowledge of the target audience and to provide appropriate educational outreach strategies to facilitate easy adoption. This becomes more difficult in areas with a high ratio of farmers to extension professionals. Jors et al. (2017) identified the lack of extension services as a barrier to the adoption of IPM by small farmers in Bolivia.

7.5 Lack of integration of indigenous practices

The goal or intention of the change agent should not be one of total disregard for indigenous practices. When working with the target audience to develop an IPM plan, it is important to study local practices to determine their effectiveness and integrate the most effective ones into the plan or trial them with an improved innovation. This shows the grower that the change agent does not consider their practices to be inferior; thus, the grower would be more willing to adopt recommended IPM strategies.

7.6 Competing sources of information or misinformation

The goal of the extension service is to provide research-based, non-biased information to the target audience to improve their livelihood. However, growers are constantly bombarded with multiple sources of information from suppliers (e.g. fertilizer and chemical company representatives), nongovernmental organizations, and the internet where some of the information may not be research-based. This information dilemma can be a difficult barrier to overcome as the grower needs to decide whose information is more trustworthy. The change agent will need to educate the grower on how to identify evidence-based information.

7.7 Access to technology

This may also be a barrier to growers with limited resources and/or those who reside in rural areas. For example, it would be futile to introduce an IPM innovation that relies on the use of the internet if growers operate in an area with unreliable internet connection.

7.8 Lack of supporting government policies

Thorburn (2015) documented the rise of the use of IPM in Indonesia through a number of intentional strategies together with a National IPM Program from 1993 to 1999, which was supported by government legislation. Subsequent deregulations resulted in the rise of input suppliers flooding the market with broad-spectrum insecticides which have since led to the demise of IPM achievements in Indonesia and an explosion of pest problems. Jors et al. (2017) mentioned that supportive government policies such as tax incentives help to speed adoption and allow for mainstreaming of IPM.

7.9 Other extenuating circumstances

The extension professional may encounter members within the social system who are facing issues that need to be addressed before they can consider adopting a new innovation. This may be the case of persons who fall in the laggards' section of the diffusion curve. Issues such as illiteracy or household food insecurity may first need to be addressed. The change agent should have an idea of the resources available in the community to direct the target audience for help. Persons who fall in this category may be willing to adopt the innovation but have other needs they perceive to be more pressing.

8 Conclusion

IPM will continue to be a major player in the production of horticultural crops for the foreseeable future. With limited reliance on pesticides, IPM provides opportunities for growers to produce crops using sustainable techniques, including biological and cultural control strategies. Growers need to receive adequate information and be able to adopt good IPM practices so that they can improve their production, lower their input costs and pesticide use, and increase their profitability. The benefits derived from effective IPM programs include billions of dollars saved in pesticide costs and the negative impacts associated with the use of conventional pesticides. It should be noted that extension educational outreach strategies are the vehicle that drives the adoption of IPM innovation. Implementation of sound IPM strategies requires a firm commitment on behalf of the extension professional and strong farmer engagement. Multiple techniques are being utilized to increase the dissemination of information, including workshops, field days, FFS, field visits, and diagnostic services, most of which have been discussed in this chapter. The use of communication technologies including emails, social media platforms (e.g. Facebook, Twitter), texts, group chats (e.g. GroupMe, WhatsApp), and plant and pest identification apps has opened new avenues for extension to expand its reach to clients, thereby increasing the potential adoption of IPM innovations.

9 Where to look for further information

Rogers (1963) Diffusion of Innovations theory has been widely used to study the adoption of integrated pest management practices in horticultural production. The exploration of other theories into why growers adopt IPM practices may provide additional insights that could influence the strategies utilized by change agents. Future research should also be directed at how artificial intelligence and technology can be used to aid grower adoptions. For instance, extensive research is being conducted on the use of 'smart phones' and how field scouts and growers can use these devices to aid in crop monitoring for pests, diseases, and physiological stress. These techniques can improve efficiency, lower input costs, and increase overall profitability. Finally, more research is needed to determine the challenges faced by the change agents in providing outreach to growers related to IPM.

A multitude of public, private, and civil society organizations are involved in the development, transfer, implementation, and evaluation of IPM innovations.

Organizations such as the Food and Agriculture Organization (FAO) of the United Nations and the Global Forum for Rural Advisory Services (GFRAS) promote a participatory or bottom-up approach to tackle issues within the agricultural innovation system. Identifying and engaging the key persons (farmers, extension professionals, educators, researchers, input suppliers, government officials, etc.) and or organizations along the spectrum from innovation to adoption of IPM technologies would increase the likelihood of such efforts being more successful than agencies operating in silos without the involvement of growers or farmers from the onset.

10 References

Anon. 2020. Internet world stats: usage and population statistics. [Online] Available at: https://www.internetworldstats.com/stats.htm [Accessed 10 January 2022].

Bently, J. W., Danielsen, S., Phiri, N., Tegha, Y. C., Nyalugwe, N., Neves, E., Hidalgo, E., Sharma, A., Pandit, V. and Sharma, D. R. 2018. Farmer responses to technical advice offered at plant clinics in Malawi, Costa Rica and Nepal. *International Journal of Agricultural Sustainability* 16(2), 187–200.

Buss, L. J. 2019. *Insect Identification Service*. Gainesville: University of Florida.

Chitere, O. P. 1998. Diffusion and adoption of farm technologies among limited-resource farmers: experiences from ICIPE/UNECA integrated pest management project in Western Kenya. *International Journal of Pest Management* 44(2), 49–52.

Comito, J., Haub, B. C. and Stevenson, N. 2017. Field day success loop. *Journal of Extension* 55(6), 3.

Food and Agriculture Organization 2021. Global farmer field schools platform - what are FFS? [Online] Available at: http://www.fao.org/farmer-field-schools/overview/en/ [Accessed 11 January 2021].

Groseta, K. J. and Myers, B. E. 2006. EDIS. [Online] Available at: https://edis.ifas.ufl.edu/pdf/WC/WC06200.pdf [Accessed 10 January 2022].

Hale, F. n.d.a Department of entomology and plant pathology - distance diagnostics: distance diagnosis through digital imaging. [Online] Available at: https://ag.tennessee.edu/EPP/Pages/TPPSN/More-about-Distance-Diagnostics.aspx [Accessed 10 January 2022].

Hale, F. n.d.b University of Tennessee institute of agriculture. [Online] Available at: https://ag.tennessee.edu/epp/pages/tppsn/more-about-distance-diagnostics.aspx [Accessed 16 May 2021].

Jors, E., Aramayo, A., Huici, O., Konradsen, F. and Gulis, G. 2017. Obstacles and opportunities for diffusion of integrated pest management strategies reported by Bolivian small-scale farmers and agronomists. *Environmental Health Insights* 11, 1178630217703390.

Kaine, G. and Bewsell, D. 2008. Adoption of integrated pest management by apple growers: the role of context. *International Journal of Pest Management* 54(3), 255–265.

LeBude, A., Fulcher, A., Dubois, J., Braman, K., Chappell, M., Chong, H. H., Derr, J., Gauthier, N., Hale, F., Klingeman, W., Knox, G., Neal, J. and Windham, A. 2017.

Experiential nursery pest management workshop series to enhance grower practice adoption. *American Society for Horticultural Science* 27(6), 772–781.

Magarey, R. D., Chappell, T. M., Trexler, C. M., Pallipparambil, G. R. and Hain, E. F. 2019. Social ecological system tools for improving crop pest management. *Journal of Integrated Pest Management* 10(1), 2.

Mullen, J. D., Alston, J. M., Sumner, D. A., Kreith, M. T. and Kuminoff, N. V. 2005. The payoff to public investments in pest-mangement R&D: general issues and a case study emphasizing integrated pest management in California. *Review of Agricultural Economics* 24(4), 558–573.

Muriithi, B. W., Gathogo, N. G., Diiro, G. M., Mohamed, S. A. and Ekesi, S. 2020. Potential adoption of integrated pest management strategy for suppression of mango fruit flies in East Africa: an ex ante and ex post analysis in Ethiopia and Kenya. Agriculture 10(7), 278. Available at: https://www.mdpi.com/2077-0472/10/7/278. Accessed January 6, 2021.

Parkinson, S. 2009. When farmers don't want ownership: reflections on demand-driven extension in Sub-Saharan Africa. *Journal of Agricultural Education and Extension* 15(4), 417–429.

Purdue University Extension-Entomology. n.d. Youth and entomology – handling insects. [Online] Available at: https://extension.entm.purdue.edu/401Book/default.php?page=handling_insects [Accessed 16 May 2021].

Rao, C. A. R., Rao, M. S., Srinivas, K., Patibanda, A. K. and Sudhakar, C. 2011. Adoption, impact and discontinuance of integrated pest management technologies for pigeon pea in South India. *Outlook on Agriculture* 40(3), 245–250.

Rebek, E. J., Schnelle, M. A. and Kelsey, K. D. 2013. Adoption of integrated pest management practices among greenhouse producers: a case study for experiential learning. *Journal of Extension* 51(6). Available at: https://tigerprints.clemson.edu/joe/vol51/iss6/21/. Accessed January 6, 2021.

Rivera, W. M., Qamar, K. M. and Crowder, L. V. 2001. *Agricultural and Rural Extension Worldwide–Options for Institutional Reform in the Developing Countries*. Rome: Food and Agriculture Organization.

Rogers, E. M. 1963. The adoption process. Part 1. *Journal of Extension*, 16–22. Available at: https://archives.joe.org/joe/1963spring/1963-1-a3.pdf. Accessed January 6, 2021.

Rogers, E. M. 2003. *Diffusion of Innovations* (5th edn.). New York: Free Press.

Saravanan, R. and Bhattacharjee, S. 2017. *Social Media for Agricultural Extension*. Hyderabad: National Institute of Agricultural Extension Management (MANAGE).

Saravanan, R., Suchiradipta, B., Chowdhury, A., Hall, K. and Odame, H. H. 2015. *Social Media for Rural Advisory Services*. Lindau: Global Forum for Rural Advisory Services.

Shiferaw, B., Kassie, M., Jaleta, M. and Yirga, C. 2014. Adoption of improved wheat varieties and impacts on household food security in Ethiopia. *Food Policy* 44, 272–284.

Skvarla, M. J. 2017. Penn state extension: a brief guide to hand lenses. [Online] Available at: https://extension.psu.edu/a-brief-guide-to-hand-lenses# [Accessed 30 November 2020].

Suber, L. 2019. North Carolina State University industry expansion solutions: what are logic models, and when should you use them? [Online] Available at: https://www.ies.ncsu.edu/blog/what-are-logic-models-and-when-should-you-use-them/ [Accessed 30 November 2020].

Tambo, J. A., Romney, D., Mugambi, I., Mbugua, F., Bundi, M., Uzayisenga, B., Matimelo, M. and Ndhlovu, M. 2021. Can plant clinics enhance judicious use of pesticides? Evidence from Rwanda and Zambia. *Food Policy* 101(102073). Available at: https://www.sciencedirect.com/science/article/pii/S0306919221000506. Accessed May 16, 2021.

Thorburn, C. 2015. The rise and demise of integrated pest management in rice in Indonesia. *Insects* 6(2), 381–408.

UF/IFAS Plant Diagnostic Center n.d. UF/IFAS plant pathology. [Online] Available at: https://plantpath.ifas.ufl.edu/misc/media/PDC/PDC_Submission_form_CC_1-18.-8.pdf [Accessed 1 December 2020].

van den Berg, H., Phillips, S., Poisot, A.-S. and Dicke, M. 2020. Leading issues in implentation of farmer field schools: a global survey. *The Journal of Agricultural Education and Extension* 27(3), 341–353.

Part 4

Case studies

Chapter 12

Practical application of integrated pest management in greenhouses and protected cultivation

Bruno Gobin, Els Pauwels and Joachim Audenaert, PCS-Ornamental Plant Research, Belgium

1 Introduction

Greenhouses provide a relatively secluded growing space for crops, but entry of pests and diseases can never be eliminated (Jarvis 1992, Albajes et al. 1999, Brownbridge 2017). The greenhouse environment proves to be optimal for fast proliferation of pests and pathogens once they gained entry. For their control, the use of biocontrol agents in integrated pest management (IPM) strategies is well established in greenhouse horticulture (Messelink et al. 2012). Augmentative and repetitive release of beneficials has provided reliable control of a number of important pests. The first great success was with spider mite control in the latter half of the last century (Hussey and Bravenboer 1971), at a moment when the spider mite *Tetranychus urticae* developed resistance to multiple traditional chemical control products (Daly et al. 1998). The more or less confined characteristic of greenhouses contributed to the success of augmentative biocontrol through the release of commercial agents, as these

http://dx.doi.org/10.19103/AS.2021.0095.14

could not escape (Fravel 2005). Growers' uptake of novel biocontrol strategies for other pests was slow at first (van Lenteren 2012). The increasing focus in the 21st century on reduction of chemical residues below official maximum residue limits (MRLs), to fulfil the demands of large retailers, resulted in an increased use of biocontrol for a variety of pests and diseases (Brownbridge 2017, van Lenteren et al. 2020). For a number of these pests and diseases, growers now have reliable micro- or macroorganisms which give adequate control within the systems used to grow high-quality commercial crops. Biocontrol agents provide a background level of control to keep the numbers of pests below damage thresholds, and chemical control can often be reserved for unexpected outbreaks or unforeseen calamities.

Bringing new young plants into a greenhouse poses a double challenge. First of all, young plants can be a source of undesirable pests and diseases that can spread throughout the greenhouse. Growers should monitor all batches of young plants meticulously upon arrival and preferably before placement in the greenhouse (Valentin 2017, Wainwright-Evans 2017). Secondly, it is important to know which chemical products have been used to treat young plants prior to their arrival at the greenhouse. Successful establishment of beneficial mites and insects on new plants will depend on previous treatments, with some residues requiring longer periods before the release of beneficials to avoid mortality (Van De Veire and Tirry 2003).

Growing crops in greenhouses requires a massive investment of money and labour that only pays off when top-quality produce or plants are sold at the end of the growing season. Growers must be able to implement reliable control strategies to prevent quality loss due to pests or diseases. IPM strategies can achieve a high level of performance only when applied in a correct, reasoned and well-informed way (Steinberg 2017). The interplay between the different essential aspects of IPM increases the chance of successful control in the greenhouse (Buitenhuis 2014). As fine-tuning IPM for a given greenhouse crop is more complicated than the traditional calendar spraying, all actors involved must understand that the level of knowledge required of growers has vastly increased. This is particularly because the greenhouse production environment proves to be highly dynamic, with interacting parameters such as crop growth, climate, light, pests and diseases and the implementation of various technical crop measures (pruning, spacing, etc.) as well as harvesting (Buitenhuis 2014, Brownbridge 2017). Fortunately, researchers have shifted their focus from highly controlled experimental set-ups – indeed essential for scientifically sound proof of concept – to more holistic demonstration trials to show pragmatic approaches to growers (Harvey et al. 2015, Wainwright-Evans 2017). Advisory and extension services have become an essential source of information for growers to keep pace with new developments. Continual communication and training of growers and operators help to speed up this process (Mechant 2014).

We will present six case studies that demonstrate how producers of protected ornamental crops in Belgium have tackled specific challenges, with the aid of research and extension, to develop reliable solutions on the grower level, within an IPM framework. We chose the cases to reflect challenges in the area of training, regulatory issues and technical aspects of IPM.

Training is key in the practical adoption of IPM strategies on farm. A greenhouse operation is like a piece of complex, well-oiled machinery where different bits need to interact smoothly. Changing any part of the operation will affect all other parts. Therefore, adopting a change in control strategy requires well-thought-out implementation and sufficient training of growers and greenhouse staff. IPM requirements, tools and techniques evolve over time. To help growers to find their way in the integration of an IPM programme, an intensified extension and training programme for ornamental plant growers in Flanders was developed and its effect was monitored (Case study 1). The programme helped growers to cope with operational changes and at the same time continue to grow qualitative crops. This has resulted in a fluent transfer of essential knowledge and practical on-farm advice to growers and a strong increase in adopted IPM strategies.

Growers must adopt greenhouse operations to novel regulatory requirements that reflect societal demands for sustainable production. We will tackle some technical challenges following changes in crop protection regulation further on. However, regulation of other – not crop protection regulated – aspects of greenhouse production can also have a strong impact on IPM strategies, a fact that regulators often forget. A major example in Flanders was the adoption of the EU Water Framework Directive, which aimed to reduce the environmental impact of farming through increased re-use of water but in doing so created a higher risk of disease spread. Case study 2 explains how growers adopted a successful solution to tackle this problem.

Apart from training and regulation, adoption of innovative and sustainable IPM strategies depends on technical success stories. The first and best-known case is the use of predatory mites to control spider mites (Case study 3), at a time that this pest became increasingly resistant to chemicals. This quickly led to the increased use of predatory mites for a number of pests, an evolution that was further triggered by the reduced availability of chemical products due to regulatory restrictions. To expand the number of target pests, specific IPM control strategies must be developed, building on a good understanding of the pest's biology and population dynamics that interact with the crop and the greenhouse environment. The commercial introduction of new predatory mite species allowed to fill these specific control needs. Broad mites are such a pest species for which adequate chemical control became problematic due to the withdrawal of a number of broad-spectrum pesticides. We were able to develop an efficient IPM strategy based on an improved understanding of the biology of the pest in the greenhouse (Case study 4).

Still, some technical challenges remain. In ornamentals, a large number of introduced beneficials leave together with the sold plants, while filling a recently sold greenhouse section with young plants always poses the risk of introducing pests. In Case study 5, we show promising results with the addition of supplementary food to keep a standing army of predators ready in the greenhouse. Finally, some pests are hard to target with either selective chemicals or biological agents; in such cases, growers must choose unavoidably harsh control methods when an outbreak occurs. This poses a threat to the success of other ongoing IPM or biocontrol strategies in the greenhouse. We present the case of mealybugs as a difficult pest, which is still lacking a cost-effective biological control method (Case study 6).

2 Case study 1: Knowledge translation, demonstration and communication

A reliable level of knowledge of growers and greenhouse operators on pest and disease control is crucial in building strong IPM strategies in the greenhouse (Jacobson 2008, van Lenteren and Nicot 2020). Extension is an important part of the Agriculture Knowledge and Information System (AKIS: https://ec.europa.eu/eip/agriculture/en/tags/akis). Such knowledge should at least comprise approaches to crop hygiene, a good knowledge of the possible pests and diseases in the crop(s) and how to monitor them in the early stages of infection, available control measures, selectivity of those measures and how to apply them correctly. As scientists and extension specialists, we have to understand the challenge this poses to growers. Researchers strive to optimise IPM strategies for problematic pests and diseases and need to discuss their results with the grower community to guarantee a thorough validation and secure a broad implementation.

Regulatory changes are frequent nowadays, with older molecules being withdrawn and novel molecules being developed (Bozzini 2017). Furthermore, with regulation being crop focused, there often is a delay between product registration for major crops and for minor crops. The first is targeted by companies for obvious commercial reasons, the latter often relies on third-party registration as the large regulatory cost is not economical for companies. The majority of greenhouse crops are minor crops, as they represent very limited total surface areas per country. International collaboration efforts exist in Europe (the Minor Uses Coordination Facility; www.minoruses.eu) to put together data from many sources to speed up minor use regulation. On top of this, there is an ever-growing group of biocontrol agents with excellent pest target qualities but which are not always compatible with chemical applications.

Fine-tuning of chemical and biological control is therefore essential (Gobin et al. 2013, Buitenhuis 2014). The recent flood of novel biostimulants adds to the challenge. Growers thus need to check which sufficiently selective chemical is available that does not affect the biocontrol agents present in the crop, or vice versa, and how long they should wait after an application before releasing new biological agents. Once they work out the best solution for their given crop/pest combination, they have to make sure that they have the right application technology (pressure, droplet size, water acidity, sachets or blowing, etc.) and good application skills. They have to calculate how long a waiting period before harvest is needed, which is especially important in edible crops. Finally, they have to monitor the effectiveness of their treatment to evaluate whether the problem is solved or whether new measures may be needed. All of that is on top of numerous other technical and commercial growing challenges, staff planning, and so on. This is a common daily practice in greenhouses all over the world but is often underestimated by stakeholders outside the horticultural sector, whose decisions can have an important effect on the sector.

In an IPM strategy, growers need a good understanding of all available control options. When the IPM directive 2009/128/EC was announced in Europe, research centres and extension services increased their focus on educating growers in the principles of IPM, boosting their knowledge before the regulation actually came into force. At the same time, various systems of continuous training of farmers and operators were set in place; in some countries, it was linked to professional licencing required to get access to professional crop protection agents. Such training focuses on the safe transport, storage, preparation, application and disposal of crop protection agents. Specifically for IPM, selectivity, use of beneficials and other aspects are included in the training. Although official legal regulatory information in each country takes precedence, such information is not always easy to read and understand. Companies supplying chemical and biocontrol agents make available clearer information via their commercial information channels. Good independent knowledge translation and demonstrations from research centres and extension services are essential to help growers choose the best solution for their specific operation. Information from extension specialists is often available in a highly usable format, taking into account local practices and sensibilities. The addition of field research results on efficacy and selectivity of certain IPM strategies in specific greenhouse crop situations in the field will add invaluable information to the legally required regulatory data on efficacy and selectivity. For biostimulants, the development of regulatory standards such as the Harmonized European Standard being developed by Technical Committee 455 of the European Committee for Standardization (CEN) will ensure minimal efficacy and safety for registered products.

Translating knowledge to growers helped the adoption of IPM and biocontrol by ornamental plant growers in the Flanders region of Belgium, where an intense demonstration and communication campaign increased knowledge and application in the year before IPM regulation 2009/128/EC came into effect. Flanders adopted an IPM system with measures defined at three levels: major, minor and recommended levels (Mechant 2014). Major measures are mandatory, minor measures can be chosen to reflect the specific on-farm situation but a minimum of 70% of minor measures should be adopted, recommended measures can be adopted at will. Examples of major measures are, among others, the regular scouting of crops for pests and diseases or the use of clean pots and recipients. Minor measures include for example frequent removal of diseased plants or provision of shelters or banker plants for beneficial organisms. Recommended measures include for example optimisation of greenhouse climate to prevent outbreaks or local pesticide application. This differentiation takes into account the huge diversity in ornamental crops and allows growers to maximally implement adequate measures useful for their specific crops.

Most growers in ornamental horticulture and tree nurseries in Flanders had already adopted the majority of good practices with regard to these IPM measures at the start of the project, with an even higher percentage in the sub-group of greenhouse ornamental horticulture. The considerably increased effort in demonstrating and communicating the various IPM measures led to an increase in adoption of IPM measures by 17% (Table 1) in greenhouses for major and minor measures. This case clearly shows the impact that a combined effort in terms of knowledge translation, demonstration and extension services can have in increasing adoption of desired practices in a sector. It is vital that this effort is continued to ensure that growers retain access to the latest knowledge and crop protection technology to maximise the success of their greenhouse operations.

Table 1 Impact of a targeted communication campaign informing growers of the possibilities and future legal requirements of IPM measures, prior to the introduction of legislation. Percentages give the percentage adoption of major measures, minor measures and recommended measures by 123 ornamental growers who were polled (of which 38 were greenhouse growers) early in the campaign and just before the end of the campaign (late)

Percentage adoption of IPM measures (Flanders region)	Total ornamental horticulture and tree nursery			Greenhouse ornamental horticulture		
No. participants	123			38		
Stage of communication campaign	Early	Late	Increase	Early	Late	Increase
Major	71%	90%	**19%**	77%	94%	**17%**
Minor	60%	84%	**24%**	66%	83%	**17%**
Recommendation	53%	70%	**17%**	61%	70%	**9%**

3 Case study 2: Use of good crop hygiene to prevent disease

Prevention has always been a cornerstone of IPM. Good crop hygiene and a preventative approach towards pest and disease control are a vital aspect of an IPM scheme. Any greenhouse operation should have good hygiene protocols in place, to prevent unintended introduction or spread of pests and diseases (Jarvis 1992, Buitenhuis et al. 2016). Tomato, cucumber and pepper greenhouses were at the forefront of hygiene protocols, most commonly aimed at eliminating the risk of virus introduction. The measures that tomato growers already had in place for the well-known tomato mosaic virus (Broadbent 1976) currently help to keep the emerging Tomato Brown Rugose fruit virus out. In addition, shifts in the climate as a result of climate change stimulate the emergence of novel threats from warmer regions. *Drosophila suzukii* is the best-known recent example (Cini et al. 2014), now posing a constant challenge, as these fruit flies easily proliferate on fruits outside greenhouses during warmer periods of the year. Maintaining clean and hygienic growing conditions is a daily challenge for growers.

Sometimes non-IPM related regulation can strongly interfere with IPM efficacy, as many different aspects of crop growing interact closely for optimised IPM control (Buitenhuis 2014). Regulators must understand that regulatory changes often have a profound effect on control strategies. The effect that the EU Water Framework Directive had on the control options of a quarantine disease is the main example.

Phytophthora ramorum is an important disease in ornamental plants, tree nurseries and in young plants (Grünwald et al. 2008). Infection with this fungus causes unattractive leaf spots and will eventually lead to wilting and premature death of the plant, causing economic losses. Furthermore, this species is considered to be a quarantine organism in many countries; thus infections occurring in a production greenhouse will halt exports of the plants produced on that site. Therefore, it is important for growers to detect any symptoms early, for instance, on novel young plants arriving, to quickly isolate and remove infected plants. Using disease-free rainwater for irrigation during the growth cycle is equally important. While this is a common practice that works well to keep the disease spreading from plant to plant, there is a catch due to the recent adoption of practices that reduce water use and its resulting environmental impact.

Growers are re-using water as much as possible for two reasons. Most growers maintain reservoirs to capture rainwater from the roofs of buildings and greenhouses, to have access to good quality irrigation water. Recycling allows growers to span increasing periods of drought, while limiting the use of scarce groundwater that is often of lesser quality or is highly regulated.

Re-using water also limits the discharge of water laced with nutrients or crop protection agents into the environment. Throughout Europe, water directives strictly regulate the maximal amount of contaminants that wastewater can contain. However, recycling water increases the risk of spreading diseases from a single contaminated plant throughout the whole greenhouse.

The use of either UV filters or slow sandbed filters (or both) is a very efficient IPM strategy to prevent *Phytophthora* or other fungal diseases from spreading via irrigation water (Werres et al. 2007, Vercauteren et al. 2013). Water that passed disinfection is then stored in a covered clean water tank to prevent accidental contamination. However, it is important to check this tank regularly for the presence of *Phytophthora*, as a control for either filter efficacy breakdown or external contamination. Detection of the presence of *Phytophthora ramorum* can be done by a laboratory test, but for frequent on-site monitoring, there is a simpler test: the leaf 'lure' test. With the use of a few netted leaves of *Rhododendron*, a plant highly sensitive to the disease, and inserted weekly for a few days into the clean water tank, the presence of the oomycete can be detected quickly if any necrotic spots are formed on the leaves (Fig. 1). The leaf test is a system with very high sensitivity, is easy to use and has a low operating cost compared to frequent water sample analysis by a laboratory (at around 100€ per test). The absence of spots clearly shows that the water is disease free and indicates that the filtering system is performing well. As soon as necrotic spots occur on the leaves, the water in the tank must be re-routed immediately over a functioning filtering or disinfection system. If contamination occurs, the exact nature of the disease should, of course, still be confirmed by a laboratory test.

(a) (b)

Figure 1 Leaves of a susceptible Rhododendron variety 'Cunninghams white' are placed in a net in water basins for a few days. Should *Phytophthora ramorum* circulate in the water tank (a), typical necrotic spots will be visible after a few days (b). This simple leaf 'lure' test is highly efficient and can be used on a day-to-day basis, reducing the need for frequent water analysis by a laboratory.

4 Case study 3: Predatory mites as a major success story

Since the first steps towards controlling pesticide-resistant spider mites with predatory mites, a wide range of predatory mites have become commercially available to control major pests in the greenhouse (Albajes et al. 1999, Brownbridge 2017, Jandricic 2017, Gobin et al. 2017, Calvo et al. 2015). All main biocontrol companies have invested in discovering and breeding potential candidate predators, testing their efficacy against a number of common pests. As with chemical pesticides, it is important for growers to choose the appropriate predator for their specific pest and cropping system, and this requires some understanding of how predators work. Predatory mites can be classified relatively easily into four main categories along the predation continuum: (1) complete specialists, (2) moderate specialists, (3) moderate generalists and (4) complete generalists (Audenaert et al. 2018). *Phytoseiulus persimilis* is the type species for complete specialism on *Tetranychus urticae*, preying on all its life stages, very active in its search for prey, highly mobile on webs and able to quickly build up large populations when prey is abundant. Because of its specialisation, it should not be introduced preventively in the absence of spider mites. Moderate spider mite specialists such as *Neoseiulus californicus* prefer feeding on spider mite nymphs but can also survive on pollen in the absence of prey. Therefore, these can be introduced preventively. A number of predatory mite species are generalists that can prey on whitefly, thrips, tarsonemid mites, spider mites or even other predatory mites, as well as pollen. The better-known species in this group are *Amblyseius andersoni, A. swirskii, Neoseiulus cucumeris, Transeius montdorensis* and *Amblydromalus limonicus.* These are well suited to being used as a police force, present and ready in the crop before pest outbreaks occur. It is generally not recommended to mix species, as they will also start feeding on each other and quickly outcompete specialists. Finally, some predatory mites such as *Euseius gallicus* are complete generalists, feeding on a wide variety of food sources including pests and pollen. Their populations can increase with just pollen as a food source and can thus be kept easily in the crop in the absence of pests. Such a large population of predators can be an asset to mitigate pests at a very early stage of infection, but availability of non-pest food sources might limit their impact on the pest, as they will readily feed on non-pest food.

Whilst predatory mites have proved a huge success in providing control of a variety of pests since the early days of IPM, changes in crop-growing techniques have led to occasional unexpected breakdowns of biocontrol systems in practice. Particularly, when energy-saving regimes such as temperature integration and dynamic lighting were implemented to save energy in greenhouses (Pollet et al. 2009, Körner and Challa 2004, Dieleman et al. 2005), some fine-tuning was needed to ensure the use of the best predator in specific greenhouse conditions. A lot of the biological parameters of predatory mites had been characterised under controlled environmental conditions, within a range that mimicked an average greenhouse temperature (e.g. Mironidis and Savopoulou-Soultani 2008, Jandricic 2017), but did not take into account the large day-to-day variations in practical energy-saving settings, similar to those in commercial greenhouses. Low day-night temperature variability tends to speed up development time (Bryant et al. 1999), whereas high-temperature variability tends to slow down development time (Hagstrum and Milliken 1991, Mironidis and Savopoulou-Soultani 2008). Different species of predatory mites show different biological responses to varying temperatures, when looking at the development time of the immature stages, fecundity and sex ratios (Van Gansbeke et al. 2013). Increased research interest in biocontrol performance under crop realistic conditions is of the essence to improve IPM strategies and to anticipate changes in efficacy when changes in greenhouse parameters occur (Van Gansbeke 2015a, b).

5 Case study 4: The detection and control of broad mites

Due to the increased adoption of IPM control strategies which considerably reduced the use of broad-spectrum pesticides, a number of previously minor pests suddenly started to damage crops. Broad mites (Tarsonemidae) are one such example (Gobin et al. 2011, Gobin and Bangels 2008). Whereas broad mites were controlled by classical organophosphates and organochlorines aimed at common pests, the switch to more selective crop protection agents allowed their populations to increase in the crop (Audenaert et al. 2009, Gobin et al. 2013). Only a limited number of selective acaricides are currently authorised in Europe to control this pest (e.g. Abamectin), while the number of permitted applications per growth season is limited in many countries to prevent the development of resistance. Furthermore, the mites typically reside in crevices of stems, leaves or buds and are thus very difficult to target with conventional greenhouse spraying equipment (Mechant et al. 2015).

On top of that, broad mites are typically difficult to detect early, due to their small size and reclusive behaviour. Broad mites prefer to reside in flower buds and move from there as soon as the buds break. The damage they cause to flowers can be commercially disastrous even when very low damage levels are seen on the leaves (Fig. 2). Counts on azalea flowers show that severe damage occurs from as few as a dozen individual broad mites per flower, resulting in azalea plants not reaching the minimal quality standards for this EU PDO (Protected Designation of Origin) product. To mitigate this, some years ago, the azalea quality monitoring board supported in-depth research into broad mite biology, spread and IPM control methods to ensure growers could meet stringent quality standards.

Figure 2 Visual representation of damage per number of broad mites (WHM) counted in open flowers. From 22 broad mites upwards, commercial quality standards are no longer met, making plants unsaleable.

To prevent broad mite outbreaks, it is important for growers to actively monitor the crop, not just for symptoms but for the broad mites themselves. Growers also need to understand the population dynamics of broad mites in the crop. When an outbreak occurs, damage symptoms become visible only after a time lag of 10–20 days as the broad mite population in that location is already declining. Often, the population in the surrounding plants is increasing rapidly by then, even though these plants lack symptoms (Fig. 3). Broad mites can be detected only by using a microscope with sufficient magnification, although modern high magnification macrolens adaptors for smartphones are an additional tool. However, detection of broad mites under a binocular microscope is more reliable, especially for early detection when numbers are low. We have developed a three-step protocol for examination under a field microscope to quickly sample plant tissue for the presence of even low numbers of broad mites (Mechant et al. 2015). Broad mites can be sampled reliably by taking a small number of plant cuttings and carefully placing them in a closed vial with 70% ethanol and then shaking this for at least 30 sec to disperse the broad mites in the alcohol. Draining the alcohol through filter paper, over a vacuum pump for speed, gives a sample that is easily assessed

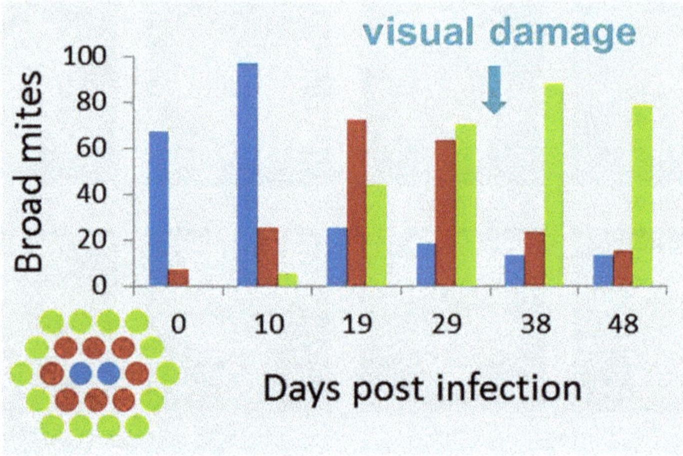

Figure 3 The presence of broad mites on potted plants adjacent to one another on given days after placing two infested potted plants (blue circles) at the centre. The arrow indicates when damage became visible on the central plants.

under a binocular microscope at 25× magnification. This method is especially useful for extension services, crop advisors or monitoring specialists in larger greenhouse operations.

To demonstrate the efficacy of biocontrol to growers, trials were conducted in a total of six commercial greenhouses spread over two growing seasons. To mitigate any risk of creating broad mite hotspots in the greenhouse, no untreated controls were used. Control efficacy was therefore compared to chemically controlled plots with the expectation that such control would be either equally good or better. Growers reserved part of the greenhouse for biocontrol of broad mites, while continuing to treat the rest of their crop for broad mites with Abamectin. The use of *Amblyseius swirskii* resulted in better control of broad mites than in the reference plots in all but one greenhouse, where it was equally good (Table 2). The degree of improvement varied greatly, between 14% and 94%, and this was enhanced when additional food was provided. Approaches to improving the population using supplementary food will be elaborated in the next section.

The use of biocontrol was about 3.5 times more expensive than the authorised chemical treatment (0.14 €/m² vs 0.04 €/m²; Table 2)). However, due to a lack of sufficient registered active ingredients with different modes of action, the risk of resistance development in broad mites was very high. Many growers indeed faced the problem of the need for more and more treatments with dwindling efficacies. Since the introduction of IPM for broad mite control, the pest can be controlled adequately as the beneficial mites are highly efficient in controlling the broad mites, while at the same time the available

Table 2 Broad mite control with *Amblyseius swirskii* in six commercial greenhouse settings, with and without added food, with release intervals (i) in weeks (wks) and an indication of efficacy as a percentage increase in efficacy compared with chemical control. The cost in euros per m² is indicated. Artemia was the commercial product Artefeed (Koppert); Pollen was the commercial product Nutrimite (Biobest)

Greenhouse	Mite	Feed	Compared to chemical control*	Cost/m²
1	swirskii 2 times i=5wks	artemia 6 times i=2 wks	+90%	€0,28
	swirskii 5 times i=3wks	None	+72%	€0,35
2	swirskii 2 times i=3wks	artemia 3 times i=2 wks	+94%	€0,21
	swirskii 2 times i=3wks	None	+94%	€0,14
3	swirskii 2 times i=2wks	pollen 4 times i=2 wks	+100%	€0,19
	swirskii 2 times i=2wks	None	+70%	€0,14
4	swirskii 2 times i=2wks	pollen 4 times i=2 wks	+57%	€0,19
	swirskii 2 times i=2wks	None	+14%	€0,14
5	swirskii 2 times i=2wks	pollen 4 times i=2 wks	+25%	€0,19
	swirskii 2 times i=2wks	None	+0%	€0,14
6	swirskii 2 times i=2wks	pollen 4 times i=2 wks	+80%	€0,19
	swirskii 2 times i=2wks	None	+20%	€0,14

* Chemical control was a tank mix of Abamectin with Hexythiazox applied two to three times at a cost of 0.03-0.04 €/m².

chemical control agents, when needed, perform better. This makes use of IPM for broad mite control worthwhile for growers, in the way that was seen some decades ago for spider mites. Once biological control is used frequently, the pest population becomes more susceptible to pesticides. However, regulatory authorities should still ensure that sufficient active ingredients with complementary modes of action remain authorised. Indeed, growers will go much further with biological control if they know they have a good fail-safe mechanism should biocontrol fail for some exceptional reason. That is the strength of IPM.

6 Case study 5: The addition of feed in the field

Commercial greenhouse operations have inherent effects on beneficials and the efficacy of IPM (Wainwright-Evans 2017). One of the main challenges in ornamental plant production is the frequent shipping out of entire batches of plants from the greenhouse, to be replaced by novel young plants (Mechant 2014, Buitenhuis et al. 2015). Growth cycles starting with young plants can vary from a few weeks to a few months, with extremes at just over a year. As the commercialised plants leave as an entire batch, any beneficial mites and insects on them depart with the plants. Emptying out parts of the greenhouse thus leaves an area devoid of predators and parasitoids, and a new batch of beneficials needs to be introduced.

One should also consider that in cross-border ornamental plant trade, the presence of beneficials can pose a risk as they can be (erroneously) taken for undesired organisms by both customs authorities and trade partners. Questions at customs clearing can generally be solved once local plant regulations officers arrive, but such proceedings can cause significant shipment delays. Furthermore, trade partners sometimes seek commercial incentives such as price reductions before accepting the shipment when insects or mites are found on plants. For further development of IPM and biocontrol in ornamental plants, it is essential to further educate consumers, trade partners and local officials too.

Another challenge for ornamental plant growers is that plants either lack flowers (green plants) or are commercialised just before the opening of flower buds. The reasons for this are to avoid flower damage during transport as well as to ensure maximal longevity of the flowers after purchase (Christiaens et al. 2016). This however means that, in contrast to tomato and pepper crops, very little pollen or nectar is available as an alternative feed source for predatory mites. Such alternative feed sources are essential to allow the population of beneficial insects to survive when prey densities are very low (Nomikou et al. 2021, He et al. 2021), thus being present as a standing army whenever a pest outbreak occurs.

A final challenge for ornamental plant growing is that tolerance for damage to leaves and flowers is extremely low (Schumacher et al. 2006, Mechant 2014). As plants and flowers are sold for decorative purposes, the entire plant or flower stem must be perfect. The same goes of course for leafy vegetables, but when only fruits are sold, some minor damage to leaves can be tolerated, as long as it does not hamper fruit development. As optimal biocontrol often involves an equilibrium between pests and beneficials, pushing this equilibrium towards minimal risk of damage can be challenging. Fortunately, the more or less closed greenhouse environment confines any added beneficials to the crop, allowing slight overdosing to avoid plant damage. Overdosing will result in faster clean-up of the pest, before the pest population reaches damage thresholds throughout the greenhouse. Pushing the equilibrium does affect the treatment costs, as a grower would add more beneficials than strictly necessary for long-term control. Equally important is thoughtful choices in other parts of crop management, in a fully integrated IPM scheme for the whole crop (van Lenteren and Nicot 2020).

Although addition of artificial or natural feed to increase predatory mite reproduction and predation works very well in laboratory trials, lab conditions are far away from greenhouse conditions (Van Gansbeke et al. 2014b). Even semi-field trials in research greenhouses offer more controlled circumstances than commercial greenhouses. Convincing growers of a new strategy in pest control often requires a thorough demonstration of novel techniques in commercial greenhouses (Messelink et al. 2014, Pijnakker et al. 2017). We were able to monitor the effect of added feed in both semi-field and commercial field conditions (table 2).

First of all, good application technology for beneficials and feed is important. Beneficials can be released near pest hotspots or evenly distributed in the whole crop by either slow-release sachets or blow application (Groot et al. 2017, van Schelt et al. 2008). Feed must be spread out broadly, to provide nutrients throughout the crop, and evenly, to avoid soiling of parts of the crop with clumps of feed (Pijnakker et al. 2017). An even distribution also reduces the risk of the development of undesired fungi on the clumps. For the same reason, distribution on a dry crop is recommended, and ideally some time before any overhead irrigation is turned on, to allow feed to penetrate within the crop.

Secondly, it is important to calculate the potential boost given to the pest species by adding feed to the crop. A number of pests, such as thrips, feed on pollen or artificial feed (Coll and Guershon 2002). The quality of alternative feed can also have a direct effect on the degree of cannibalism in the predator (Van Gansbeke et al. 2014a). Low-quality feed will not suffice for the predatory mites; if there is a lack of other feed, they will turn to cannibalism. Still, the benefits of maintaining a patrolling population of predators generally outweigh the potential benefit to the pest species (Van Gansbeke et al. 2014b). To convince growers that feed addition improved pest control, we set up large-scale pilot trials at commercial greenhouses, with part of the greenhouse receiving feed and part without. Monitoring of pests and predators showed that good predator populations were established and pests were adequately controlled compared to control plots without feed.

A population of predatory mites benefits from adding feed to the crop. In semi-field cage trials in the greenhouse, the predator density of *Amblyseius swirskii* could be increased significantly by adding feed (Van Gansbeke et al. 2016). On two common ornamental crops, Chrysanthemum and Ivy (Hedera helix), adding pollen of *Typha angustifolia* or a dry artificial rearing diet at 500 g/ha dose rates significantly increased the number of *Amblyseius swirskii* per leaf throughout the observation period. In Ivy, this was demonstrated both at near constant and at more commercially realistic variable temperatures. Cysts of Artemia brine shrimps at a similar dose rate also provided a good food source in one trial under constant conditions, but were more susceptible to clumping in conditions of high humidity, resulting in the rapid growth of fungi during a trial at a grower's greenhouse. This might have affected their performance as a food source in another trial. Even when *A. swirskii* can use Artemia as an alternative food source for short-term survival, it is impossible to rear multiple generations on this food source (Van Gansbeke et al. 2015a,b). Nevertheless,

Figure 4 Field trial of thrips control on Chrysanthemum in a near-commercial setting. Non-crop areas separate different plots where predatory mites are released with or without additional food.

these experiments showed that feed addition has a high potential to sustain predatory mite populations and to reduce the need for repeated release of predatory mites. To convince growers of the potential, we took this approach one step further in demonstration trials in commercial greenhouse settings (Fig. 4). We aimed to demonstrate three things: predator population establishment, improved pest control and the cost-effectiveness of adding food to the crop. We focused on the four main pests - spider mites, thrips, whitefly and broad mites - with good results and enhanced control when food was added. Figure 5 shows that efficacy was obtained in spider mite control with *Neoseiulus californicus* in young Chrysanthemums, prior to the start of flowering.

7 Case study 6: Mealybugs as difficult pests compromising integrated pest management strategies

The majority of plant pests in ornamental plant greenhouses can be adequately controlled through an IPM strategy with the main focus on biocontrol. However, biocontrol is still at risk from outbreaks of persistent pests that require harsh chemical treatments. Mealybugs are some of those pests compromising growers' efforts to increase non-chemical control (Audenaert and Verhoeven 2020). Doing

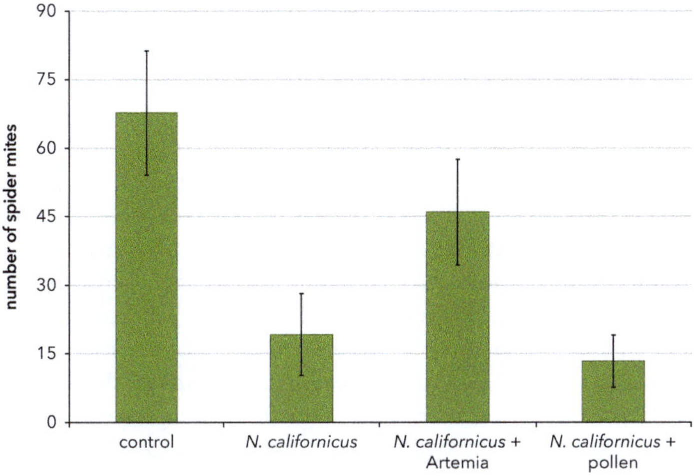

Figure 5 The number of spider mites in a semi-field trial in Chrysanthemum comparing untreated control plots with plots treated with *Neoseiulus californicus* as such and with *N. californicus* and added feed (Artemia cysts or Typha pollen).

both direct damage caused by sucking sap from the plant and indirect spoiling through excretion of honeydew and the growth of sooty mould, they render plants unsaleable. Furthermore, mealy bugs are difficult to treat chemically, as their protective waxy layers deter spray droplets from actually reaching the insect. The required applications of harsher chemicals and adjuvants to dissolve wax can have negative side effects on biocontrol agents aimed at spider mites, thrips or whiteflies. Even when mealybugs have a close endosymbiotic relationship with a number of bacteria, it was shown that reduction of these symbionts did not affect mealybug fecundity (Parkinson et al. 2016), nor insecticide tolerance or host plant preference in *P. citri* (Parkinson et al. 2018).

Mealybug control relies on regular crop monitoring, positive identification of the species and good knowledge of the IPM tools at hand. The three predominant mealybug species in European greenhouse horticulture are *Planococcus citri, Pseudococcus longispinus* and *Pseudococcus viburni*, and easy identification tools for growers exist for all these species. Other species, such as *Planococcus ficus* and *Phenacoccus madeirensis*, are more important in greenhouses in Mediterranean climates. Monitoring can be done by hand; by searching leaf shafts, crevices, branch nodes and other seclusive areas on the plants; or by checking for the presence of winged males on yellow sticky traps. For *P. citri, P. longispinus* and *P. viburni*, adding female sex pheromone lures to the yellow traps greatly increases the capture rate of males. Though pheromone trapping is very good for monitoring, it cannot be used reliably for mass trapping. Therefore, weekly monitoring is essential for an optimal IPM strategy to control mealybugs. Two efficient biocontrol agents *Cryptolaemus montrouzieri* and *Chrysoperla carnea* are well suited to controlling mealybugs. Our preliminary experiments show high efficacy (>90% reduction) against *P. citri* and good efficacy against *P. longispinus* (>75%) and *P. viburni* (>65%). However, the cost of these beneficials is currently too high for crop-wide application but they should be directed at specifically localised hotspots within the greenhouse. Crop-wide application of *Cryptolaemus* or *Chrysoperla* will be too expensive, but a grower can treat around 50 hotspots with about 15 predators per m^2 for less than 100€. In comparison, a cover spray of the crop with selective chemicals and added adjuvants will be up to 300€ per ha. Indeed, the biological control options for some of the more challenging pests are still too expensive for regular use, and research into improving the efficiency of both the breeding of beneficials and their application and population development in the greenhouse, should improve cost-effectiveness in the future.

8 Conclusion

In this chapter, we presented a number of case studies that show how well-thought-out IPM strategies ensure adequate control of most pests in ornamental greenhouses. As IPM evolves further, with less and less availability of chemistry and ever-increasing use of biological control, sufficient training and support of growers must be available to ensure uptake of innovative methods. On the other hand, regulators must take into account that regulatory changes can have a profound effect on greenhouse operations and should check the level of impact and availability of mitigating measures beforehand. Predatory mites as a success story were a main driving force to get IPM into the greenhouses. A good understanding of the biology of pests allowed further development of adequate IPM strategies, such as the one for broad mite control. To increase the efficacy of predatory mites even further, as well as to keep a standing army in the absence of pests, the addition of supplementary food is useful. Finally, work is still ongoing to develop reliable IPM strategies for some hard to control pests such as mealybugs. Overall, IPM in ornamental greenhouse crops has matured over the past decades to become a reliable and sustainable control strategy, and future additions to the IPM toolbox will strengthen this even further.

9 Future trends in research

Over the last few decades, greenhouses have evolved considerably with increased height and screening, increased photoperiod with artificial lighting, climate control aimed at energy saving, increased automation, and so on, we believe that greenhouse technology will evolve at an even greater pace over the next decades. Ambitious global societal goals on reduction in CO_2 emissions, water use, environmental risks, food safety and operator safety will drive innovation in greenhouse production. Greenhouse horticulture should look forward and address these challenges as a sector, to find practical, feasible and economical solutions for each and all of these ambitions. Even though pest control or IPM is just one aspect of these challenges, it will closely interact with changes in all other aspects of greenhouse horticulture. From a practical point of view, we present three foreseeable major changes and their effect on IPM: the increased use of LED lighting in more complex lighting strategies, the increase in the degree of automation and the move towards multilayer growing.

Increased lighting in ornamental crops has allowed increased production of vegetables and ornamentals during winter periods, prolonged growth and harvest seasons or allowed production to be focused on specific festive days such as Valentine's Day. Traditionally growers prolonged daylight periods in fairly continuous bouts through the use of high-pressure sodium lamps at around 100 μmols per m^2 light energy at plant level. This would add around 30% to the incoming natural daylight on a dark and cloudy winter day as well as adding top heat to the apex of the plants. The recent development of LED grow lights with increased efficiency in blue, red and far-red light spectra provides a highly interesting alternative to high-pressure sodium lights: low energy consumption, the possibility to optimise the spectrum for specific plant production and less limitations in on/off switching, as no lamp heating period is needed. This allows growers to switch lights on intermittently in direct response to grid fluctuations in energy pricing. Lamps can also be placed nearer to, or in between, the crop as less heat is generated, at the same time being advantageous for energy efficiency but somewhat disadvantageous for certain ornamental crops preferring top heat, such as cut roses. Intermittent use of specific light spectra is becoming the norm in greenhouse horticulture, and any effects of this transition on plants, fungi and arthropods should be taken into account when building an IPM programme.

By using specific light spectra, both the plant's growth and its resilience to pests and diseases can be actively steered (Ballare 2014, Vanninen et al. 2010). Additionally, different light spectra have the potential to affect the virulence of fungal diseases. This opens up a whole new toolbox for IPM. Especially the red:far-red ratio affects both the jasmonic and salicylic acid pathways involved in immune responses in the plant. UV B has a more direct effect on plant phenolics, which have a role in deterring leaf-eating insects, in addition to direct effects on pests and diseases (Johansen et al. 2017). Fungal diseases such as Botrytis are known to have light receptors for specific light frequencies and intelligent use of light spectra can induce phenotypic variation in sporulation and formation of sclerotia (Meng et al. 2020). Many predators are known to be positively affected by photoperiod extension with artificial lighting (Gilkeson and Hull 1986, Gillespie and Quiring 2005). The effect that spectrum-specific lighting and the switch towards LED spectra have on the use of predators and parasitoids has received some scientific attention since the paper by Johansen et al. in 2011. In practice, pest development and biocontrol both work under LED light regimes, even in the absence of daylight. However, we do not yet fully understand how the pest-predator interactions are affected by different spectral ratios. Photoperiod seems more important than light spectrum for some predators (Labbe et al. 2017), but wavelength can influence development times (Wang et al. 2013) or recruitment of predators (Ogino et al. 2016). We believe that future understanding would help improve biocontrol in illuminated crops.

The trend towards increased automation of crop cultivation in greenhouses, to limit labour costs, will have an immediate effect on IPM strategies. While climatisation, fertigation, lighting and ventilation are commonly automated, modern ornamental plant greenhouses have fully robotised processes starting from potting young plants, to moving these into the greenhouse, to spacing the plants mid growth and to order picking when plants reach commercial stages. Such operational optimisation allows for much larger crop areas and crop densities, limiting the need for and possibility of operator entry into the greenhouse, to walk amongst the crop. Application of crop protection, whether chemical or biological, can also be automated. The main challenge lies with the early detection of pest and disease outbreaks. Growers quickly identify aberrant growth or colour when walking through crops, and this skill needs to be transferred to digitalised monitoring systems. Recent technological advances have allowed for crop monitoring with RGB-, hyper- and multispectral cameras, as well as thermography, sensors for chlorophyllic fluorescence and 3D scanning of plants (Mahlein 2016). While mainly developed to monitor growth and quality parameters, further development towards disease and pest monitoring is a reality. Camera-based observation of crops and traps will prove an essential tool for IPM, but still requires practical optimisation towards much needed monitoring resolution and density, to have a resilient system at a reasonable cost.

Ever-increasing automation and energy-efficient LED lights have opened the way for even denser crop growing in multilayer systems and plant factories (Shamshiri et al. 2018). The reduced spacing between layers and the height of stacked layers will make direct monitoring and intervention by growers near impossible. Such systems will require even more creative IPM approaches, from monitoring to pest control, but also offer some advantages (Roberts et al. 2020). The highly controllable and sealed environment of a plant factory limits the risk of entry of pathogens compared to classic greenhouses (Goodman and Minner 2019). Many of these systems now aim to implement short growth cycles from disease-free young plants in smaller units. However, upscaling in operation size and length of growth cycle will increase the risk of pathogens emerging and proliferating in the system. There will be a specific need for monitoring strategies and optimised IPM application solutions for plant factory systems, while highly automated multi-layered systems will require starting with a clean crop. Practical techniques to ensure clean starts (i.e. Buitenhuis et al. 2016) need to be further improved through research, while at the same time breeding for disease and pest resistance must be taken to a higher level.

Finally, we must appeal to government agencies to positively stimulate innovation to tackle the many challenges facing commercial greenhouse growers. Research questions must be tackled at different knowledge levels, from academia to practical research, to ensure both in-depth understanding of biocontrol and pragmatic on-farm solutions for commercial crops. Both are needed to further develop sustainable horticulture for a demanding society. At the same time, growers must be stimulated to adopt novel sustainable control strategies, without abolishing old and proven practices prematurely. Novel techniques take more than a few seasons to become fully validated in practical circumstances. Only then can they be applied with confidence in commercial operations.

10 Where to look for further information

The IOBC (https://www.iobc-global.org/): The International Organisation For Biological And Integrated Control is a tremendous source of information due to its global network of scientists working on many different aspects of biological control and integrated pest and disease management as well as the publication of books and the dedicated journal 'Biocontrol'. it is organised in regional sections with active working groups that meet every couple of years to discuss ongoing research and development.

IOBC WPRS (https://www.iobc-wprs.org/expert_groups/wg.html): the West Palearctic Regional Section has two dedicated working groups on greenhouse biocontrol and IPM, which attract participants from all continents. Both groups report ongoing research in their respective IOBC/WPRS Bulletins:

1. the working group 'Integrated Control in Protected Crops, Mediterranean Climate' with a focus on greenhouse cultivation with lower needs for heating and lighting;
2. the working group 'Integrated Control in Protected Crops, Temperate Climate' with a focus on greenhouse cultivation with high heating and lighting requirements.

Extension Services: Practical and hands-on information on IPM of different crops and growing areas can be obtained from local private or public extension

Figure 6 Extension services and their communication play a major role in AKIS (Agriculture Knowledge and Information Systems) by providing information in highly readable and useable formats and taking local circumstances into account.

services. These often provide easy to read information and courses with a focus on practical use within the greenhouse (Fig. 6) and can show novel techniques through demonstration trials.

AKIS (https://ec.europa.eu/eip/agriculture/en/focus-groups): The Agriculture Knowledge and Information system of the EU works on several topics, including IPM, through a number of focus groups.

11 References

Albajes, R., Gullino, M. L., van Lenteren, J. C. and Elad, Y. (Eds.) (1999). *Integrated Pest and Disease Management in Greenhouse Crops*. Kluwer Publishers, Dordrecht. 545 pp.

Audenaert, J., Verhoeven, R., Leus, L., Van Huylenbroeck, J., Witters, J., Jaques, J., Flors, V., Baroffio, C., Minguely, C., Winkler, K. and van Holstein-Say, R. (2018). Monitoring, Identification and control of mites: The IPM approach. PCS publications 60 pp.

Audenaert, J. and Verhoeven, R. (2020). Wolluisbeheersing 2.0 (Mealy bug control 2.0, in dutch) PCS Publications. 62 pp.

Audenaert, J., Vissers, M., Haleydt, B., Verhoeven, R., Goossens, F. and Gobin, B. (2009). Acaricides and predatory mites against the begonia mite, *Polyphagotarsonemus latus* (Acari: Tarsonemidae), on Hedera helix. *Comm. Agri. Appl. Biol. Sci.* 74(1): 217–224.

Audenaert J., Vangansbeke D., Verhoeven R., De Clercq P., Tirry L. and Gobin B. (2014). Predation efficiency of predatory mites from different climatic origin under variable climates in Belgian greenhouses. IOBC-WPRS Bulletin 102, 7–13.

Ballare, C. L. (2014). Light regulation of plant defense. *Annu. Rev. Plant Biol.* 65: 335-363.

Bozzini, E. (2017). *Pesticide Policy and Politics in the European Union: Regulatory Assessment, Implementation and Enforcement*. Springer.

Broadbent, L. (1976). Epidemiology and control of tomato mosaic virus. *Annu. Rev. Phytopathol.* 14(1): 75-96.

Brownbridge, M. (2017). Biological control in greenhouse IPM systems: where we've been, where we are, and where we need to go? *IOBC WPRS Bull.* 124: 1-10.

Bryant, S. R., Bale, J. S. and Thomas, C. D. (1999). Comparison of development and growth of nettle-feeding larvae of Nymphalidae (Lepidoptera) under constant and alternating temperature regimes. European Journal of Entomology 96: 143-148.

Buitenhuis, R. (2014). Systems approach: integrating IPM in the production system. *IOBC WPRS Bull.* 102: 37-43.

Buitenhuis, R., Brownbridge, M., Brommit, A., Saito, T. and Murphy, G. (2016). How to start with a clean crop: biopesticide dips reduce populations of *Bemisia tabaci* (Hemiptera: Aleyrodidae) on greenhouse poinsettia propagative cuttings. *Insects* 7(48): 13 pp. doi:10.3390/insects7040048.

Buitenhuis, R., Murphy, G., Shipp, L. and Scott-Dupree, C. (2015). *Amblyseius swirskii* in greenhouse production systems: a floricultural perspective. *Exp. Appl. Acarol.* 65(4): 451-464.

Calvo, F. J., Knapp, M., van Houten, Y. M., Hoogerbrugge, H. and Belda, J. E. (2015). *Amblyseius swirskii*: what made this predatory mite such a successful biocontrol agent? *Exp. Appl. Acarol.* 65(4): 419-433.

Christiaens, A., De Keyser, E., Pauwels, E., De Riek, J., Gobin, B. and Van Labeke, M. C. (2016). Suboptimal light conditions influence source-sink metabolism during flowering. *Front. Plant Sci.* 7: 249.

Cini, A., Anfora, G., Escudero-Colomar, L. A., Grassi, A., Santosuosso, U., Seljak, G. and Papini, A. (2014). Tracking the invasion of the alien fruit pest *Drosophila suzukii* in Europe. *J. Pest Sci.* 87(4): 559-566.

Coll, M. and Guershon, M. (2002). Omnivory in terrestrial arthropods: mixing plant and prey diets. *Annu. Rev. Entomol.* 47: 267-297.

Daly, H., Doyen, J. T. and Purcell, A. H. III (1998). Chapter 14. *Introduction to Insect Biology and Diversity* (2nd edn.). Oxford University Press, New York, NY, pp 279-300.

Dieleman, J. A., Meinen, E. and Dueck, T. A. (2005). Effects of temperature integration on growth and development of roses. *Acta Hortic.* 691(691): 51-58.

Fravel, D. R. (2005). Commercialization and implementation of biocontrol. *Annu. Rev. Phytopathol.* 43: 337-359.

Gilkeson, L. A. and Hill, S. B. (1986). Diapause prevention in Aphidoletes aphidimyza (Diptera: Cecidomyiidae) by low-intensity light. *Environ. Entomol.* 15(5): 1067 -1069.

Gillespie, D. R. and Quiring, D. M. J. (2005). Diapause induction under greenhouse conditions in two populations of Dicyphus hesperus (Hemiptera: Miridae). *Biocontrol Sci. Technol.* 15(6): 571-583.

Gobin, B., Audenaert, J., Vissers, M., Van Delsen, B., Vlaeminck, M. and Pauwels, E. (2013). Broad mite control in woody ornamentals: developing an integrated pest management solution. *Acta Hortic.* 990: 47-53. II International Symposium on Woody Ornamentals of the Moderate Zone (ed. by J Van Huylenbroeck, MC Van Labeke & K Van Laere), ISHS, Leuven, Belgium.

Gobin, B. and Bangels, E. (2008). Field control of strawberry mite *Phytonemus pallidus* IOBC. *Bull* 39: 97–100.

Gobin, B., Pauwels, E., Mechant, E. and Audenaert, J. (2017). Integrated control of broad mites in ornamental plants under variable greenhouse conditions. *IOBC WPRS Bull.* 124: 119–124.

Gobin, B., Vissers, M. and Pauwels, E. (2011). Integrated pest management solutions for the control of *Polyphagotarsonemus latus* in ornamentals: from trial to practice. *IOBC WPRS Bull.* 68: 47–50.

Goodman, W. and Minner, J. (2019). Will the urban agricultural revolution be vertical and soilless? A case study of controlled environment agriculture in New York City. *Land Use Policy* 83: 160–173.

Groot, T. M. V., van Houten, Y. M., Hoogerbrugge, H. and Lenferink, K. O. (2017). Improved predatory mite release system. *IOBC WPRS Bull.* 124: 165–170.

Grünwald, N. J., Goss, E. M. and Press, C. M. (2008). *Phytophthora ramorum*: a pathogen with a remarkably wide host range causing sudden oak death on oaks and ramorum blight on woody ornamentals. *Mol. Plant Pathol.* 9(6): 729–740.

Hagstrum, D. W. and Milliken, G. A. (1991). Modeling differences in insect developmental times between constant and fluctuating temperatures. *Ann. Entomol. Soc. Am.* 84(4): 369–379.

Harvey, J. A., Malcicka, M. and Ellers, J. (2015). Integrating more biological and ecological realism into studies of multitrophic interactions. Ecological Entomol. 40: 349–352.

He, X., Kiær, L. P., Jensen, P. M. and Sigsgaard, L. (2021). The effect of floral resources on predator longevity and fecundity: a systematic review and meta-analysis. *Biological Control* 153. doi:10.1016/j.biocontrol.2020.104476.

Hussey, N. W. and Bravenboer, L. (1971). Control of pests in glasshouse culture by the introduction of natural enemies. In: Huffaker, C. B. (Ed) *Biological Control*. Plenum Press, New York, pp 195–216.

Jacobson, R. (2008). Towards a robust IPM programme for organic tomato. *IOBC WPRS Bull.* 32: 87–90.

Jandricic, S. E. (2017). Potential for *A. limonicus* to control whitefly in greenhouse ornamental crops under cool weather conditions: preliminary tests in a commercial greenhouse. *IOBC WPRS Bull.* 124: 119–124.

Jarvis, W. R. (1992). Managing diseases in greenhouse crops. *Plant Dis.* 73(3): 190–194.

Johansen, N. S., Tadesse, B. A., Suthaparan, A., Stensvand, A., From, P. J. and Gadoury, D. M. (2017). Nighttime treatments of ultraviolet (UV) light targeting powdery mildews also suppress the two-spotted spider mite (*Tetranychus urticae*). *Phytopathology* 107: 77–77.

Johansen, N. S., Vanninen, I., Pinto, D. M., Nissinen, A. I. and Shipp, L. (2011). In the light of new greenhouse technologies: 2. Direct effects of artificial lighting on arthropods and integrated pest management in greenhouse crops. *Ann. Appl. Biol.* 159(1): 1–27.

Körner, O. and Challa, H. (2004). Temperature integration and process-based humidity control in chrysanthemum. *Comput. Electron. Agric.* 43(1): 1–21.

Labbe, R., Gagnier, D. and Shipp, L. (2017). Light supplementation improves the short-season establishment of *Dicyphus hesperus* (Knight) on commercial tomato crops. *IOBC WPRS Bull.* 124: 107–112.

Mahlein, A. K. (2016). Plant disease detection by imaging sensors – parallels and specific demands for precision agriculture and plant phenotyping. *Plant Dis.* 100(2): 241–251.

Mechant, E. (2014). *Integrated Pest Management in Ornamental Horticulture* (in Dutch). PCS Publications, 41 pp.

Mechant, E., Luypaert, G., Van Delsen, B., Witters, J., Van Huylenbroeck, J. and Gobin, B. (2015). Development and validation of a three-step detection protocol for broad mites (*Polyphagotarsonemus latus*) in pot azalea (*Rhododendron simsii* hybrids) Entomologia Experimentalis et Applicata 156: 99–104.

Meng, L., Mestdagh, H., Ameye, M., Audenaert, K., Höfte, M. and Van Labeke, M.-C. (2020). Phenotypic variation of *Botrytis cinerea* isolates is influenced by spectral light quality Front. *Plant Sci.* 11.

Messelink, G. J., Bennison, J., Alomar, O., Ingegno, B. L., Tavella, L., Shipp, L., Palevsky, E. and Wäckers, F. L. (2014). Approaches to conserving natural enemy populations in greenhouse crops: current methods and future prospects. *BioControl* 59(4): 377–393.

Messelink, G. J., Sabelis, M. W. and Janssen, A. (2012). Generalist predators, food web complexities and biological pest control in greenhouse crops. In: *Integrated Pest Management and Pest Control - Current and Future Tactics*, pp 191–214.

Mironidis, G. K. and Savopoulou-Soultani, M. (2008). Development, survivorship, and reproduction of *Helicoverpa armigera* (Lepidoptera: Noctuidae) under constant and alternating temperatures. *Environ. Entomol.* 37(1): 16–28.

Nomikou, M., Sabelis, M. W. and Janssen, A. (2010). Pollen subsidies promote whitefly control through the numerical response of predatory mites. *BioControl* 55(2): 253–260.

Ogino, T., Uehara, T., Muraji, M., Yamaguchi, T., Ichihashi, T., Suzuki, T., Kainoh, Y. and Shimoda, M. (2016). Violet LED light enhances the recruitment of a thrip predator in open fields. *Sci. Rep.* 6: 32302.

Parkinson, J., Gobin, B. and Hughes, W. O. H. (2016). Heritability of symbiont density reveals distinct regulatory mechanisms in a tripartite symbiosis. *Ecol. and Evol.* 6(7): 2053–2060.

Parkinson, J. F., Gobin, B. and Hughes, WOH (2018). Symbionts in excess? No correlation between symbiont density and the ability of mealybug hosts to exploit plant sepcies or tolerate insecticide stress. Agricultural and Forest Entomology 20:451–460

Pijnakker, J., Arijs, Y., Van Gansbeke, D. and Wäckers, F. (2017). A food supplement for the predatory mite *Iphiseius degenerans* (Berlese) in sweet pepper crops. *IOBC WPRS Bull.* 124: 159–164.

Pollet, B., Steppe, K., Dambre, P., Van Labeke, M. C. and Lemeur, R. (2009). Temperature integration of *Hedera helix* L.: quality aspects and growth response. *Sci. Hort.* 120(1): 89–95.

Roberts, J. M., Bruce, T. J. A., Monaghan, J. M., Pope, T. W., Leather, S. R. and Beacham, A. M. (2020). Vertical farming systems bring new considerations for pest and disease management. *Ann. Appl. Biol.* 176(3): 226–232.

Schumacher, S. K., Marsh, T. L. and Williams, K. A. (2006). Optimal pest control in greenhouse production of ornamental crops. *Agr. Econ.* 34(1): 39–50.

Shamshiri, R. R., Kalantari, F., Ting, K. C., Thorp, K. R., Hameed, I. A., Weltzien, C., Ahmad, D. and Mojgan Shad, Z. (2018). Advances in greenhouse automation and controlled environment agriculture: A transition to plant factories and urban agriculture. *Int. J. Agric. Biol. Eng.* 11(1): 1–22.

Steinberg, S. (2017). Innovative strategies that enhance the cost-effectiveness, reliability and efficacy of arthropod-based biocontrol in greenhouse crops. *IOBC WPRS Bull.* 124: 11–16.

Valentin, R. (2017). Using Biological Control as a first line of defence from 'Start to Finish' in greenhouse vegetables and ornamental production. *IOBC WPRS Bull.* 124: 147–150.

Van De Veire, M. and Tirry, L. (2003). Side effects of pesticides on four species of beneficials used in IPM in glasshouse vegetable crops: "worst case" laboratory tests. *IOBC WPRS Bul.* 26: 41–50.

Van Gansbeke, D., De Schrijver, L., Spranghers, T., Audenaert, J., Verhoeven, R., Nguyen, D. T., Gobin, B., Tirry, L. and De Clercq, P. (2013). Alternating temperatures affect life table parameters of *Phytoseiulus persimilis, Neoseiulus californicus* (Acari: Phytoseiidae) and their prey *Tetranychus urticae* (Acari: Tetranychidae). *Exp. Appl. Acarol.* 61(3): 285–298.

Van Gansbeke, D., Nguyen, D. T., Audenaert, J., Verhoeven, R., Deforce, K., Gobin, B., Tirry, L. and De Clercq, P. (2014a). Diet-dependent cannibalism in the omnivorous phytoseiid mite *Amblydromalus limonicus. Biol. Control* 74: 30–35.

Van Gansbeke, D., Nguyen, D. T., Audenaert, J., Verhoeven, R., Gobin, B., Tirry, L. and De Clercq, P. (2014b). Food supplementation affects interactions between a phytoseiid predator and its omnivorous prey. *Biol. Control* 76: 95–100.

Van Gansbeke, D., Audenaert, J., Nguyen, D. T., Verhoeven, R., Gobin, B., Tirry, L. and De Clercq, P. (2015a). Diurnal temperature variations affect development of a herbivorous arthropod pest and its predators. *PLoS ONE* 10(4): e0124898.

Van Gansbeke, D., Nguyen, D. T., Audenaert, J., Verhoeven, R., Gobin, B., Tirry, L. and De Clercq, P. (2015b). Prey consumption by phytoseiid spider mite predators as affected by diurnal temperature variations. *BioControl* 60(5): 595–603.

Van Gansbeke, D., Nguyen, D. T., Audenaert, J., Gobin, B., Tirry, L. and De Clercq, P. (2016). Establishment of Amblyseius swirskii in greenhouse crops using food supplements. Systematic & Applied Acarology 21: 1174–1184.

van Lenteren, J. C. (2012). The state of commercial augmentative biological control: plenty of natural enemies, but a frustrating lack of uptake. *BioControl* 57(1): 1–20.

van Lenteren, J. C., Alomar, O., Ravensberg, W. J. and Urbaneja, A. (2020). Biological control agents for control of pests in greenhouses. In: Gullino, M., Albajes, R. and Nicot, P. (Eds) *Integrated Pest and Disease Management in Greenhouse Crops. Plant Pathology in the 21st Century* (vol. 9). Springer, Cham.

van Lenteren, J. C. and Nicot, P. C. (2020). Integrated pest management methods and considerations concerning implementation in greenhouses. In: Gullino, M., Albajes, R. and Nicot, P. (Eds) *Integrated Pest and Disease Management in Greenhouse Crops. Plant Pathology in the 21st Century* (vol. 9). Springer, Cham.

Vanninen, I., Pinto, D. M., Nissinen, A. I., Johansen, N. S. and Shipp, L. (2010). In the light of new greenhouse technologies: 1. Plant-mediated effects of artificial lighting on arthropods and tritrophic interactions. *Ann. Appl. Biol.* 157(3): 393–414.

van Schelt, J., Tetteroo, A., Hoogerbrugge, H., Veenman, R. and Bolckmans, K. (2008). The release of beneficials in greenhouses with an air blower, a new wind in biocontrol. *IOBC WPRS Bull.* 32: 245–248.

Vercauteren, A., Riedel, M., Maes, M., Werres, S. and Heungens, K. (2013). Survival of *Phytophthora ramorum* in Rhododendron root balls and in rootless substrates. *Plant Pathol.* 62(1): 166–176.

Wäckers, F. and van Rijn, P. (2005). Food for protection: an introduction. In: Wäckers, F. L., van Rijn, P. C. J. and Bruin, J. (Eds) *Plant-Provided Food for Carnivorous Insects:*

A Protective Mutualism and Its Applications. Cambridge University Press, New York, pp 1–14.

Wainwright-Evans, S. (2017). Integration of biological controls into production systems: success in the real world. *IOBC WPRS Bull.* 124: 87–88.

Wang, S., Tan, X. L., Michaud, J. P., Zhang, F. and Guo, X. (2013). Light intensity and wavelength influence development, reproduction and locomotor activity in the predatory flower bug *Orius sauteri* (Poppius) (Hemiptera: Anthocoridae). *BioControl* 58(5): 667–674.

Werres, S., Wagner, S., Brand, T., Kaminski, K. and Seipp, D. (2007). Survival of *Phytophthora ramorum* in recirculating irrigation water and subsequent infection of Rhododendron and Viburnum. *Plant Dis.* 91(8): 1034–1044.

Chapter 13

Practical applications of integrated pest management in horticultural cultivation: the cases of protected tomato and outdoor *Brassica* production

Richard H. Binks, FreshTec Agricultural Consultancy Limited, UK

1 Introduction

Vegetable crops are cultivated throughout the world using protected and outdoor production systems (Welbaum, 2015). Global vegetable production was recorded at 1.04 billion tons in 2010, a 33% rise since 2000 (Leskovar et al., 2014). The majority of the world's vegetable production (approx.

http://dx.doi.org/10.19103/AS.2021.0095.15

75%) is located in Asia, with China, India and the USA accounting for 60% of the world's vegetable production (Leskovar et al., 2014). Over the next 50 years, farmers throughout the world face enormous challenges from biotic and abiotic factors. This is due to the rapid rise in the size of the human population, rising bio-fuel demands, shrinking land and water resources and significant changes in the global climate. According to the Food and Agriculture Organisation of the United Nations (FAO), the world's population is currently growing by approximately 1.1% per year, and if current trends continue, according to the medium-variant projection, the world's population is projected to reach 9.7 billion by 2050 (Molotoks et al., 2021). In addition to these startling figures, world poverty is also on the rise, with the number of undernourished people in the world reaching an estimated 821 million in 2017 which is around one in nine people (Molotoks et al., 2021). With the demand for urban and industrial land, there are limited possibilities to increase fertile crop production areas. Farmers will therefore have to develop sustainable and more environmentally friendly production technologies. This will enable them to extend production into new marginal production areas and better manage the impacts of food production (Leskovar et al., 2014). In addition to the environmental changes they face worldwide, farmers face serious challenges from pest insects. These include pests associated with indoor- and outdoor-cultivated vegetable crops.

Over the past 50 years, integrated pest management (IPM) has been one of the strategies available to farmers to sustainably and holistically reduce the impact of pest insects on crops and maintain resilient and environmentally friendly cropping systems (Stenberg, 2017; Dent and Binks, 2020). The adoption of IPM strategies should enable vegetable producers to meet the needs of the human population while reducing their ecological footprint (Vallad et al., 2018). IPM in vegetable crops involves the use of cultural techniques, including host-plant resistance, biological control, environmental management techniques and, as a last resort, selective plant protection products (Bird, 2014; Dent and Binks, 2020).

This chapter will outline two case studies where IPM strategies and techniques have been implemented to control pest insects in indoor and outdoor vegetable cropping situations. In particular, the case studies will highlight the impact that biological control has had on protected tomato production as well as the benefits that biological systems can have on outdoor vegetable crops, for example in *Brassica* production.

2 Integrated pest management in protected tomato production

Tomato (*Solanum lycopersicum* L.) belongs to Solanaceae (nightshade family) and is one of the world's major horticultural products, both fresh and processed.

(Ramasamy and Ravishankar, 2018). This is primarily due to the way in which tomatoes can be consumed, for example as a fresh product, or processed for their juice or puree (Costa and Heuvelink, 2018; Flint, 2018). Global tomato production was estimated to be 117 million tons in 2018 (75% fresh and 25% processed), with China, the European Union (EU) including the Netherlands, Spain and UK, India, the USA and Turkey contributing approximately 70% of this production across an area of 4.8 million hectares (Vallad et al., 2018; Ramasamy and Ravishankar, 2018). Tomatoes are produced throughout the world including in (sub) tropical climates, but mainly in temperate zones where summers are long and there is winter precipitation (Costa and Heuvelink, 2018).

Tomato crops are grown under protection (glasshouses or plastic structures) or in open-field conditions. Tomato plants are short-lived perennials and are normally maintained in greenhouses for up to a year or more, depending on the environmental conditions and on the purpose of the crop. Tomato crops are grown outdoors in soil (polythene-mulched fields) or indoors, in soil, or in 'soil-less' cultures with drip irrigation (Kubota et al., 2018). The term 'hydroponic production' refers to substrate-based 'soil-less' culture or water cultures, such as by using the nutrient film technique (NFT), which is used less frequently in commercial settings than substrate-based 'soil-less' culture (Kubota et al., 2018) (Fig. 1).

3 Integrated pest management strategies in protected tomato crops: resistant varieties, physical and cultural control

An IPM programme in any cropping system should consider non-chemical control methods in the first instance, as many of the pest insects that affect

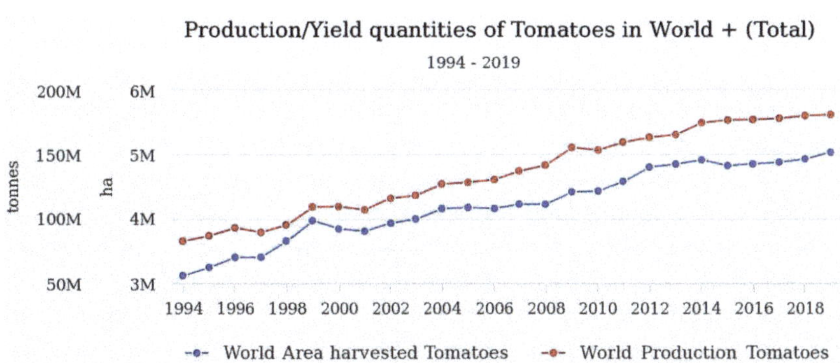

Figure 1 Global tomato production (Global area harvested and global production). Source: FAOSTAT 2021.

tomato crops can develop resistance to insecticides (Csizinszky et al., 2005). In the case of tomato, this will include the use of crop varieties which exhibit resistance or tolerance to pests. Plant breeders focus on developing disease resistance, particularly virus resistance traits within tomato varieties. Several key viruses are transmitted by insects and arthropod vectors (Picó et al., 1996; Csosnek and Latterot, 1997; Stout et al., 2018; Vallad et al., 2018). These include the tomato yellow leaf curl virus (TYLCV), which is transmitted by the tobacco whitefly (*Bemisia tabaci*) and which is a serious disease of tomato crops grown in Mediterranean countries, the Middle East and Africa (Csizinszky et al., 2005). The tomato spotted wilt virus (TSWV) belongs to the genus *Tospovirus* and is transmitted mechanically, but more importantly, can be transmitted by thrips including western flower thrips (*Frankliniella occidentalis*) (Rosellò et al., 1996; Vallad et al., 2018). In the case of TYLCV, tolerant varieties are cultivated (van der Velden et al., 2012).

Cultural control strategies for both these viruses include changing the planting schedule of the crop to avoid periods of peak activity by the insect vectors and eliminating primary or secondary virus sources (i.e. weeds). Physical barriers to prevent weed ingress into the greenhouse can also be deployed (Bonsignore and Vacante, 2017).

In greenhouses in Southern Europe, the use of mesh screens located in the ventilation openings can help to prevent the ingress of pests into tomato crops (Berlinger et al., 2002; van der Velden et al., 2012; Bonsignore and Vacante, 2017). The use of ultraviolet (UV)-reflective plastic or straw has been shown to reduce the numbers of whiteflies and thrips landing on crops (Ramasamy and Ravishankar, 2018; Vallad et al., 2018). Crops can also be protected from thrips by using UV-absorbing plastic films (Ramasamy and Ravishankar, 2018). In addition to their use for insect monitoring and scouting, sticky cards can act as passive insect traps.

Manipulation of adult insect behaviour has been exploited through the use of semiochemicals, functioning as signalling compounds. Methyl isoniconate is the active ingredient in Lurem-TR (Koppert Biological Systems, Berkel en Rodenrijs, the Netherlands) and is an attractant for both male and female western flower thrips (WFT) as well as other thrips species, and is used by them to locate host plants (Teulon et al., 2017; Mouden et al., 2017). This attractant is used as a lure in conjunction with blue sticky card traps (Mouden et al., 2017; Ramasamy and Ravishankar, 2018). By using this technique in conjunction with introductions of the predatory bug *Orius laevigatus,* populations of thrips can be reduced in crops. Large sticky 'roller' traps can be deployed in greenhouses to mass-trap mobile insects such as adult whiteflies, thrips and leaf miners (Ramasamy and Ravishankar, 2018; Dent and Binks, 2020). The crop environment is important, particularly the relative humidity in the greenhouse, when controlling diseases in tomato crops, but can also be important when trying to reduce certain pests, for example spider mites and thrips.

Monitoring and trapping of insects, in particular lepidopteran pest insects, can be achieved using sex pheromones. Tomatoes can be affected by a number of lepidopteran species depending on the country of tomato production. The common species that can be found in protected tomato crops in Northern Europe include the silver Y moth (*Autographa gamma*), tomato moth (*Laconombia oleracea*) and the tomato looper (*Chrysodeixis chacites*) which is found in Southern Europe but can be found across Europe because it is a migratory species (Molotievskiy et al., 2019). Other species which can affect tomatoes include the *Spodoptera* (armyworms), for example, the black cutworm (*Spodoptera littoralis*), and the corn earworm (*Helicoverpa armigera*) (Bonsignore and Vacante, 2017; Vallad et al., 2018). The tobacco cutworm (*Spodoptera litura*) can also be a serious problem in netted greenhouses in Asia (Ramasamy and Ravishankar, 2018). Sex pheromones are commercially available for all these species in many countries and can be used for monitoring and mass trapping.

4 Integrated pest management strategies in protected tomato crops: chemical and biological control

Over the past decade, there has been a sharp decline in the use of synthetic insecticides in protected tomato crops. This has primarily been due to the success of biological strategies to control noxious pest insects and the impact that insecticides can have on bumble bees which are used to pollinate protected tomato crops. Insecticide use alone to control the two-spotted spider mite (*Tetranychus urticae*) in protected crops is estimated to be valued at (US) $1 billion (George et al., 2015). Additional reasons for the decline in use include the discontinuation of reliable insecticides, the increase in insect resistance in insects such as thrips, aphid and caterpillar species, limited pesticide options for greenhouse use since the enclosed area increases risks to human health, environmental concerns, the exposure of the applicator to the chemicals and consumer trends desiring chemical-free produce (Amundson, 2012). The careful use of insecticides must be considered as part of any IPM strategy, with decisions being made using information derived from monitoring insect levels in the crop. The choice of insecticide can then be made based on its selectivity for the specific pest and also its low impact on natural enemies and bees (Matthews, 2018).

Current efforts are very much focused on developing robust biological solutions including the use of natural enemies in IPM strategies. However, the use of biopesticides, including biorational products, may also hold some promise, as these products are reported as having the potential for reduced occurrence of pest resistance, a higher amenability to combined use with biological control, and lowered environmental persistence with reduced product residues (George et al., 2015). According to Grant et al. (2010), biopesticides can be defined as mass-produced, biologically based products for crop protection. Products include microbial natural enemies, as well as insecticides derived from plants or insect-mating pheromones. However, biopesticides are not without their limitations. These limitations include having lower and slower overall kill rates than conventional chemical pesticides, being more sensitive to environmental degradation and often having a requirement for direct pest contact that necessitates high spray volumes and more informed application procedures (Pavela, 2014). In organic production, growers use natural predators as well as natural repellents (e.g. garlic, onion, zinnia, marigold and nasturtium) (Dorais and Schwarz, 2018). The use of pyrethrum is also employed when pest populations reach economically damaging levels. Microbial insecticides such as *B. thuringiensis* (Bt) and the nuclear polyhedrosis virus (NPV) can be effective against young caterpillars, for example, the corn earworm *Helicoverpa armigera* (Popham et al., 2016; Bonsignore and Vacante, 2017).

The biological IPM methods which are practised in protected tomato crops versus those grown in open fields are fundamentally different. Firstly, augmentation biological control (mass-reared natural enemies) is far more advanced in greenhouses than in field-grown tomato crops. This is historical and can be traced back to the late 1920s when E.R. Speyer identified parasitism of whitefly scales by the parasitoid *Encarsia formosa* in tomato crops (Hussey and Scopes, 1985). By 1929, Speyer had managed to mass produce *Encarsia* and distribute these parasitoids to seven countries around the world, including France and Canada (Hussey and Scopes, 1985; Helyer et al., 2014). It was not until the late 1960s that these practices were fully commercialised (Dent and Binks, 2020). Today, natural enemies are mass produced around the world by over 500 companies (Dent and Binks, 2020) and are introduced into glasshouse crops in more than 20 countries (van Lenteren, 2000; Dent and Binks, 2020). Secondly, protected, and in particular, glasshouse-grown tomato crops are in general of higher market value and require higher production levels than field crops. This means that more money can be spent on control measures, and in particular, the release of natural enemies. In recent years, the release of natural enemies into outdoor settings has become more popular and easier to carry out. This has included the release of predatory mites using unmanned aerial vehicles (UAVs) to control two-spotted spider mites (*Tetranychus urticae*) in field-grown tomato crops (Iost Filho et al., 2020; P. Banzato, pers. comm., 10 November 2019). The third benefit of using natural enemies in protected crops is that they are confined to the crop under production, whereas natural enemies released into fields disperse into surrounding fields and margins, and therefore, their effectiveness in the crop can be reduced (Vallad et al., 2018). In field-grown crops, the use of conservation biological control techniques has been shown to be beneficial, for example, the use of flower strips to encourage natural enemies in the crop.

In tomato crops, both specialist and generalist natural enemies are used to control pest insects. Specialist natural enemies are highly adapted to controlling their prey, such as the whitefly parasitoid *Encarsia formosa* and the predatory mite *Phytoseiulus persimilis* (Table 1.) which is used to control two-spotted spider mites. These insects normally have to be introduced repeatedly through the growing season and their introduction needs to be timed correctly. Alternatively, generalist predators establish quickly in a crop and can feed on a variety of food sources. The generalist predators *Nesidiocoris tenuis* and *Macrolophus pygmaeus* (Table 1) are particularly well adapted to tomato crops and can survive on plant sap and pests which include whiteflies, aphids, thrips, spider mites and small lepidopteran species (Vallad et al., 2018; van der Ent et al., 2017). These predators are normally introduced at the beginning of the cropping cycle and even if pests are not present in the crop, artificial food sources such as pollen or the eggs of the flour moth *Ephestia kuehniella* can be introduced (van der Ent et al., 2017).

In recent years, the most threatening and damaging pest insect of tomatoes meant for processing and fresh consumption is the South American tomato leaf miner (*Tuta absoluta*). This pest was first detected outside South America in 2006 but is now distributed in most of the tomato-growing areas of Europe, Africa and Asia (Campos et al., 2017; Mansour et al., 2018). Various strategies can be deployed to control this pest, but this can be difficult due to

Table 1 Commercially applied IPM programme for tomato crops

Pests	IPM programme
Trialeurodes vaporariorum	*Encarsia formosa; Macrolophus pygmaeus; Eretmocerus eremicus; Nesidiocoris tenuis*
Tetranychus urticae	Chemical control; *Phytoseiulus persimilis; Feltiella acarisuga*
Aculops lycopersici *Polyphagotarsonemus latus* *Liriomyza bryoniae; L. trifolii; L. huidobrensis*	*Dacnusa sibirica; Diglyphus isaea*
Noctuid spp.	*Bacillus thuringiensis* (Bt)
Aphid spp.	*Aphidoletes aphidimyza; Aphelinus abdominalis, Aphidius matricariae; Aphidius ervi*
Tuta absoluta	*Macrolophus pygmaeus; Nesidiocoris tenuis*

Adapted from Dent and Binks, 2020.

the endophytic habit of the larvae, which are protected in the leaf mesophyll or inside fruits. In addition, there have been many reports of insecticide resistance in *T. absoluta* to several key active ingredients, including the insect's ability to detoxify spinosad (Cocco et al., 2013; Guedes et al., 2019). Management strategies include the use of microbial insecticides such as Bt, the use of biological control agents (e.g. mirid predators and egg parasitoids) and mass trapping. An alternative control method which relies on the use of synthetic pheromones is mating disruption. Mating disruption involves the saturation of the atmosphere with a synthetic pheromone. This alters the ability of male moths to locate females and mate, thereby causing a long-term reduction in the population density. When pheromone dispensers containing a 90:10 blend of (3E,8Z,11Z)-tetradecatrien-1-yl acetate and (3E,8Z)- tetradecadien-1-yl acetate were introduced into greenhouses at a rate of 1000 dispensers/ha (60 g of active ingredient (a.i.)/ha), leaf damage was significantly reduced and infestations of *T. absoluta* were reduced throughout the growing season (Cocco et al., 2013). This mating disruption pheromone is now registered as Isonet® T in several countries and is sold as a vapour-releasing product formulation in a slow-release dispenser (micro capillary container – Shin Etsu Chemical Co. Ltd), and is proving to be a valuable component in IPM strategies against this pest.

5 Integrated pest management strategies to control particular insect pests of tomato crops: plant-feeding mites and whiteflies

5.1 Plant-feeding mites

Plant-feeding mites (*Acari*) are pests of many crops throughout the world and can be extremely problematic in tomato crops. The principal species that feed on tomato crops include the two-spotted spider mite *Tetranychus urticae* (Kock), the tomato red spider mite *Tetranychus evansi* (Baker and Pritchard), tomato russet mite *Aculops* lycopersici (Masee) and the broad mite *Polyphagotarsonemus latus* (Banks) (Table 1) (Vallad et al., 2018; van der Ent et al., 2017). Under favourable conditions, the lifecycles of the tomato russet mite and broad mite can be completed in under a week at 24–27°C. Spider mites can complete their life cycle in a similar time scale at higher temperatures (30°C). Mites cause damage to plants by using their piercing and sucking mouthparts to puncture cells and feed on sap. Spider mites can also cause webbing and leaf surface chlorosis when in high numbers. Broad mites feed on epidermal cells and cause bronzing on stems and lower leaf surfaces. Young expanding leaves fail to develop, and appear narrow. The stems of young plants are also swollen and appear to be russeted and greying-green in colour

(Bonsignore and Vacante, 2017). The undersides of tomato plant leaves which are colonised by russet mites appear silvery and chlorotic.

For over 40 years, the use of the specialist predatory mite *Phytoseiulus persimilis* has been at the centre of biological control strategies against spider mites. This predatory mite is used worldwide to control spider mites that affect tomato as well as cucumber and sweet peppers (Perdikis et al., 2008). As soon as the first spider mites are detected in and around plants and within hotspots, predatory mites are normally introduced. It is usually difficult to find *Phytoseiulus* within the first 2–3 weeks of application. It is the second generation that is more easily detected, and the generation that delivers a higher level of spider mite control. The presence of glandular trichomes on tomato plants reduces the ability of *P. persimilis* to control spider mites (Perdikis et al., 2008). For this reason, *P. persimilis* mites are released when plants are infested with spider mites (Perdikis et al., 2008). The rearing of the predators for four generations on tomato instead of bean has also been shown to considerably increase the performance and effectiveness of the predator on tomato (Perdikis et al., 2008). The control of russet mites (*Aculops lycopersici*) and the broad mite (*Polyphagotarsonemus latu*) on tomato using predatory mites remains difficult due to the issues with glandular trichomes (Bonsignore and Vacante, 2017). Instead, growers use synthetic chemistry, including wettable sulphur, to reduce the incidence of these pests in crops (Hubert, pers. comm., 5 February 2021).

5.2 Whiteflies

Whiteflies remain a serious problem in protected tomato crops. The principal species affecting tomato crops are the greenhouse whitefly *Trialeurodes vaporariorum* (Westwood) and the tobacco whitefly (*Bemisia tabaci* Gennadius). Both species develop from an egg through four larval instars and finally into an adult insect. Adult females can lay between 100 and 300 eggs during the 3–6 weeks that they are alive (Vallad et al., 2018; van der Ent et al., 2017). At 25°C, *B. tabaci* can develop from an egg to an adult in approximately 20 days, and development of *T. vaporariorum* takes 21 days.

Plants are affected in three different ways. Firstly, whiteflies are phloem feeders and use their piercing-sucking stylets to penetrate plants. When high numbers of whitefly adults feed on plants, the loss of sap can seriously affect the plant. Secondly, whiteflies excrete honeydew, which can then provide an ideal growth medium for sooty moulds to proliferate on leaf surfaces. Finally, and probably the most serious problem, is the ability of whiteflies to reduce tomato crop yields through their ability to transmit viruses. As one of the world's most important agricultural pests, *B. tabaci* is known to transmit over 60 plant viruses (Denholm et al., 1998). By far the most important virus problem in greenhouse- and field-grown tomato crops is the tomato yellow leaf curl virus

(TYLCV) which is transmitted by *B. tabaci* (Perdikis et al., 2008; Vallad et al., 2018). The virus is acquired by both immature and adult whiteflies. The TYLCV is transmitted through a persistent circulation mechanism (whiteflies acquire the virus during feeding and then transmit the virus after a latent period of 8–24 hours). In addition, adult whiteflies can hold on to the TYLCV throughout their life (Vallad et al., 2018). Plant symptoms include stunting and distortion and yellowing on tomato plant leaves. *B. tabaci* nymphs can also cause irregularities in fruit ripening through feeding.

The issue of insecticide resistance in whitefly populations remains a global challenge and has been a major driver for tomato breeders to develop resistant varieties (Lapidot and Friedmann, 2002) and for growers to rely on robust biological control solutions to control whitefly. The use of the parasitoid *Encarsia formosa* remains the backbone of biological control programmes against the glasshouse whitefly (*T. vaporariorum*) since its introduction in 1926. Female *Encarsia* wasps lay their eggs in the third and fourth instar larvae of whitefly (Vallad et al., 2018). *Encarsia* spp. activity can be recognised by the presence of black 'scales' on leaves (Helyer et al., 2014). The black scales are the pupal stages of the parasitoid and are formed inside the pupae of the whitefly. Females can lay up to 300 eggs at a rate of 12–15 per day under ideal conditions.

Other beneficial insects include another parasitoid wasp, *Eretmocerus mundus*. This wasp species is more tolerant of higher temperatures, which hinder *Encarsia,* and is also more tolerant than *Encarsia* to pesticides. *Eretmocerus* is also effective against *Bemisia tabaci* (Helyer et al., 2014; van der Ent et al., 2017). Growers of protected tomato crops are also reliant on the predatory mirid bugs *Macrolophus pygmaeus* and *Nesidiocoris tenuis* (Table 1). These predators have been used to control glasshouse whitefly and tobacco whitefly since the early 1990s. These are key predators in the fight against pests such as whitefly in greenhouses, particularly in tomato crops, as they have the ability to live on plants that have glandular trichomes (Sanchez et al., 2018). Although these two species of mirid bugs are effective predators, the main disadvantage of using them is that they can also be a pest, and can, if left unmanaged in a greenhouse, cause damage to plants, in particular, tomato fruits of the cherry tomato type (Helyer et al., 2014; Calvo et al., 2009).

6 Integrated pest management methods to control particular insect pests of tomato crops: aphids and leaf miners

6.1 Aphids

Aphids are small soft-bodied insects which feed on plant sap and live on plants in dense colonies. Aphids generally have pear-shaped bodies and lack segmentation or division in the head, thorax or abdomen (Rabasse and Wyatt,

1985). Their legs and antennae are slender and at the rear there are a pair of tubular wax glands known as cornicles. Aphids in temperate regions pass through an egg stage and four to five nymphal stages before reaching adulthood (Blackman and Eastop, 2000), whereas in warmer regions, aphids tend to be parthenogenetic and viviparous; that is, adults reproduce without mating and give birth to nymphs rather than eggs. This enables aphid populations to increase rapidly in warmer conditions. In some instances, aphid species such as the melon cotton aphid (*Aphis gossypii*) can tolerate the higher temperatures which can be prevalent in greenhouses in Southern Europe. Aphids are known to be *r*- strategists (maximal rate of increase) because they are very well adapted to exploiting a new temporary habitat through rapid population increase. Depending on the environmental conditions, adults may be wingless (apterae) or winged (alatae). Infestations in greenhouses in spring are normally brought about by alate aphids entering through greenhouse vents. Aphids may come from overwintering hosts, or later in the season in Southern Europe, they may fly between summer host-plant species (Rabasse and van Steenis, 1999). Normally, aphids are more problematic in cooler temperate regions than in subtropical climates where temperatures can be too high for many species (Vallad et al., 2018). On entering the greenhouse, aphids develop from small isolated foci, and from these foci, populations develop rapidly with overlapping generations, colonising neighbouring plants. In ideal conditions, that is, constant temperatures and in the absence of natural enemies, aphid populations can grow exponentially (Rabasse and van Steenis, 1999) so that aphids increase in number by a fixed proportion each day. The value *r* is used to measure this rate of increase, which is dependent on the host plant and temperature.

Aphids can be found on the underside of tomato leaves, but if conditions are favourable, they can be found on all plant parts. The four most important species that affect protected tomatoes are the polyphagous species – the peach-potato aphid *Myzus persicae* (Sulzer), the melon cotton aphid *Aphis gossypii* (Glover), the potato aphid *Macrosiphum euphorbiae* (Thomas) and the glasshouse potato aphid *Aulacorthum solani* (Kaltenbach) (van der Ent et al., 2017). Damage to tomato crops caused by these species can be severe and has three aspects. Firstly, aphids are sap feeders and therefore excrete honeydew; which in turn leads to the development of sooty moulds on plants. Secondly, aphid feeding can cause wilting, necrosis and spotting on leaves. Thirdly, aphids can be vectors of damaging plant viruses, and particularly in tomato crops, they can transmit the tobacco etch virus (TEV), potato Y virus (PYV) and cucumber mosaic virus (CMV) (Vallad et al., 2018).

In modern protected tomato production, aphid control is brought about by the use of specialist parasitoids and insecticides. The most commonly used beneficial insects are the parasitoid wasps *Aphidius ervi*, *Aphelinus abdominalis* and the midge *Aphidoletes aphidimyza* (Table 1). Aphids can sometimes be present very soon after planting, and if so, control measures should commence immediately. *Aphidius ervi* will parasitise the potato aphid (*M. euphorbiae*) and the glasshouse potato aphid (*A. solani*). *Aphidius ervi* can also parasitise the English grain aphid *Sitobion avenae* which is used in banker plant systems. The banker plant system provides supplementary food resources and sites for oviposition (Pijnakker et al., 2020). The system uses a plant species different from the crop and provides the parasitoid wasp species with alternative (non-pest) prey and/or plant-based food resources. In addition, adult wasps are active at temperatures as low as 10°C (van der Ent et al., 2017). This makes *A. ervi* a useful predator in tomato crops grown in cooler climates. Sometimes, secondary parasitoids, known as hyperparasitoids, can affect the ability of *A. ervi* to work effectively in the glasshouse (Vallad et al., 2018). The gall midge *Aphidoletes aphidimyza* can be used to control aphid populations in protected tomato crops. Female midges actively seek and are attracted to the honeydew produced by their prey (Perdikis et al., 2008). The larvae of this predator inject a paralysing toxin into the aphid which then initiates pre-digestion of the aphid's body contents. Generally, the females prefer to lay eggs in dense aphid colonies or nearby, as the larvae move for only short distances. This predator has been shown to be effective against many species of aphid, but in tomato crops, its ability to control *Myzus persicae* is well documented (Perdikis et al., 2008). Other predators including chrysopids, for example, the green lacewing (*Chrysoperla carnea*) and syrphids, for example, *Sphaerophoria rueppellii* and *Miridae*, can be used effectively against aphids in the glasshouse.

6.2 Leaf miners

Three species of leaf miner are associated with tomato crops. They are the tomato leaf miner *Liriomyza bryonae* (Blanchard), which is the most dominant species in greenhouses, the vegetable leaf miner *L. sativae* (Blanchard) and the American serpentine leaf miner *L.trifolii* (Burgess). Adult leaf miners are small yellow-and-black-coloured flies. Females lay their eggs on the surface of leaves. Eggs are laid on the underside of leaves by *L. huidobrensis* and the larvae mine near the midrib and lateral veins of leaves (Weintraub and Horowitz, 1995). The other two species, *L. sativae* and *L. trifolii*, oviposit on the upper side of leaves and the larval mines do not follow a specific pattern. Three larval stages feed on mesophyll tissue and then exit the leaf and drop to the soil, or in the case of artificial substrate production between the folds of the plastic membrane, to pupate. Before pupating, larvae cut a sickle-shaped exit hole in the leaf with their mouthparts and after approximately one hour, the larva exits the leaf (van der Ent et al., 2017). A small proportion of larvae remain attached to the leaf where they pupate. As well as leaf-mining damage (Fig. 2), the adults puncture leaves with their ovipositors to feed on plant juices.

In protected tomato production, leaf miners are controlled using commercially produced parasitoids. The two most widely used are *Diglyphus isaea* (Walker) and *Dacnusa sibirica* (Telenga). *D. isaea* is an ectoparasite, which means that it lays its eggs beside the host. *D. sibirica* is, on the other hand, an endoparasite - the eggs are laid inside the host (van der Ent et al., 2017). *D. sibirica* is most effective in the early season due to its ability to control leaf miners at lower temperatures. Later in the season, when temperatures are higher (>20°C), the use of *D. isaea* predominates. Female wasps lay 1–5 eggs in each mine, with one egg completing its development. The females paralyse their hosts with venom, preferring to lay their eggs on or near the second or third instar larvae. The females then pierce leaf miner larvae with their ovipositors and feed on the fluids that are exuded. There are three larval stages. The older larvae leave their dead host and crawl back into the mine to pupate. Pupation occurs between the upper and lower epidermis of the leaf, which is held apart by six pillars of frass

Figure 2 Tomato leaf miner (*Liriomyza bryoniae*) damage on tomato leaf (Photo courtesy of Richard H. Binks).

Figure 3 *Diglyphus isaea* pupa (Photo courtesy of Koppert Biological Systems).

(Fig. 3) - this stops pupal dehydration. The parasitic wasps then exit the mine by making a hole through the upper leaf epidermis. In practice, high levels of control can be gained from the introduction of these parasitoids. Sampson and Walker (1998) showed that using pest thresholds for *D. isaea* release could be used to improve early parasitoid establishment in commercial crops. An action threshold of one new mine per plant per week was implemented to trigger the release of *D. isaea* three times at seven-day intervals (Liu et al., 2011). Liu et al. found that *D. isaea* established faster and the number of parasitoids released and control costs were reduced (Liu et al., 2011).

Researchers have also investigated the use of entomopathogenic nematodes (*Heterorhabditis* and *Steinernema* species) to infect and kill leaf miner larvae, the pre puparia and early puparia (Williams and MacDonald, 1995; Tomalak et al., 2005; Liu et al., 2009; Garcia-del-Pino et al., 2018). Infective stages of nematodes sprayed on *Liriomyza*-infested plants enter leaf mines via the punctures created by female flies. However, the efficacy of the nematodes does depend on pest species, the development stages of the pest, concentration of nematodes and importantly, the environmental conditions (Xuan et al., 2018). In greenhouse and laboratory studies, leaf miner mortality can be as high as 85-97% at high humidity (Xuan et al., 2018). The use of nematodes on a commercial scale to control *Liriomyza* species still remains challenging because of their low mortality and sensitivity to low humidity.

7 Integrated pest management for tomato crops: conclusions

IPM systems in protected tomato crops are very advanced and include a range of measures to control noxious pest insects. These techniques and strategies need to be promoted more so as to rationalise and reduce pesticide inputs (Arnó i Pujol et al., 2009). In general, pest control in protected tomato crops is technically possible. However, there are some big challenges which will need addressing. These include research focusing on whiteflies and their ability to transmit viruses such as the TYLCV that can cause up to 100% yield losses in tomato crops (Prasad et al., 2020). Studies on mites, including the tomato russet mite (*Aculops lycopersici*) and western flower thrips (*Frankliniella occidentalis*) as secondary pests in Mediterranean tomato production will also be important (Arnó i Pujol et al., 2009). Finally, a better understanding of the potential of plant health and the development of biopesticide technology as a means to reduce the dependency on chemical pest control will be needed.

8 Integrated pest management in outdoor *Brassica* production

Brassica vegetable crops belong to the Brassicaceae (=Cruciferae) family which contains over 1800 species from more than 100 genera worldwide. This family includes many important vegetable, oil and field crops (Dixon, 2007; Welbaum, 2015). Brassicaceae contain more than 40 species of both annual and biennial plants but also include perennial herbaceous plants and small shrubs of Old World origin (Welbaum, 2015). A full review of the origin and diversity of *Brassica* crops has been covered in detail by Sadowski and Kole (Sadowski and Kole, 2016) as well as Dixon (2007). This case study will cover vegetable crops in the genus *Brassica*, albeit many of the pest insects that affect vegetable crops will affect canola oil crops, for example, *Brassica napus* and *Brassica campestris*.

The majority of commonly grown *Brassica* vegetables include a range of popular crops which form part of the human diet. These include cabbage, broccoli, cauliflower, Brussels sprouts, kale, collards and kohlrabi (Table 2) (Welbaum, 2015). *Brassica* vegetable crops are grown worldwide, with the exception of some tropical areas. Among these, cabbage is most widely produced, followed by cauliflower and broccoli. The origin of this group is believed to be in Eastern Mediterranean countries (Dogru et al., 2018). According to the FAO, the global vegetable *Brassica* production is over 70 million tonnes per year, with China producing over 30 million tonnes (FAOSTAT, 2021). India, the Russian Federation, Republic of Korea, Ukraine and Turkey are also significant producers, with Spain being the largest European producer (Dogru et al., 2018). Brassica crops are in general grown in cool climates and require deep, fertile sandy or silty loam soils. Vegetable *Brassica* can either be directly seeded using precision seeders, or where early production is required and where temperatures and moisture stress are common, crops are established by transplanting young plants. Growers

Table 2 Economically important crops in the *Brassicaceae*.

Brassicaceae	Common name
Armoracia rusticana	Horseradish
Sinapis alba	White mustard
Brassica juncea	Leaf mustard
Brassica napus (Napobrassica group)	Rutabaga
B. napus (Pabularia group)	Siberian kale
Brassica oleracea (Acephhala group)	Kale, collard
Brassica oleracea (Albogiabra group)	Chinese kale
Brassica oleracea (Botrytis group)	Cauliflower
Brassica oleracea (Capitata group)	Cabbage
Brassica oleracea (Gemmifera group)	Brussels sprouts
Brassica oleracea (Gongylodes group)	Kohlrabi
Brassica oleracea (Italica group)	Broccoli
Brassica oleracea (Costata group)	Tronchuda cabbage
Brassica napus (Chinensis group)	Chinese cabbage (non-heading), pak choi
B. rapa (Pekinensis group)	Chinese cabbage (heading), pe-tsai
B. rapa Perviridis group)	Spinach mustard
B. rapa (Rapifera group)	Turnip
B. rapa (Ruvo group)	Broccoli raab, rapini
Lepidium sativum	Garden cress
Crambe maritima	Sea kale
Nasturtium officinale	Watercress
Raphanus sativus	Radish
Wasabia japonica	Japanese horseradish

Adapted from Welbaum, 2015.

may produce their own transplants or purchase these from specialist plant propagation businesses.

9 Integrated pest management methods to control insect pests of *Brassica* crops: aphids and cabbage root fly

9.1 Aphids

The principal species of aphid found on *Brassica* crops around the world are the foliage-feeding aphids *Brevicoryne brassicae* (cabbage aphid), *Lipaphis pseudobrassicae* (turnip aphid), *Myzus ascalonicus* (shallot aphid), *Myzus persicae* (peach-potato aphid), *Macrosiphum euphorbiae* (potato aphid), *Aphis maidiradicis* (corn root aphid) and the root-feeding aphids *Pemphigus populitransversus* and *Smynthurodes betae* (Collier and Finch, 2017). The most economically important species are *B. brassicae* (cabbage aphid) and *M. persicae* (peach-potato aphid). A range of control options are available to growers, which include chemical, cultural and biological control options. These options and strategies will be discussed in relation to the important aphid species. Aphid infestations can cause direct feeding damage, which leads to leaves becoming yellow and curled (Gratwick, 1992), as well as being crop contaminants (Collier and Finch, 2017). *M. persicae* can also transmit important plant viruses such as the turnip mosaic virus (TMV) and cauliflower mosaic virus (CMV).

Developing resistant cultivars by using conventional breeding methods has proven to be challenging, primarily due to the lack of resistance sources within cultivated and even wild relatives of the Brassicaceae (Bhattacharya, 2019). The use of insecticides is still the most common method of controlling aphids in *Brassica* crops. Insecticides come in the form of seed treatments, granule formulations used at sowing or as foliar treatments. Different mechanisms of insecticide resistance (e.g. modified acetylcholinesterase (MACE) resistance to carbamates and knockdown (kdr and super-kdr) resistance to pyrethroid insecticides) have become prominent in *M. persicae* due to the intensive use of insecticides over many years (Bass et al., 2014; Prince and Chandler, 2020). *M. persicae* has been shown to be resistant to most classes of insecticides including organophosphates, carbamates, pyrethroids, cyclodienes and neonicotinoids (Bass et al., 2014; Foster and Devine, 2017). To date, insecticide resistance in *B. brassicae* has only been reported in Pakistan (Ahmad and Akhtar, 2013). The occurrence of insecticide resistance has led to the need to investigate and identify integrated and more sustainable forms of control to manage aphids in *Brassica* vegetable crops. The use of biorational products and biopesticides have become much more common - not just in organic production but increasingly in conventional vegetable *Brassica* production. These include the use of neem, horticultural oils and soaps, as well as pyrethrins. The efficacy of these products has been shown to be improved through improved application technology as well as by using wetting agents and adhesives such as those based on polyterpenes (Daniel et al., 2018). The use of entomopathogenic fungi (EPF), for example, *Beauveria bassiana*, to control pest insects including aphid species that affect *Brassica* crops is now of great interest. Recent work conducted by Prince and Chandler showed in field studies that when the product Botanigard ES - *Beauveria bassiana* (LAM International, USA) was sprayed twice (at seven-day intervals) against *B. brassicae* on cabbage plants, the treatment reduced aphid numbers by 73% (Prince and Chandler, 2020). Questions still remain about the use of these products in outdoor situations, as their efficacy can be affected by fluctuating temperatures and humidity. In addition, further work is required to investigate their mode of action, which will lead to improvements in application timing and determining the frequency of applications (Prince and Chandler, 2020).

In several countries including the UK, data collected from a network of suction traps are used by growers and agronomists to determine the flight patterns of key species of aphids (Collier and Finch, 2017). Day-degree models based on published data have also been developed (Collier and Finch, 2017; Cividanes and Santos-Cividanes, 2012). Treatment thresholds to determine the timing of insecticide sprays have been employed in several countries and are becoming more important, as spraying crops on a prophylactic basis is no longer acceptable. Efforts have been made to develop thresholds in vegetable *Brassica* crops but these are time-consuming to develop and can be inflexible, which has led to poor uptake by growers. However, thresholds are used by growers in the USA; for example, when *B. brassicae* infestations reach 1–2%, then the plants are treated (Collier and Finch, 2017).

Crop covers made of fine-mesh netting are used in *Brassica* seedling production to protect plants from aphids (Daniel et al., 2018). However, crop covers are expensive and are only used to protect high-value crops. Incidental aphid control can be provided when *Brassica* crops are covered to protect them against the cabbage root fly (*Delia radicum*) (Collier and Finch, 2017). Reflective and straw crop mulches have also proved to be useful (Daniel et al., 2018). Predatory natural enemies including lacewings and ladybirds, as well as parasitoids, play a vital role in the suppression of aphid populations in field crops. For aphid control in many field-grown crops, the emphasis has moved towards the development of conservation biological control (CBC) strategies. These strategies are designed to counteract habitat loss and environmental disturbance associated with intensive crop production. The CBC systems conserve natural enemies and thus contribute to pest suppression. In-field flowering strips have the benefit of encouraging naturally-occurring parasitoids and predators into crops (Hommes and Herbst, 2014; Ben-Issa et al., 2017; Ribeiro and Gontijo, 2017; Begg et al., 2017). For example, work carried out by Kienegger showed that cabbage aphid numbers could be reduced significantly by the presence of flowering strips in broccoli crops (Kienegger et al., 2003). The use of flowering and companion plants including *Alyssum* has also been investigated (Brennan, 2016). Trap crops and living mulches have been shown to be useful tools in the fight against aphids, particularly in organic cropping systems (Sarkar et al., 2018; Piñero and Manandhar, 2015). Augmentative biological control of aphids is practised in protected *Brassica* crops, particularly in propagation units where organic vegetable transplants are being produced prior to being despatched to the field. The release of predators such as lacewing larvae and eggs (*Chrysopa* spp.) into outdoor crops has been undertaken, particularly in vegetable crops, as far back as the mid-1970s (Beglyarov and

Smetnik, 1977; McEwen et al., 2001). There still is, however, a need to overcome the practical and economic constraints associated with releasing insects into crops on a field scale (Sigsgaard, 2005; Hance et al., 2017).

9.2 Cabbage root fly

The cabbage root fly (*Delia radicum* L.) is a serious pest of vegetable *Brassica* crops in Europe and North America (Gratwick, 1992). In Western Europe and North America, economic losses due to cabbage root fly damage have been estimated to be up to $100 million in some years (Shuhang et al., 2016).

Female flies can lay more than 60 eggs in their lifetime, which are laid mainly in the soil close to the stems of *Brassica* plants (Finch, 1971). Damage is caused by larvae which feed usually on the plant roots and lower parts of the stem and can affect any stage of plant growth, although normally, young plants are affected in the seedbed or shortly after transplanting. Female cabbage root flies sometimes lay eggs on the aerial parts of a plant and the larvae which develop subsequently can tunnel into the buttons or spears respectively, leading to rejections by retailers and processors. Flies complete 1–5 generations per year depending on temperature and location (McKinlay, 1992). In areas where the weather is particularly warm, flies will aestivate in the pupal stage (Finch and Collier, 1985). There are a range of approaches for monitoring and forecasting infestations of cabbage root fly in the field to support growers and their advisors in making treatment decisions (Collier et al., 2020)

Control strategies aimed at *D. radicum* include the use of insecticides, but insecticide use to control cabbage root fly is limited on some crops, therefore growers need to employ other methods. Cultural control strategies include ploughing or discing in crop residues after harvest to attempt to reduce the emergence of the new fly generation (Daniel et al., 2018); it is suggested that agronomic practices such as reducing irrigation during periods of oviposition might reduce egg hatch, as eggs are sensitive to desiccation (Daniel et al., 2018). Egg-laying by *D. radicum* can be disrupted by undersowing and/or intercropping with non-host plants (Finch and Collier, 2005), while straw and plastic mulches can be useful in reducing cabbage root fly infestations (Daniel et al., 2018).

Physical measures to control *D. radicum* can also be employed. Fine mesh netting will prevent female flies from accessing the crop to lay eggs, although the flies are able to lay their eggs on or through the mesh where plants are in contact with the mesh (Witkowska et al., 2018). The timing of the application and removal of the mesh also needs careful management to avoid 'trapping' cabbage root flies under the net (Witkowska et al., 2018). Blackshaw and co-workers have shown that vertical barriers can reduce the dispersal of *D. radicum* females into *Brassica* crops (Blackshaw et al., 2012).

Over the past few years, there has been an increased interest in the use of biorational products including the use of entomopathogenic fungi to control *D. radicum* (Razinger et al., 2017; Daniel et al., 2018; Hutchison, 2020). The use of entomopathogenic nematodes has also been investigated by various groups (Beck et al., 2014; Kapranas et al., 2020). Some evidence is available to show that garlic extracts have some insecticidal activity against cabbage root fly larvae under glasshouse conditions (Jukes et al., 2005). As discussed earlier in this case study, the use of conservation biological control and landscape management approaches requires further investigation (Josso et al., 2013; Kergunteuil et al., 2015). In particular, the use of so-called 'push–pull' techniques to manipulate cabbage root fly behaviour in and around crops could be beneficial in deterring oviposition by cabbage root flies in commercial *Brassica* crops (Lamy et al., 2018).

10 Integrated pest management methods to control insect pests of *Brassica* crops: lepidopteran pests

There are several species of Lepidoptera whose larvae feed on vegetable *Brassica* crops. These include the small white butterfly (*Pieris rapae*), the large white butterfly (*Pieris brassicae*), the cabbage looper (*Trichoplusia ni)*, the cabbage moth (*Mamestra brassicae*) and the garden pebble moth (*Evergestis forficalis*). In general, the larvae of these species cause feeding damage to plants and can contaminate them with their frass. Most species will have two or more generations per year. By far the most economically important and widespread lepidopterous pest that affects vegetable *Brassica* crops worldwide is the diamond back moth (DBM) (*Plutella xylostella* L.) and this section will focus on this pest. It was estimated in 2013 that *P. xylostella* damage, including that associated with yield losses and the cost of control in canola crops, costs the world economy US$4–5 billion annually (Furlong et al., 2013). The DBM has an extensive global distribution and is a particular problem in tropical regions where the pest can complete up to 14 generations in a year (Wainwright et al., 2020; CABI, 2021). The main challenge with this pest has been driven by the overuse of, and the consequent increased selection pressure for resistance to, a wide range of chemical and biological insecticides. As well as being a multi-generational pest species, DBM can migrate large distances and rapidly colonise suitable hosts. For example, DBM populations originating in Mexico were identified in summer-grown Canadian canola crops (Gurr et al., 2018). Over the past three decades, there has been a move to introduce more selective insecticides which are more compatible with natural enemies. In the 1990s, the launch of insecticides such as spinosad, emamectin and indoxacarb were heralded, but their overuse led to IPM strategies being dropped in many countries (Furlong et al., 2013). Many of the new insecticides were introduced sequentially and in spray mixtures with less selective older compounds to which the insect population had already become resistant. This resulted in the rapid development of resistance to the newer insecticides. One example where DBM resistance to spinosad was identified was in the Hawaiian Archipelago. The Insecticide Resistance Action Committee (IRAC)-sponsored programme aided by the University of Hawaii's extension advisors banned the use of spinosad and replaced its use with monthly rotations of emamectin benzoate and indoxacarb until DBM populations recovered their susceptibility to spinosad (Dent and Binks, 2020). Other examples where insecticide resistance has developed in populations of DBM are summarised by Furlong et al. (2013). The problem of insecticide resistance has therefore led to many decades of investigation into more sustainable and integrated methods of control.

A variety of novel strategies have been investigated to control and mitigate the effects of DBM populations on vegetable *Brassica* crops. These have included the use of conservation biological techniques with the aim of encouraging predators and parasitoids of DBM (Gurr et al., 2018). Intercropped trap crops can also be useful to reduce egg production by female DBM (Gurr et al., 2018) Researchers in China have recently shown that faba beans that have been sprayed with aqueous extracts of cruciferous plants rich in glucosinolates can be used as a 'dead-end' trap crop, that is, where the trap crop, being a non-host crop, fails to support the DBM offspring (Zhu et al., 2020). Non-host 'barrier' crops such as sorghum and maize can also aid in reducing the influx of DBM into crops (Daniel et al., 2018). The use of biopesticides such as *Bt kurstaki* can be effective against small larvae and these products are used in particular in organic farming systems. In addition, the entomopathogenic fungus *Beauveria bassiana* and neem can also be effective (Daniel et al., 2018).

The development and use of transgenic Bt-*Brassica* crops has shown some promise. Here, researchers have introduced Bt-toxin genes which enable *Brassica* lines including cauliflower and cabbage to suppress DBM populations (Furlong et al., 2013). Due to legislative barriers, this promising technology and the release of transgenic vegetable *Brassica* crops designed to combat DBM and insecticide resistance continue to be evaluated by countries around the world.

Area-wide pest management strategies have received a lot of attention over the past decade. Traditionally, the sterile insect technique (SIT) uses radiation or chemicals to sterilise mass-produced immature males which are then released to outcompete wild male moths. This subsequently reduces the likelihood of successful wild mating (Brown et al., 2019). Mass release of sterile γ-irradiated DBM has proven to be unsuccessful to date (Brown et al., 2019). However, technology that involves the development of self-limiting strains of DBM may hold some promise. One such technique is known as 'release of insects with a dominant lethal' (RIDL). Here, insect lines are genetically engineered such that female larvae die without the addition of an antidote, that is, tetracycline, in their diet; this switches off expression of the lethal gene. Such insects can be mass-reared in the laboratory and released, with the female progeny of successful mating with wild moths dying subsequently (Brown et al., 2019). Male-only release of individuals which are homozygous for the lethal gene may aid in suppressing populations. The system is self-limiting, that is, male moths pass on a self-limiting gene that prevents their female offspring from reaching adulthood. This reduces the number of females able to reproduce, and so with repeated releases of male moths, the pest population in the release area shrinks. As this method is self-limiting and does not persist in the environment, it is unlike gene drive approaches, which are designed to persist in the environment long-term (Shelton et al., 2020). The development of the *P. xylostella* strain OX4319L has allowed the production of male-only cohorts of moths for field release; that is, wild-type females that mate with these self-limiting males will not produce viable female progeny (Shelton et al., 2020). To date, greenhouse experiments and field work have shown that this technology works. However, further field studies will be required to demonstrate its potential to provide pest suppression and resistance management benefits in the field (Shelton et al., 2020). The potential benefits of RIDL over SIT include a reduction in the ratio of sterilised to wild insects released, and therefore reduced production costs. This is achieved by reducing the fitness costs that can result from using the high levels of radiation which are required to sterilise male insects. In Asia and Africa, where DBM is a serious problem, biological strategies have been employed to reduce the reliance on chemicals to control DBM. These have included the use of host-plant resistance, preservation and augmentation of natural enemies, introduction of appropriate parasitoids and predators, use of pathogens, manipulation of insect/host semiochemistry, cultural manipulation, the use of selective insecticides and improved spray application. All these approaches have seen success and failure but remain as the main focus of future strategies (Grzywacz et al., 2010).

11 Integrated pest management for *Brassica* crops: conclusions

A conceptual framework for IPM has been described by Stenberg. He states that a 'holistic' science of IPM has not been developed – '*there is a need to develop and formulate general principles for synergistically combining traditional and novel IPM actions to improve efforts to optimize plant protection solutions.*' Many research programmes are focused on single plant protection methods, but to be truly integrated, these studies should be designed to study the compatibility and optimisation of actions associated with two or more pest management elements (Stenberg, 2017). One example which relates to this case study is the use of biological control to support plant breeding strategies. It was suggested that biological control agents that decrease the differential pest fitness between resistant and susceptible cultivars should delay the evolution of counter-resistance in the pest (Gould et al., 1991). This concept has been demonstrated with Bt-broccoli expressing the *Cry1Ac* toxin against DBM. In this study, DBM evolved counter-resistance to Bt-broccoli, which developed more slowly in the presence of the spotted lady beetle (*Coleomegilla maculata*) (Liu et al., 2014).

Finally, sustainable IPM systems will be required to effectively control the three main groups of *Brassica* pests covered in this case study. In particular, research on DBM will need to continue at a rapid pace as this pest threatens most of the world's *Brassica* production, particularly in developing countries and in small-scale farming operations (Grasswitz, 2019). The development of resistant transgenic crops and studies into their interactions with parasitic and predatory enemies will undoubtedly continue to be of interest when developing IPM solutions to control aphids in *Brassica* (Bhattacharya, 2019). Strategies involving the development of ecologically friendly methods such as mating disruption, conservation biocontrol and landscape management techniques will hopefully ensure sustainable vegetable *Brassica* production in the future. A truly integrated approach will also need to be developed where the components of IPM programmes for different pests, pathogens and weeds, link effectively with one another. With the problems that insecticide resistance poses to growers and farmers around the world, lessons may also be learnt from organic production through the use of cultural and biological control solutions, and in particular, through improved use of biorational products in IPM systems.

12 Future trends in research

Integrated pest management systems in protected vegetable cropping, in particular those using augmentative insect-release systems, have been in place around the world for over 40 years (van Lenteren et al., 2018). With the advent of commercial biological control systems in the 1970s, the use of beneficial insects to control pests revolutionised IPM systems and led to reductions in the use of synthetic insecticides in protected vegetable crops. In outdoor vegetable cropping arenas, IPM strategies have been used but have not been as widely implemented, in part due to the availability and use of cheap insecticides. This situation is now changing, with a greater emphasis on more holistic and joined-up approaches to control pest insects in crops around the world. The key drivers here include a greater emphasis on the environmental agenda, the loss of active ingredients as well as the costs associated with developing new and environmentally sustainable pesticides. Complete and integrated pest insect control solutions in outdoor vegetable crops will need to be developed to ensure that vegetable production remains viable in many countries. One of the most important groups of vegetable crops in the world is that belonging to the Brassicaceae family. *Brassica* vegetable crops remain important in many countries as many species can be grown in a range of environmental conditions and can be grown all year round. The health benefits from edible *Brassica* crops are also helping to enhance their popularity in particular through their nutritional value, anticancer, antioxidant and anti-inflammatory properties (Sanlier and Guler, 2018). Vegetables of *Brassica*, though, are under attack from a whole array of pest insects. Future strategies will need to concentrate on controlling key pests such as aphids, cabbage root fly and DBM if crops are to remain economically viable. Therefore, sustainable IPM solutions need to be developed. Protected vegetable crops, including tomatoes grown under protection, are also not without their problems; in particular their vulnerability to viruses which are transmitted by insect vectors such as the whitefly. The integration of robust pest insect controls coupled with crop breeding strategies will be key to ensuring future tomato crop viability and production on a global scale. The development of insecticide resistance in whitefly is a key driver here, and therefore, robust IPM strategies are paramount in the fight against this pest and the harmful viruses that it transmits. With global trade comes the problem of movement of pest insects. In tomato production, the spread of the tomato leaf miner (*Tuta absoluta*) has been a major problem in many countries, including those in Europe, over the past decade. This pest is now controlled through integrated strategies which include mating disruption, insect trapping, the release of macro biological control agents and biopesticides.

Learnings from IPM strategies developed for indoor crops can in the future be translated to outdoor cropping scenarios. This will take time and a significant shift in the way that growers, farmers and agronomists think about controlling pest insects in their crops. A move to adopting organic farming practices in conventional systems may well be useful. The term 'organic fusion farming' may be used here, as this term describes the mixing of both conventional farming practices with those practised in organic farming. The conceptual holistic IPM framework described by Daniel et al. (2018) may act as an important future reference point for many growers, agronomists and IPM practitioners. In this scenario, pest management is achieved through blending direct and curative methods with indirect and preventative strategies. The reliance on nature conservation and the enrichment of biodiversity coupled with the use of more biologically based control measures may enable growers of outdoor vegetable crops to become less reliant on 'direct' chemical options, only using these where absolutely necessary. However, this concept is very much experimental and would need to be validated on farms in the future.

Funding streams that cover IPM strategies are now becoming more available through government initiatives, enabling more joined-up thinking across research and farming communities. An example where decision support systems (DSS) are being developed to assist with IPM practices is in the EU-funded IPM decisions project. https://www.ipmdecisions.net. Technology is playing more of a central and important role in IPM systems in vegetable crops. This is as diverse as using machine learning to identify insects on traps in greenhouses and in the field (Eliopoulos et al., 2018; Preti et al., 2021) through to using unmanned aerial vehicles (UAVs) or 'drones` to dispense beneficial insects into crops (Del Pozo-Valdivia et al., 2021; Zhan et al., 2021). Further investments and developments from private companies will ensure that relevant technology will form part of farming landscapes in the future. More specifically, researchers in Europe have been actively developing IPM strategies through the Fly IPM project, with the aim of developing more robust and IPM-focused strategies against vegetable crop root fly pests (https://projects.au.dk/c-ipm/news/nyhed/artikel/getting-to-the-root-of-root-flies/). On a global scale, the FAO continues to actively promote IPM through its three regional programmes in Asia, the Near East and West Africa as well as through several stand-alone national projects. Under these programmes and projects, the FAO provides assistance in capacity building and policy reform and facilitates collaboration among ongoing national IPM programmes. http://www.fao.org/agriculture/crops/core-themes/theme/pests/ipm/en/. A common theme to these programmes is the use of farmer schools. Farmer IPM school programmes are common in Asia and Africa but are not common in Europe. The adoption of such schemes in Europe and in western farming communities may be important to develop in the future so that IPM techniques and strategies can be showcased and further developed at a practical level on-farm.

Finally, the global farming landscape will change dramatically over the next century and farmers will need to adapt to these changes. Through the development and implementation of scientifically robust IPM strategies, farmers on the global stage will be well positioned to ensure that pest insect problems are effectively controlled and that adverse impacts on the environment and natural landscapes are minimised.

13 Where to look for further information

13.1 Global websites

- http://www.fao.org/pest-and-pesticide-management/ipm/integrated-pest-management/en/.
- https://www.cabi.org.
- https://www.epa.gov/safepestcontrol/integrated-pest-management-ipm-principles.
- https://www.usda.gov/oce/pest/integrated-pest-management.

13.2 European websites

- https://ec.europa.eu/food/plants/pesticides/sustainable-use-pesticides/integrated-pest-management-ipm_en.
- https://voluntaryinitiative.org.uk/.
- https://ahdb.org.uk/.

13.3 Literature

- Dent, D. and Binks, R. H. (2020). *Insect Pest Management*, 3rd edition. CAB International, Wallingford, UK.
- Kogan, M. and Heinrichs, E. A. (Eds) (2019). *Integrated Management of Insect Pests: Current and Future Developments*. Burleigh Dodds Science Publishing.
- Onstad, D. W. and Crain, P. (Eds.). (2019). *The Economics of Integrated Pest Management of Insects*. CABI.
- Rapisarda, C. and Cocuzza, G. E. M. (Eds.). (2017). *Integrated Pest Management in Tropical Regions*. CABI.
- Vacante, V. and Kreiter, S. (Eds.). (2017). *Handbook of Pest Management in Organic Farming*. CABI.

14 References

Ahmad, M. and Akhtar, S. (2013). Development of insecticide resistance in field populations of *Brevicoryne brassicae* (Hemiptera: Aphididae) in Pakistan. *Journal of Economic Entomology* 106(2), 954–958.

Amundson, S. K. (2012). *Cultural Techniques to Improve Yield and Cost Efficiency of Greenhouse Grown Tomatoes*.

Arnó i Pujol, J. et al. (2009). Implementation of integrated pest management programs in European greenhouse production areas: Trends and constraints. EU ENDURE Project – Univesitat de Lleida, Lleida, Spain.

Bass, C., Puinean, A. M., Zimmer, C. T., Denholm, I., Field, L. M., Foster, S. P., Gutbrod, O., Nauen, R., Slater, R. and Williamson, M. S. (2014). The evolution of insecticide resistance in the peach potato aphid, *Myzus persicae*. *Insect Biochemistry and Molecular Biology* 51, 41–51.

Beck, B., Spanoghe, P., Moens, M., Pollet, S., Temmerman, F. and Nuyttens, D. (2014). Improving control of the cabbage root fly with entomopathogenic nematodes. *Aspects of Applied Biology* 122, 125–134.

Begg, G. S., Cook, S. M., Dye, R., Ferrante, M., Franck, P., Lavigne, C., Lövei, G. L., Mansion-Vaquie, A., Pell, J. K., Petit, S., Quesada, N., Ricci, B., Wratten, S. D. and Birch, A. N. E. (2017). A functional overview of conservation biological control. *Crop Protection* 97, 145–158.

Beglyarov, G. A. and Smetnik, A. I. (1977). Seasonal colonization of entomophages in the USSR. In: *Biological Control by Augmentation of Natural Enemies*, pp. 283–328. Springer, Boston, MA.

Ben-Issa, R., Gomez, L. and Gautier, H. (2017). Companion plants for aphid pest management. *Insects* 8(4), 112.

Berlinger, M. J., Taylor, R. A. J., Lebiush-Mordechi, S., Shalhevet, S. and Spharim, I. (2002). Efficiency of insect exclusion screens for preventing whitefly transmission of tomato yellow leaf curl virus of tomatoes in Israel. *Bulletin of Entomological Research* 92(5), 367–373.

Bhattacharya, S. (2019). Brassica-aphid interaction: challenges and prospects of genetic engineering for integrated aphid management. *Physiological and Molecular Plant Pathology* 108, 101442.

Bird, C. (2014). *The Fundamentals of Horticulture: Theory and Practice*. Cambridge University Press, Cambridge.

Blackman, R. L. and Eastop, V. F. (2000). *Aphids on the World's Crops: An Identification and Information Guide* (2nd edn.). John Wiley & Sons Ltd.

Blackshaw, R. P., Vernon, R. S. and Prasad, R. (2012). Reduction of *Delia radicum* attack in field brassicas using a vertical barrier. *Entomologia Experimentalis et Applicata* 144(2), 145–156.

Bonsignore, C. P. and Vacante, V. (2017). 15 Pest management in organic vegetable greenhouses. Handbook of pest management in. *Organic Farming* 374.

Brennan, E. B. (2016). Agronomy of strip intercropping broccoli with alyssum for biological control of aphids. *Biological Control* 97, 109–119.

Brown, Z. S., Jones, M. S. and Mumford, J. D. (2019). Economic principles and concepts in area-wide genetic pest management. *The Economics of Integrated Pest Management of Insects*, 96–120.

CABI (2021). *CABI Invasive Species Compendium*. Available at: https://www.cabi.org/isc/ (accessed on 11 March 2021).

Calvo, J., Bolckmans, K., Stansly, P. A. and Urbaneja, A. (2009). Predation by *Nesidiocoris tenuis* on *Bemisia tabaci* and injury to tomato. *BioControl* 54(2), 237–246.

Campos, M. R., Biondi, A., Adiga, A., Guedes, R. N. C. and Desneux, N. (2017). From the western Palaearctic region to beyond: *Tuta absoluta* 10 years after invading Europe. *Journal of Pest Science* 90(3), 787–796.

Cividanes, F. J. and Santos-Cividanes, T. Md (2012). Predicting the occurrence of alate aphids in Brassicaceae. *Pesquisa Agropecuária Brasileira* 47(4), 505 –510.

Cocco, A., Deliperi, S. and Delrio, G. (2013). Control of *Tuta absoluta* (Meyrick) (Lepidoptera: Gelechiidae) in greenhouse tomato crops using the mating disruption technique. *Journal of Applied Entomology* 137(1–2), 16–28.

Collier, R., Mazzi, D., Folkedal Schjøll, A., Schorpp, Q., Thöming, G., Johansen, T. J., Meadow, R., Meyling, N. V., Cortesero, A. M., Vogler, U., Gaffney, M. T. and Hommes, M. (2020). The potential for decision support tools to improve the management of root-feeding fly pests of vegetables in western Europe. *Insects* 11(6), 369. doi: 10.3390/insects11060369.

Collier, R. H. and Finch, S. (2017). IPM case studies: brassicas. In: *Aphids as Crop Pests* (2nd edn.), pp. 578–586.

Costa, J. M. and Heuvelink, E. P. (2018). The global tomato industry. In: *Tomatoes*, pp. 1-26. CABI, Wallingford, UK.

Csizinszky, A. A., Shuster, D. J., Jones, J. B. and van Lenteren, J. C. (2005). Crop protection. In: *Tomatoes* (vol. 13). CABI Publishing.

Czosnek, H. and Latterot, H. (1997). A worldwide survey of tomato yellow leaf curl viruses. *Archives of Virology* 142, 1391–1406.

Daniel, C., Collier, R., Niggli, U. and Koller, M. (2018). Chapter 17 Pest management in organic field vegetables in temperate areas. In: *Handbook of Pest Management in Organic Farming*, p. 443.

Del Pozo-Valdivia, A. I., Morgan, E. and Bennett, C. (2021). In-field evaluation of drone-released lacewings for aphid control in California organic lettuce. *Journal of Economic Entomology* 114(5), 1882–1888.

Denholm, I., Cahill, M., Dennehy, T. J. and Horowitz, A. R. (1998). Challenges with managing insecticide resistance in agricultural pests, exemplified by the whitefly *Bemisia tabaci. Philosophical Transactions of the Royal Society of London. Series B: Biological Sciences* 353(1376), 1757–1767.

Dent, D. and Binks, R. H. (2020). *Insect Pest Management* (3rd edn.). CAB International, Wallingford, UK.

Dixon, G. R. (2007). *Vegetable Brassicas and Related Crucifers (No. 14).* CABI Publishing.

Dogru, S. M., Balkaya, A. and Kurtar, E. S. (2018). A comprehensive perspective on Brassica vegetable crops grown in Turkey. *Acta Horticulturae*, 1202, 1-7.

Dorais, M. and Schwarz, D. (2018). Organic tomato. *Tomatoes* 27, 337.

Eliopoulos, P., Tatlas, N. A., Rigakis, I. and Potamitis, I. (2018). A "smart" trap device for detection of crawling insects and other arthropods in urban environments. *Electronics* 7(9), 161.

FAOSTAT (2021). Food and Agriculture Organization of the United Nations (FAO). *FAOSTAT Database*. Available at: http://www.fao.org/faostat/en/#data/QC (accessed on 13th February 2021).

Finch, S. (1971). The fecundity of the cabbage root fly *Erioischia brassicae* (Bouché) under field conditions. *Entomologia Experimentalis et Applicata* 14(2), 147–160.

Finch, S. and Collier, R. H. (1985). Laboratory studies on aestivation in the cabbage root fly (*Delia radicum*). *Entomologia Experimentalis et Applicata* 38(2), 137–143.

Finch, S. and Collier, R. (2005). Answers to questions concerning the 'appropriate/inappropriate landing' theory of host-plant finding. *IOBC/WPRS Bulletin* 28(4), 103–121.

Flint, M. L. (2018). *Integrated Pest Management for Tomatoes* (4th edn.). University of California Statewide Integrated Pest Management Project, Division of Agriculture and Natural Resources Publication, p. 3274.

Foster, S. P. and Devine, G. (2017). 19 Insecticide resistance. *Aphids as Crop Pests* 426.

Furlong, M. J., Wright, D. J. and Dosdall, L. M. (2013). Diamondback moth ecology and management: problems, progress, and prospects. *Annual Review of Entomology* 58, 517–541.

Garcia-del-Pino, F., Morton, A. and Shapiro-Ilan, D. (2018). Entomopathogenic nematodes as biological control agents of tomato pests. In: *Sustainable Management of Arthropod Pests of Tomato*, pp. 269–282. Academic Press.

George, D. R., Banfield-Zanin, J. A., Collier, R., Cross, J., Birch, A. N. E., Gwynn, R. and O'Neill, T. (2015). Identification of novel pesticides for use against glasshouse invertebrate pests in UK tomatoes and peppers. *Insects* 6(2), 464–477.

Gould, F., Kennedy, G. G. and Johnson, M. T. (1991). Effects of natural enemies on the rate of herbivore adaptation to resistant host plants. *Entomologia Experimentalis et Applicata* 58(1), 1–14.

Grant, W. P., Chandler, D., Bailey, A., Greaves, J., Tatchell, M. and Prince, G. (2010). *Biopesticides: Pest Management and Regulation*. CABI Publishing.

Grasswitz, T. R. (2019). Integrated pest management (IPM) for small-scale farms in developed economies: challenges and opportunities. *Insects* 10(6), 179.

Gratwick, M. (1992). *Crop Pests in the UK*. Chapman & Hall, London, pp. 36–41.

Grzywacz, D., Rossbach, A., Rauf, A., Russell, D. A., Srinivasan, R. and Shelton, A. M. (2010). Current control methods for diamondback moth and other brassica insect pests and the prospects for improved management with lepidopteran-resistant Bt vegetable brassicas in Asia and Africa. *Crop Protection* 29(1), 68–79.

Guedes, R. N. C., Roditakis, E., Campos, M. R., Haddi, K., Bielza, P., Siqueira, H. A. A., Tsagkarakou, A., Vontas, J. and Nauen, R. (2019). Insecticide resistance in the tomato pinworm *Tuta absoluta*: patterns, spread, mechanisms, management and outlook. *Journal of Pest Science* 92(4), 1329–1342.

Gurr, G. M., Reynolds, O. L., Johnson, A. C., Desneux, N., Zalucki, M. P., Furlong, M. J., Li, Z., Akutse, K. S., Chen, J., Gao, X. and You, M. (2018). Landscape ecology and expanding range of biocontrol agent taxa enhance prospects for diamondback moth management: a review. *Agronomy for Sustainable Development* 38(3), 1–16.

Hance, T., Fatemeh, T. K. and Manaut, F. (2017). 20 biological control. *Aphids as Crop Pests* 549.

Helyer, N., Cattlin, N. D. and Brown, K. C. (2014). *Biological Control in Plant Protection: A Colour Handbook*. CRC Press.

Heuvelink, E. P. (2018). *Tomatoes – Crop Production Science in Horticulture* (2nd edn.). CAB International, Wallingford, UK.

Hommes, M. and Herbst, M. (2014). Supervised control of aphids and caterpillars on white cabbage and impact of flower strips. *Supervised Control. of Aphids and Caterpillars on White Cabbage and Impact of Flower Strips* 107, 43–49.

Hussey, N. W. and Scopes, N. (1985). *Biological Pest Control: The Glasshouse Experience*. Cornell University Press, Ithaca.

Hutchison, I. K. (2020). *Biological Control of the Cabbage Root Fly (Delia radicum) Using an Entomopathogenic Fungus (Metarhizium brunneum)*. Master's thesis, Norwegian University of Life Sciences, Ås.

Iost Filho, F. H., Heldens, W. B., Kong, Z. and de Lange, E. S. (2020). Drones: innovative technology for use in precision pest management. *Journal of Economic Entomology* 113(1), 1–25.

Josso, C., Le Ralec, A., Raymond, L., Saulais, J., Baudry, J., Poinsot, D. and Cortesero, A. M. (2013). Effects of field and landscape variables on crop colonization and biological control of the cabbage root fly *Delia radicum*. *Landscape Ecology* 28(9), 1697–1715.

Jukes, A., Collier, R. and Finch, S. (2005). Attempts to reduce cabbage root fly infestations by applying garlic and other 'natural' products. *Comptes Rendus de la Réunion* 28(4), 45–52.

Kapranas, A., Sbaiti, I., Degen, T. and Turlings, T. C. J. (2020). Biological control of cabbage fly *Delia radicum* with entomopathogenic nematodes: selecting the most effective nematode species and testing a novel application method. *Biological Control* 144, 104212.

Kergunteuil, A., Dugravot, S., Danner, H., van Dam, N. M. and Cortesero, A. M. (2015). Characterizing volatiles and attractiveness of five brassicaceous plants with potential for a 'push-pull' strategy toward the cabbage root fly, Delia radicum. *Journal of Chemical Ecology* 41(4), 330–339.

Kienegger, M., Kromp, B. and Kahrer, A. (2003). The effect of strips of flowers on pests and beneficial arthropods in adjacent broccoli plots. *IOBC WPRS Bulletin* 26(3), 61–70.

Kubota, C., de Gelder, A. and Peet, M. M. (2018). Greenhouse tomato production. In: *Tomatoes*, pp. 276–313. C.A.B. International Publishing, Wallingford, UK.

Lamy, F., Dugravot, S., Cortesero, A. M., Chaminade, V., Faloya, V. and Poinsot, D. (2018). One more step toward a push-pull strategy combining both a trap crop and plant volatile organic compounds against the cabbage root fly *Delia radicum*. *Environmental Science and Pollution Research International* 25(30), 29868–29879.

Lapidot, M. and Friedmann, M. (2002). Breeding for resistance to whitefly-transmitted geminiviruses. *Annals of Applied Biology* 140(2), 109–127.

Leskovar, D. I., Crosby, K. M., Palma, M. A. and Edelstein, M. (2014). Vegetable crops: linking production, breeding and marketing. In: *Horticulture: Plants for People and Places* (vol. 1), pp. 75–96. Springer, Dordrecht.

Liu, T., Kang, L., Lei, Z. and Hernandez, R. (2011). Hymenopteran parasitoids and their role in biological control of vegetable *Liriomyza* leafminers. In: *Recent Advances in Entomological Research*, pp. 376–403. Springer, Berlin, Heidelberg.

Liu, T. X., Kang, L., Heinz, M. and Trumble, J. (2009). Biological control of *Liriomyza* leafminers: progress and perspective. *Biocontrol News and Information* 30(1), 1R.

Liu, X., Chen, M., Collins, H. L., Onstad, D. W., Roush, R. T., Zhang, Q., Earle, E. D. and Shelton, A. M. (2014). Natural enemies delay insect resistance to Bt crops. *PLoS ONE* 9(3), e90366.

Mansour, R., Brévault, T., Chailleux, A., Cherif, A., Grissa-Lebdi, K., Haddi, K., Mohamed, S. A., et al. (2018). Occurrence, biology, natural enemies and management of *Tuta absoluta* in Africa. *Entomology Genetics* 38(2), 83–112.

Matthews, G. A. (2018). *A History of Pesticides*. CABI Publishing.

McEwen, P., New, T. R. and Whittington, A. E. (2001). *Lacewings in the Crop Environment*. Cambridge.

McKinlay, R. G. (Ed.) (1992). *Vegetable Crop Pests*. Springer.

Molotievskiy, M. N., Toderaş, I., Iurcu-Străistaru, E. and Moldovan, A. (2019). A review of the major pest insects of tomato crops in the Republic of Moldova. *Buletinul Academiei de Ştiinţe a Moldovei. Ştiinţele Vieţii* 337(1), 7–18.

Molotoks, A., Smith, P. and Dawson, T. P. (2021). Impacts of land use, population, and climate change on global food security. *Food and Energy Security* 10(1), e261.

Mouden, S., Sarmiento, K. F., Klinkhamer, P. G. and Leiss, K. A. (2017). Integrated pest management in western flower thrips: past, present and future. *Pest Management Science* 73(5), 813–822.

Pavela, R. (2014). Limitation of plant biopesticides. In: *Advances in Plant Biopesticides*, pp. 347–359. Springer, New Delhi.

Perdikis, D., Kapaxidi, E. and Papadoulis, G. (2008). Biological control of insect and mite pests in greenhouse solanaceous crops. *European Journal of Plant Science and Biotechnology* 2(1), 125–144.

Picó, B., Díez, M. J. and Nuez, F. (1996). Viral diseases causing the greatest economic losses to the tomato crop. II. The tomato yellow leaf curl virus–a review. *Scientia Horticulturae* 67(3–4), 151–196.

Pijnakker, J., Vangansbeke, D., Duarte, M., Moerkens, R. and Wäckers, F. L. (2020). Predators and parasitoids-in-first: from inundative releases to preventative biological control in greenhouse crops. *Frontiers in Sustainable Food Systems* 4, 245.

Piñero, J. and Manandhar, R. (2015). Effect of increased crop diversity using trap crops, flowering plants and living mulches on vegetable insect pests. *Trends in Entomology* 11, 91–109.

Popham, H. J. R., Nusawardani, T. and Bonning, B. C. (2016). Introduction to the use of baculoviruses as biological insecticides. In: *Methods in Molecular Biology*, pp. 383-392. Humana Press, New York, NY.

Prasad, A. et al. (2020). Tomato yellow leaf curl virus: impact, challenges and management. *Trends in Plant Science* 25(9), 897–911.

Preti, M., Verheggen, F. and Angeli, S. (2021). Insect pest monitoring with camera-equipped traps: strengths and limitations. *Journal of Pest Science* 94(2), 203–217.

Prince, G. and Chandler, D. (2020). Susceptibility of *Myzus persicae*, *Brevicoryne brassicae* and *Nasonovia ribisnigri* to fungal biopesticides in laboratory and field experiments. *Insects* 11(1), 55.

Rabasse, J. M. and Wyatt, I. J. (1985). Biology of aphids and their parasites in greenhouses. In: Hussey, N. W. and Scopes, N. (Eds), *Biological Pest Control: The Glasshouse Experience*, pp. 66–73. Cornell University Press, Ithaca.

Rabasse, J. M. and van Steenis, M. J. (1999). Biological control of aphids. In: *Developments in Plant Pathology*, pp. 235–243. Springer, Dordrecht.

Ramasamy, S. and Ravishankar, M. (2018). Integrated pest management strategies for Tomato under protected structures. In: *Sustainable Management of Arthropod Pests of Tomato*, pp. 313-322. Academic Press.

Razinger, J., Žerjav, M., Zemljič-Urbančič, M., Modic, Š, Lutz, M., Schroers, H. J., Grunder, J., Fellous, S. and Urek, G. (2017). Comparison of cauliflower-insect-fungus interactions and pesticides for cabbage root fly control. *Insect Science* 24(6), 1057-1064.

Ribeiro, A. L. and Gontijo, L. M. (2017). Alyssum flowers promote biological control of collard pests. *BioControl* 62(2), 185–196.

Roselló, S., Díez, M. J. and Nuez, F. (1996). Viral diseases causing the greatest economic losses to the tomato crop. I. The Tomato spotted wilt virus–a review. *Scientia Horticulturae* 67(3-4), 117–150.

Sadowski, J. and Kole, C. (Eds.) (2016). *Genetics, Genomics and Breeding of Vegetable Brassicas*. CRC Press.

Sampson, C. and Walker, P. (1998). *Improved Control of* Liriomyza bryoniae *Using an Action Threshold for the Release of* Diglyphus isaea *in Protected Tomato Crops*. Mededelingen-Faculteit Landbouwkundige en Toegepaste Biologische Wetenschappen Universiteit, Gent (Belgium).

Sanchez, J. A., López-Gallego, E., Pérez-Marcos, M., Perera-Fernández, L. G. and Ramirez-Soria, M. J. (2018). How safe is it to rely on *Macrolophus pygmaeus* (Hemiptera: Miridae) as a biocontrol agent in tomato crops? *Frontiers in Ecology and Evolution* 6, 132.

Sanlier, N. and Guler, S. M. (2018). The benefits of *Brassica* vegetables on human health. *Journal of Human Health Research* 1, 1–13.

Sarkar, S. C., Wang, E., Wu, S. and Lei, Z. (2018). Application of trap cropping as companion plants for the management of agricultural pests: a review. *Insects* 9(4), 128.

Shelton, A. M., Long, S. J., Walker, A. S., Bolton, M., Collins, H. L., Revuelta, L., Johnson, L. M. and Morrison, N. I. (2020). First field release of a genetically engineered, self-limiting agricultural pest insect: evaluating its potential for future crop protection. *Frontiers in Bioengineering and Biotechnology* 7, 482.

Shuhang, W., Voorrips, R. E., Steenhuis-Broers, G., Vosman, B. and van Loon, J. J. A. (2016). Antibiosis resistance against larval cabbage root fly, *Delia radicum*, in wild Brassica-species. *Euphytica* 211(2), 139–155.

Sigsgaard, L. (2005, November). Biological control of arthropod pests in outdoor crops–the new challenge. In: *Proceedings of the International Workshop. Implementation of Biocontrol in Practice in Temperate Region–Present and Near Future*, pp. 153–167. Research Centre, Flakkebjerg, Denmark.

Stenberg, J. A. (2017). A conceptual framework for integrated pest management. *Trends in Plant Science* 22(9), 759–769.

Stout, M. J., Kurabchew, H. and Leite, G. L. D. (2018). Host-plant resistance in tomato. In: *Sustainable Management of Arthropod Pests of Tomato*, pp. 217–236. Academic Press.

Teulon, D. A. J., Davidson, M. M., Perry, N. B., Nielsen, M. C., Castañé, C., Bosch, D., Riudavets, J., Van Tol, R. W. H. M. and de Kogel, W. J. (2017). Methyl isonicotinate–a non-pheromone thrips semiochemical–and its potential for pest management. *International Journal of Tropical Insect Science* 37(2), 50–56.

Tomalak, M., Piggott, S. and Jagdale, G. B. (2005). *Glasshouse Applications. Nematodes as Biocontrol Agents*. CABI Publishing, Oxon, pp. 147–166.

Vallad, G. E., Messelink, G. and Smith, H. A. (2018). Crop protection: pest and disease management. In: *Tomatoes* (vol. 13). CABI Publishing.

van der Ent, S., Knapp, M., Klapwijk, J., Moerman, E., van Schelt, J. and de Weert, S. (2017). *Knowing and Recognizing: The Biology of Pests, Diseases and Their Natural Solutions*. Koppert Biological Systems.

van der Velden, N. J. A., Suay, R., Urbaneja, A., Giorgini, M., Ruocco, M., Poncet, C. and Lefèvre, A. (2012). Recent developments and market opportunities for IPM in greenhouse tomatoes in Southern Europe; consequences for advanced IPM toolboxes and greenhouse engineering (No. 12-077). LEI, Part of Wageningen UR.

van Lenteren, J. C. (2000). A greenhouse without pesticides: fact or fantasy? *Crop Protection* 19(6), 375–384.

van Lenteren, J. C., Bolckmans, K., Köhl, J., Ravensberg, W. J. and Urbaneja, A. (2018). Biological control using invertebrates and microorganisms: plenty of new opportunities. *BioControl* 63(1), 39–59.

Wainwright, C., Jenkins, S., Wilson, D., Elliott, M., Jukes, A. and Collier, R. (2020). Phenology of the diamondback moth (*Plutella xylostella*) in the UK and provision of decision support for brassica growers. *Insects* 11(2), 118.

Weintraub, P. G. and Horowitz, A. R. (1995). The newest leafminer pest in Israel, *Liriomyza huidobrensis*. *Phytoparasitica* 23(2), 177–184.

Welbaum, G. E. (2015). *Vegetable Production and Practices*. CAB International, Wallingford, UK.

Williams, E. C. and MacDonald, O. C. (1995). Critical factors required by the nematode *Steinernema feltiae* for the control of the leafminers *Liriomyza huidobrensis*,

Liriomyza bryoniae and *Chromatomyia syngenesiae*. *Annals of Applied Biology* 127(2), 329–341.

Witkowska, E., Moorhouse, E. R., Jukes, A., Elliott, M. S. and Collier, R. H. (2018). Implementing integrated pest management in commercial crops of radish (*Raphanus sativus*). *Crop Protection* 114, 148–154.

Xuan, J. L., Liu, W. X., Zhang, Y. B., Cheng, X. Q., Guo, J. Y. and Wan, F. H. (2018). Interactions between *Diglyphus isaea* and *Neochrysocharis formosa* (Hymenoptera: Eulophidae), two parasitoids of agromyzid leafminers. *Biological Control* 126, 45–52.

Zhan, Y., Chen, S., Wang, G., Fu, J. and Lan, Y. (2021). Biological control technology and application based on agricultural unmanned aerial vehicle (UAV) intelligent delivery of insect natural enemies (Trichogramma) carrier. *Pest Management Science* 77(7), 3259–3272.

Zhu, J. Y. et al. (2020). A new pest management strategy: transforming a non-host plant into a dead-end trap for the diamondback moth. *Pest Management Science*. doi: 10.1002/ps.6126.

Chapter 14

Practical application of integrated pest management to control cabbage root fly in vegetables

Louis Lippens, PCG vzw – Vegetable Research Centre Kruishoutem, Belgium; Sander Fleerakkers, PSKW vzw – Research Station for Vegetable Production Sint-Katelijne-Waver, Belgium; Femke Temmerman, Inagro vzw, Belgium; and Annelies De Roissart, HOGENT University of Applied Sciences & Arts, Belgium

1 Introduction

In Northwestern Europe, the cabbage root fly (CRF) (or cabbage maggot) (*Delia radicum* L.) (Diptera: Anthomyiidae) is considered an important pest of *Brassica* vegetables (Mesmin et al., 2019). Cabbage root flies lay most of their eggs in the soil in proximity to the stem of *Brassica* crops. Eggs hatch after 3-5 days. The larval development consists of 3 instars that feed on roots (internally and externally) and stem tissue (internally). Larval development requires 18–22 days and is followed by a pupal period of approximately 12–25 days in summer. Since this is the most common overwintering stage, the pupal period is prolonged for 5–8 months during winter when the pupae are in diapause

http://dx.doi.org/10.19103/AS.2021.0095.16

(Collier and Finch, 1983). Pupation takes place in the soil near the base of the stem (Finch and Skinner, 1973). Adult flies measure 6–7 mm and feed on the nectar from flowering plants to mature eggs. In Flanders (Belgium), generally three generations occur each year. The first generation of adult flies appears in spring by the beginning of April. The second generation is present by the end of June, the third by the beginning of September. In Flanders, the first generation of flies causes the most severe crop damage, mainly due to the sensitivity of the crop in its early stages and the high pest pressure in April and May (pers.comm. Vegetable Research Centre, Kruishoutem).

Cabbage root flies are known to attack *Brassica* vegetable crops including broccoli, Brussels sprouts, cabbage, radish and cauliflower (McKinlay, 1992). Early in the season, larvae feeding on plant roots reduce seedling vigour while their susceptibility to infection by secondary pathogens increases (Van Herk et al., 2017). Later in the season, the larvae enter the roots, causing damage and eventually, crop loss. After entering the plant, they move to the main root to mature. Plant death often co-occurs with periods of drought and water stress. More problems are observed on light soils (Capinera, 2008). In some specific crops, CRFs directly damage the marketable product. In these crop types, a little damage can cause high losses (Capinera, 2008). In Brussels sprout crops, for example, the third generation of flies lays eggs on the sprout buttons instead of near the stem base. The emerging larvae consequently feed on the edible buds. Since Chinese cabbage does not have a visible stem, eggs are laid between the leaf bases near the ground, and larvae feed through the soft leaf veins. For cruciferous root vegetables like turnip or radish, the tunnels made by the feeding larvae cause severe damage to the edible part (pers. comm. Vegetable Research Centre, Sint-Katelijne-Waver), making it unattractive.

Pest damage due to the CRF causes considerable economic losses in the *Brassica* vegetable industry. In Flanders, this industry produces 258 755 tons of cabbage on a yearly basis with a production value of more than 76 million euro. More than 1800 growers produce cabbages on an area of 10 347 ha (Table 1). Financial losses caused by a reduction in quality and production weight occur at between two and six percent when the cabbages are treated with conventional chemical substances. Without treatment, losses of more than 90% are observed, corresponding to 1.5 million euro (Volckaert et al., 2018; Maes and Lippens, 2019).

The control of CRF is a challenging task for several reasons, the first of which is their hidden development. The eggs are very small and deposited in the soil. Therefore, they are difficult to observe in the field (Bligaard et al., 1999), which makes it hard to determine whether and when to apply insecticides. Root fly larvae themselves live in the soil or in the plant tissue, hidden from non-systemic insecticide sprays. Although adult flies are exposed, they need to acquire a sufficient dose of insecticide through direct or indirect (on treated

Table 1 Economic data on the Brassica crop production in Flanders (Belgium) (VBT, 2018–2020)

Crop	Number of growers	Area (ha)	Production weight (ton)	Production value (x1000 eur)
Brussels sprout	483	2.450	54.550	25.989
Cauliflower	819	5.745	118.384	38.547
Broccoli	74	250	3.267	2.802
Red cabbage	122	240	11.700	2.578
White cabbage	172	450	27.600	4.329
Savoy cabbage	164	300	13.600	na
Chinese cabbage	na	47	1504	na
Turnip	na	370	na	na

surfaces) contact. Since most of them are only active in the crop during a limited timeframe of the day, effective control of the adult flies is not easily achieved (Collier et al., 2020). Insect control, therefore, preferably takes place before the larvae enter the roots (Leger and Riga, 2009).

In general, seed treatments, granule treatments and drench treatments are expected to be more efficient than foliar sprays. In Flanders, the application of a preventive drench treatment with pesticides containing spinosad is the most common practice. However, practical research indicates that this treatment does not always result in adequate control of the pest (Maes and Lippens, 2019; Volckaert et al., 2018) (see Section 5). The occurrence of induced resistance against insecticides due to the intense use of a limited number of substances complicates the control of CRF (Finch and Collier, 2000). The development of spinosad resistance has been observed in several laboratory and field populations of insect pest species (*D. melanogaster* (fruit fly), *Liriomyza trifolii* (American serpentine leafminer), *Helicoverpa armigera* (cotton bollworm), and *Frankliniella occidentalis* (western flower thrips) (Sparks et al., 2012)). Due to the intense use of spinosad against several generations of CRF each year, the risk of resistance development and its interference with root maggot control in Flanders is high.

In addition to the struggle by growers to control CRF with a limited number of pesticides, the increasing demands of consumers and policy makers for residue-free vegetable products that are cultivated in a sustainable manner (Frank, 2010; Van Lenteren et al., 2018) also create a certain pressure. The European Union ('Sustainable Use of Pesticides Directive'-EC 2009), as well as the Belgian government (Joas and Cotillon, 2009) advise the implementation of integrated pest management (IPM) to control agricultural pests. The IPM method offers an approach to achieve pest control by the use of a combination of alternative pest management techniques as a substitute for chemical control, to minimise the amount of pesticide needed (Collier et al., 2020). IPM contains a mix of cultural (e.g. crop rotation, trap cropping, companion planting), physical (e.g. row covering) and biological (e.g. the release of entomopathogenic nematodes, fungi or natural enemies) control methods that result in a decrease of harmful side-effects of pesticides on the environment and on human health (Radcliffe et al., 2008; Collier et al., 2020).

While some of these alternative strategies, like the release of natural enemies, are well-developed in a greenhouse environment (e.g. the release of predatory mites in tomato) their use outdoors is challenging and needs further research and refinement (Van Lenteren, 2012). In Flanders, the implementation of a monitoring system, netting, entomopathogenic nematodes (EPN) and selective insecticides are the most commonly studied components of an IPM strategy for CRF. The practical efficacy of these techniques for cabbage production has been studied since 1998 in vegetable research centres in Flanders.

2 Developing a decision-support system to predict pest outbreaks

Understanding pest insect phenology is an important aspect of IPM. By monitoring and forecasting pest populations, the timing of critical events in their lifecycle can be determined. This allows for optimal timing of control actions and reduces the negative impact of unnecessary treatments (Collier et al., 2020). In Flanders, a decision-support system for vegetable pests and diseases was formulated in 1998. The association of Flanders' vegetable auctions (LAVA) decided to invest in this scientifically based warning system for vegetable producers in order to minimise pesticide applications. Advice is given on a regular basis to the vegetable growers by e-mail (Martens and Plovie, 2007). The main crops that are monitored are cauliflower (*Brassica oleracea* var. *botrytis*), Brussels sprout (*Brassica oleracea* var. *gemmifera*), leek (*Allium porrum*) and carrot (*Daucus carota*). An overview of the pests and diseases monitored for this warning system is listed in Table 2.

The existing warning system merely informs growers about the root fly pressure present in their crops. This information is specifically valuable to growers of Brussels sprouts, where losses caused by the second (and sometimes third) generation of CRF are more severe and field application of pesticides is often necessary (pers.comm. with Rudi Vandewiele (advisor to cabbage growers for TACO cvba)). Nonetheless, a decision-support system based on threshold levels or forecasting of pest numbers is a crucial part of good agricultural practice that includes IPM, and has proven to be useful for the control of many other pests like carrot fly (*Psila rosae*) and thrips (*Thrips tabaci*) (Collier et al., 2020). In Flanders, a new day-degree model is currently being developed to predict the appearance of *T. tabaci* in different vegetables.

Table 2 Overview of the decision-support system for vegetable pests and diseases in Flanders

Crop	Pests monitored	Diseases monitored	Monitoring system
Cauliflower	Cabbage root fly, diamondback moth, *Thrips tabaci Pegohylemia fugax*, other caterpillars, aphids, slugs	*Mycosphaerella brassicicola*, *Alternaria sp.*, *Albugo candida*, *Erysiphe cruciferarum* and *Peronospora parasitica*	Bell-shaped traps, prisma-shaped pheromone trap, visual assessment
Brussels sprout	Cabbage root fly, diamondback moth, other caterpillars, aphids, slugs, *Thrips tabaci*	*Mycosphaerella brassicicola*, *Alternaria sp.*, *Albugo candida*, *Erysiphe cruciferarum* and *Peronospora parasitica*	Bell-shaped traps, blue sticky traps, prisma-shaped pheromone trap, visual assessment
Leek	*Thrips tabaci*, leek moth (*Acrolepiopsis assectella*), leek miner fly (*Phytomyza gymnostoma*), *Delia antiqua*	*Phytophthora porri*, *Puccinia allii*, *Alternaria porri*, *Colletotrichum circinans*	Blue and yellow sticky plates, pheromone trap, visual assessment
Carrot	Carrot fly		

In the future, knowledge on the phenology of the CRF could be used to inform growers about the ideal timing to apply alternative pest management techniques such as physical barriers (see Section 3) and when to release biological control agents like EPN (see Section 4).

2.1 Monitoring of cabbage root fly

In the decision-support system for vegetable pests and diseases in Flanders, the presence of the CRF is monitored during the growth season (from mid-April until the end of October) on at least 10 *Brassica* fields (cauliflower and Brussels sprout). The fields are situated in the main vegetable-growing regions in Flanders (West-Flanders, East-Flanders and Antwerp). The monitoring approaches used in these crops are summarised in Table 2. The number of adult flies captured by bell-shaped traps is recorded six times a month (Fig. 1A). Bell-shaped traps (Van Iperen BV) combine a yellow sticky trap with a chemical lure that attracts adult flies. For the last 20 years, ethyl isothiocyanate has been used as chemical attractant. Recently, a new lure, n-butyl, has been shown to be more effective in attracting adult flies to the trap. An additional advantage of n-butyl is the slower release-rate, resulting in the practical benefit that no lure replacement is needed during crop cultivation. In addition to the number of adult flies, oviposition is monitored using felt traps (Olbis les ateliers.ch) placed at the base of a number of plants (Fig. 1B). Female root flies deposit their eggs in the trap. The number of eggs on the trap is counted six times a month. Following the above-mentioned counts, a general message is compiled

Figure 1 Monitoring method for adult cabbage root flies with bell-shaped traps (a) and oviposition with felt traps (b).

regarding the pest pressure on the monitored fields and sent to 350 cauliflower growers and 120 Brussels sprouts growers.

3 The use of row covers in crop protection

The use of row covers is a preventive crop protection measure used to exclude vegetable pests used in organic and integrated farming. This approach generally achieves a good level of control against the CRF (Ludwig and Meyhofer, 2016). It enables passive control of flying pest insects of vegetables by creating a physical barrier (Simon et al., 2014). The disadvantage of this method is the additional production cost, including the labour to apply and remove the covers, and the cost of the materials used (Witkowska et al., 2018). Care must be taken not to 'trap' pest species below the cover. Since the covers exclude natural enemies as well, this might result in an increase rather than a decrease in pest damage (Collier et al., 2020). Despite these disadvantages, the use of row covers is a valuable method. It is a reliable technique to exclude CRF from the crop, and most types of cover additionally protect the crop from game and/or frost damage (Andersen et al., 2006; Witkowska et al., 2018).

The key factors associated with row covering and which determine the level of pest control are choice of material, mesh size and timing of coverage. Most of these factors have an impact on the microclimate under the net by affecting air movement, light quality and quantity, air temperature and relative humidity. Indirectly, they can therefore moderate crop growth, development and yield (Simon et al., 2014).

3.1 Choice of material

Different types of row covers can be used to protect crops from CRF. The three most commonly used fabrics are non-woven fleece row covers, climate net[1] with transparent polyethylene straps and insect-proof netting. In temperate regions, non-woven covers are most suited in spring as they protect the crops from frost. This type of fabric is less suited for summer covering because the raised temperature and relative humidity below the cover can result in a detrimental microclimate, causing increased disease pressure, weed development and a risk of heat stress to the plants. On the contrary, climate nets can be used year-round due to their more open, breathable structure. This type of row cover is made of transparent polyethylene straps and is primarily meant to protect from adverse weather conditions (mainly cold temperatures). An additional benefit of this type of net is their durability. Their longer lifespan makes them a more sustainable option than non-woven fabrics that deteriorate quickly. Their disadvantage is the purchasing cost (Table 3). Insect-proof nets are an even more costly alternative but with even greater durability. Their heavy weight can be disadvantageous for good crop growth. As a result, companies are now producing new types of lightweight, fine-mesh insect-proof nets. Instead of having woven monofilaments, these new materials are made using a new technique which results in knitted monofilaments. The resulting nets not only weigh less, but are also more user friendly, since they are less rigid (pers.com m. Vegetable Research Centre, Sint-Katelijne-Waver).

3.2 Mesh size of insect nets

The success of using row covers will depend on the choice of mesh size, as most often, more than one pest species needs to be considered (Collier et al., 2020). In general, effectiveness against CRF increases with decreasing mesh size. Unfortunately, row covers with a smaller mesh size reduce ventilation. This might result in adverse climatic conditions under the cover and favour the development of fungal disease and weed growth (Martin et al., 2013). To protect cabbage crops against vertebrate, aphid, lepidopteran and dipteran pests, a mesh size of 1.35 mm is advised (Ludwig and Meyhofer, 2016).

3.3 Timing of covering

In Flanders, it is observed that CRFs deposit their eggs on plants growing in module trays prior to transplanting. Therefore, it is important to cover up the trays

Table 3 Characteristics of cover fabrics used to protect crops against cabbage root fly (adapted from Spiessens et al., 2019)

Fabric	Guide price (€/m²)	Mesh size (mm)	Weight (g/ m²)
Non-woven fleece	0.07-0.14	-	12-30
Climate net	0.30-0.35	-	38
Insect-proof netting	0.5-1.4	<0.2- >1.35	45-130

to guarantee adequate plant protection. At the time of planting, in April, high numbers of CRF are usually captured in the bell-shaped traps. Consequently, the plants need to remain covered following transplanting in the field. While, ten years ago, a covering period of four weeks in the field ensured sufficient pest control in cauliflower, today, this is no longer adequate. Increased pest pressure caused by a prolonged oviposition period makes it necessary to cover the cabbage plants during the entire cultivation period (Spiessens et al., 2019).

4 Entomopathogenic nematodes as biocontrol agents

Over the past two decades, several laboratory and field trials examined the potential of nematodes as a biocontrol agent for several pest insects. In 1985, Morris argued that spray-application of entomopathogenic nematodes (EPN) could result in the control of CRF since their eggs hatch in the upper soil layer where they are in contact with the nematodes (Finch, 1993). Slug-parasitic (*Phasmarhabditis*) and entomopathogenic (EPN - *Steinernema* and *Heterorhabditis*) nematodes were shown to be effective and hence are produced commercially (Grewal et al., 2005).

4.1 Biology and ecology of entomopathogenic nematodes

Many nematodes live in close association with agricultural pest organisms like insects, mites and molluscs. The *Steinernemid* and *Heterorhabdid* species are pathogenic to insects, contributing to the natural regulation of insect populations. Due to their pathogenicity, they are particularly interesting when inundatively released as a biocontrol agent for pest insects (Grewal et al., 2005). Their pathogenicity is caused by symbiotic gut bacteria belonging to the genus *Photorhabdus* (*Heterorhabdit* nematodes) and *Xenorhabdus* (*Steinernemid* nematodes) (Boemare et al., 1993). In this partnership, the infective juveniles (IJs: the single free-living stage of *Steinernema* and *Heterorhabditis*) of the nematode perform the task of host-seeking while the gut bacteria kill the insect host. The IJ enters a suitable host insect through the mouth, anus, spiracles or by penetration of the intersegmental membranes of the cuticle (Peters and Ehlers, 1994). Once it has arrived in the haemocoel, cells of the bacterial symbiont are released. The bacteria multiply and cause death of the insect within 24–48 h. The nematodes develop to the adult stage and reproduce within the host cadaver. Depending on the available resources, one or more generations occur in the deceased insect body.

4.2 Factors affecting the success of biocontrol

The species, temperature, water content, timing of release, application method and rate are key factors pertaining to the EPNs that determine whether the control of CRF is successful or not.

4.2.1 EPN species

Different strains of *S. carpocapsae* and *S. feltiae* are most often tested in laboratory and field-trials against the cabbage root fly. In most of these studies, *S. feltiae* strains appeared to be slightly better suited than *S. carpocapsae* for control of the CRF (Schroeder et al., 1996; Beck et al., 2014).

4.2.2 Temperature

Temperature is one of the most important parameters limiting the success of EPN in temperate regions. In temperate regions like Belgium, the CRF causes most of the damage in early spring, when the soil temperature is low (approximately 10°C). The failure or success of a biocontrol strategy with EPN will therefore largely depend on their performance at low temperatures. Chen et al. (2003) analysed the impact of soil temperature on the control of cabbage root fly by IJs of different nematode species (Chen et al., 2003). In this study, *S. feltiae* was the only species able to kill cabbage root fly larvae at 10°C. However, both its development and the time needed for penetrating the larvae were significantly slowed at this temperature.

4.2.3 Water content of the soil

Desiccation is an important abiotic factor affecting the survival of EPN. When free water disappears from the soil, the movement of nematodes, and thus, their host-seeking behaviour are hampered (Glazer, 2002). Further desiccation of the environment withdraws water from the nematode body, causing its death.

4.2.4 Timing and method of application

Since CRF larvae are in the soil for a limited period, the timing of nematode application has to be optimal. More active strains might therefore achieve higher infection rates (Beck et al., 2014). The application method used has an impact on the level of pest control achieved (Chen et al., 2003). The EPN should enter the soil as quickly as possible, avoiding exposure to ultraviolet (UV) radiation and desiccation. Applying nematodes at the soil surface, as close as possible to the larvae, appears to be the most successful method (Chen et al., 2003).

4.2.5 Application rate

In general, increasing the application rate of EPN increases the level of control achieved. On the other hand, bearing the cost of IJs in mind, application rates should be kept as low as possible.

4.3 Challenges in use of application technologies

Most EPN are applied by using standard spray equipment in the field, based on the irrigation system in greenhouses or by drench treatment. All these techniques are often used to apply pesticides without significant modification. However, different EPNs might require a different application method in order to maximise the level of control (Grewal et al., 2005). Losses during transfer from the mixing tank to the host insect must be minimised. Research aiming to improve and optimise the application systems is certainly needed, and might result in the more widespread use of EPNs as biological control agents (Grewal et al., 2005). Some of the practical difficulties linked to nematode application technology are listed below.

4.3.1 Temperature

The aqueous solutions in which nematodes are applied should have the ideal temperature (4–30°C). Temperatures exceeding 30°C are detrimental for the EPNs. Oxygen concentration decreases with increasing temperature, inactivating the nematodes. Black irrigation hoses, sometimes used to deposit nematodes, might warm up considerably and reduce the biological control potential of EPN (Grewal et al., 2005).

4.3.2 Sedimentation

Since nematodes are heavier than the solution in which they are applied, sedimentation on the bottom of the tank should be expected. Care must be taken to guarantee an equal distribution over the treated crop. Increasing the viscosity of the solution can decrease the sedimentation rate of the nematodes (Grewal et al., 2005).

4.3.3 Compatibility with pesticides

In general, nematodes are relatively resistant to fungicides and herbicides. Insecticides might, on the contrary, impose negative effects on the EPNs applied (Patel and Wright, 1996). More research that explores the impact of pesticides on nematodes is certainly needed.

4.3.4 Shear forces

During application, EPNs are exposed to relatively high shear forces (e.g. in the pumps, while passing through filters and nozzles and hitting the crop). These forces might be harmful. Research by Nilsson and Gripwall (1999) indicates that this harmful effect on the nematodes is partially caused by a rise in temperature of the tank liquid after its passage through the pump. However, care must be taken to minimise mechanical stress. The pressure in the tank should be limited to 5 bar while sieves and filters should have a mesh size of at least 300 μm, and nozzles with apertures >500 μm are recommended (Grewal et al., 2005).

5 Key issues in chemical control

To reduce the impact of chemical crop protection on the environment and health, the range of insecticides approved for application is quite limited in several European countries. The previous decades show a sequence of different methods and products being used, although drench applications remain the most common method of treatment. With the introduction of products based on spinosad as drench application, the use of other methods appeared less necessary, although they were still used. However, with changing pest pressure and the loss of certain products, spinosad is no longer sufficient to control damage when pressure from the CRF is high (pers.comm. advisor to cabbage growers). This makes the control of CRF a challenging task for growers.

In Flanders, a whole range of commonly used products was recently prohibited. Since 1 June 2017, the active substance (a.s.) fipronil, and since 15 March 2017, the a.s. dimethoate can no longer be used. Until 2020, chlorpyrifos granules applied in the field were often used (in combination with a drench treatment) to extend the period of protection for a few weeks. In 2020, authorisation for the application of products containing chlorpyrifos also ceased (Maes and Lippens, 2019). Since then, chemical crop protection to protect cabbage plantlets against cabbage root fly in the field is limited to a few products which gain an emergency permit for 120 days. In 2020, the granule treatments Force 1,5 GR (1,5% tefluthrin) and Sherpa 0,8 GR (0,8% cypermethrin) gained such a temporary permit. So did Verimark (200 g/L cyantraniliprole, SC), an alternative to products containing spinosad.

Cabbage growers in Flanders are strongly advised to take preventive measures (application of a drench treatment or covering the crop) against CRF, especially in regions with high pest pressure. Preventive treatment is necessary for three reasons: (1) the observation that CRF eggs are laid on the seed trays, meaning that adequate plant protection is needed even before planting; (2) the detection of high numbers of CRF on the sticky traps during the period of planting in spring; and (3) the limited possibilities for growers to protect the crop once planted. It is a common practice in Flanders to treat *Brassica* plantlets with a drench application of products containing spinosad before they are transplanted. Despite this treatment, regional losses of 50% can occur when pest pressure is high (Maes and Lippens, 2019). From a series of field trials (see Sections 6-8), it is known that drench applications can protect the plants for a period of four to five weeks. However, the larvae of the cabbage root fly can still cause considerable damage to the roots, with the risk of yield or quality loss, once the effects of the application have faded. At that time, applying products containing spinosad to the plant base or application of a granule treatment might be a final option.

A major advantage of both the drench and granule treatments is their limited environmental impact, as there is no risk of drift to off-field areas. Both the drench application and the granules are applied to the exact location where damage by the CRF takes place.

Since 2017, products containing cyantraniliprole have been granted a yearly temporary permit and can be alternated with products containing spinosad as a drench treatment. This allows growers to reduce the risk of resistance build-up against spinosad. The risk of resistance build-up against products used frequently to control CRF has been reported in the past. In 2012, Blackshaw et al. reported resistance of CRF larvae to chlorpyrifos in Canada. Possibly a more relevant case is the report of resistance to spinosad amongst certain populations of *T. tabaci* in Israel (Lebedev et al., 2012), a problem that is increasing in certain Flemish regions as well. Further research is being conducted regarding this last case.

6 Case study 1: row covers as a preventive measure against *Delia radicum*

Chinese cabbage is a preferred host of the CRF. The larvae feed through the leaf veins, causing visual damage to the marketable product and increasing the risk of secondary infections. If not too severe, damage can be eliminated by removing the outer leaves. For brassicaceous root vegetables like turnip, swede or radish, damage cannot be removed since the larvae create tunnels in the root, which is the edible part.

Even a low incidence of CRF can cause high economic losses in the above-mentioned crops. Therefore, chemical protection can be insufficient to prevent damage. An alternative is the use of row covers. With proper use and the right choice of material, a high level of protection can be achieved.

Row covers are not always easy to use. Growers need extra time to install and remove the netting. Certainly, on large fields of brassicaceous crops, the technique is not standard practice. However, on small fields of crops such as turnip and Chinese cabbage, row covers can be a valuable option. Especially in organic agriculture, where approved pest control measures against cabbage root fly are scarce, the use of insect netting or climate nets might provide adequate protection in these crops and can be considered an important crop-protection technique for these vegetables.

Field trials were performed to assess the impact of different types of row cover on turnip and Chinese cabbage production. The results of these trials were used to inform growers about the potential of row covering to protect their crops against the cabbage root fly.

6.1 Material and methods

6.1.1 Field trials

All trials were performed on fields of the Research Station for Vegetable Production in Sint-Katelijne-Waver, Belgium. The impact of row covers on CRF damage and yield was assessed on turnip and Chinese cabbage (Table 4). Turnip was cultivated as a conventional crop, while Chinese cabbage was cultivated in an organic production system. The trials were performed in a randomised block experimental design with three replicates. Different types of row cover were used (see Table 4 for the specifications). The distance between the straps of the Rova climate net is larger compared to the Howicover climate net, which results in a more open structure and a higher mesh size. The plot size was 1.75 m x 8 m.

6.1.2 Assessments

At harvest, CRF incidence is assessed by counting the number of damaged turnips and Chinese cabbages. For turnip, 100 plants were assessed per plot, and 20 were assessed for Chinese cabbage. Additionally, the plants were weighed to calculate yield.

6.1.3 Statistical analysis

Data were analysed by performing ANOVA, followed by Duncan's multiple-range test, with a significance level of 0.05.

6.2 Results and discussion

The results of pest incidence and yield for four trials evaluating row covers are presented in Tables 5 and 6.

Table 4 Overview of the trials

Crop	Year of trial	Planting or sowing date	Harvest date
Turnip	2018	March 26 2018	June 1 2018
Turnip	2019	May 2 2019	June 24 2019
Chinese cabbage	2019	August 15 2019	October 22 2019
Chinese cabbage	2020	April 2 - April 24 - May 15 2020	June 4 - June 16 - June 29 2020

6.2.1 Turnip 2018

Plants from all the treatments were highly damaged. The damage was probably caused by emerging CRFs from the pupae present in the soil. However, it is remarkable that the number of damaged plants was highest in the control treatment without row covers. This might indicate that the row covers avoided additional damage caused by CRF from nearby fields.

6.2.2 Turnip 2019

The best results in this trial were achieved with the insect nets with the smallest mesh size. Moreover, the yield was highest under these nets. The broad-meshed stretch-net did not prevent damage.

6.2.3 Chinese cabbage 2019

In this trial, all types of insect nets were sufficient to control the CRF. The climate net also provided adequate protection. Again, covering up the crops resulted in a higher yield in almost all treatments.

6.2.4 Chinese cabbage 2020

All materials tested gave significantly better protection than the control. Likewise, with the trial on turnip in 2019, the best results were obtained with the row covers with the smallest mesh size. Row covers resulted in an increased yield of Chinese cabbage (40–70% increase in weight/vegetable) (Table 5).

6.2.5 Conclusion

Row covers with a mesh size of 1.35 mm x 1.35 mm or lower can prevent a considerable amount of damage by CRF. However, the best protection in turnip and Chinese cabbage crops is obtained by row covers with a finer mesh size (< 0.8 mm x 0.8 mm). Densely woven climate nets can provide adequate protection as well.

Covering up the crop always resulted in a higher yield in the trials. It is likely that higher temperatures under the nets stimulate plant production as well as mineralisation of nitrogen in the soil. This is advantageous for vegetables with a short cropping cycle, such as Chinese cabbage and turnip.

6.2.6 Success factors

Besides the right choice of material, the level of protection is determined by other factors. The row covers need to be well installed, avoiding any access

Table 5 Incidence of damage due to cabbage root fly (CRF) in four different trials (percentage of plants damaged by CRF) assessing different types of row cover. Covers are arranged according to decreasing mesh size, if applicable (treatment means followed by the same letter are not statistically significantly different)

			Percentage of plants damaged by cabbage root fly			
Material	Type of cover	Mesh size (mm)	Turnip 2018	Turnip 2019	Chinese cabbage 2019	Chinese cabbage 2020
Control	Pigeon net	5 x 7	73 d	96 c	75 c	97 b
Stretchnet	Anti-hail net	4 x 5	43 abc	77 c	35 b	
Rova	Climate net	n/a	39 ab	38 b	12 a	
Howicover	Climate net	n/a	46 bc	25 ab	2 a	20 a
Ornata plus 135	Insect net	1.35 x 1.35	52 bc	20 ab	2 a	33 a
Insectprotect 1.3	Insect net	1.3 x 1.3			7 a	43 a
Ornata addu 80100	Insect net	0.8 x 1	52 bc	14 ab	7 a	2 a
Ornata plus 80	Insect net	0.8 x 0.8	50 bc	8 a	5 a	35 a
Insectprotect 0.8	Insect net	0.8 x 0.8			5 a	42 a
Ornata addu 5080	Insect net	0.5 x 0.8				18 a
Ornata Light	Insect net	0.6 x 0.66	53 c	5 a	8 a	15 a
Fleece cover	Non-woven fleece	n/a	31 a	19 ab		

Table 6 Yield data from four different trials assessing different types of row cover. Covers are arranged according to decreasing mesh size, as in Table 4 (treatment means followed by the same letter are not statistically significantly different)

	Production			
Material	Turnip 2018 (kg/100m²)	Turnip 2019 (g/vegetable)	Chinese cabbage 2019 (g/vegetable)	Chinese cabbage 2020 (g/vegetable)
Control	111 d	192	871 d	d b
Stretchnet	220 ab	230 bcd	1065 bc	
Rova	117 d	245 bc	1155 ab	
Howicover	123 d	216 cd	995 cd	1012 a
Ornata plus 135	201 abc	211 cd	1134 abc	1143 a
Insectprotect 1.3			1205 ab	957 a
Ornata addu 80100	170 c	271 b	1220 ab	1108 a
Ornata plus 80	217 ab	269 b	1072 bc	1005 a
Insectprotect 0.8			1238 a	1142 a
Ornata addu 5080				1097 a
Ornata Light	232 a	328 a	1102 abc	1140 a
Fleece cover	189 bc	276 b		

points for insects. A good rotation is essential to prevent the presence of pupae in the soil. Covering up the trays of transplants before planting is another important factor that prevents egg-laying on young plants. Removal of row covers needs to be avoided, especially when the pest pressure is high. If necessary, it is better to take off the covers when pest activity is lower.

7 Case study 2: the use of entomopathogenic nematodes against *Delia radicum*

The CRF is a devastating pest in spring crops (April–June) of cauliflower. In organic crops, a module drench treatment using spinosad or the application of row covers must be used to protect the plants. Crops are most susceptible during the first few weeks after planting, before their root systems have developed. However, the oviposition rate remains high in the following weeks and continued protection is needed.

In this field trial, we tested the efficacy of biological control with entomopathogenic nematodes (EPN). *S. feltiae* is one of the most promising biological control agents for field application against CRF larvae. Previous field research revealed that the application method should be optimised in order to increase control for effective protection of cauliflower plantings. Two *S. feltiae* formulations, Capirel®(Koppert Biological Systems) and Nemacaps (Agrocaps SPRL), and an adjuvant were tested. In addition, two treatment schemes were tested: a single pre-planting module application was compared with a module application followed by a supplementary application in the field. The same dose rate, 100 000 EPN per plant, was used for all treatments.

7.1 Materials and methods

7.1.1 Trial site

The trial was carried out at the experimental organic farm belonging to Inagro in Rumbeke-Beitem, Belgium. The soil type is sandy loam. On this site, CRF pressure during the spring has been increasing in recent years.

7.1.2 Trial design

The trial was organised as a randomised complete block design with four replicates of each treatment. Plot size was 12 m long and 3 m wide, with four plant rows.

7.1.3 Cauliflower planting

The cauliflower plants were transplanted on 8 April, 2020. Cauliflowers were planted at twice the normal density within rows, 28 x 70 cm, to allow for destructive assessments during the cropping period. After these assessments, plant rows were thinned out to the normal density to obtain a normal yield assessment. Immediately after planting, the whole plot was covered with anti-bird netting to prevent damage by pigeons.

7.1.4 Treatments

The reference treatment was a module drench treatment with spinosad (Tracer®, Corteva Agriscience). Before planting, spinosad was applied to the modules using a spray boom at an application rate of 12 mL/1000 plants. Shortly after spraying, the plant containers were watered until saturated.

The IJs of *S. feltiae* were applied at different times and using different methods. The first treatment was a pre-planting module drench application with Capirel® (Koppert Biological Systems) (for this trial, a similar product, Entonem, was used, with a smaller package size). For this treatment, a knapsack sprayer was used with Teejet XR 110 04 nozzles. The filter in the tank sprayer was removed at filling. For each tray container of 150 plants, a volume of 0.5 L of water with 15×10^6 IJs was sprayed (= dose rate of 100 000 EPN/plant). For Treatment 7, the adjuvant was added at 0.1% to the EPN suspension (2.5 mL on 2.5 L water).

In Treatments 4, 6 and 8 a new experimental formulation of encapsulated *S. feltiae* nematodes was tested: Nemacaps® produced by Agrocaps SPRL (Gedinne, Belgium). For the pre-planting module application, the capsules were spread manually onto the plant modules at a rate of 3000 capsules per tray (= 20 capsules/plant) and then sprayed with water.

Because rainfall was lacking following the EPN treatments, the field was irrigated on 10 April, two days after planting, and on 23 April, the day after the second EPN application in the field.

After planting, a second application of nematodes in the field was scheduled 10 days after planting for Treatments 5 to 8. To apply Capirel, with or without adjuvant, 150 mL EPN suspension/plant was poured manually using a cup to the base of each plant. The nematode capsules were sprinkled at a dose rate of 20 capsules/plant at the base of the plants and then covered with soil using a hoe. The field was irrigated the following day. The control plots were not treated with insecticides or EPN suspensions (Table 7).

7.1.5 Assessments

To monitor CRF activity in the field, 40 egg-laying traps (felt collars - Olbisles ateliers.ch) were distributed in the outer plant rows of the trial field. The CRF eggs deposited in these traps were counted and removed on a weekly basis.

Assessments of maggot density and damage to the roots were made 2, 3, 5, 6 and 8 weeks after planting. For each assessment, 6 plants in the outer plant rows of each plot were dug up and the numbers of CRF larvae and pupae around / in the roots and stem of each of these plants were counted.

In the inner two plant rows of the plot (i.e. 60 plants), all dead plants were removed from the field on a weekly basis. The roots of these plants were visually inspected to determine the cause of death. Plants that died due to CRF infestation were recorded as such.

The damage rate to the roots was graded on a scale of 1 to 5:

Table 7 Overview of the treatments applied in the trial on the use of entomopathogenic nematodes to control cabbage root fly larvae (MDT = module drench treatment; FT = field treatment; Timing A = just before planting; Timing B = 10 days after planting)

N°	Treatment	Dose	Active substance	Application method	Timing
1	Control (untreated)	-	-	-	-
2	Spinosad MDT	12 mL/1000 plants	spinosad (480 g/l)	Pre-planting module drench application	A
3	Entonem MDT	100 000 IJs/ plant	*Steinernema feltiae*	Pre-planting module drench application	A
4	Nemacaps MT	100 000 IJs/ plant	*S. feltiae*	Pre-planting module sprinkle application;	A
5	Entonem MDT + FT	100 000 IJs/ plant	*S. feltiae*	Pre-planting module drench; Soil drench at the base of the plants	A B
6	Nemacaps MT + FT	100 000 IJs/ plant	*S. feltiae*	Pre-planting module sprinkle application; Field application at the base of the plants	A B
7	Entonem with adjuvant MDT + FT	100 000 IJs/ plant	*S. feltiae*	Pre-planting module drench application; Soil drench at the base of the plants	A B
8	Entonem MDT + Nemacaps FT	100 000 IJs/ plant	*S. feltiae*	Pre-planting module drench; Field application at the base of the plants	A B

1 = no damage
2 = slight attack (superficial damage), the stem still firm
3 = moderately affected (larvae only at the bottom of stems)
4 = very degraded (larvae high in stem), stem strongly hollowed out
5 = completely degraded (larvae in stalk to cabbage), stem almost completely hollowed out

After harvest, the roots of the 12 remaining stems per plot were pulled out and assessed for damage to the roots using the same scale.

7.1.6 Statistical analysis

The results were processed using the AGROVA-R statistical package developed by Inagro in R and validated with SPSS (ANOVA WITH UNTREATED, Tukey's test). Values followed by the same letter are not significantly different (P =0.05).

7.2 Results and discussion

Egg-laying by CRF females started in the second week after planting. Therefore, the pressure was low during the first two weeks after planting. From the end of April until mid-June, egg-laying was high (Fig. 2). During May, more than 15 eggs per trap per week were recorded.

To evaluate the efficacy of the treatments, the build-up of damage was monitored during the first eight weeks after planting. Weekly or every two weeks, we counted the number of dead or wilted plants damaged by CRF larvae on a marked area of 60 plants. In addition, we inspected 6 randomly-selected plants per plot for root damage and the number of larvae present in and around the roots and stem. These assessments were made on five occasions. On 17 June, the trial was finished because of insufficient marketable yield. Just before clearing the crop, final root damage assessments were made. The results are presented in Tables 8 and 9.

Plants that died as a result of CRF damage appeared from three to four weeks after planting. An average of three larvae per plant were found on the roots of untreated plants (Table 8 and Fig. 3). In the following five weeks, this number increased to eight larvae per plant. All EPN treatments reduced the larval density on the roots and stems of the plants, but at most assessment times, this reduction was not statistically significant. Only at the time of the last assessments did two treatments show statistically significant differences from the untreated control: for Treatment 7, two applications of Capirel with adjuvant, and for Treatment 8, a Capirel module drench application followed by a field application of Nemacaps. Treatments 5 and 6 showed no significant

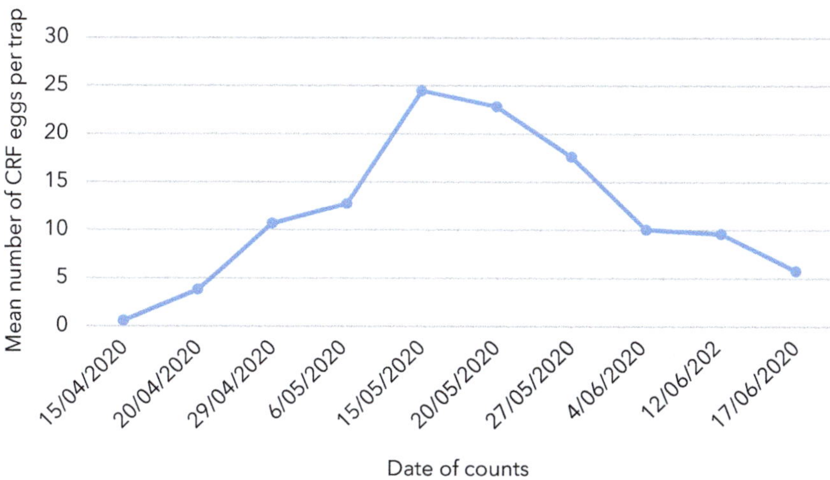

Figure 2 Mean number of eggs laid per plant in the cauliflower trial during the cropping period (planting date 8 April 2020). Eggs were laid in felt egg traps.

reductions in the number of larvae but they did slow down the progression of the root damage (Table 9). At the time of the assessments in May, i.e. two and three weeks after the second EPN treatment, the root damage of these treated plants was significantly lower compared to the control.

Two EPN doses of 100 000 IJs per plant (Treatments 5 to 8) resulted finally in 7% to 12% less plant loss compared to the control. However, 25% to 30% of the plants still died in the field, which is too many. This result was due to insufficient continuous protection provided by the EPN under the high pressure of CRF during the whole cropping period. The drought also contributed to the plant loss. We noticed, for example, that eggs were laid on the stem as well as at the base of the plant. In this case, the larvae are more difficult for the nematodes to reach because they bore directly into the stem.

The protection provided by a spinosad module treatment proved to be sufficient to limit damage accumulation and plant loss during the first six weeks after planting. After six weeks, however, many plants died nonetheless. This was due to insufficient continuous protection in combination with unfavourable growing conditions (drought).

7.3 Conclusion

The infestation of cabbage root fly larvae in cauliflower spring cultivation was high but at the time of planting, 8 April, egg-laying by CRF was still low. The population of cabbage root larvae only started to build up from three weeks after planting. For the module drench treatments, applied just before planting,

Table 8 Effect of the treatments on the number of cabbage root fly larvae per plant, found in and around the stem and roots of the cauliflower plants (planting date 8 April 2020) (treatment means followed by the same letter are not statistically significantly different). WAP = weeks after planting

No	Object	22/4/20 2 WAP	29/4/20 3 WAP	6/5/20 4 WAP	14/5/20 5 WAP	3/6/20 8 WAP
1	Control (untreated)	0	3.2 a	2.6 a	5.2 a	8.3 a
2	Spinosad MDT	0	0.0 c	0.2 c	0.9 b	2.7 c
3	Entonem MDT	0	2.0 ab	1.7 ab	4.2 a	6.7 ab
4	Nemacaps MT	0	1.1 bc	1.7 ab	4.0 a	6.9 ab
5	Entonem MDT + FT	0	1.2 abc	1.2 b	2.8 a	6.5 ab
6	Nemacaps MT + FT	0	2.4 ab	1.8 ab	3.4 a	6.4 ab
7	Entonem with adjuvant MDT + FT	0	2.7 ab	2.4 a	3.3 a	5.7 b
8	Entonem MDT + Nemacaps FT	0	1.9 ab	1.5 ab	4.1 a	5.9 b
Statistical test applied			Kruskal-Wallis	Tukey	Tukey	Tukey
C.V. (%)			26.2	16.2	15.3	6.9
P-value			< 0.01**	< 0.001***	< 0.001***	< 0.001***

Table 9 Effect of the treatments on cumulative damage to the roots of cauliflower plants caused by cabbage root fly larvae (planting date 8 April 2020) (treatment means followed by the same letter are not statistically significantly different). The damage is scored on a scale from 1 (= no damage) to 5 (= roots completely degraded and stem almost completely hollowed out)

No	Object	22/4 2 WAP	29/4 3 WAP	6/5 4 WAP	14/5 5 WAP	3/6 8 WAP	17/6 10 WAP
1	Control (untreated)	1.0	2.4 a	3.3 a	3.8 a	4.9 a	4.9 a
2	Spinosad MDT	1.0	1.0 c	1.2 d	1.7 c	2.2 b	3.9 b
3	Capirel MDT	1.0	2.2 ab	2.4 c	3.4 ab	4.8 ab	4.7 a
4	Nemacaps MT	1.0	1.7 bc	2.5 bc	3.3 ab	4.8 ab	4.8 a
5	Capirel MDT + FT	1.0	1.7 ab	2.6 bc	2.6 bc	4.9 a	4.5 ab
6	Nemacaps MT + FT	1.0	2.0 ab	2.3 c	2.8 ab	4.7 ab	4.5 ab
7	Capirel with adjuv. MDT + FT	1.0	1.9 ab	3.1 ab	3.2 ab	4.8 ab	4.8 a
8	Capirel MDT + Nemacaps FT	1.0	2.1 ab	2.3 c	3.0 ab	4.7 ab	4.7 a
Statistical test applied			Tukey	Tukey	Tukey	Kruskal-Wallis	Tukey
C.V. (%)			15.5	11.3	14.5	8.9	7.0
P-value			< 0.01**	< 0.001**	< 0.001**		< 0.01**

Figure 3 Cabbage root fly larvae at the roots of plants three weeks after planting (left) and inside the stem five weeks after planting (right) on 30 April and 14 May, respectively.

this means that there is a time lag between the protective period and the pest presence in the field. *S. feltiae* nematodes, sprayed at a dose rate of 100 000 IJs per plant on the plant modules, proved to persist in effective numbers for two weeks. However, during this period, there was a lack of larvae as prey in the field.

All EPN treatments, with an additional field application two weeks after planting, when egg-laying was increasing, could slow down damage progression and reduce the plant loss caused by CRF to some extent. However, the control was finally insufficient because of the continued high pest pressure.

A module drench treatment with spinosad provides sufficient control during the first six weeks after planting. However, when the high CRF pressure continues under conditions that inhibit the formation of new roots, plants can still die after this period of adequate protection.

8 Case study 3: chemical control of *Delia radicum*

Despite the implementation of a decision support system in Flanders, chemical intervention is often necessary to prevent severe damage by CRF. At the vegetable research centres in Flanders, trials to evaluate the efficacy of different products and methods of application against CRF have been performed for over ten years. Recent trials have mainly focused on the use of pre-planting drench applications with products containing spinosad, granular pesticide supplementation and alternative drench applications.

8.1 Materials and methods

8.1.1 Materials

Products with the active ingredient spinosad (Tracer®, Conserve Pro™, etc.) (Corteva Agriscience) are commonly used in pre-planting drench applications in Flanders. In regions with high pressure from CRF, additional treatments with granules are often necessary to prevent heavy losses. Until 2019, these granules often contained chlorpyrifos (Pychlorex, etc.) but since their withdrawal, alternative granules, such as Force 1,5GR (1,5% tefluthrin, GR) (Syngenta) and Sherpa (0,8% cypermethrin, GR) (Corteva Agriscience) have been tested (Table 10). The Force 1,5GR granule received a temporary permit for application in 2020, for 120 days, at a dose rate of 10 kg/ha. Thus, in 2020, this granule was tested at 0.4 g/plant, which would be the maximum dose rate found in practice. Granules were added manually within one hour after planting. Growers, who only use one granulate per field, use a granulate disperser, which releases a dose of granules on the plants at the moment of planting. Due to the risk of resistance build-up by the flies against spinosad, Verimark (200 g/L cyantraniliprole, SC) (FMC Chemical S.P.R.L.) was also tested as a drench application to the module trays.

8.1.2 Assessment methods

The damage caused by CRF was assessed according to the EPPO Guideline PP 1/009(3) 'Guideline for the efficacy evaluation of insecticides on *Delia radicum*.' During assessment, the percentage of damaged plants and plants lost due to CRF was counted. At harvest, the cabbages were weighed (to determine the yield) and sorted according to their quality (to determine the percentage of marketable cabbages). Shortly after harvest, a final assessment was done on the roots. After washing the roots, the damage caused by CRF was scored according to four classes, between 0 and 3 (Fig. 4). With these scores, a root damage index was calculated according to the Townsend-Heuberger formula:

$$\text{Percent damage} = \left(\Sigma\left(n * v\right) / \left(I * N\right)\right) * 100$$

V = class of damage defined as degree of infestation by CRF, according to Fig. 4
i = highest class of damage (3 in this case)
n = number of plants in each class
N = total number of plants

8.1.3 Trial setup

The field trials were set up in a research field in Kruishoutem, Belgium, in a randomised block design with at least 3 replications.

8.1.4 Statistical analysis

If the residuals were not homogeneous, a transformation was performed. The means were compared by analysis of variance. If significant differences (α=0.05)

Table 10 Products tested and trial details for trials on insecticides between 2018 and 2020

Treatment	Active ingredient	Dose rate	2018	2019	2020
Untreated			x	x	x
Tracer (drench)	spinosad	0.012 mL/plant	x		x
Verimark (drench)	cyantraniliprole	0.015 mL/plant	x		x
Pychlorex (granule)	chlorpyrifos	1 g/plant	x		
Force (granule)	tefluthrin	0.236 g/plant	x		
Sherpa (granule)	cypermethrin	0.42 g/plant	x		
Tracer + Pychlorex			x	x	
Tracer + Force				x	x
Tracer + Sherpa				x	x
Verimark + Force				x	x
Verimark + Sherpa				x	x
Planting date			18/04/20	15/04/20	24/04/20
Harvesting date				24/06/20	7/07/20
Variety			Clarina	David	David

Figure 4 Classification of roots of cauliflowers at harvest; CRF = cabbage root fly (Volckaert et al., 2019).

occurred, post hoc Tukey's test was performed to find significant differences between means.

8.2 Results and discussion

In 2018, the main research question was whether combining a drench application and granules increased plant protection. With 61% of the untreated plants dying because of CRF damage (Table 11), it was clear that protection was necessary in order to grow a profitable crop. Every treatment reduced the number of dead plants. Both with regard to dead plants and to plants infested by CRF, it was clear that the combination of Tracer and Pychlorex showed the greatest efficacy. During this trial, the pre-planting drench applications proved to be effective for up to four to six weeks from planting. Any damage caused at that point might not cause additional loss of plants, but could reduce the quality or yield of the crop.

Table 11 Results of the insecticide trial in 2018 when plants were assessed on 20 June (treatment means followed by the same letter are not statistically significantly different)

	Assessment of the plants (20/06/18)	
Object	Percentage dead due to CRF	Percentage infested by CRF
Untreated	61 a	31 ab
Tracer	11 bc	35 ab
Verimark	9 bc	41 ab
Sherpa	14 bc	59 a
Force	8 bc	28 ab
Pychlorex	24 b	46 ab
Tracer + Pychlorex	3 c	18 b
P-value	<0.01	0.0359
transformation	*y*=*sqrt*(*x*+0.5)	-

Table 12 Results of the insecticide trial in 2019 when plants were assessed on 24 June 2019 and the roots were assessed on 2 July (treatment means followed by the same letter are not statistically significantly different)

	Assessment of the plants (24/06/19)		Assessment of the roots (2/07/19)		Harvest (24/06/19)
Object	Percentage dead due to CRF	Percentage infested by CRF	Percentage of undamaged roots	Root damage index	Percentage of marketable plants
Untreated	64 a	30 a	0	91 a	24 b
Tracer + Pychlorex	0 b	6 b	8	47 d	99 a
Tracer + Force	6 b	14 ab	0	59 cd	98 a
Tracer + Sherpa	1 b	10 ab	0	71 bc	98 a
Verimark + Force	1 b	15 ab	3	58 cd	99 a
Verimark + Sherpa	2 b	21 ab	0	78 ab	94 a
P-value	<0.01	0.0312	0.0745	<0.01	<0.01
transformation	-	-	-	-	-

In 2019, the trial focused on different combinations of drench applications and granules. The damage caused by CRF was severe, with 64% of untreated plants lost (Table 12). All the combinations tested showed good efficacy and reduced plant losses and infestation. The effectiveness of the treatments was obvious when looking at the percentage of marketable plants. Pychlorex was the most effective granule treatment in all respects. Sherpa appeared to be the least effective product in terms of protecting the roots.

2020 was the first year that growers could no longer use chlorpyrifos against CRF. That year, a range of possible treatments were tested (Table 13). Although the losses of untreated plants were as high as in previous years, no clear differences were seen between the treatments. Even the added value of the granule treatments was not observed, as Tracer and Verimark alone were as effective as the treatments with both a drench treatment and a granulate. The added value of the granules became clear when assessing root damage, especially the efficacy of Force, which was similar to the observation made in 2019. This could indicate that the effect of the drench applications faded after a couple weeks. By that time, the cabbage plants could tolerate some root damage, without dying or losing yield. Although the roots suffered some damage, it was not that prevalent at harvest.

8.3 Conclusion

This series of trials showed that intervention is necessary for growers to protect their crop from CRF. The efficacy of drench applications was confirmed in these trials but they had limited persistence. The trials in 2018 and 2020 showed that after four to five weeks of protection, larvae were able to cause root damage, with varying consequences at harvest. By adding granule treatments, the period of protection could be extended. Although none of the granules tested was as efficient as Pychlorex, they were able to reduce the damage and yield losses significantly.

9 Conclusion and future trends

The reduction in authorised chemical crop protection products will force growers to rely increasingly on IPM strategies using a variety of alternative techniques. Practical research in Flanders indicates that EPNs and row covers are promising techniques to control CRF. These methods can be implemented as part of a pest management strategy to reduce the amount of chemical insecticides used. However, although promising, uptake by growers is limited due to a lack of knowledge about the efficacy and practical application of these alternative techniques in a commercial setting. To guarantee increased uptake by the growers, further refinement of such methods is necessary. For the EPNs,

Table 13 Results of the trial in 2020 when plants were assessed on 26 June 2020, harvested on 6 July, and the roots were assessed on 13 July (treatment means followed by the same letter are not statistically significantly different)

	Assessment on the plants (26/06/20)		Assessment on the roots (13/07/20)		Harvest (06/07/20)
Object	Percentage dead due to CRF	Percentage infested by CRF	Percentage of undamaged roots	Root damage index	Percentage of marketable plants
Untreated	58 a	28	2 b	74 a	35 b
Tracer + Force	0 b	20	43 a	22 c	69 ab
Tracer + Sherpa	0 b	16	10 b	41 bc	92 a
Verimark + Force	0 b	13	40 a	23 c	77 a
Tracer	1 b	15	0 b	60 ab	88 a
Verimark	0 b	15	5 b	48 abc	85 a
P-value	<0.01	0.3897	<0.01	<0.01	<0.01
transformation	$y=\log(x+1)$	-	-	-	-

the following aspects of the technique should be studied further: avoidance of desiccation of the EPN after application, prolongation of the lifespan of the EPN, development of more virulent strains, development of cold-resistant strains suitable for use in early spring in temperate regions and improved timing of application. For row covers, the microclimate in Flemish conditions and the optimal timing of covering should be further investigated for different net types, to optimise the use of this technique. For some crops and in some circumstances, the use of row covers favours other pests, such as aphids or whitefly, while protecting the crop from CRF. Further research should be done to investigate how to avoid these pests, by optimising the covering technique or by combining row covers with other practices. Although the above-mentioned techniques are separately less efficient than chemical control, the future lies in the application of a combination of them in order to obtain an integrated approach.

Practical research centres, such as the vegetable research centres in Flanders, play an important role in gathering this knowledge. They validate new pest-control approaches in an open-field context, thereby building bridges between academic research and vegetable growers. Additionally, they inform the growers about the new methods to ensure that they are incorporated into the vegetable production system correctly, and advise growers on their pest control strategy. Additionally, knowledge on pest density and phenology is essential and will remain important in the future.

Since most of the non-insecticidal methods of pest control are less efficient compared with chemical control, the combined use of different techniques will be necessary to control CRF adequately. Furthermore, CRF is part of a pest complex consisting of many insect species and pathogens, all with their specific impact on crop health and yield. Future research should therefore apply an integrated approach and focus on the cropping system as a whole, considering the impact of a combination of control measures on the pest complex and crop yield. In this more integrated approach, accurate information on pest densities and phenology is likely to be very important. A decision-support system where the abundance of root-feeding flies is monitored and forecasted, guiding growers in their decision-making, will become more important in the future when the number of insecticidal control measures further diminishes. Alternative control methods should be integrated with the decision-support system and possibilities to automate the system through the use of, for example, insect cameras and simulation models, should be explored further.

10 Where to look for further information

A standard introduction to the use of IPM against cabbage root fly, both monitoring methods and control methods:

- Glazer, I. 2002. Survival biology. In: R. Gaugler (Ed.), *Entomopathogenic Nematology*. CAB International, Wallingford, UK, pp. 169–187.
- Collier, R., Mazzi, D., Schjoll, A. F., Schorpp, Q., Thoming, G., Johansen, T. J., Meadow, R., Meyling, N. V., Cortesero, A. M., Vogler, U., Gaffney, M. T. and Hommes, M. 2020. The potential for decision support tools to improve the management of root-feeding fly pests of vegetables in Western Europe. *Insects* 11(6), 369.
- Chen, S. L., Han, X. Y. and Moens, M. 2003. Biological control of *Delia radicum* (Diptera: Anthomyiidae) with entomopathogenic nematodes. *Applied Entomology and Zoology* 38**,** 441–448.
- Capinera, J. L. 2008. *Encyclopedia of Entomology*. Springer Science & Business Media, Heidelberg, p. 2061.

Several institutions and organisations are actively working on how to tackle this and other similar problems and supporting growers in their regions and countries, including

- Commonwealth Agricultural Bureaux International (www.cabi.org).
- Research centre for agriculture in West Flanders (Belgium) (www.inagro.be).
- Provincial research centre for vegetables in East Flanders (Belgium) (www.pcgroenteteelt.be).
- Research station for vegetables in Antwerp (Belgium) (www.proefstation.be).
- Research centre AgroFoodNature at HOGENT in Ghent (Belgium) (https://www.hogent.be/onderzoekscentra/agrofoodnature/).
- The School of Life Sciences, a research centre for sustainable agriculture, horticulture and food security (United Kingdom)(www.warwick.ac.uk/fac/sci/lifesci/wcc).

11 Acknowledgements

The writing of this chapter was realized with the financial support of the Flemish government, the Department of Agriculture and Fishery, Flanders Innovation and Entrepreneurship (HBC.2019.2861) and CCBT.

12 References

Andersen, C. L., Hazzard, R., Van Driesche, R. and Mangan, F. X. 2006. Alternative management tactics for control of phyllotreta cruciferae and phyllotreta striolata (coleoptera: chrysomelidae) on Brassica rapa in Massachusetts. *Journal of Economic Entomology* 99(3), 803–810.

Beck, B., Brusselman, E., Nuyttens, D., Moens, M., Temmerman, F., Pollet, S., Van Weyenberg, S. and Spanoghe, P. 2014. Improving the biocontrol potential of entomopathogenic nematodes against Mamestra brassicae: effect of spray application technique, adjuvants and an attractant. *Pest Management Science* 70(1), 103–112.

Blackshaw, R. P., Vernon, R. S. and Prasad, R. 2012. Reduction of Delia radicum attack in field brassicas using a vertical barrier. *Entomologia Experimentalis et Applicata* 144(2), 145–156.

Bligaard, J., Meadow, R., Nielsen, O. and Percy-Smith, A. 1999. Evaluation of felt traps to estimate egg numbers of cabbage root fly, Delia radicum, and turnip root fly, Delia floralis in commercial crops. *Entomologia Experimentalis et Applicata* 90(2), 141–148.

Boemare, N.E., Akhurst, R.J. and Mourant, R.G. 1993. DNA Relatedness between *Xenarhabdus* spp. (*Enterobacteriaceae*), symbiotic bacteria of entomopathogenic nematodes, and a proposal to transfer *Xenorhabdus luminescens* to a new genus, *Photorhabdus* gen. Nov. *International journal of systemic and evolutionary microbiology*, 43(2), 249–255.

Capinera, J. L. 2008. *Encyclopedia of Entomology*. Springer Science & Business Media, Heidelberg, p. 2061.

Chen, S. L., Han, X. Y. and Moens, M. 2003. Biological control of delia radicum (diptera: anthomyiidae) with entomopathogenic nematodes. *Applied Entomology and Zoology* 38(4), 441–448.

Collier, R., Mazzi, D., Schjoll, A. F., Schorpp, Q., Thoming, G., Johansen, T. J., Meadow, R., Meyling, N. V., Cortesero, A. M., Vogler, U., Gaffney, M. T. and Hommes, M. 2020. The potential for decision support tools to improve the management of root-feeding fly pests of vegetables in western Europe. *Insects* 11(6), 369.

Collier, R. H. and Finch, S. 1983. Completion of diapause in field populations of the cabbage root fly (*Delia radicum*). *Entomologia Experimentalis et Applicata* 34(2), 186–192.

Finch, B. S. and Skinner, G. 1973. Distribution of cabbage root flies in Brassica crops. *Annals of Applied Biology* 75(1), 1–14.

Finch, S. 1993. Integrated pest-management of the cabbage root fly and the carrot fly. *Crop Protection* 12(6), 423–430.

Finch, S. and Collier, R. H. 2000. Integrated pest management in field vegetable crops in northern Europe - with focus on two key pests. *Crop Protection* 19(8–10), 817–824.

Frank, S. D. 2010. Biological control of arthropod pests using banker plant systems: past progress and future directions. *Biological Control* 52(1), 8–16.

Glazer, I. 2002. Survival biology. In: Gaugler, R. (Ed.), *Entomopathogenic Nematology*. CAB International, Wallingford, UK, pp. 169–187.

Grewal, P. S., Ehlers, R.-U. and Shapiro-Ilan, D. I. 2005. *Nematodes as Biocontrol Agents*. CABI Publishing, Wallingford, UK.

Joas, R. and Cotillon, A.-C. 2009. *Development of Guidance for Establishing Integrated Pest Management (IPM) Principles*. Final report, European Commission, Munich, Germany.

Lebedev, G., Abo-Moch, F., Gafni, G., Ben-Yakir, D. and Ghanim, M. 2012. High-level of resistance to spinosad, emamectin benzoate and carbosulfan in populations of thrips tabaci collected in Israel. *Pest Management Science* 69(2), 274-277.

Leger, C. and Riga, E. 2009. Evaluation of marigolds and entomopathogenic nematodes for control of the cabbage maggot delia radicum. *Journal of Sustainable Agriculture* 33(2), 128–141.

Ludwig, M. and Meyhofer, R. 2016. Efficacy of crop cover netting against cabbage pests and their natural enemies and relevance of oilseed rape. *Journal of Plant Diseases and Protection* 123(6), 331–338.

Maes, T. and Lippens, L. 2019. Plantbakbehandeling en granulaat combineren loont tegen koolvlieg. *Proeftuinnieuws* 19, 26–27.

Martens, K. and Plovie, N. 2007. A pest and disease warning system for cole crops and leek in flanders: a practical approach. *Integrated Protection of Field Vegetables IOBC/wprs Bulletin* 30, 75–82.

Martin, T., Palix, R., Kamal, A., Deletre, E., Bonafos, R., Simon, S. and Ngouajio, M. 2013. A repellent net as a new technology to protect cabbage crops. *Journal of Economic Entomology* 106(4), 1699–1706.

McKinlay, R. G. 1992. *Vegetable Crop Pests*. Macmillan Press, London, p. 406.

Mesmin, X., Vincent, M., Tricault, Y., Estorgues, V., Daniel, L., Cortesero, A. M., Faloya, V. and Le Ralec, A. 2019. Assessing the relationship between pest density and plant damage: a case study with the belowground herbivore delia radicum (diptera: anthomyiidae) on broccoli. *Applied Entomology and Zoology* 54(2), 155–165.

Nilsson, U. and Gripwall, E. 1999. Influence of application technique on the viability of the biological control agents *Verticillium lecanii* and *Steinernema feltiae*. *Crop Protection* 18(1), 53–59.

Patel, M. N. and Wright, D. J. 1996. The influence of neuroactive pesticides on the behaviour of entomopathogenic nematodes. *Journal of Helminthology* 70(1), 53–61.

Peters, A. and Ehlers, R. U. 1994. Susceptibility of leatherjackets (tipula paludosa and tipula oleracea; tipulidae; nematocera) to the entomopathogenic nematode steinernema feltiae (tipula-paludosa and tipula-oleracea, tipulidae, nematocera). *Journal of Invertebrate Pathology* 63(2), 163–171.

Radcliffe, E. B., Hutchinson, W. D. and Cancelado, R. E. 2008. *Integrated Pest Management: Concepts, Tactics Strategies and Case Studies*. Cambridge University Press, Cambridge, UK. 550p.

Schroeder, P. C., Ferguson, C. S., Shelton, A. M., Wilsey, W. T., Hoffmann, M. P. and Petzoldt, C. 1996. Greenhouse and field evaluations of entomopathogenic nematodes (nematoda: heterorhabditidae and steinernematidae) for control of cabbage maggot (diptera: anthomyiidae) on cabbage. *Journal of Economic Entomology* 89(5), 1109–1115.

Simon, S., Komlan, F. A., Adjaito, L., Mensah, A., Coffi, H. K., Ngouajio, M. and Martin, T. 2014. Efficacy of insect nets for cabbage production and pest management depending on the net removal frequency and microclimate. *International Journal of Pest Management* 60(3), 208–216.

Sparks, T. C., Dripps, J. E., Watson, G. B. and Paroonagian, D. 2012. Resistance and cross-resistance to the spinosyns - a review and analysis. *Pesticide Biochemistry and Physiology* 102(1), 1–10.

Spiessens, K., Caluwé, S. and Temmerman, F. 2019. Afdekken tegen koolvlieg. *Proeftuinnieuws* 15, 29–31.

Van Herk, W. G., Vernon, R. S., Waterer, D. R., Tolman, J. H., Lafontaine, P. J. and Prasad, R. P. 2017. Field evaluation of insecticides for control of cabbage maggot (diptera: anthomyiidae) in Rutabaga in Canada. *Journal of Economic Entomology* 110(1), 177–185.

Van Lenteren, J. C. 2012. The state of commercial augmentative biological control: plenty of natural enemies, but a frustrating lack of uptake. *BioControl* 57(1), 1–20.

Van Lenteren, J. C., Bolckmans, K., Kohl, J., Ravensberg, W. J. and Urbaneja, A. 2018. Biological control using invertebrates and microorganisms: plenty of new opportunities. *BioControl* 63(1), 39–59.

Volckaert, A., Maes, T. and Venneman, S. 2018. Koolvliegbestrijding in bloemkool 1ste vrucht anno 2018. *Proeftuinnieuws* 7, 28–29.

Volckaert, A., Cap, N., Buysens, S. and De Reycke, L. 2019. Control of cabbage root fly in cauliflower – a practical point of view. *IOBC-WPRS Bulletin* 142, 73–78.

Witkowska, E., Moorhouse, E. R., Jukes, A., Elliott, M. S. and Collier, R. H. 2018. Implementing integrated pest management in commercial crops of radish (Raphanus sativus). *Crop Protection* 114, 148–154.

Index

www.ingramcontent.com/pod-product-compliance
Ingram Content Group UK Ltd.
Pitfield, Milton Keynes, MK11 3LW, UK
UKHW021824100926
8556IPUK00003B/149

9 781786 767530